INSTITUTION OF CIVIL ENGINEERS

Current and future trends in bridge design, construction and maintenance 2
Safety, economy, sustainability and aesthetics

Proceedings of the international conference organized by the Institution of Civil Engineers, and held in Hong Kong on 25-26 April 2001

Edited by P C Das, D M Frangopol and A S Nowak

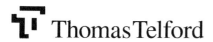

Conference organized by the Conference Office of the Institution of Civil Engineers.

Organizing Committee:

Prof J R Casas, Technical University of Catalonia, Spain
Mr R D Csogi, Greenman-Pedersen, Inc., USA
Dr P C Das (co-chairman), Highways Agency, UK
Prof M C Forde, University of Edinburgh, UK
Prof D M Frangopol, University of Colorado, USA
Prof H Furuta, Kansai University, Japan
Prof J M Ko, The Hong Kong Polytechnic University
Dr J S Kuang, Hong Kong University of Science and Technology
Ir Dr C K Lau, The Hong Kong Institution of Engineers
Mr K Leung, Maunsell Consultants Asia Ltd, Hong Kong
Prof A S Nowak (co-chairman), University of Michigan, USA
Mr D Poineau, SETRA, France
Prof P Thoft-Christensen, Aalborg University, Denmark

Published by Thomas Telford Publishing, Thomas Telford Ltd, 1 Heron Quay, London E14 4JD.
URL: http://www.thomastelford.com

First published 2001

Distributors for Thomas Telford books are
USA: ASCE Press, 1801 Alexander Bell Drive, Reston, VA 20191-4400, USA
Japan: Maruzen Co. Ltd, Book Department, 3–10 Nihonbashi 2-chome, Chuo-ku, Tokyo 103
Australia: DA Books and Journals, 648 Whitehorse Road, Mitcham 3132, Victoria

A catalogue record for this book is available from the British Library

ISBN: 0 7227 3091 6

© The Institution of Civil Engineers, unless otherwise stated 2001

All rights, including translation, reserved. Except as permitted by the Copyright, Designs and Patents Act 1988, no part of this publication may be reproduced, stored in a retrieval system or transmitted in any form or by any means, electronic, mechanical, photocopying or otherwise, without the prior written permission of the Publishing Director, Thomas Telford Publishing, Thomas Telford Ltd, 1 Heron Quay, London E14 4JD.

This book is published on the understanding that the authors are solely responsible for the statements made and opinions expressed in it and that its publication does not necessarily imply that such statements and/or opinions are or reflect the views or opinions of the publishers or of the organizers. While every effort has been made to ensure that the statements made and the opinions expressed in this publication provide a safe and accurate guide, no liability or responsibility can be accepted in this respect by the authors or publishers.

Printed and bound in Great Britain by MPG Books, Bodmin

Preface

In order to celebrate the achievements of the bridge engineering profession at the end of the Millennium, with the help of a number of eminent bridge engineers and academics from around the world, The Institution of Civil Engineers has been organizing a series of international bridge conferences. The broad purpose of these conferences is to enable the profession to take stock of the current developments in bridge engineering, and also to look forward to the future at this historic juncture.

The first conference in the series was held in Singapore in October 1999 and the second in Hong Kong in April 2001. This book is a compilation of the papers presented at the Hong Kong conference.

The 20th century has seen major expansions of the transport networks in most countries of the world accompanied by the construction of most of the bridges that exist today. Even in the closing years of the century vigorous activities were taking place in every aspects of bridge engineering – design, construction, management and research.

A number of major bridges with record-breaking spans and outstanding aesthetic appeal have been built around the world in recent years. A few more are in the pipeline. There have been many recent advances in the techniques of construction and materials. These have been accompanied by the development of a number of national and international codes and standards aimed at producing more durable and reliable structures. Increasing attention is being paid to whole-life performance and sustainability.

Regarding the management of the existing bridge stocks, many countries are now facing the problem of having to deal with large numbers of deteriorated or otherwise sub-standard bridges. Considerable efforts are being made to improve the methods of inspection, assessment and strengthening so that the scarce resources can be utilized more effectively. A number of bridge management systems are now available, with new ones still being developed.

Wide-ranging research and development activities are being carried out at universities, national laboratories and other institutions. Innovations are taking place in most areas, particularly in the use of fibre-reinforced plastics, non-destructive testing and smart monitoring. In addition, assisted by research, rapid advances are being made towards the application of reliability analysis and whole-life costing to bridges.

This book provides a contemporary account of the important developments in all these areas, written by expert bridge engineers, academics and senior government officials from some eighteen countries. It is divided into six main parts covering the following broad topics:

- Design and construction
- Maintenance practice
- Design theory
- Management
- Assessment
- Research

The papers contain the details of major new bridge construction projects from around the world and the more general state-of-the-art developments in bridge engineering, particularly relating to bridge maintenance and management. It is hoped, therefore, that this volume will prove useful to all sectors of the bridge engineering community.

In conclusion, the editors would like to express their gratitude to the authors of the individual papers for their enthusiastic co-operation in preparing this publication.

Parag C Das
Dan M Frangopol
Andrzej S Nowak

Contents

SECTION 1: Plenary papers

Bridge management in Hong Kong: the selection of appropriate techniques
P.C. Wong, C.Y. Wong and J. Darby — 3

Maintenance strategies for bridge stocks: cost-reliability interaction
D. M. Frangopol, J.S. Kong and E.S. Gharaibeh — 13

Strategies for the management of post-tensioned concrete bridges
R.J. Woodward, D.W. Cullington and J.S. Lane — 23

Stonecutters bridge design competition - aesthetic consideration
M.C.H. Hui and C.K.P. Wong — 33

Long term fatigue behaviour of steel girders with welded attachments under highway variable amplitude loading
M. Sakano — 43

A landmark structure over the Charles River in Boston, Massachusetts
V. Chandra, A. Ricci, K. Donington and P. Towell — 52

Optimal scheduling for bridges based on life-cycle evaluation
D. De Leon and A. H-S. Ang — 64

SECTION 2: Design and construction

The hot dip galvanized wires for bridge cables
J.Y. Kang — 71

Genoa Harbour crossing
A. Farooq, G.W. Davies and M. Grassi — 82

The planning and design of viaduct construction in Route 9 between Tsing Yi and Cheung Sha Wan
N. Hussain — 92

Recent achievements in Polish bridge engineering
W. Radomski — 102

The bridge on the Wadi Kuf Valley – Libya. Rehabilitation and maintenance project.
E. Codacci-Pisanelli — 112

Metsovitikos Bridge – a towerless suspension bridge
A. Paul and I. Wilson 122

KCRC West Rail Viaducts – design development
N. Hussain and A. Crockett 131

The alternative design of the West Rail Viaducts
N.J. Southward and J.H. Cooper 141

Construction of KCRC's West Rail Viaducts
H. Boyd, T. Gregory and N. Thorburn 153

Construction of a steel girder bridge rigidly connected to concrete piers with perfobond plates
H. Hikosaka, K. Akehashi, Y. Sasaki, K. Agawa and L. Huang 166

Twinning of Jindo Grand Bridge, Republic of Korea
M.J. King, W.J. Kim and C.Y. Cho 175

The design of the Stonecutters Bridge, Hong Kong
S. Withycombe, I. Firth and C. Barker 185

SECTION 3: Maintenance practice

Safety and performance of an active load control system for bridges
T. Atkinson, P. Brown, J. Darby, T.A. Ealey, J.S.Lane, J.W. Smith and Y.Zheng 197

Replacement of steel and composite bridges under traffic
R. Saul and S. Hopf 207

Reconstruction of the Lions Gate Bridge
M.J. Abrahams and J.K. Tse 215

Kingston Bridge phase 1 strengthening
M. Collings and I. Telford 225

Acoustic emission – a tool for bridge assessment and monitoring
J.R. Watson, P.T. Cole, S. Yuyama and D. Johnson 236

The strengthening and refurbishment of Westfield Pill Bridge, Pembrokeshire
D.T. Gullick 246

Monitoring system for fatigue crack propagation by image analysis
K. Tateishi, T. Hanji and M. Abe 256

Health monitoring system for bridge structures based on continuous stress measurement
N. Horikawa, H. Namiki and T. Kusaka 261

Bridge inspection in steel road bridge based on real measurement
A. Koshiba, M. Abe, T. Sunaga and H. Ishii 268

Installation of advance warning system at highway structures which are susceptible to flooding
P.C. Wong, C.Y. Wong, L.H.Y. Ho and Y.H. Leung 274

Rehabilitation of Tsing Yi South Bridge, Hong Kong
P.C. Wong, C.Y. Wong and F. Kung 282

Post tentioning of steel beam using high strength steel plate
M. Sakano and H. Namiki 292

Vibration induced fatigue of overhead sign structures on elevated highway bridges
K. Yamada, T. Ojio, S. Lee, Z. Xiao and S. Yamada 299

SECTION 4: Design and theory

Design for durability – a maintenance engineer's viewpoint
R.J. Feast 309

Review of design thermal loading for steel bridges in Hong Kong
F.T.K. Au, L.G. Tham and M. Tong 318

Decision support system for bridge aesthetic design using immune system
H. Furuta, M. Hirokane and K. Ishida 328

SECTION 5: Management

Optimum maintenance strategies for trunk road bridges in England and Italy
P.C. Das and L. Pardi 341

Inspection and maintenance of Hong Kong's long span bridges
J.D. Gibson 349

How effective is bridge posting in enhancing reliability?
S.B.A. Asantey and F.M. Bartlett 359

Life cycle cost of post-tensioned T-section girder bridges
T. Yoshioka, S. Ogawa, C. Wu and T. Sugiyama 369

West Rail Viaducts – an overview
C. Calton and S. Lo 379

Life cycle cost analysis of bridges where the real options are considered
Y. Koide, K. Kaito and M. Abe 387

Safety management of highway structures
A.J. Wingrove 395

Staged investigations of bridges
D. Pearson-Kirk 406

Verification of girder distribution factors and dynamic load factors by field testing
A.S. Nowak and J. Eom 414

Seismic retrofitting of bridges in New York
A.H. Malik   424

SECTION 6: Assessment

Current and future trends in heavy haulage bridge assessment process
S.N. Sergeev, G. Sobol and C.C. Candy   439

The use of reliability-based assessment techniques for bridge management
R.J. Lark and K.D. Flaig   453

Assessment of fatigue damage in the Tsing MA Bridge under traffic loadings by finite element method
T.H.T. Chan, L. Guo and Z.X. Li   463

Some outcomes from load testing of small span bridges in Western Australia
I. Chandler   473

Sensitivity of steel bridge fatigue life estimates to fatigue crack modelling
T.D. Righiniotis and M.K. Chryssanthopoulos   484

SECTION 7: Research

Full strength joints for precast reinforced concrete units in bridge decks
S.R. Gordon and I.M. May   497

Vibration and impact studies of multi-girder steel bridge in laboratory
L. Yu and T.H.T. Chan   507

Fundamental study on application of carbon fiber reinforced polymer strips to a notched steel member
H. Suzuki   517

Three dimensional modelling of masonry arch bridges
P.J. Fanning and T.E. Boothby   524

Full scale testing of high performance concrete bridge beams with in-situ slabs
D.J. Doyle and D.L. Keogh   534

Categorizaton of damaged locations on concrete bridge structures by a neural network
L. Bevc and I. Peruš   544

A qualitative and quantitative comparative study of seismic design requirements in bridge design codes
M.M. Bakhoum and S.S. Athanasious   554

Plenary papers

Bridge Management in Hong Kong: the selection of appropriate techniques.

P.C. WONG, C.Y. WONG
Highways Department, Hong Kong SAR Government, Hong Kong

J DARBY
Mouchel, West Byfleet, UK

Abstract

Bridge managers throughout the world share the same general objectives. However, the management techniques that are most appropriate for their use will depend upon local factors. In particular, the age and nature of bridge stocks will vary widely. This paper will review available techniques and relate their application to the circumstances in Hong Kong.

1   Introduction

The bridge stock in Hong Kong is relatively new, when compared with the stock of structures in Europe and North America. Bridge managers are therefore fortunate not to have the resource and prioritisation problems encountered by managers of older structures that are weak and/or deteriorating fast. Opportunity can therefore be taken to review all available techniques, and to develop management procedures that will ensure that structures continue to perform satisfactorily in the future.

Despite the young age of the bridge stock, it is important for Hong Kong that its bridge assets are managed effectively. Structures convey large volumes of traffic critical to the economy, with little opportunity for diversionary routes in centres of dense population. Many structures are difficult to access, aesthetics and environment are of political importance, and the marine environment has the potential to influence a high proportion of structures.

2   Objectives of Bridge Management

The general objectives of Bridge Managers are similar throughout the world, even if the techniques appropriate in the variable situations differ widely. Defining objectives may appear unnecessary, because many activities appear self-evident. Nevertheless, to do so does focus the mind upon the real choices to be made. All adopted techniques must deliver a particular objective, and be the most efficient means of delivering that objective.

The following four objectives are proposed: -

Performance: To maintain structures in service with minimum disruption, and to the standards required by the local society.

Prediction: To understand and monitor the stock of structures in sufficient detail to enable effective forward planning.
Funding: To ensure that the funds society provides produce maximum value.
Social: To take account of wider social responsibilities, such as safety and sustainability.

Figure 1. Island Eastern Corridor. A typical multi-span concrete viaduct with a marine influence.

3   The Hong Kong Bridge stock.

The Hong Kong Government Highways Department owns and maintains 1969 structures supporting pedestrian or vehicular loading, as shown in Table 1.

'Bridge' Type	No. of structures
Road Bridges	810
Footbridges	564
Subways	338
Underpasses	50
Nullah decks. (covered waterways.)	27
Culverts	180
Total 'Bridges'	**1969**

Table 1. Pedestrian and vehicular loaded structures maintained by Hong Kong Government Highways Department.

In addition, the Department maintains 732 ancillary structures, such as sign gantries, noise barriers and noise enclosures.

The number of bridge structures is not large, but this must be balanced against the above average number of spans and span length for each bridge, as demonstrated by the following statistics: -

- 5 road bridges have between 50 and 80 spans.
- 38 road bridges have between 20 and 50 spans.

- 69 road bridges have between 10 and 20 spans.
- 167 road bridges have between 5 and 10 spans
- 123 footbridges have 5 or more spans.
- 11 road bridges have a largest span over 70m.
- 61 road bridges have a largest span between 40m and 70m

The main materials used for each structure type are shown in Table 2

'Bridge' Type	% By Main Material		
	Concrete	Steel	Masonry
Road Bridges (810)	92%*	5%	3%
Footbridges (564)	82%	15%	3%
Subways (338)	100%	-	-
Underpasses (50)	100%	-	-
Nullah decks(27)	100%	-	-
Culverts (180)	97%	-	3%

Table 2: Main materials used in the Hong Kong bridge stock.

*Concrete road bridges are 42% reinforced concrete, 30% precast prestressed concrete, and 28% insitu prestressed concrete. The precast prestressed concrete structures may also be post-tensioned in addition to those that are insitu prestressed concrete.

The age profile for all bridge and ancillary structures may be judged from the following Table 3, which show them to be remarkably new with the exception of the nullah decks: -

Age of Structure	Structures excluding Nullah Decks. (2674)	Nullah Decks (27)
Under 40 years	98.5%	37%
Under 30 years	92.6%	26%
Under 20 years	79.6%	22%
Under 10 years	49.4%	15%

Table 3: Age of Bridge and ancillary highway structures in Hong Kong.

Figure 2. Extensive noise enclosure to a 4 lane highway in the New Territories.

These statistics demonstrate that the most common structures in Hong Kong are relatively new multi-span concrete viaducts. Of equal significance is the fact that these structures are mostly within a city environment, resulting in heavy usage, restricted access, and limited availability of alternative routes. The ancillary structures also represent valuable assets, and include extensive noise barriers and enclosures. The investment must be maintained.

4 Comparison between bridge stocks of Hong Kong and the UK

Management techniques must be designed to meet the needs presented by a particular stock of structures. Consideration of techniques in use in other parts of the world must therefore be accompanied by an appreciation of the nature of the structures to which they are applied.

The most fundamental property of a bridge stock is age, and figure 3 illustrates the comparison between Hong Kong Highways Department structures and those of three UK bridge owners. In the UK the earliest forms of transport was by local roads, hence local Highway Authorities together maintain the largest bridge stock. These authorities are represented by Oxfordshire, a typical County. 5% of Oxfordsshire's structures were constructed before 1800, and 27% before 1900. Structures have been built at a fairly steady rate since 1900, due to both expansion of the network and replacement.

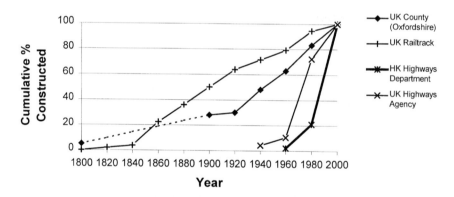

Figure 3   Comparison between the age of the Hong Kong Highways Department structures and those of UK Bridge Owners.

Railtrack has stock of over 80,000 structures, 40,000 of which support the railway. 50% of those still in service are over 100 years old, being built between 1800 and 1900 when most of the Railway network was constructed. A high proportion of the structures built since 1900 are replacements of earlier structures, particularly those that were metal structures subjected to corrosion and fatigue.

The UK Highways Agency is responsible for Trunk Roads and Motorways, a network that has expanded rapidly since the 1960's, resulting in a much younger stock. Thus only 4.4%%

were constructed before 1940, and 10.5% before 1960. However, 72.5% are still over 20 years old.

The bridge stock owned by the Hong Kong Government Highways Department even newer than that of the UK Highways Agency, with only 2.1% constructed before 1960 and only 20.9% over 20 years old.

Figure 4   Main Materials used for bridge structures in the UK and Hong Kong.

Materials used for bridge construction above all reflect the date of construction. This is illustrated by a comparison between the materials shown in Figure 4 and the dates of construction in Figure 3. The older bridge stocks of UK Counties and Railtrack comprise a high proportion of masonry structures, but with Railtrack having a significant proportion of metal structures to meet their need for larger spans and heavy loads, often in locations with difficult access.

The newer bridge stocks of the Highways Agency and the Hong Kong Highways Department are remarkably similar, with very high proportions of concrete. Prestressing has now enabled this material to achieve the long spans that were previously only possible with steel.

5 World Trends in Bridge Management

It may be seen that countries such as the UK have bridge stocks that have been in service for many years, a high proportion of which were not designed to current standards. Loading requirements and traffic volumes continue to increase, whilst material deterioration is accelerating with age. These factors combine to present a significant problem to bridge managers that is compounded by tight restraint on the funding of maintenance and refurbishment works.

Only a few decades ago bridge maintenance comprised little more than painting of steelwork and masonry pointing. The evolving situation now requires much more sophisticated techniques for analysis, monitoring and prioritisation. Many of these techniques are still developing. They require and deserve official support for further progress to be made. Nevertheless, during this development period bridge managers therefore have a bewildering choice. Whilst it is right to support promising development, most decisions will inevitably be based upon methods that managers believe to be proven, reliable and cost effective. Above all, the selected methods must meet specific bridge management objectives, and must correspond to the specific problems of the bridge stock.

At the heart of any maintenance process is an information system to record and manipulate asset information. Computing hardware has developed so fast that its capacity is no longer a significant restraint in the selection of a process. However, the software systems in use still vary considerably in their sophistication. The most basic systems are essentially inventories of the bridge stock and track inspection programmes. Data on condition and works prioritisation may also be included, and the most sophisticated systems will extend to include a predictive function.

If the assertion that the techniques must be tailored to meet the needs of the bridge stock and the specific objectives of the client is accepted, then the information system itself must also be designed to meet the same specific needs. The information system must supply all of the data required to support the selected techniques, but at the same time should avoid wasteful collection and storage of information that is not essential. The particular inspection process in use must also be reflected in the database design.

This need to match the bridge management system to the techniques and inspection process appropriate for a particular bridge stock is recognised by many bridge managers. They are in the best position to recognise the needs arising from the type and age of their structures, and the influence of financial and political constraints upon their decisions. Standard bridge management products are most unlikely to meet all of the requirements that result, which perhaps explains the rather limited number of applications for such products. Managers generally prefer systems customised to meet their particular needs.

6 Techniques appropriate for the Hong Kong Bridge Stock

6.1 Inspection of Structures

Inspection of structures has been undertaken for many years. In the UK some authorities have the records of inspections undertaken over a hundred years ago. However, these inspections have become progressively more detailed particularly since the 1970's, and Hong Kong is no exception. Currently, detailed inspections of structures in Hong Kong are conducted at 2-

yearly intervals. In addition to the detailed inspections, brief inspections are also carried out at 6-monthly intervals to check for obvious defects which might lead to safety problems, loss or restriction of use of the structure. Furthermore, special inspections of structures are also carried out when the need arises.

Inspections have increased the understanding that bridge engineers have of their stock. This has opened up the opportunity to further refine the inspection process by varying the frequency according the characteristics of particular structures, thereby increasing efficiency. Hong Kong is considering the following categories of inspection:

Inspection Type	Scope of Inspection	Frequency
Opportune	Brief inspection after receiving a public complaint or report of defects from staff.	
Superficial	Brief Inspection from ground level to note obvious defects, accident damage, graffiti etc.	3 monthly or 12 monthly according to structure type and location
Benchmark	Detailed inspection of all parts of a structure, including minor testing, to set benchmark and determine future inspection and maintenance strategy.	Once in a lifetime, unless there is reason to repeat the inspection.
Custom	Planned inspections of particular part(s) of a structure, to varying levels of detail, including testing as required. These scheduled inspections are designed to meet the needs of a particular structure, and are frequently determined after a Benchmark inspection.	Variable, according to a schedule drawn up for a particular structure. Revised if new circumstances arise
Special	One off inspection arising from a particular problem discovered during opportune, superficial, Benchmark or Custom inspections, or due to notification. Also includes inspection arising from potential deficiencies of a structure type. Inspections include testing as required.	Varies according to nature of problem.

Table 4   Types of Inspection proposed for Hong Kong

The inspection regime outlined in Table 4 delivers the following benefits : -

- Very frequent superficial inspection of vulnerable structures, enabling the high standards of appearance demanded in Hong Kong to be maintained.
- The needs of particular structures to be reflected in their inspection schedule, whilst minimising the collection of unnecessary data for structures not expected to deteriorate.
- A framework for special inspections to deal with unforeseen problems, such as those associated with a particular type of construction.

6.2 Material Deterioration and Condition Monitoring

Engineers have historically adopted new materials for bridge construction soon after they have been developed, generally with the incentive of increasing the potential span. Unfortunately, many of these new materials have also led to a progressively lower life. Thus in the UK the historical trend was as follows: Masonry arches, Cast Iron, Wrought iron, steel, reinforced concrete, post-tensioned concrete to pre-tensioned concrete. Many of these material types consequently require refurbishment or replacement at the present time. It is confidently anticipated that this trend towards lower life has been reversed by the most recent developments.

In Hong Kong the material of greatest interest is concrete. The UK faces major expenditure for concrete repair and maintenance. Figure 3 illustrates that the Hong Kong bridge stock is following on approximately 20 years behind the construction of the UK Trunk and Motorway network, and with similar materials. This raises the question as to whether similar problems will arise in Hong Kong.

Hong Kong has in its favour the fact that de-icing salts are not introducing chlorides to the structures, and the later construction has resulted in the use of more durable materials. Nevertheless, chlorides from the marine environment are certainly affecting some structures, and the range of that influence remains unclear. Carbonation is also a risk, and corrosion due to either cause may be accelerated in the warm climate. It is therefore proposed that the Hong Kong bridge stock will be closely monitored for indications of these potential forms of degradation. Where signs of deterioration or contamination are detected, early remedial measures can then be put in hand that minimise both cost and disruption consequences.

6.3 Load Capacity and Construction Type

The UK has incurred high expenditure to assess and strengthen structures, and to check for potential defects in particular construction types. To what extent is this liability also applicable to Hong Kong?

Hong Kong increased the loading standard as recently as 1993, and therefore some 70% of structures were designed to the earlier standard despite the young age of the bridge stock. The increase in loading standard was to make way for future development, and it did not imply that the designed loading was insufficient. In fact the new increased loading has not been reached to date. Furthermore, many large span structures and those checked for HB loading will not be affected by the change. A programme of checks is to be commenced, and 16t weight limits are already in place on some structures. Extensive problems are not considered likely because of the age of the bridge stock and relatively small effect of loading changes. Much is known about the materials and loading of structures, and any sub-standard structures will be strengthened as required. There is therefore no need to adopt sophisticated techniques such as reliability analysis as a means of coping with uncertainty and prioritising the programme.

Special Inspection programmes in the UK have shown that the grouting of post-tensioned structures is often defective. In a few cases this has resulted in replacement of the structure, although in most cases the defects have been found to be at an acceptable level. There is evidence that the problem is worldwide. Hong Kong has a large stock of such structures, and has already undertaken checks on the effectiveness of grouting. There is no evidence to date of any serious problems, although the inspection programme will be extended to ensure that no problems arise in the future.

6.4 Selection of maintenance and remedial measures

The key activity for all bridge managers is to select the most appropriate maintenance and remedial measures, and to undertake them at the right time in the life of a structure. If managers is able achieve these aims, their management will be at maximum efficiency.

Bridge managers always try to meet these aims, but the difficulty comes in balancing conflicting benefits. Should the inexpensive early repair be selected, or the more expensive solution that will also remove the need for future repair or disruption? How long will materials last? and what weighting should be given to future traffic disruption in the consideration. These difficult decisions increasingly require some framework against which they can be decided. Management priorities can then be defined, and decisions taken in an ordered and auditable way. Whole life costing is increasingly the framework within which maintenance and management decisions are taken, and the Hong Kong Highways Department proposes to consider its relevance to the Hong Kong situation.

Sustainability is also an aspect that is likely to influence all aspects of construction in the future, including bridge management decisions. As global warming accelerates, and consequences for mankind become more evident, political pressure to minimise harmful emissions and the use of scarce resources is likely to increase. Professional Engineers have a moral responsibility to provide guidance on these issues in their particular areas of expertise. The data and tools required to assess sustainability accurately in the bridge management field are not yet available, but this aspect will be reviewed in the light of developments.

6.5 Bridge Management System

The Hong Kong Highways Department already has computerised bridge management system that records details of its stock of structures. Networked to all potential users, it provides the means of viewing bridge drawings and records, as well as inspection reports and the state of the inspection programme.

Development of bridge management systems must keep abreast of bridge management initiatives. In Hong Kong, this particularly applies to the changes proposed to the inspection process and condition monitoring.

7 Conclusions

The Highways Department of the Government of Hong Kong has an extensive stock of structures to maintain that are vital to the economy. They are relatively new, inspected regularly, and providing good service with relatively little disruption.

The objective of the Highways Department is to maintain this level of service as the bridge stock become older. The potential problems can be anticipated by reviewing problems encountered by countries with similar bridge stocks that are of greater age. Condition monitoring and whole life costing techniques will therefore be adopted to ensure that any necessary interventions are both cost effective and avoid undue disruption. The inspection process will be designed to meet specific local needs.

Bridge Management in Hong Kong will thus be adapted to include appropriate techniques. This will ensure that the stock of structures continues to provide the infrastructure so essential to the health of the economy.

Acknowledgements

The authors would like to express thanks to the Director of Highways, Highways Department of Hong Kong SAR Government for permission to publish this paper.

MAINTENANCE STRATEGIES FOR BRIDGE STOCKS: COST - RELIABILITY INTERACTION*

DAN M. FRANGOPOL[1], JUNG S. KONG[2], and EMHAIDY S. GHARAIBEH[3]
[1] Professor, , Department of Civil, Environmental, and Architectural Engineering, University of Colorado, Boulder, CO 80309-0428, USA
[2] Graduate Research Assistant, Department of Civil, Environmental, and Architectural Engineering, University of Colorado, Boulder, CO 80309-0428, USA
[3] Assistant Professor, Department of Civil Engineering, Mu'tah University, Al-Karak, Jordan; formerly Graduate Research Assistant, Department of Civil, Environmental, and Architectural Engineering, University of Colorado, Boulder, CO 80309-0428, USA

INTRODUCTION

In bridge engineering, the financial resources do not keep pace with the growing demand for bridge maintenance, rehabilitation, and replacement. For this reason, it is imperative that the best possible use of existing resources should be achieved. The current bridge management systems try to meet this goal. However, as indicated in Frangopol and Das (1999), currently available bridge management systems have important limitations. One of the most severe limitations is that bridge reliability is not directly incorporated in bridge management. Consequently, so far, the best possible use of financial resources has been only a subject of continuing concern to bridge engineers and a dream for bridge managers. This dream can only come true by using reliability-based bridge management (Frangopol and Das 1999, Thoft-Christensen 1999, Das 2000, Frangopol *et al.* 1999, 2000a).

The next generation of bridge management systems has to use powerful reliability, optimization, and life-cycle engineering tools to predict the performance of bridges and the costs associated with bridge interventions including both agency and user costs, to find the optimal lifetime strategy based on benefit/cost methodologies, to implement this strategy at both network and project levels, and to update the optimal strategy as more information becomes available. This paper is based on a recent contribution of the authors published by Thomas Telford and presented at the Conference on Bridge Rehabilitation in the U.K. (Frangopol *et al.* 2000b). It describes those portions of the current effort which have direct bearing on the bridge performance prediction from the viewpoint of a reliability approach and on the integration of whole life costing with lifetime reliability. A particular emphasis is placed upon cost-reliability interaction in bridge management.

---

* A modified version of this paper was published in *Bridge Management 4*, Edited by M.J. Ryall, G.A.R. Parke, and J.E. Harding, Thomas Telford, London, 2000, and presented at the *Conference on Bridge Rehabilitation in the U.K.*, The Institution of Civil Engineers, London, October 2-3, 2000.

BRIDGE RELIABILITY STATES

Currently available bridge management systems, including Pontis (Thompson et al. 1998) and BRIDGIT (Hawk and Small 1998), use condition states to define condition of bridge elements at any given point in time. These condition states, largely based on visual inspections, indicate relative health of bridge elements but do not identify their specific reliability. Recently, Frangopol and Das (1999) and Thoft-Christensen (1999) proposed five bridge reliability states. These states are indicated in Fig. 1 along with the associated reliability indices. The service life of bridges is a progression of reliability states from excellent ($\beta \geq 9.0$) to unacceptable ($\beta < 4.6$). As indicated in Wallbank et al. (1999), Das (1999), and Frangopol et al. (1999), the justification for carrying out essential maintenance (such as major repairs) is that without it the element will be unsafe, and the justification for preventive maintenance (such as painting, silane treatment) is that if it is not done at the time it will cost more at a later stage to keep the element from becoming critical. The attributes for reliability and associated maintenance actions are also indicated in Fig. 1. It should be mentioned that preventive maintenance work should be considered as a package of actions (such as silane treatment, deck waterproofing, expansion joint replacement, extraction of contaminants). It is expected that the cost of these packages (called options in Fig. 1) should increase with the decrease in the reliability state of the bridge.

RELIABILITY STATE

5	4	3	2	1

RELIABILITY INDEX

$\beta \geq 9.0$	$9.0 > \beta \geq 8.0$	$8.0 > \beta \geq 6.0$	$6.0 > \beta \geq 4.6$	$4.6 > \beta$

ATTRIBUTE FOR RELIABILITY

EXCELLENT	VERY GOOD	GOOD	FAIR	UNACCEPTABLE

MAINTENANCE ACTION

PREVENTIVE 5	PREVENTIVE 4	PREVENTIVE 3	PREVENTIVE 2	ESSENTIAL 1
OPTION 5a	OPTION 4a	OPTION 3a	OPTION 2a	OPTION 1a
OPTION 5b	OPTION 4b	OPTION 3b	OPTION 2b	OPTION 1b
OPTION 5c	OPTION 4c	OPTION 3c	OPTION 2c	OPTION 1c
• • •	• • •	• • •	• • •	• • •

Figure 1. Definition of Reliability States, Attributes, and Maintenance Actions

Using this approach, maintenance actions are selected in response to distinct changes in the reliability states. In this manner, bridge reliability is directly incorporated in bridge management and all limitations associated with current Markovian-based bridge management systems can be relaxed.

FIRST REHABILITATION TIME

In order to estimate the number of bridges of a particular type requiring first rehabilitation (i.e., first essential maintenance) each year in the future it is necessary to capture the propagation of uncertainties from the time of construction to the time of failure (i.e., time at which the reliability index β downcrosses the target level 4.6). This time is called first rehabilitation time or rehabilitation rate. There are two cases to be investigated: (a) first rehabilitation time assuming no preventive maintenance has been done, and (b) first rehabilitation time assuming preventive maintenance has been done. The random variables associated with no maintenance action scenario are: initial target reliability index B_o, time of damage initiation T_I, and reliability index deterioration rate A. Five additional random variables have to be introduced in order to characterize the preventive maintenance scenario: time of first application of preventive maintenance T_{PI}, time of reapplication of preventive maintenance T_P, duration of preventive maintenance effect on reliability T_{PD}, deterioration rate of reliability index during preventive maintenance effect Θ, and improvement in reliability index (if any) immediately after the application of preventive maintenance Γ. The assumed PDFs of all the eight random variables for steel/concrete composite bridges are indicated in Kong and Frangopol (2001).

FIRST REHABILITATION TIME AFTER NO MAINTENANCE

Using Monte-Carlo simulation it is possible to generate the PDF of the reliability index of a group of steel/concrete composite bridges at any point in time. Fig. 2 compares the PDFs of the first rehabilitation time (i.e., rehabilitation rate) for steel/concrete composite bridges provided by experts in 1997 [i.e., triangular distribution (20, 35, 50) where 20, 35, and 50 represent the lowest, mode, and highest age (in years), respectively, Maunsell Ltd. and Transport Research Laboratory 1998], in 1998 (i.e., the logistic distribution characterized by the parameters 35.9 years and 6.2 years, Maunsell Ltd. and Transport Research Laboratory 1999), and the one obtained through complex reliability analysis computations using Monte-Carlo simulation and quadratic fitting. It is interesting to note that: (a) the target reliability index was not specified in 1997 and 1998; (b) in the reliability analysis carried out in 1999 the target reliability index was specified as 4.6; (c) the modes of the three distributions in Fig. 2 are approximately the same.

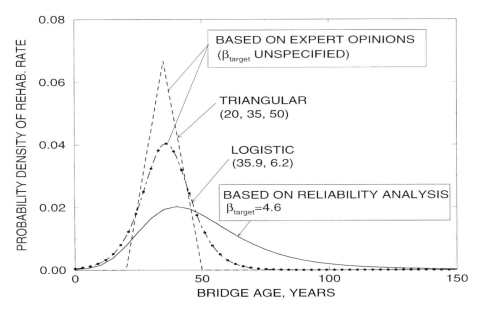

Figure 2. Probability Density of First Rehabilitation Time for Steel/Concrete Composite Bridges Assuming No Maintenance

FIRST REHABILITATION TIME AFTER PREVENTIVE MAINTENANCE
The reliability-based procedure developed for finding the PDF of bridge rehabilitation rate for steel/concrete composite bridges assuming no preventive maintenance was extended to the case of preventive maintenance by adding the random variables associated with the preventive maintenance actions.

EFFECT OF PREVENTIVE MAINTENANCE
The beneficial effect of preventive maintenance is clearly demonstrated on a stock of 100 steel/concrete composite bridges built at the same time (i.e., all bridges have the same age) by examining the results presented in Fig. 3. As indicated in Figs. 3 (a) and (b), the expected total number of bridges in reliability states 1 (i.e., $\beta < 4.6$), and 1 and 2 (i.e., $\beta < 6.0$) decreases considerably when preventive maintenance has been done. Figs. 6(c) and (d) indicate the expected total number of bridges in reliability states 1 to 3 (i.e., $\beta < 8.0$), and (d) 1 to 4 (i.e., $\beta < 9.0$), respectively. As shown, after about 48 years none of the 100 bridges is in the reliability state 5 even if preventive maintenance has been done.

In general, individual bridges in a bridge stock have different ages and their reliability is time dependent. In this case, it is also possible to predict the number of bridges in each reliability state over a prescribed time interval. As an example, Fig. 4 shows the time variation of the expected number of bridges in each reliability state from a stock of 389 steel/concrete composite bridges built during 1965-1974 (see Table 1, Maunsell Ltd. 2000) assuming no maintenance has been done.

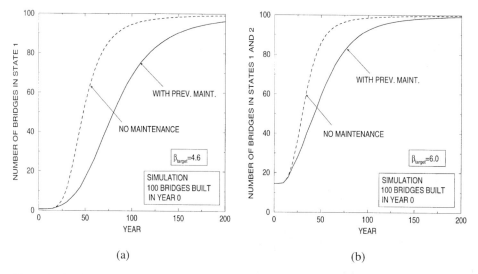

Figure 3. Stock of 100 Steel/Concrete Composite Bridges Built in Year 0. The Effects of no Maintenance and Preventive Maintenance on the Time Variation of the Expected Total Number of Bridges in Reliability States: (a) 1 (i.e., $\beta < 4.6$), and (b) 1 and 2 (i.e., $\beta < 6.0$)

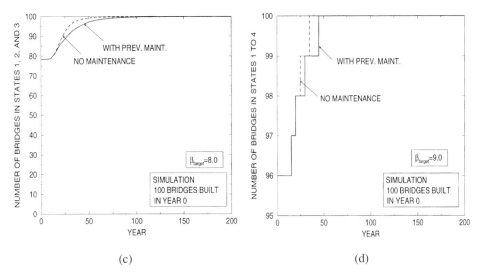

Figure 3. (continued) Stock of 100 Steel/Concrete Composite Bridges Built in Year 0. The Effects of no Maintenance and Preventive Maintenance on the Time Variation of the Expected Total Number of Bridges in Reliability States: (c) 1 to 3 (i.e., $\beta < 8.0$), and (d) 1 to 4 (i.e., $\beta < 9.0$)

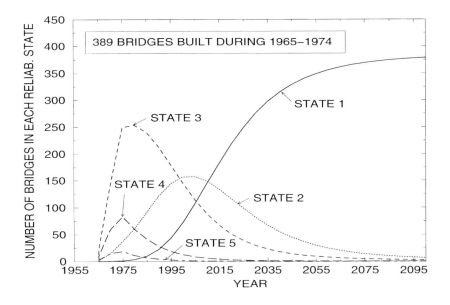

Figure 4. Expected Number of Bridges in Each Reliability State from a Stock of 389 Steel/Concrete Composite Bridges

Table 1. Steel/Concrete Composite Bridge Stock Built from 1965 to 1974

Year (1)	Number of Bridges Built (2)	Year (1)	Number of Bridges Built (2)
1965	37	1970	53
1966	8	1971	88
1967	32	1972	29
1968	25	1973	32
1969	74	1974	11

The study of the effect of various preventive maintenance actions on the time variation of the expected number of bridges in each reliability state from a large stock of bridges with different ages is a formidable task. However, advances in computing power in recent years have made solutions to this problem possible (Kong and Frangopol 2001).

COST-RELIABILITY INTERACTION

In general, the present value of cost is evaluated based on the assumption that the maintenance cost is independent on its effect on the system reliability level. However, in reality, cost varies according to the maintenance method selected and the improvement in system reliability level. Maintenance expenses invested in the past influence the current system reliability and the maintenance cost requested in the future. This interaction between maintenance costs and system reliability over lifetime is considered in Fig. 5 by using three different relationships between the unit rehabilitation cost and system reliability improvement.

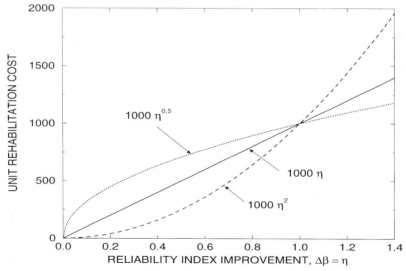

Figure 5. Unit Rehabilitation Cost Versus Reliability Improvement

Fig 6 shows the present value of the expected cumulative unit rehabilitation cost for the case in which only one essential maintenance is applied during the service life of a bridge group. It is clear that the present value of the expected cumulative unit rehabilitation cost is dependent on the cost function selected.

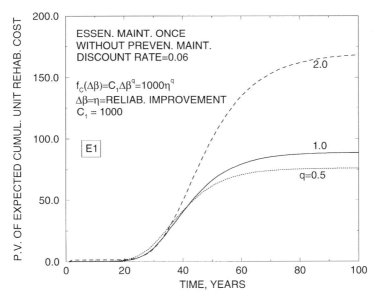

Figure 6. Present Value of Expected Cumulative Unit Rehabilitation Cost; Essential Maintenance Applied Once During the Service Life of a Bridge Group

COST-BENEFIT ANALYSIS

In order to determine the optimum maintenance strategy for each bridge type, total maintenance cost profiles for the future years have to be obtained. The following measures are appropriate to assess the effectiveness of preventive maintenance:

$$B_1 = C_{NPM} - C_{WPM} \qquad (1)$$

$$B_2 = C_{NPM} - (C_{WPM} - C_{PM}) \qquad (2)$$

where C_{NPM} = present value of expected cumulative unit maintenance cost assuming no preventive maintenance has been done; C_{WPM} = present value of expected cumulative unit maintenance cost assuming preventive maintenance has been done; C_{PM} = present value of expected cumulative unit maintenance cost of preventive maintenance; B_1 = expected cumulative unit benefit cost of using preventive maintenance including the cost of preventive maintenance; and B_2 = expected cumulative unit benefit cost of using preventive maintenance excluding the cost of preventive maintenance.

Considering a planning horizon of 50 years, Fig. 7 (Miyake and Frangopol 1999) shows the effects of both user cost and discount rate of money r on B_1. From this figure it is clear that the benefit of using preventive maintenance is increasing with user cost and is affected by the discount rate. The data considered in computations associated with Figs. 7 are taken from the strategic review of bridge maintenance costs (Maunsell Ltd. and Transport Research Laboratory 1998).

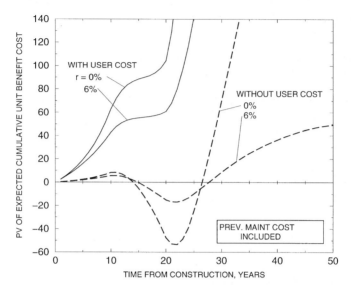

Figure 7. Effects of User Cost and Discount Rate on the Present Value of Expected Unit Benefit Cost for Steel/Concrete Composite Bridges

CONCLUSIONS
Bridge management based on lifetime reliability and whole life costing is considered to be the next generation of bridge management systems. These systems are based on reliability states instead of condition states. In order to provide realistic cost estimation the cost-reliability interaction has to be considered. Maintenance actions are selected in response to distinct changes in reliability states. In this manner, bridge reliability is directly incorporated in bridge management. Application of benefit/cost analysis to reliability-based bridge management decision making guides the selection of the optimum maintenance strategy in the face of uncertainties and fiscal constraints. Further research is needed for the successful implementation of the proposed reliability-based bridge management methodology in connection with probabilistic modeling of (a) bridge group deterioration, (b) effects of essential and preventive maintenance actions on bridge reliability, and (c) whole life assessment and costing. More data expected in the future will reduce uncertainty and, in turn, provide optimal bridge maintenance strategies with greater confidence.

ACKNOWLEDGEMENTS
The partial financial support of the U.K. Highways Agency and the U.S. National Science Foundation is gratefully acknowledged. Fruitful discussions with Dr. Parag Das of the U.K. Highways Agency are also gratefully acknowledged. The opinions and conclusions presented in this paper are those of the writers and do not necessarily reflect the views of the sponsoring organizations.

REFERENCES

Das, P.C. (1999). "Prioritization of Bridge Maintenance Needs," *Case Studies in Optimal Design and Maintenance Planning of Civil Infrastructure Systems*, D.M. Frangopol, ed., ASCE, Reston, Virginia, 26-44.

Das, P.C. (2000). "Reliability Based Bridge Management Procedures," *Bridge Management 4*, M.J. Ryall, G.A.R. Parke, and J.E. Harding, eds., Thomas Telford, London, 1-11.

Frangopol, D.M., and Das, P.C. (1999). "Management of Bridge Stocks Based on Future Reliability and Maintenance Costs," *Current and Future Trends in Bridge Design, Construction, and Maintenance*, P.C. Das, D.M. Frangopol, and A.S. Nowak, eds., Thomas Telford, London, 45-58.

Frangopol, D.M., Gharaibeh, E.S., Kong, J.S., and Miyake, M. (2000a). "Optimal Network-Level Bridge Maintenance Planning Based on Minimum Expected Cost," *Journal of the Transportation Research Board*, TRR, No. 1696, Vol. 2, 26-33.

Frangopol, D.M., Kong, J.S., and Gharaibeh, E.S. (2000b). "Bridge Management Based on Lifetime Reliability and Whole Life Costing: The next generation," *Bridge Management 4*, M.J. Ryall, G.A.R. Parke, and J.E. Harding, eds., Thomas Telford, London, 392-399; also "Optimum Maintenance Strategy: Steel Bridges," Conference on Bridge Rehabilitation in the U.K., The Institution of Civil Engineers, London, U.K., October 2-3, 2000.

Frangopol, D.M., Thoft-Christensen, Das, P.C., Wallbank, E.J., and Roberts, M.B. (1999). "Optimum Maintenance Strategies for Highway Bridges," *Current and Future Trends in Bridge Design, Construction, and Maintenance*, P.C. Das, D.M. Frangopol, and A.S. Nowak, eds., The Institution of Civil Engineers, Thomas Telford, London, 541-550.

Hawk, H., and Small, E.P. (1998). "The BRIDGIT Bridge Management System." *Structural Engineering International*, IABSE, **8**(4), 309-314.

Kong, J.S. and Frangopol, D.M. (2001). "Bridge Life-Cycle Safety Management," University of Colorado, Boulder (in progress).

Maunsell Ltd. (2000). "Optimum Maintenance Strategies for Different Bridge Types: Bridge Data," Vol. 3, *Final Report to the Highways Agency*, London.

Maunsell Ltd. and Transport Research Laboratory (1998). "Strategic Review of Bridge Maintenance Costs: Report on 1997/98 Review," *Draft Report*, The Highways Agency, London.

Maunsell Ltd. and Transport Research Laboratory (1999). "Strategic review of bridge maintenance costs: Report on 1998 Review," *Final Report*, The Highways Agency, London.

Miyake, M., and Frangopol, D.M. (1999). *Cost-Based Maintenance Strategies for Deteriorating Structures,* Report No. 99-5, *SESM Research Series,* Department of Civil, Environmental, and Architectural Engineering, University of Colorado, Boulder, Colorado.

Thoft-Christensen, P. (1999). "Estimation of Bridge Reliability Distributions," *Current and Future Trends in Bridge Design, Construction, and Maintenance*, P.C. Das, D.M. Frangopol, and A.S. Nowak, eds., Thomas Telford, London, 15-25.

Thompson, P.S., Small, E.P., Johnson, M., and Marshall, A.R. (1998). "The Pontis Bridge Management System," *Structural Engineering International*, IABSE, **8**(4), 303-308.

Wallbank, E.J., Tailor, P., and Vassie, P.R. (1999). "Strategic Planning of Future Maintenance Needs," *Management of Highway Structures*, P.C. Das, ed., Thomas Telford, London, 163-172.

STRATEGIES FOR THE MANAGEMENT OF POST-TENSIONED CONCRETE BRIDGES

Authors Dr. Richard Woodward, Dr. David Cullington & Mr. John Lane
TRL Ltd, Crowthorne, UK.

Introduction

The idea of prestressing concrete was first applied by Freyssinet at the beginning of the last century but it was not until the middle of the century that the first post-tensioned bridges were constructed. The first structures built were the Oued Fodda bridge in Algiers in 1936 and the Oelle bridge in Westfalia in 1938 both of which used pretensioning in the form of bars. The use of prestressed concrete in bridge construction didn't really take off until after the end of the Second World War. The construction of five portal frame bridges on the Marne between 1947 and 1950 enhanced the reputation of prestressed concrete construction and encouraged its use in other countries.

The first post-tensioned bridge built in the UK was Nunns bridge in Lincolnshire which was completed in 1947 with a span of 22.5m. In Switzerland the first railway bridge was constructed in 1943 and the first road bridge followed in 1950. The first major prestressed concrete bridge in the United States was the Walnut Lane Bridge in Philadelphia which was constructed in 1950.

Over the last fifty years post-tensioned concrete has become the preferred method of construction for bridges in the 30m to 50m span range. Its use has brought many benefits to the economy and construction of large bridges, particularly the extended range of possibilities open to designers, the use of various combinations of precast and in-situ concrete and the use of cantilevered methods of construction. Many thousands of structures have been constructed worldwide[1].

There is now a need to ensure that these structures are properly maintained in a safe and serviceable condition. This is particularly true for post tensioned concrete construction, as there has always been concern that voids could be formed in ducts during grouting resulting in increased risk of oxygen, moisture, carbon dioxide and other deleterious materials entering the ducts and destroying the protective environment around the tendons. This paper will describe the inspection procedures in current use, how the data is interpreted, in terms of both the current and future load carrying capacity, and where there is evidence of deterioration, the options available to the engineer. These range from do nothing to monitoring, regrouting partially grouted ducts, strengthening and replacement. The nature of post-tensioned inspections is such that decisions have to be taken on incomplete information as it is only possible to inspect a small sample of ducts. The paper will focus on the factors that need to be taken into account in the decision making process and is based on research undertaken at TRL over the last twenty five years. The paper puts forward strategies that could be adopted for the management of structures of this type based on the experience gained in the UK.

During the 1950s and early 1960s post-tensioned concrete bridges gave satisfactory performance. However the collapse of two footbridges during the 1960's gave an indication of the problems that could arise. Both were of segmental post-tensioned construction and both collapsed without warning under self-weight.

The first problems with a major structure, were found during the early 1970s when longitudinal cracks were observed on some of the I-beams of a beam and slab bridge built in 1961. They followed the profile of the tendons and holes drilled into the ducts disclosed the presence of voids and water. The cracks were attributed to expansion of the water on freezing. Holes were also drilled into the ducts in a neighbouring bridge that did not show any external evidence of trouble. This investigation revealed that 62% of the ducts contained voids that were continuous along the whole of their length and two ducts were empty.

Subsequent investigations were undertaken on twelve bridges built between 1958 and 1977[2]. They differed in respect of type of bridge, prestressing system, designers and contractors and none of the bridges examined showed any external indication of trouble. However voids were found in over 50% of the ducts examined and were present in all but one of the structures. The size of the voids varied considerably and they were often of sufficient size to expose the tendons. The distribution of voids was uneven. In continuous structures they were usually concentrated at high points over the piers. In general ducts were dry and there was usually a thin film of cement paste over the wires. There was no evidence of serious corrosion with the exception of slight pitting on a strand in a virtually empty duct and it was concluded that the presence of some grout in a duct provided an environment in which corrosion did not take place.

During the 1980s problems were found in an increasing number of post-tensioned bridges. In 1981 leakage through the joints, and between precast and in-situ beams was observed under a ten span post-tensioned bridge built in 1960 and the following year problems with a large segmental bridge were found during a principal inspection. Further investigations revealed serious corrosion of the tendons which was caused by road salt percolating through the deck waterproofing. The most serious event to-date occurred in December 1985 when a single span segmental post-tensioned bridge collapsed without warning[3]. The structure consisted of precast units post-tensioned together both longitudinally and transversely and chlorides had penetrated to the joints and corroded the tendons.

A sample of nine segmental bridges was subsequently examined and voids were found in seven of the structures and severely corroded tendons were found at two locations in one of the bridges. It was concluded that despite widespread deficiencies in grouting, the tendons were generally in good condition.

Despite the findings of the survey the number of structures with problems continued to increase and during the late 1980s and early 1990s about a dozen post-tensioned structures were found with serious corrosion problems and they required either major refurbishment or replacement.

There has also been a steady increase in problems reported abroad. In Japan, longitudinal cracks that followed the line of the ducts were observed in a post-tensioned bridge built in 1956. When it became redundant and was demolished 65% of the ducts were classified as fully grouted, 25%

were classified as over half full and the remainder were less than half full. About 10% of the tendons showed some corrosion but only 3% showed loss of section. In 1992 a post-tensioned bridge across the River Scheld in Belgium collapsed. It had been built in 1956 and collapse was due to fracture of the hinge joint at one end of the deck, which was caused by corrosion of the post-tensioning wires. Problems have also been reported in France, Germany and Italy[4].

This experience to-date has shown that the problem of tendon corrosion in post-tensioned bridges is increasing not just in the UK but also in the rest of Europe. Isolated voids in themselves are not a major problem. However if chlorides can penetrate to the tendons through vulnerable areas such as anchorages, joints or cracks in a structure then serious corrosion can occur. The difficulty of detecting corrosion and the absence of visual evidence of deterioration in most cases means that careful inspections and investigations are required.

Current Management of Post-Tensioned Bridges in the UK

Until 1992, the management of post-tensioned concrete bridges was no different to the management of other types of bridge structure. However the bridge authorities in the UK were concerned that problems on structures of this type were often being found by accident, for example when other work was being carried out on a structure. An investigation of an apparently minor problem sometimes revealed serious corrosion of the tendons that would otherwise have gone unnoticed. This led in 1992 to the then UK Department of Transport, announcing a five year programme of Special Inspections of its stock of existing post-tensioned concrete bridges. Following the announcement, the Department of Transport's Highways Agency developed a number of Standards and Advice Notes to assist Maintaining Agents manage the inspection programme. The areas addressed were as follows:

- Prioritisation for inspection (BD 54/93)[5]
- Planning and execution of the inspections (BA 50/93)[5]
- Strengthening, repair and monitoring (BA 43/94)[5]
- Assessment of concrete structures affected by steel corrosion (BA 51/95)[5]

Condition of post-tensioned bridges in the UK

Before the programme of Special Inspections started and in the first few years of its existence, the TRL had a commission from the Highways Agency to provide support for the programme. Initially, this was to review available methods of inspection including non-destructive techniques, collecting data on the findings of the programme in its early stages and disseminating this information to inspectors. Later, the primary task was to gather copies of the inspection reports, enter the important data on a database, and provide feedback to the HA on the progress of the programme and its outcome. When this work ended, there were 447 bridges on the database and full entries for 281 of these. An analysis of the data shows that overall, the condition of the bridges was found to be surprisingly good considering the problems that preceded the inspections and that led to its implementation.

Based on the sample of 447, voids were found in over 40% of the bridges inspected but only those that are sufficiently widespread to constitute an un-bonded system or contain water, chlorides or both are considered significant. The potential for water ingress in the future may be related to the leakage to atmosphere. About 20% of bridges contain large voids or ungrouted tendons. Of these, about 14% have a high incidence of occurrence (3% of the total sample). About 30% of bridges contain voids large enough to allow the tendons to be at least partially exposed. These observations relate to the condition at the inspected position and not the extent or volume of the voids. They also denote the worst case found on a bridge.

Based on year of construction, the earlier bridges in the sample are more inclined to have larger voids. In the 1950s and 60s, just over 50% of the bridges had small voids at worst (the remainder having larger voids). In the 1970s, the figure is 70%. For the 1980s and 1990s, there are relatively few bridges in the sample, but of these very few have significant voids detected. The late 1960s and early 1970s were the most productive years to date in the UK for the construction of post-tensioned bridges.

Compared with voids, the detected tendon corrosion is much less severe and more isolated in occurrence, almost 90% of bridges containing minor corrosion at worst (often just surface discoloration). Sometimes this can be widespread but more substantial corrosion is generally isolated. As expected, corrosion is more prevalent in older bridges, the 1950s being the worst with nearly one-quarter of the sample containing substantial corrosion. Very little substantial corrosion has been found in bridges from the late 1970s onwards.

The observations relating to year of construction must be treated with caution as they are based on small samples in each five-year period adopted in the analysis.

Other observations are based on the smaller number of 281 bridges inspected for which full entries are available on the database. For example, anchorage inspections were carried out in about 65% of these bridges, almost 50% of which were reported to contain minor corrosion at worst. Of the remaining 15%, 12% do not contain serious corrosion.

The presence of water, sometimes also with chlorides in the grout, is reported in 13% of the 281 bridges. Many of these are connected with corrosion classified as moderate or more. Most of the 13% have voids present in the ducts, a factor that arguably assists in water movement and suggests the potential for future deterioration.

In terms of overall condition, 3% of bridges are classified as having significant defects and a further 4% as needing attention. Over 50% are classified as being in good condition with the balance having minor problems. For the future, the important factors are likely to be the rate of deterioration of the post-tensioning systems and the frequency of inspection necessary to ensure continuing bridge safety.

Future strategy

In general, the results of the programme of Special Inspections are therefore showing that although voids are commonly found in post-tensioning ducts, very few bridges show evidence of serious corrosion.

The long-term integrity of these structures relies on the corrosion protection afforded by the grout and it is assumed that in well grouted ducts, the life of this corrosion protection will exceed the service life of the structure. However, the time for which this corrosion protection is effective is likely to be reduced in ducts containing voids and, as indicated by the problems that have occurred in the UK, may in some cases be less than the service life. As the procedures used for post-tensioning are similar throughout the world and many post-tensioned bridges have been designed and built by consultants and contractors with international experience, it is likely that the condition of structures in the UK is not unrepresentative of that in other countries. These structures should therefore be managed such that those where problems are likely to occur are identified and appropriate action is taken.

Developing a strategy that allows this to be done is far from straight forward and it is recommended that a staged approach is adopted as follows:

- classification of structures according to risk
- detailed inspection of structures deemed to be most at risk
- development of a management plan.

A more detailed flowchart showing the actions that could be taken is shown in Figure 1 and a description of the process is given below.

Classification

Classification of structures according to risk should be based on factors that might affect the susceptibility of bridges to corrosion of the tendons, such as type of construction, age of the structure, traffic loading, design and environment, and the consequences should this occur, such as degree of redundancy. The type of construction should consider whether details are present that might provide easy access for deleterious materials into the duct. Details that should be considered include:

- joints (construction joints in in-situ concrete and joints between in-situ and precast concrete and between precast elements.)
- deck slab
- detailing at anchorages
- duct linings
- deck waterproofing
- drainage systems.

All of these details will affect the protection given to the tendons. Other factors that will affect the durability of the structure, include the use of sea-dredged aggregates or admixtures containing chlorides, location of the structure, design and traffic loading.

The location of the structure is a major factor. For example, structures in a marine environment or in countries where de-icing salts are used during winter are more at risk than structures in a dry mild environment. Structures subjected to harsh extremes of climate are more at risk than those in moderate climatic areas.

The detailed design of the structures should also be considered. For example, the class of structure, ie whether cracks or tension are allowed under full design loading. If cracking is allowed, there is a reduction in the degree of protection provided to the tendons.

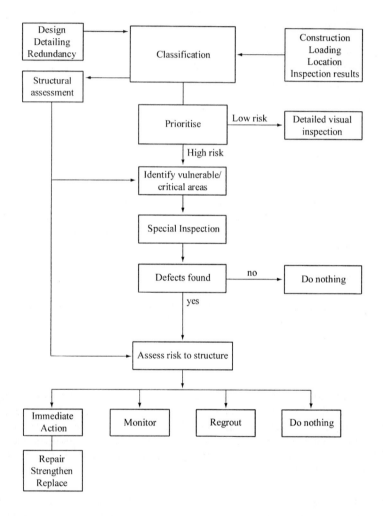

Figure 1: Staged Approach to Management of Post Tensioned Bridges

Traffic loading may accelerate deterioration, particularly if it causes cracks to open and close as it will make it easier for deleterious materials to gain access to the ducts. Such problems are more acute on structures used by overloaded vehicles.

The degree of redundancy also needs to be considered. Structures where failure of one member would not cause complete collapse, are less at risk than those with no redundancy. Structures that are particularly vulnerable, are those that use segmental construction or which rely on post-tensioned tie down tendons.

Finally the results of previous inspections should be examined to determine whether there is any evidence of distress. The reports should be reviewed for any evidence of water leakage through the structure, ingress of chlorides, rust staining or cracking, particularly along the line of the ducts.

Prioritisation

This information should be collected together and used to provide an initial classification of structures according to risk and to identify those areas on each structure that are deemed to be most at risk. The results of structural assessments should also be taken into account where available. For structures that are most at risk, a structural assessment should be undertaken to determine the load carrying capacity, whether it exceeds current design requirements and if so by how much. In addition, an assessment should be made of the sensitivity of the structure to loss of prestress.

Detailed inspections

Structures with the highest risk should be subjected to a special inspection (ie visual plus testing) with particular emphasis being paid to the areas that were classified as being most vulnerable to tendon corrosion. The inspections should focus on whether there is any evidence of distress such as water leakage through the structure, cracking along the length of the duct and rust staining.

The only reliable method of checking the actual condition of the tendons is to expose them and inspect directly. This can be done either by drilling or coring into ducts but needs to be done selectively as it damages the structure, there is a risk of damage to the tendons and it disturbs the environment inside the ducts. Once tendons have been exposed the opportunity should be taken to measure the volume of any voids present, check whether there is leakage between the ducts and the exterior, and take samples of grout for analysis.

The extent of testing required and the number of ducts that should be inspected, is a matter of engineering judgement. If a small number of ducts are inspected at locations where corrosion is most likely to occur and they are found to be fully grouted with no evidence of tendon corrosion or chloride ingress, then the risk to the structure is likely to be small. However if voids and corrosion are found, then the probability of such conditions occurring elsewhere on the structure is significantly increased and further investigations are required.

As far as the overall management of the bridge stock is concerned, sufficient inspections should be undertaken to give a reliable indication of the overall condition of the bridge stock and to enable structures where further action is required to be identified.

Management Plan

Where defects are found it is necessary to formulate a management plan for the structure. A structural assessment should be undertaken if one has not already been carried out and the findings should be taken into account in developing the plan. The options available are do nothing, re-grout voided ducts and monitoring. Where there is deemed to be a risk to the structure then more positive action should be taken.

For structures where there is no evidence of any problems and voids, if present, are dry and there is no evidence of corrosion, do nothing is the appropriate action. Such structures should continue to be regularly inspected and the decision revised should any evidence of deterioration be detected.

Where there are extensive voids re-grouting should be considered, both to improve the protection of the tendons and restore the structural advantage that bonded tendons have over un-bonded tendons. The need to re-grout voided ducts is a matter of judgement and the decision on whether or not to recommend it, will generally depend on the vulnerability and importance of the tendons and the practicality of accomplishing the task. Small isolated dry voids pose less of a problem than large voids containing exposed tendons, leakage to atmosphere, dampness and corrosion. Wholly un-grouted tendons are potentially vulnerable and because they are un-bonded may lead to a lower assessed capacity for the member. Moreover, a fracture will affect the tendon along its whole length because re-anchorage cannot occur. Where chlorides are present, re-grouting cannot be relied upon to prevent further corrosion.

Re-grouting of voided ducts has been carried out in the UK to a small but apparently growing extent. This may be because the management of the structures inspected in the Programme of Special Inspections is beginning to lead to activity on the ground. Another factor may be that engineers were initially reluctant to undertake this work before there was sufficient experience to ensure a satisfactory outcome.

It has been shown that re-grouting is a practical exercise for some types of bridge, ducts and voids provided access is satisfactory. To date, it has generally entailed drilling a large number of holes into the concrete to gain access to the ducts, to act as injection or venting points. Some engineers feel this is undesirable because of the damage caused.

When a structure or member contains relatively few tendons, each one is likely to be structurally important. This may not be the case when there are a great many tendons. In addition, when there are many tendons, re-grouting may be impractical because of the scale of the work. It should be borne in mind that the inspection may only have covered a small proportion of the tendons and relatively little may be achieved by re-grouting just these.

Where re-grouting is impractical but the tendons are vulnerable to corrosion and fracture or the structure is in an acceptable condition at present but there is concern that there is a risk of corrosion and possibly tendon fractures in the future, monitoring should be considered. The overall effect of a single wire fracture is extremely small. There is no measurable increase in overall deformation of the structure and although there might be a small local change in strain it is extremely difficult to detect. To do so would require strain gauges to be mounted at all locations where wire fractures might occur and continuous monitoring is required in order to distinguish between strain changes due to wire fractures and other effects such as temperature changes.

Another option is acoustic monitoring. This requires acoustic transducers to be positioned on a structure and then monitored to detect the energy released as a wire fractures. It is possible to determine the location of events from the time of arrival of the acoustic signals at different transducers. The equipment used and the associated analysis software must be able to distinguish between wire fractures and other acoustic events such as traffic noise and vehicles crossing expansion joints. Trials at TRL have demonstrated that acoustic monitoring can be used for this purpose but specialist equipment that has been specifically developed for detecting wire fractures should be used[6].

An alternative approach is to install corrosion monitoring equipment in selected ducts. This would enable the environment within the ducts to be monitored continuously and any changes that might lead to corrosion could be identified. This has the disadvantage that only discrete locations are being monitored rather than the whole structure, but they could represent a worst case scenario, since the environment within the ducts will have been disturbed during the installation process.

For some structures immediate action might be required. In the worst case this might entail closure of the structure. In other cases weight or lane restrictions or propping might be sufficient. A decision then needs to be taken to determine whether the structure should be repaired or replaced and this should be based on the whole life cost of the options available. This should take account of all costs including both traffic management and disruption to users.

Where remedial action is the preferred option, it is necessary to apply measures that reduce or prevent future corrosion and replace any prestress that has been lost. The installation of external tendons can be used for this purpose although care is required to ensure that the structure is not overstressed. Measurement of concrete stress at selected locations should be considered using one of the techniques that have been developed in recent years.

Conclusions

Post-tensioned concrete bridges have given good service over the last fifty years. However where problems do occur they can lead to failure.

The procedures described in this paper should enable a stock of post-tensioned bridge to be managed to provide a safe and economic service for the community.

If properly managed post-tensioned concrete bridges will remain a popular and economic form of construction for spans in the range 30m to 50m.

Acknowledgements

The work described in this paper is based on research undertaken by TRL on behalf of the UK Highways Agency.

References

1. Sriskandan K. 1989. *Prestressed concrete road bridges in Great Britain: a historical survey.* Proc Instn Civ Engrs, Part 1, 86, Apr, 269-303.

2. Woodward R J, 1981. *Conditions within ducts in post-tensioned prestressed concrete bridges.* TRRL Laboratory Report, LR 980. Transport and Road Research Laboratory, Crowthorne, Berkshire.

3. Woodward R J and F Williams. *Collapse of Ynys-y-Gwas bridge, West Glamorgan.* Proc Instn Civ Engrs, Part 1, Vol 84. Aug 1988. pp635-669.

4. Storrar, D B, 1993. *The management of post-tensioned grouted-duct bridges owned by the Department of Transport.* Proceedings of fifth international conference on structural faults and repair, Edinburgh, 29 June 1993. Editor M C Forde. Engineering Technics Press, Edinburgh.

5. Design Manual for Roads and Bridges. Vol 3 Section 1: Part 2. BD 54/93. *Post-tensioned concrete bridges. Prioritisation of Special Inspections.* Vol 3 Section 1: Part 3. BA 50/93. *Post-tensioned concrete bridges. Planning, organisation and methods for carrying out Special Inspections.* Vol 3 Section 3: Part 2. BA 43/94. *Strengthening, repair and monitoring of post-tensioned concrete bridge decks.* Vol 3 Section 4: Part 13. BA 51/95. *The assessment of concrete structures affected by steel corrosion.* Department of Transport. London

6. Cullington D W, MacNeil D, Paulson P and J Elliot. (2001). Continuous acoustic monitoring of grouted post-tensioned concrete bridges. NDT&E International 34. pp 95 – 105.

Stonecutters Bridge Design Competition – Aesthetic Consideration

MICHAEL C H HUI, CHRIS K P WONG
Major Works Project Management Office, Highways Department,
The Government of the Hong Kong Special Administrative Region

Abstract

Highways Department conducted an international design competition for Stonecutters Bridge. The competition attracted renowned bridge designers from all over the world to submit a variety of proposals. These proposals contained many innovative ideas and imaginative concepts. Different bridge forms, pylon shapes, cable and deck arrangements were proposed by the competitors. The designs expressed the aesthetic qualities in the perception of both the road users as well as observers from different locations in the harbour and the city. Some of the competitors explained in detail the themes of their designs in the competition entries.

This paper first describes the aesthetic evaluation criteria adopted in the assessment of competition entries. It then describes the aesthetic design evolution process of those Stage 2 entries and how the major comments given by the judges in Stage 1 were addressed in the Stage 2 submissions.

Introduction

Stonecutters Bridge will have a span exceeding 1000m and will be situated in the famous Hong Kong Harbour. To conduct an international design competition was seen to be the right step for procuring the conceptual design of the record-breaking bridge located in this unique maritime setting. The design competition did not only aim at obtaining a competent design for the longest span cable-stayed bridge in the world, it also provided a good opportunity to look for a design that could make the new Stonecutters Bridge stand out amongst the world's long span bridges and become a fitting landmark in Hong Kong.

The competition was conducted in two stages, with 27 entries in Stage 1 whittled down to 5 in Stage 2. The entries were assessed by two committees, i.e. the Technical Evaluation Committee and Aesthetic Evaluation Committee, each of which comprised international and local experts as judges. At Stage 1, the 5 design shortlisted were selected purely based upon the marks given by the two committees. At Stage 2, the two committees also evaluated the schemes. They then submitted the marks and evaluation reports of the 5 finalists to an Executive Panel who selected the winner by voting.

The design competition attracted the world's leading bridge designers to put up a great variety of proposals. The design concept and aesthetic considerations varied very much but a common goal was to provide a technically sound and aesthetically pleasing design for this landmark structure.

Aesthetic Evaluation Criteria

It is accepted that it is not possible to lay down clear-cut rules for appearance in the same way as for more technical matters, and that the aesthetic evaluation therefore inherently is a subjective assessment. However, a bridge is a structure, which has to satisfy a need, get built and be durable, maintainable and economical, and at the same time it also has to fit into its environment and be elegant. In a properly balanced and integrated approach to its design, the appearance of a bridge takes its rightful place alongside function, structure/construction and economy.

While it is inappropriate to restrict the work of the Aesthetic Evaluation Committee to the consideration of a set of fixed evaluation criteria, some guidelines were given. These guidelines were not meant to be prescriptive, but to help ensure that all aspects of visual excellence were considered and evaluated, irrespective of design approach.

The aesthetic evaluation was under two headings both carrying equal weight in the evaluation. Under each heading a number of subheadings were considered. Below is an evaluation guide to some important aspects of the design:

- *The bridge in the landscape (carried 50% weighting):*
1) Relationship to the site
2) Scale and harmony in the context of the environment
3) Relationship to the adjacent environment
4) Relationship to the adjoining viaducts
5) Cultural context

- *The bridge itself (carried 50% weighting):*

Fundamental aspects:
1) Elegance, simplicity - scale and proportion
2) Consistency of design, unity of appearance of the various parts – form and harmony

Other aspects:
3) Expression of function
4) Expression of structural behaviour
5) Expression of construction technique
6) Quality of details
7) Colour and tone - texture
8) Durability and maintenance, materials, surfaces and weathering
9) Bridge furniture – drainage, parapets, signage, lighting etc.

Aesthetic Design Evolution Process of the 5 Finalists

Five proposals were selected to enter Stage 2 of the competition. The competitors were required to develop their schemes in much more detail to cover all aspects of the evaluation criteria. In addition to the reports and drawings, the competitors were required to submit 2 physical models of their bridges. The first one is of a scale of 1:2000 showing the bridge in its entirety. The second one is a smaller model of a scale of 1:4000 built on a model base block supplied by Highways Department. This smaller model can be inserted into a background model of the surrounding area to facilitate the judges to assess the appearance of the bridge in the landscape. In order to facilitate design evolution, the competitors were informed of the judges' comments on their Stage 1 proposals. In refining their schemes, the competitors were requested to address these comments. The following is a description of their aesthetic design considerations. It also describes how the competitors addressed the judges' major comments during the design evolution process.

Proposals Given Honourable Mentions

A cable-stayed bridge with A-shaped towers and stainless steel cladding at the top

It is a cable-stayed bridge with A-shaped towers. The cables are arranged in a semi-fan configuration. The bridge deck consists of a concrete box in the side spans and 52m into the main span. The remainder of the main span adopts a double steel box system with an open structure in the middle. The tower top is designed as a steel core covered by stainless cladding.

The A-shaped pylon has a steel core at the pylon top covered by corrugated steel cladding. The pylon top will be equipped with devices providing illumination through the corrugated steel system. Because the bridge will form the gate to one of the most important and busiest harbours in the world, the designer considered it given extra meaning by designing the pylons to act symbolically as navigation signs for ships. According to international rules of marine traffic, the top of the right pylon will send green light while the top of the left pylon will send red light when ships enter the harbour. The idea of providing a lighting system at the top was commended by the judges but the use of green and red light did not impress the judges. The major comments from the judges include:
- The pylon shape was the strong part of the design but the lower part of the pylon might benefit from further refinement.
- The cable connections at the pylon top is unresolved and must be shown fully.

In the Stage 2 submission, the designer modified his scheme by thickening the leg of the pylon, which, in the opinion of the judges, ruined the proportions. Furthermore, the design of the pylon top was not considered co-ordinated as the details seemed to compete.

The design was submitted by Consulting KORTES Ltd. in association with Finroad Ltd.

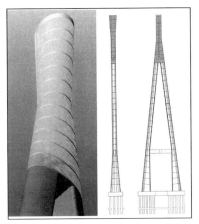

Fig. 1 - Stage 1 Pylon Shape

Fig. 2 - Stage 2 Pylon Shape

A cable-stayed bridge with inverted Y-shaped towers and "deviators"

It is a cable-stayed bridge with inverted Y-shaped towers. The cables are arranged in a semi-fan in the longitudinal direction of the bridge. In the transverse direction, the stay cables in pairs are attached to the outside of the girder and the two cables are pulled together through a deviator to a central position approximately halfway towards the tower. The bridge deck consists of a concrete box girder in the side spans and 100m into the main span, and a closed steel box in the remainder of the main span.

The design was primarily driven by the use of deviators to achieve a special three-dimension cable net system with a central node. Such a cable net solution was proposed to reduce wind drag and to control cable-vibration. At the same time, the designer claimed that the stay cable system would offer tremendous aesthetic opportunities due to its three dimensional nature. At Stage 1, the designer proposed to use a dark colour for the upper portion of the cables and a light colour for the lower portion. The dark-colour upper cable plane will be prominent in daylight while it will nearly disappear at night. In contrast, the lower portion of the cable arrangement has the reverse effect. The major comments from the judges include:

- The different colour on each side of the deviator is visually confusing.
- It is not apparent that substantial benefits are achieved by the inclusion of a cable deviator.
- As a consequence of the long vertical pylon top, the pylon legs appear to be spread very wide at the base.

Fig. 3 - Stage 1 Scheme

In the Stage 2 submission, the designer modified the proposed solution to maintain a constant cable colour throughout to avoid visual confusion. The designer instead adopted an aesthetic lighting strategy to highlight the fan-like quality of the upper cables at night time. For other aspects, the designer maintained the presence of the deviators and very wide pylon base. In the opinion of the judges, the presence of deviators, although interesting, does not have appreciable aesthetic justification. Furthermore, the very wide pylon base does not impart a feeling of strength and this is further undermined by the emphasis put on the deviators as illustrated in the night time view.

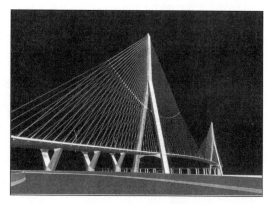

Fig. 4 - Stage 2 Scheme

The scheme was submitted by HNTB team. The bridge architect was Wolf Architecture.

The Second Runner-up Scheme

It is a cable-stayed bridge with H-shaped towers without cross beams above bridge girder level. The cables are arranged in a semi-fan configuration in two vertical planes. The bridge deck consists of a closed steel box girder over the entire length of the bridge with ballast concrete inside the box girder in the side spans.

Fig. 5 - Polygonal Pylon Shape

The designer's idea was to develop a design that represented a step forward both in architectural style and in engineering concept – a bridge that was elegant and a style that was new to both Hong Kong and to the world. The designer explained in his submission that Stonecutters Bridge would form a sculptural portal of grand style and dignity. It would create a monumental gateway for ships as they enter and depart the harbour. For motorists, it would be a ceremonial connection between Stonecutters Island and Tsing Yi.

Tapered twin towers would form landmark gateposts at bridge-ends. No crossbeam was proposed above the roadway to obscure the view of the towers, nor to diminish their symbolism. During the design process, the designer considered many tower shapes: single pole, inverted-Y, and diamond, amongst others. The designer felt that all popular forms would no longer arouse the public and that for the world's longest cable-stayed span bridge, replication of another bridge was unbefitting. Although the double pole towers had been used before, the designer considered that none were of such a magnificent scale. The towers were proposed to be of polygonal shape to provide aesthetically pleasing architectural features. The sculpted towers would reach upward illustrating power and strength. At the top of each tower, a lantern of gold-coated glass would catch sunlight and reflect it in all directions. At night, the lanterns would mark the tower tops.

The judges commended that the bridge was elegant and impressive. They considered that the vertical and elegant slender pylons would sit comfortably in the landscape and would fit well with the existing major bridges nearby. Other major comments include:
- The connection between the anchor piers and the adjoining viaduct piers needs refinement.
- The relationship with the adjoining viaduct should be given further consideration.

In the Stage 2 submission, the designer accordingly made revisions to address the judges' concerns. However, the effect of the protruding anchor piers prompted much discussion, some members felt them to be an unnecessary detail while others saw them as symbolizing a gateway to the bridge.

The design was submitted by T.Y. Lin International in association with Gensler Architecture Design & Planning Worldwide and other team members.

Fig. 6 - Stage 1 Back-Span Arrangement

Fig. 7 - Stage 2 Back-Span Arrangement

The First Runner-up Scheme

It is a cable-stayed bridge with A-shaped pylons. The cables are arranged in a fan configuration. The bridge girder consists of a concrete box girder in the side spans and 75m into the main span, and a closed steel box girder in the remainder of the main span.

At the outset of the proposal, the designer stated the theme "Heavenwards" by referring to the characteristics of Chinese calligraphic characters. The two slight curved legs of the 300m pylon of this bridge symbolize the two legs of the Chinese calligraphic characters "Man" (人) and "Sky" (天 which gave rise to the working title "Heavenwards".

Fig. 8
"Heavenwards" theme

The scheme is characterized by inwardly curved A-shaped pylons with a damper box at the tower top. It was the scheme which had undergone most changes in the aesthetic design after taking into account the comments given by the judges in Stage 1. Major comments include:
- The pylon top was commented to be over-designed in the way it attempts, in an overly literate gesture, to mirror the stroke in the Chinese character for 'Man'.
- The size of the damper box seems exaggerated.
- The side span piers distract from the statement made by the strong pylon design.

In Stage 2, the simulation of "Man" and "Sky" at the pylon top was played down. The large yellow damper box was changed to a grey colour and its size was also trimmed down. The wide spread legs of the side span piers were also changed to bend inwards in order to match the approach viaducts. Overall speaking, the judges considered that the pylon itself was elegant but expressed a concern that the appearance of the curved legs of the pylons were not in harmony with the surroundings.

The design was submitted by Scott Wilson Hong Kong Ltd. in association with Leonhardt Andra and Partners and other team members. The bridge architect was Planungsgruppe Professor Laage.

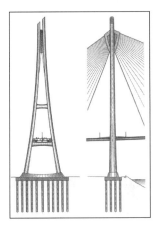

Fig. 9 - Stage 1 Pylon and Damper Box Arrangement

Fig. 10 - Stage 1 Back-Span Arrangement

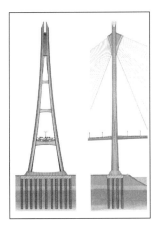

Fig. 11 - Stage 2 Pylon and Damper Box Arrangement

Fig. 12 - Stage 2 Back-Span Arrangement

The Winning Design

It is a cable-stayed bridge with tapered cylindrical pylons. The cables are arranged in a semi-fan configuration. The bridge deck consists of twin box concrete girders in the side spans and 24m into the main span and, and twin box steel girders in the remainder of the main span. The two boxes are interconnected with lateral members.

Fig. 13 - A View from under side of the bridge of the Winning Scheme

The designer came up with a bold yet simple scheme – a record span cable stayed bridge with two single tapered cylindrical pylons. The use of a steel top tower section would not only enable rapid construction but also create a sense of a "modern-look" for the bridge which is to be built in the twenty first century. Apart from the bridge itself, the designer also carefully considered the relationship with its surroundings. The bridge defines the entrance to Hong Kong's busiest port area, one of the world's leading sea ports, characterized by enormous activity and dynamic growth. With the elegant towers and the structural clarity of the bridge, the designer believes that the bridge will be an attractive and dynamic contribution to Hong Kong both during daytime and at night.

The judges considered the design an impressive and powerful scheme. They were of the view that the built landscape in the surroundings comprised predominantly vertical forms in a dense configuration with a mountainous backdrop and the busy harbour within which the single vertical pylons would be compatible. The pylons and structure as a whole would complement and enhance the setting. The elegant bridge would fit well and belong with the existing three major bridges nearby. The bridge silhouette will be recognizable from all view points – the shape does not change significantly according to the view.

The judges also commended the split deck with a crossbeam curved in profile. The view from below the bridge deck was considered to be magnificent from both the sea and from the ground level around the container terminals. The proposed configuration, along with the simple lines of the towers, would offer beneficial opportunities for dramatic lighting of the bridge.

Despite the above commendation, some concerns were raised after the Stage 1 evaluation:
- The competitors were cautioned about the possible joss stick analogy in the pylon design.
- The lighting design at the top of the pylon appears exaggerated.

As a matter of interest, in refining the scheme at Stage 2, the designer consulted a Feng Shui expert who advised that for the joss stick analogy to be relevant, the bridge would require three towers. Nevertheless, the lighting feature at the pylon top was reduced in scale and intensity, lessening its impact for a more subtle appearance. The bridge was in fact likened to a dragon, the lighting features marking the dragon's eyes, the curving approach viaducts its tail. The dragon would guard the channel entrance offering a safe haven for those within. The Feng Shui expert commended the design and suggested that with the eyes to light the harbour, it would be prettier and luckier than before.

The winning design was submitted by a team comprising Halcrow Group Limited, Flint & Neill Partnership, Dissing + Weitling and Shanghai Municipal Engineering Design Institute.

Fig. 14 - Stage 1 Pylon and Lighting Design

Fig. 15 - Stage 2 Pylon and Lighting Design

Conclusion

Before deciding to go ahead with the design competition, there were worries that the room for the variance of the appearance of the bridge could be limited as its basic form might be driven by the technical requirements. On the contrary, the design competition has attracted designers to submit a variety of proposals with very different shapes. These design proposals have provided inspiring and imaginative ideas. It can be seen that apart from the technical side, the competitors have spent a lot of effort in coming up with proposals with high aesthetic qualities. The winning design is highly praised for its suitability in serving as an icon in Hong Kong. The outcome of the competition is promising.

Acknowledgement

This paper has been published with the permission of the Director of Highways, the Government of the Hong Kong Special Administrative Region.

Long Term Fatigue Behaviour of Steel Girders with Welded Attachments under Highway Variable Amplitude Loading

SAKANO, MASAHIRO
Kansai University, Osaka, Japan

1. Introduction

Long life fatigue behaviour is of great importance in predicting the remaining life of existing bridges as well as in designing new bridges, because the service life of highway and railway bridges is usually expected to be so longer than 100 years. The author previously investigated the long life fatigue behaviour of transverse stiffener joint and web penetration joint in steel railway bridges [1],[2] and of transverse stiffener joints in steel highway bridges [3] under variable amplitude loading.

In this study, the long life fatigue behaviour of steel highway bridges is investigated through variable amplitude fatigue tests using steel girder specimens with web gusset welded joints and flange attachment welded joints under computer-simulated variable amplitude loading with the number of loading cycles more than 10 to 100 million cycles.

2. Fatigue Test

2.1 Specimens

Figures 1-3 show three types of plate girder specimens with web gusset joint and flange attachments; a longitudinal hanger piece and a patch plate to suspend maintenance facilities. Fatigue cracks developed in these structural details are very fatal to break up the bottom flange of girders.

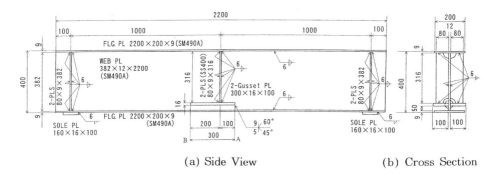

(a) Side View   (b) Cross Section

Fig. 1   Web Gusset Specimen (WG Specimen)

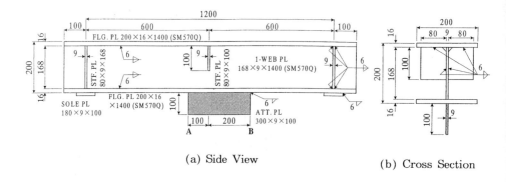

Fig. 2 Hanger Piece Specimen (HP Specimen)

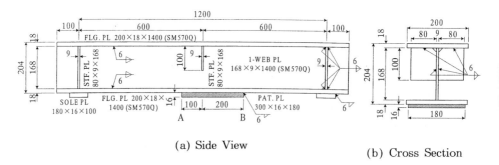

Fig. 3 Patch Plate Specimen (PP Specimen)

2.2 Variable Amplitude Loading

Figure 4 shows an example of the variable amplitude bending moment fluctuation generated by the computer simulation for highway live loading [4]. Figures 5-8 show examples of the variable amplitude bending moment fluctuation and its moment range histogram used in fatigue tests on web gusset specimens (Figs. 5 and 6) and flange attachment specimens (Figs. 7 and 8). Moment ranges higher than 85 % of the maximum value are cut off due to the loading capacity of fatigue testing machine, and moment ranges lower than 20 % (for web gusset specimens) or 30 % (for flange attachment specimens) of the maximum value, which corresponds to the cut-off- limit for variable amplitude stresses recommended by JSSC [5], are cut off in order to save time for fatigue tests.

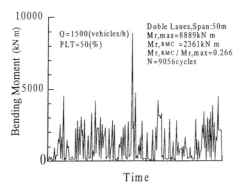

Fig. 4 Variable Amplitude Loading Fluctuation generated by Computer Simulation

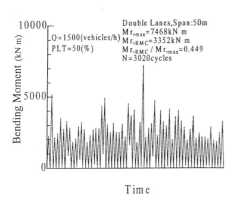

Fig. 5 Loading Fluctuation (WG Specimen)

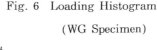

Fig. 6 Loading Histogram (WG Specimen)

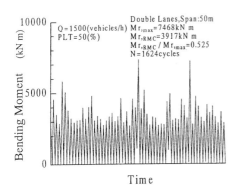

Fig. 7 Loading Fluctuation (HP&PP Specimens)

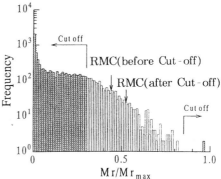

Fig. 8 Loading Histogram (HP&PP Specimens)

3. Fatigue Behaviour of Welded Girders

Tables 1-3 give fatigue test results obtained in this study. $\Delta \sigma_1$ represents the maximum principal stress range in the web plate, which is calculated from bending moment and shearing force at the section. $\Delta \sigma$ represents stress range in the bottom flange, which is calculated from bending moment at the section. Nd is fatigue crack detection life defined as the number of stress cycles until fatigue cracks are detected. The surface length of detected cracks is represented by 2bd. Nf is fatigue failure life defined as the number of stress cycles until the bottom flange fails.

Table 1 Variable Amplitude Fatigue Test Results (WG Specimen)

Specimen No.	Load Range ΔP_{max} (kN)	Section	Stress Range $\Delta \sigma_1$ (MPa)			Fatigue Life (Mcycles)		Crack Length
			max	eq	min	Nd	Nf	2bd (mm)
1	235	A	80.5	36	19.2	7.4	23.6	10
		B	73.6	33	17.5	10.3	26.1	7,5
2	196	A	67	30	16	9.7	31.7	12
		B	59.6	26.7	14.2	>50.0		-

Table 2 Variable Amplitude Fatigue Test Results (HP Specimen)

Specimen No.	Load range ΔP_{max} (kN)	Section	Stress Range $\Delta \sigma$ (MPa)			Fatigue Life (Mcycles)		Crack Length
			max	eq	min	Nd	Nf	2bd (mm)
1	118	A	47	24.6	16.8	4.59	24.7	10
		B	35.9	18.8	12.9	26.6	76.3	7
2	98	A	33.8	17.7	12.1	16.5	60.3	20
		B	33	17.3	11.8	6.1	44.6	16

Table 3 Variable Amplitude Fatigue Test Results (PP Specimen)

Specimen No.	Load range ΔP_{max} (kN)	Section	Stress Range $\Delta \sigma$ (MPa)			Fatigue Life (Mcycles)		Crack Length
			max	eq	min	Nd	Nf	2bd (mm)
1	176	A	63.9	33.5	22.8	65.1	86.6	30
		B	49.3	26	17.7	109.7	153.9	13

3.1 Crack Initiation and Propagation Behaviour

Figures 9-14 show fatigue cracks and fracture surfaces for each type of specimen. In the WG specimen, fatigue cracks are initiated at the toe of turn-round fillet weld on web plate near the edge of gusset plate, and propagate in the direction perpendicular to the maximum principal stress until breaking out the bottom flange. In the HP specimen, fatigue cracks are also initiated at the toe of turn-round fillet weld but on the attachment-side near the edge of the attachment, and propagate through fillet welds to penetrate into the bottom flange and the web plate. In the PP specimen, fatigue cracks are initiated at the root of transverse fillet weld, and propagate through fillet welds to penetrate into the bottom flange and the web plate too.

Fig. 9 Fatigue Crack (WG Specimen)

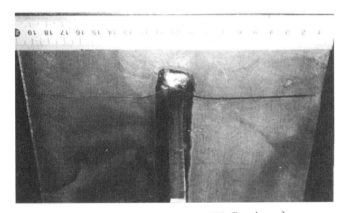

Fig. 10 Fatigue Crack (HP Specimen)

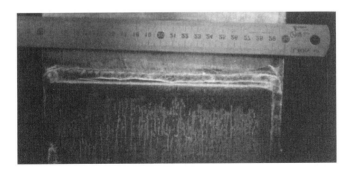

Fig. 11 Fatigue Crack (PP Specimen)

Fig. 12 Fracture Surface (WG Specimen)

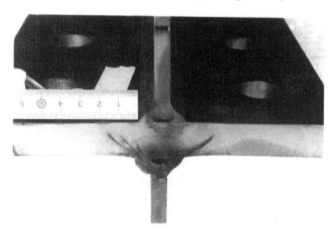

Fig. 13 Fracture Surface (HP Specimen)

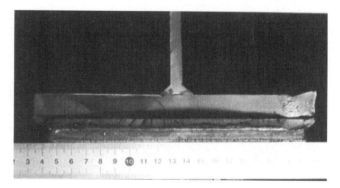

Fig. 14 Fracture Surface (PP Specimen)

3.2 Fatigue Strength

Figures 15-17 show S-N relationships for three types of specimens. Constant amplitude fatigue test results [6]-[8] are also plotted in these figures. Fatigue failure life of web gusset joints plotted against the equivalent nominal principal stress range satisfies the class F fatigue design curve with a sufficient margin, which is one rank higher than the class G recommended by JSSC [5]. Fatigue failure life of hanger piece joints plotted against the equivalent nominal bending stress range do not satisfy the class G fatigue design curve recommended by JSSC [5] in the low stress region. Fatigue failure life of patch plate joints plotted against the equivalent nominal bending stress range satisfies the class E fatigue design curve with a sufficient margin in the low stress region, which is one rank higher than the class F recommended by JSSC [5].

4. Conclusions

The principal results obtained through this study were as follows:
(1) Fatigue failure life of web gusset joints and patch plate joints plotted against the equivalent nominal principal stress range satisfies the one rank higher fatigue class recommended by JSSC.
(2) Fatigue failure life of longitudinal hanger piece joints plotted against the equivalent nominal bending stress range satisfies the one rank lower fatigue class recommended by JSSC.

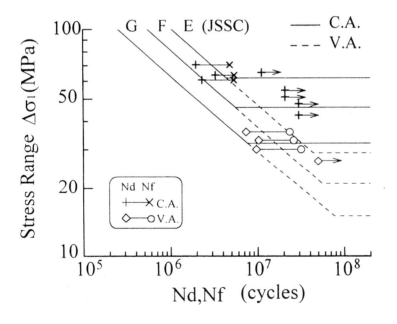

Fig.15 S-N Diagram for WG Specimen

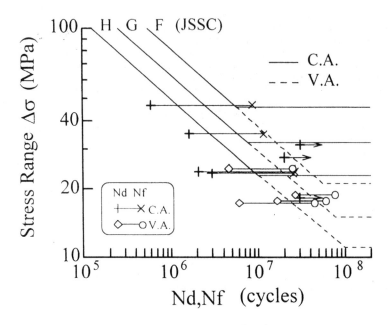

Fig.16 S-N Diagram for HP Specimen

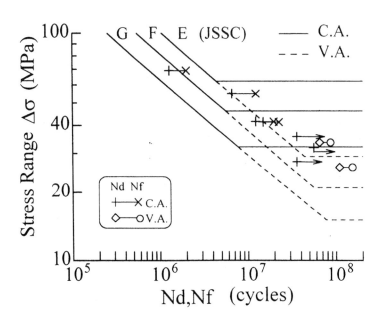

Fig.17 S-N Diagram for PP Specimen

References

[1] Sakano,M. and Wahab,M.A.: Fatigue strength of welded transverse stiffener joints under variable amplitude loading, International Journal of Pressure Vessels and Piping, Vol.75, No.15, pp.1037-1045, 1998.
[2] Sakano,M. and Wahab,M.A.: Fatigue strength of railway floor beams with web penetration, Proc. of the International Conference on Mechanics of Structures, Materials and Systems (MSMS'99), Wollongong, Australia, pp.83-89, 1999.
[3] Sakano,M., Yonemoto,E, Kano,K. and Mikami,I.: Fatigue strength of vertical stiffener joints in steel highway bridges, Journal of Structural Mechanics and Earthquake Engineering, No.612/I-46, pp.31-43, 1999.
[4] Sakano,M., Mikami,I. and Hori,K.: Simultaneous loading factors for fatigue assessment of urban expressway bridges, Journal of Structural Engineering, Vol.41A, pp.855-863, 1995.
[5] Japanese Society of Steel Construction: Recommendations for fatigue design of steel structures, 1993.
[6] Sakano,M., Hohzumi,M., Shimora,T. and Mikami,I.: Long term fatigue strength of web gusset joint in floor-beam to main-girder connection, Steel Construction Engineering, Vol.5, No.18, pp.31-40, 1998.
[7] Sakano,M., Mikami,I., Yonemoto,E. and Sakuragi,D.: Long term fatigue behavior of patch-plate-type flange attachment joint, Steel Construction Engineering, Vol.3, No.9, pp.67-74, 1996.
[8] Sakano,M. and Yonemoto,E.: Long term fatigue behavior of plate girder bottom flanges with longitudinal out-of-plane attachments, Steel Construction Engineering, Vol.7, No.28, pp.1-10, 2000.

A LANDMARK STRUCTURE OVER THE CHARLES RIVER IN BOSTON, MASSACHUSETTS

VIJAY CHANDRA, P.E.
Senior Vice President
Parsons Brinckerhoff Quade & Douglas, Inc.
New York, New York, USA

ANTHONY RICCI, P.E.
Chief Bridge Engineer, CA/T Project
Massachusetts Turnpike Authority
Boston, Massachusetts, USA

KEITH DONINGTON, P.E.
Project Engineer, CA/T Project
Bechtel/Parsons Brinckerhoff
Boston, Massachusetts, USA

PAUL TOWELL, P.E.
Senior Engineer, CA/T Project
Bechtel/Parsons Brinckerhoff
Boston, Massachusetts, USA

Introduction

Boston's immense Central Artery/Tunnel (CA/T) Project consists of many kilometers of tunnels, six major interchanges, and two long-span parallel crossings of the Charles River, one connecting to Storrow Drive and the other a cable-stayed bridge connecting to I-93 called the Leonard B. Zakim Bunker Hill Bridge (Bunker Hill Bridge). The project's keystone is the Bunker Hill Bridge, featuring a landmark hybrid cable-stayed bridge that will be the first of its type in the US. The river crossing will provide virtually the only access to Boston from the north, straddling the Charles River in a historic area where Paul Revere began his famous ride and the Battle of Bunker Hill took place. Accordingly, there is great interest in ensuring that the bridge features a distinctive design while also providing a dramatic gateway to the city's downtown.

This paper summarizes the authors' first-hand knowledge of the development of this unique bridge from concept to construction.

Design Constraints

Major physical constraints for the Bunker Hill Bridge include:
- Existing Orange Line subway ventilation building adjacent to the south main pier
- Existing Orange Line tunnel alignment traversing below the bridge
- Existing 0.92-meter-diameter waterline located below the south main pier
- Steep 5 percent grade entering a tunnel at the south end of the area, tying into a three-level interchange at the north end
- Existing Charles River lock and dam system abutting the bridge on the east side

Bridge Type Selection

The Bunker Hill Bridge had to meet the objectives of numerous state and federal regulatory agencies, including the Federal Highway Administration (FHWA). The bridge design had to present sound engineering solutions to numerous site constraints while also meeting community expectations that the structure create a distinctive "signature" on Boston's skyline. To fulfill these goals, the project team conducted a bridge type study (design charette) and assembled a multidisciplinary team of experts in structural engineering, inspection and maintenance, highway design and engineering, urban planning, construction, environmental engineering, architecture, cost control, and scheduling.

Over 16 bridge types, ranging from trusses to suspension bridges, were studied and seven of them were short listed. The short listed types included an arch, a truss, and five types of cable-stayed bridges. After an evaluation matrix was prepared for the seven bridges, the concept put forward by Dr. Christian Menn was finally selected as the most appropriate bridge for the I-93 crossing of the Charles River, providing a "signature bridge" for Boston (see Fig. 1).

Fig. 1: Leonard B. Zakim Bunker Hill Bridge model

Preliminary designs were then prepared for steel, concrete, and hybrid alternates for the asymmetrical cable-stayed structure. Although each alternate posed unique challenges, a committee composed of international bridge experts concurred with the project team that only the hybrid alternate should be pursued. Due to the relatively short south back span (compared to the main span), the back spans feature "heavy" cast-in-place post-tensioned concrete construction to counterbalance the "light" main span constructed of steel floor beams and edge girders with a precast concrete composite deck. The bridge will be the first hybrid cable-stayed structure built in the US.

Site-Specific Seismic Study

The Bunker Hill Bridge will be a lifeline to Boston. Like all major interstate highway structures, it is considered "important" for seismic considerations. This means that the bridge should be serviceable after a design earthquake, sustaining only minor damage. Therefore, due to the structure's critical role and its unusual design features, a site-specific seismic study was undertaken.

Evaluating seismic sources in the New England area is based on the latest developments in seismic source zone characterization and attenuation relationships for the eastern US. Response spectra for both 2 percent and 5 percent damping for the 500-year operating design earthquake and 2,000-year maximum design earthquake return periods were developed for the bridge.

Bridge Configuration

The shape of the tower piers and the cable arrangement (main span cables splayed out from the tower with back span cables centered in the median) evolved principally from technical considerations with aesthetics playing an important role. The proximity of existing double-deck ramps at the south end, which need to remain in service during construction of the new bridge, dictated that the back span cables be anchored in the median and not splayed out similar to the main span cables.

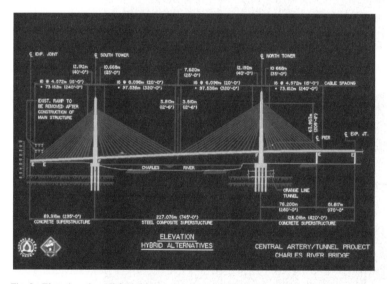

Fig. 2: Elevation view of the bridge

The Bunker Hill Bridge features a 56.4-meter-wide, five-span hybrid superstructure with a main span of 227 meters; two south back spans of 34.2 and 39.6 meters; and two north back spans of 51.8 and 76.2 meters (see Fig. 2). The tower piers are inverted Y shapes (see Fig. 3). The tops of the south and north towers rise 89.9 and 98.5 meters from their foundations, respectively. The towers include ladders inside their hollow legs to provide access for inspection and maintenance to the cable anchorages as well as the aviation beacons on top. The back spans consist of multi-cell concrete box girders, 3 meters deep and 38.4 meters wide. Main structural elements include a 3-meter-wide central spline beam with internal floor beam diaphragms at 4.6 meters on center, framed with four secondary webs (see Fig. 4). The spline beam in turn is supported by a single plane of cables spaced at 4.6 meters.

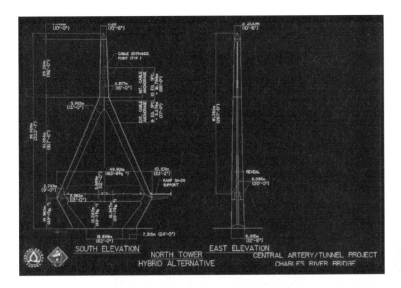

Fig. 3: North tower view

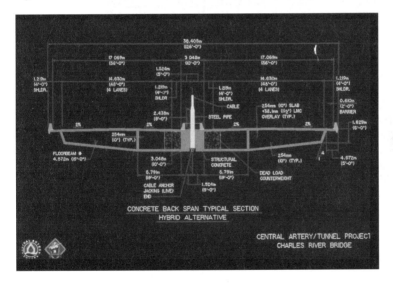

Fig. 4: Typical back span section

The main span consists of precast concrete deck panels acting compositely with longitudinal steel box edge girders and transverse steel floor beams (at 6.1 meter centers) by means of cast-in-place closure strips (see Fig. 5). The box edge girders are supported by cables anchored on the outside web at 6.1-meter intervals. On the main span side, the two-lane ramp (SA-CN) is carried on floor beam extensions cantilevered on the east side of the main line deck. Open-grid fiberglass closure panels partially cover the underside of the main span superstructure to enclose utilities, creating a more aesthetic underbelly while also providing access for inspection and maintenance. Lightweight precast concrete deck panels are used for the ramp to minimize eccentric dead loads. On the back spans, the ramp is a single cell concrete box girder, supported on piers that are independent from the cable-stayed structure with roadway joints at the tower interfaces.

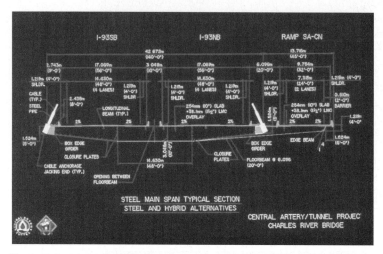

Fig. 5: Typical main span section

Foundation and Drilled Shaft Design

The tower foundations consist of footings on 2.4-meter drilled shafts (14 at the south tower and 16 at the north tower) designed to carry a working load of 2,270 tonnes each. At the north tower, the footing and the supporting drilled shafts straddle the Orange Line tunnel (see Fig. 6).

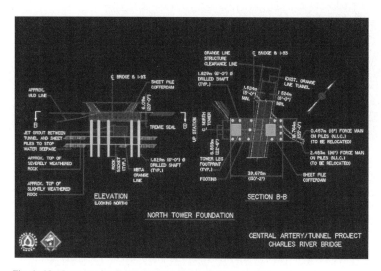

Fig. 6: North tower pier foundation straddling Orange transit line

Measures were taken to ensure that the tunnel is not adversely impacted by foundation construction activities. The drilled shafts closest to the Orange Line tunnel are placed outside a 1.5-meter buffer zone. The effect of lateral forces transmitted from the drilled shafts through the surrounding soil to the Orange Line tunnel was investigated. As a result, the shafts closest to the tunnel were installed within 2.7-meter-diameter isolation casings. Meanwhile, the south tower foundation is very close to the existing subway ventilation building on the west side, while also bridging a 0.9-meter-diameter waterline. The three drilled shafts closest to the ventilation building are also isolated within a 2.7-meter-diameter isolation casing. The gap between the casing and the drilled shaft is unfilled. The south tower foundation slab is 47 meters long by 12.2 meters wide and 4.6 meters thick. The north tower foundation slab is 42.1 meters long by 16.8 meters wide and 4.6 meters thick.

Strut at Tower Piers

The change in direction of the tower legs at the deck level produces large tension forces in the tower strut. The strut also serves as the transition from the main span composite steel superstructure to the post-tensioned concrete box girder back spans. Additionally, compressive forces from the cables, which are carried in the spline beam of the back spans, needed to be transferred to the edge box girders of the main span and vice versa. The edge girders, in turn, are then attached to the main span cables. As a result, special attention was focused on strut design, while shear lag effects and tension stresses in the concrete were also carefully evaluated.

In addition, the imbalance of bending, shear, and axial load forces in the main span and back spans under different loading combinations produce torsion, bi-axial bending, and bi-axial shear stresses in the tower strut. Due to its critical structural nature, the strut was post-tensioned in stages to a total jacking force of 27,700 tons. Limiting principal tensile stresses to pre-determined values under the working loads was an important design consideration for this element.

Post-Tensioned Concrete Back Spans

Major challenges for the south back span design and detailing were the physical overlap of the plan area of the proposed bridge at the south end, with an adjoining tunnel that is part of the project as well as the proximity of the existing Storrow Drive double deck ramps connecting to I-93, which had to be maintained in service during construction. At the north back span, numerous existing, temporary, and future ramps posed unique challenges for construction. At the south interface, the design solution features an early termination of the main line bridge deck (by approximately 15.2 meters), while extending the central spline beam the required length as a cantilever to receive and anchor the last three cables (see Fig. 7). Heavyweight ballast concrete with a density of 4,000 kg/m^3 was placed in the box girder cells within the last three floor beam bays to counteract the local reduction in superstructure weight due to the early termination of the roadway. The cantilever extension of the spline beam is housed in a vault that is accessible for inspection and maintenance of the last three stay cables. To reduce the impact of north back span construction to ramp traffic by limiting ramp closures and detours, final design and detailing of the north back span was conducted based on the incremental launching construction method. However, after a detailed evaluation, the contractor proposed—and the project team accepted—a value engineering alternative to cast-in-place the north back span using falsework with accommodations for the various ramps.

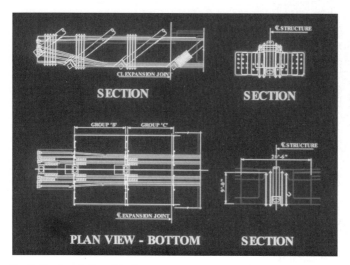

Fig. 7: South end of south back span

Steel Edge Girders and Floor Beams

The edge girders are asymmetric steel box sections with inclined bottom flanges and inclined fascia web. Typical edge girder sections are 18.3 meters long, supported by three stay cables. To achieve a full moment connection between the tower and the edge girders at the strut level, a base plate connection with 36-millimeter-diameter, 1,030 Mpa post-tensioning bars is used. Floor beams, spaced at 6.1 meters longitudinally, will span 42.7 meters between box edge girders. A separate floor beam cantilevers approximately 13.7 meters to carry two lanes of traffic outside the east plane of the stay cables. A longitudinal edge beam provided at the fascia of the cantilevered section will distribute truck loads to multiple floor beams.

The main span composite superstructure erection starts with the box edge girders (in 18.3 meter lengths) being placed and field spliced, followed by the cantilever floor beams and the east fascia edge beam. The first floor beam between the edge girders is then brought in and connected to the edge girders. Then the stringers connecting to the floor beam are installed. Next, the precast deck slabs closest to the previously erected section are installed and first stage stressing of the main span and back span stay cables is performed. The floorbeam and next set of precast panels are then erected and corresponding stay cables are stressed to their first stage stressing. This is repeated for the third floor beam and third set of precast panels, after which longitudinal post-tensioning is installed followed by the placement of transverse closure strip concrete over the floor beams. After the closure strip concrete has attained a strength of 24 Mpa, the slabs are longitudinally post-tensioned. Then the longitudinal closure pours are made over the edge girders and stringers, and second stage stay cable adjustments are made to all cables in the unit.

Fig. 8: "Iso-tension" method of stressing cable stays in progress

Stay Cables

Project design documents required stay cable strands to be either greased and sheathed or "Flo-fil" epoxy coated. Flexibility was provided concerning the type of anchorage—wedge or socket or wedge/socket—that could be utilized. After the project was bid, the Atkinson/Kiewit joint venture, the successful bidder, proposed using ungrouted stay cables employing the Freyssinet "Iso-tension" stressing method (see Fig. 8), which was accepted by the project. This will be the first US project to use ungrouted stays and the Iso-tension stressing method. The contractor also successfully proposed using co-extruded high density polyethylene (HDPE) pipe with a spiral bead to reduce potential stay cable rain/wind induced vibration (see Fig. 9).

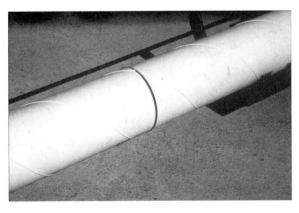

Fig. 9: Co-extruded stay cable with spiral bead

The Iso-tension method of stressing allows the stay cables to be installed one strand at a time. Based on the calculated installation force for the stay cable, the force required for the first or reference strand is calculated. This force must account for the weight of the HDPE sheathing, which is temporarily supported by the single strand and structural deformations that will occur due to stay stressing. After the reference strand is installed, each subsequent strand is installed to match the force in the reference strand. Forces are measured by load cells on the reference strand and the strand being stressed, with electronic control stopping the stressing operation when forces are equal. As strands are added, the weight of the HDPE sheathing is shared equally by the strands, slightly reducing the force in the reference strand. Additionally, deformations of the deck and tower reduce the force in the reference strand as each subsequent strand is stressed. When the stressing operation is complete, the force in each strand multiplied by the number of strands equals the calculated installation force. Tensioning of stay cables is performed at the superstructure level.

The final dead load force in the stays range from 1,330 to 7,100 KN, and the live load force in the stays range from 60 to 1,330 KN.

Cable Anchorages at Towers

The vertical leg at the tower top varies from 4.9 meters square at its base to 3.2 meters square just beneath the peak. Because of the limited room to anchor cables, a prefabricated steel anchor box is built into the tower, acting compositely with the exterior concrete by means of shear connectors. The cables are anchored by bearing at the inner end of structural pipe sections built into an anchorage stiffener (see Fig. 10). This detailing offers the following advantages:
- Reduced torsional moment due to closer transverse spacing of the cables
- Improved geometry control of the cable anchorages
- Elimination of complicated forming of the inside walls
- Elimination of post-tensioning in the tower cross section
- Ready access for inspection and maintenance

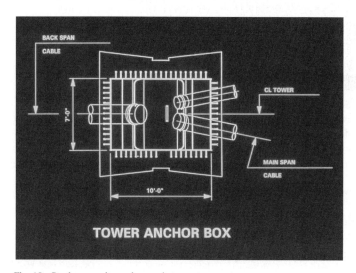

Fig. 10: Steel composite anchorage in tower top

However, torsion in the tower leg due to the cantilevered ramp on one side posed a challenge. The east side cables of the main span carry a 30 percent greater load than the west side cables. This is mitigated by using a lightweight concrete deck slab in the cantilever and offsetting the geometric centerline of the back span and main span cables by 75 millimeters with respect to the tower centerline.

To avoid external cable anchorages and their related maintenance issues in the inclined legs of the tower, a non-uniform cable spacing scheme with internal anchorages is used. This entails gradually increasing vertical spacing from the standard 1.7 meters for the uppermost ten spaces up to 2.9 meters for the lowermost cable pair.

Girder-to-Cable Anchorages

The cable anchorages on the main span box edge girders are mounted on the outside webs and detailed as a pipe assembly bolted to the side of the girder (see Fig. 11). The cables are then passed through the anchor pipe, with the cable anchor bearing against the lower end of the pipe. The pipe is connected to a base plate with a single web plate.

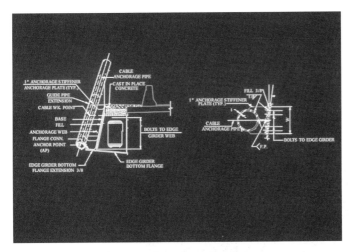

Fig. 11: Stay cable anchorage to superstructure in main span

This detail was selected due to its visual appeal over typical box-type cable anchorages; fabrication and erection considerations; and for allowing easy access for inspecting all critical welds and bolts.

Aerodynamic Evaluation

Wind tunnel tests of both the sectional and aeroelastic models were performed for the final structure as well as for intermediate construction stages. Vortex excitation occurred at about 128 kph, within criteria, while flutter speed was measured at 715 kph, well above the requirement of 210 kph (see Fig. 12). Smoke flow visualization tests also indicated that wind flows were not significantly altered by changes to the deck section, such as deck openings and open mesh closure panels on the underside.

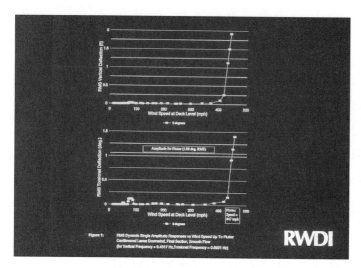

Fig. 12: Result of aerodynamic wind tunnel study

The assessment of the potential for cable vibration, considering the use of co-extruded HDPE pipe with a spiral bead to reduce rain/wind vibration and ungrouted stay cables, resulted in the need for cross-ties to offset any potential galloping. After a study of the Freyssinet viscoelastic dampers was conducted by RWDI and Construction Technology Laboratories, it was concluded that providing dampers at all lower anchorages (see Fig. 13), coupled with a cross-tie arrangement for longer cables, would best meet project needs.

Fig. 13: Typical viscoelastic damper detail

Construction

The cast-in-place back spans were constructed on falsework concurrent to tower construction. The tension strut at the tower piers was post-tensioned in stages. Afterwards, the superstructure of the main span was erected in a cantilever fashion, as previously described (see Fig. 14). As sections of the main span were erected, stay cables were installed and tensioned.

Fig. 14: Bridge construction status as of February 2001

Construction of the Bunker Hill Bridge began in September 1997. As of March 2001, the total cost of the bridge, including change orders and value engineering savings, stands at $95.1 million. The superstructure mid-span closure at the main span was completed in April 2001, and by mid 2001 the bridge is expected to be substantially complete with vehicle barriers, high performance concrete overlay, and utilities.

Conclusion

Boston, in the forefront of the American Revolution over two centuries ago, is now in the forefront of another revolution—in the field of cable-stayed bridge technology. A highly complicated structure, unique in the world, has been successfully designed and is nearing completion. New technologies and innovations have become hallmarks of the Leonard B. Zakim Bunker Hill Bridge.

Credits

Owner:	Massachusetts Turnpike Authority
Management Consultant:	Bechtel/Parsons Brinckerhoff (B/PB)
Bridge Concept:	Dr. Christian Menn
Type Study and Preliminary Engineering:	PB (of B/PB)
Final Design:	HNTB
Construction Management:	B/PB
Contractor:	Atkinson/Kiewit, joint venture
Cable Stay and Post-tension Supplier:	Freyssinet
Structural Steel Supplier:	Grand Junction Steel

OPTIMAL SCHEDULING FOR BRIDGES BASED ON LIFE-CYCLE EVALUATION

David de Leon[1] and Alfredo H-S. Ang [2]
[1] Instituto Mexicano del Petroleo. Competencia de Ingeniería Civil
Eje Lázaro Cárdenas
152, Col. San Bartolo Atepehuacan
Mexico City, 07730. Mexico
Ph. Number (52-5) 333-7322, Fax (52-5) 368-4529, E-mail: dleon@www.imp.mx
[2] University of California, Irvine
International Director, ASCE

ABSTRACT

Recently, formulations to support decisions on the design and maintenance of bridges are based on the life-cycle cost assessment of alternative programs. In this paper, an approach to determine the optimal schedule for maintenance works for a typical bridge is described in a conceptual manner. The formulation considers the balance between safety and cost and both aspects are treated as random variables. The results may be used to target optimal scheduling for the tasks of preventive maintenance of a typical bridge.

KEYWORDS: Reliability, availability, optimal maintenance schedule, time to failure, repair time, life – cycle cost.

INTRODUCTION

To be effective, risk management of bridges requires the quantitative assessment of the likelihood and consequences of undesirable events that may cause fatalities, injuries, damage, interruption of traffic, etc. These assessments provide elements to the operators and managers of bridges to discriminate, in a cost-effective way, among a set of alternative measures to anticipate, prevent and mitigate the occurrence and consequences of those undesirable events according to the specific risks and available resources for the bridges under their care.
Optimal maintenance and life – cycle costing of bridges have been previously addressed (4, 5, 7, 8).
However, cost functions for each loss component have not been formulated. Furthermore, the indirect loss for bridge failure has not yet been included into the design or maintenance criteria.
Given the enormous economic importance of major bridges, the incorporation of cost functions and the estimation of the reliability, maintainability and availability associated with the functionality of a bridge becomes specially relevant to generate, later on, cost-effective measures

that help to improve the risk–benefit balance of the maintenance program of the bridge. Recent progress on risk and reliability assessment (1, 3, 6) may be implemented for the incorporation and adaption of modern probabilistic tools towards the generation of a decision - making criteria. However, for the purpose of comparing the cost-effectiveness of alternative maintenance programs, the performance of the structure as well as the economic consequences of failure need to be assessed and properly combined (2).

The expected life-cycle cost may be estimated in terms of the future (repair and deferred availability) costs and of the global functionality of the bridge. The future costs must be expressed in present value as the causing events occur at several times in the future. If several alternatives for the maintenance of bridges are assumed, a comparative criterion for optimal maintenance scheduling and decisions may be devised on the basis of minimum expected life-cycle cost.

In this work, the above mentioned issues are explored and a RAM (reliability, availability and maintainability) - based approach for selecting an optimal maintenance schedule is formulated.

In the near future, integrated risk assessment may be fully implemented to lead bridge authorities towards optimal decisions on areas as: strengthening, inspection, protection and other safety measures.

DESCRIPTION OF THE PROPOSED FORMULATION

The operating cost proposed by Goble (6)

$$C_o = (C_f + C_m) * L \tag{1}$$

where C_o = operating cost, C_f = failure cost, C_m = maintenance cost and L = operating life, is re-written in probabilistic terms and composed of the failure costs C_f^L and the maintenance costs C_m^L that are expected during the service life of the bridge. For maintenance schedule j:

$$E(C_f^L)_j = E\{\sum_{i=1}^{nf} C_{fj}(\Delta t f_i) PVF(t_i)[1 - A(\Delta t f_i)]\} \tag{2}$$

where $C_{fj}(\Delta t f_i)$ and $A(\Delta t f_i)$ are the failure cost and availability associated with the time to failure $\Delta t f_i$. Also, the present value factor, of expenditures required at time t_i, is expressed in terms of the discount rate r:

$$PVF(t_i) = 1/(1+r)^{t_i} \tag{3}$$

These costs may be estimated from Monte Carlo simulation, for the service life L of the bridge, according to the potential occurrence of the failure or maintenance events. If Δt_j is the prescribed constant period for maintenance of the bridge, according to schedule j and $\Delta t f_{ij}$ is the random time to failure (modeled from a history of the bridge's performance), a failure (and its subsequent repair) occurs whenever a simulated value of $\Delta t f_{ij}$ (a realization of the random variable) is less than Δt_j and, for this case, the maintenance cost $C_m(\Delta tf_{ij})$ is 0. On the other hand, when $\Delta tf_{ij} > \Delta t_j$ a maintenance event occurs and the corresponding failure cost $C_f(\Delta tf_{ij})$ is not 0.

A trial of the simulation process consists of two sets of failure times Δtf_{ij} and repair times Δtr_{ij} being generated, according to predetermined distributions, and subsequently added up to reach the bridge service life:

$$\sum_{i=1}^{nf}(\Delta tf_{ij} + \Delta tr_{ij}) + nm_j (\Delta t_j) \approx L \qquad (4)$$

Once all the failure, repair and maintenance events are accommodated into the service life L, the life-cycle failure as well as the maintenance costs are accumulated and the total life-cycle cost is estimated for the maintenance schedule j.

After several trials of the simulation process are completed, the average of the life-cycle cost $E(C^L_f)_j$ is estimated for maintenance schedule j.

Finally, the optimal maintenance schedule will be the one corresponding to the minimum expected life-cycle cost.

Conceptually, it is expected that, as the maintenance period decreases, the cost of maintenance rises and the cost of failure decreases. Conversely, for a maintenance period large enough, the maintenance cost decreases whereas the cost of failure goes up. On the basis of the above, the corresponding conceptual cost functions for the expected life-cycle costs may be outlined. See Fig. 1 for sketches of these functions.

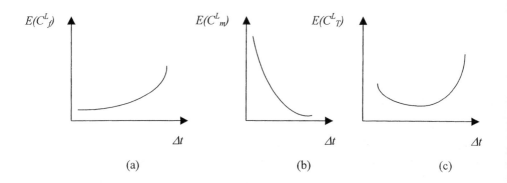

Figure 1. Conceptual cost functions for maintenance.

An optimal scheme will correspond to the combination of bridge performance and maintenance schedule that minimizes the total expected life-cycle cost.

COST FUNCTIONS FOR FAILURE AND MAINTENANCE

The shape of the cost functions corresponding to the alternative events of failure and repair or maintenance may also be plotted. For every realization of $\Delta t f_{ij}$, the costs will result as may be seen in Fig. 2.

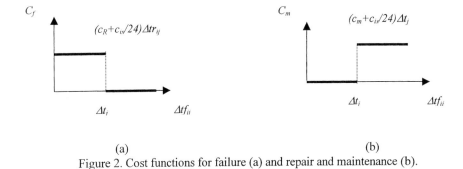

(a)   (b)
Figure 2. Cost functions for failure (a) and repair and maintenance (b).

Where:
c_R and c_{is} are the repair cost (per hour) and the loss associated with the interruption of service on the bridge (per day) in case that a failure occurs and major repairs are required. Similarly, c_m is the maintenance cost.

CONCLUSIONS

A probabilistic approach to generate optimal maintenance schedules was outlined for bridges. Similar schemes may be developed to integrate risk assessment with cost estimations and produce cost-benefit models that may be used by managers for cost-optimal decision-making.

ACKNOWLEDGEMENTS

The authors thank the IMP and PEMEX for the information that served as a basis for the application of the proposed formulation.

REFERENCES

1.- Ang, Alfredo H-S. and Tang, W. H., 1984, *"Probability Concepts in Engineering Planning and Design"* Vol. II – Risk, Reliability and Decisions. John Wiley and Sons, New York.
2.- Ang, A. H-S. and De León, D., 1997, "Determination of optimal target reliabilities for design and upgrading of structures". *Structural Safety*, **19**, 1, Elsevier Science Ltd. The Netherlands, pp. 91- 103.
3.- Goble, William M., 1992, *"Evaluating Control Systems Reliability. Techniques and Applications"*. Resources for Measurement and Control Series. Instrument Society of America. North Carolina, U. S.
4.- Kong, J. S., Frangopol, D. M. and Gharaibeh, E. S., 2000, "Life Prediction of Highway Bridges with or without Preventive Maintenance", PMC2000-300, Notre Dame, U. S. A.

5.- Lin, K. Y. and Frangopol, D., 1996, " Reliability-based optimum design of reinforced concrete girders". *Structural Safety*, Elsevier, **18**, 2-3, 123-150.

6.- Locks, Mitchell O., 1995, *"Reliability, Maintainability and Availability Assessment"*. 2nd. Edition, ASQC Quality Press, Milwaukee, Wisconsin, U. S.

7.- Mori, Y. and Ellingwood, B. R., 1994, "Maintaining reliability of concrete structures. I: Role of inspection/repair". *Journal of Structural Engineering, ASCE*, **120**, 3, Mar., 824-845.

8.- Mori, Y. and Ellingwood, B. R., 1994, "Maintaining reliability of concrete structures. I: optimum inspection/repair". *Journal of Structural Engineering, ASCE*, **120**, 3, Mar., 846-862.

Design and construction

The hot dip galvanized wires for bridge cables

JIANG YONG KANG
Shanghai Shenjia Metal Products Co., Ltd, Shanghai, P.R.C.

Abstract: This paper introduces types and main quality requirements of galvanizing wires used for bridge cables, process of galvanized wire and engineering application for galvanized wire. The paper describes in focus the comparison of domestic wires with the international standard and with the quality of the real product imported. It is considered that the standard level and the real product level of galvanizing wires used for cables of bridges entered in the world's advanced area. In 2000, 55 bridges in the whole nation used domestic wires for the construction.

Key words: Bridge cable   Hot dip galvanizing wires   Quality   Application

1. Preface

The loss brought by the corrosion of metals to the national economy is considerably astonishing, upon the statistics, metallic materials discarded every year due to the corrosion on the world corresponds to one third of the annual output. It is more important that the value and influence of the structure damages resulted from the metallic corrosion is far more than their own values of metals. Based on the protection layer of prestressed steel wire and strand, it could be divided into nonmetallic protection layer (referred as coating) and metallic protection layer (referred as plating) two kinds. Referring to nonmetallic protection coatings, there are mainly plastic, epoxy resin and anti-corrosion grease three kinds of nonmetallic coatings. But the anti-aging of the nonmetallic coating isn't solved yet, and its mechanics properties are much lower than that of metallic plating (Epoxy coated and filled strand was used on cable stayed bridges inside the U.S). Therefore, the application scope is limited. As an excellent plating of the anti-corrosion of wires, the galvanizing was confirmed commonly by the manufacturing industry. Because the zinc (or zinc-aluminum alloy) plating has a good result in the anti-corrosion of the metal and has matured technology facilities, therefore it is widely used by prestressed steel. In view of the "hydrogen embrittlement" dispute of prestressed structure, the spreading is obstructed once, and the conclusion is made until Federation Internationale de la Procontrainte (F.I.P) reported it on December 1992. "The galvanizing processes which may be used for prestressing steels, notably hot dip galvanizing and sherardising, do not cause hydrogen embrittlement. Electroglavanizing is known to introduce the risk of hydrogen embrittlement and should not be used for prestressing steels." "And it should not be touched directly with the cement mortar just blended". [1] After then France declared the world's first product standard, "Steel products--Hot-dip galvanized prestressing smooth wires and strands " (NFA35-035) on December 1993, it determines the hot-dip galvanizing is the unique plating method. Our nation issued successively the standard of hot-dip galvanized steel wires for bridge cables (GB/T17101-1997). It makes the anti-corrosion technology of prestressed steel of our nation enter into the line of world's advanced level. Now, zinc-aluminum (Galfan) plated wire is developing by Shanghai ShenJia Metal

Products Co., Ltd. our company set about the establishment of a new type industrial company for utilizing over forty years production experiences of wires and strands of Shanghai iron and steel industry and for introducing European advanced technology facilities. Since the putting into production in 1994, it has accepted the fabrication task of wires used for cables of 55 bridges in the whole nation, the total amount is nearly over fifty thousand tons, thus it makes our company's galvanizing wires and strands used for bridge cables enter into the world's advanced line. In the past, galvanizing wires for cables all depended upon import. After entering into the 1980s, there are several suspension bridges with large spans built or being at design and construction. For example, <u>Santou Bay</u> Bridge with a main span of 452 m, Guangdong Humen Bridgewith a main span of 888 m(This bridge was the first large suspension bridge in China), Hubei Xiling Yangzi River Bridge with a main span of 900 m, Yichang Yangzi River Bridge with a main span of 960 m, Jiangyin Yangzi River Bridge with a main span of 1385 m.(The bridge is ranked first in length in Asia and fourth in the world). With reference to cable stayed bridge, The Shanghai Xupu Bridge is a cable-stayed bridge with twin towers at both ends. The length of the main span of the bridge is 590 m. The main cables for this bridge, provided by Shanghai Shenjia Metal Products Co., Ltd, were the first hot dipped galvanized wire used for this type of application that were manufactured by a Chinese company.The Jin Wu Bridge is a cable-stayed bridge with a main span of (100+125) m. The main cable represents the first use of galvanized strand of 15.24mm dia. in China.The Qian Jian No.3 Bridge is a cable-stayed bridge with twin towers and one side. The length of the main span of the bridge is (160+160) m. The main cable of the bridge is the biggest for this type of bridge in China. The main cable is composed of 451 wires of 7 mm. dia. galvanized wire. The Nanjing Yangtze River No.2 Bridge is a cable-stayed bridge with twin towers on both ends. The length of the main span of the bridge is 628 m. The main span of the bridge is ranked third in the world.

2. Types of galvanizing wires used for bridge cables

According to the design style of pulling cables of bridges, they could be divided into pulling cable for cable stay and cable for suspension bridge two types (see Table 1). In cable stay bridge the pulling cable is the direct load bearing structure of pulling the bridge surface. Therefore it must have the requirement of low stress relaxation. There are two kinds of wires used for the cable suspension bridge: the wire used for the main cable is the main structure that bears the bridge surface load, although it hasn't the requirement of low stress relaxation, but requirements of the straightness and ductility of wires are higher for ensuring a good shape of cable manufacture and a sufficient ductility during the supporting of suspension cables. The " ductility " of wires is different from the "plasticity". The "plasticity" index is measured with the "elongation" and the "reduction of area". The " ductility " index is measured with the "torsion times", "bending times" and "wrapping (generally is 3D mandrel)". Some selections could be found in property indexes of various countries (see Table 2). The hanger of suspension bridge could be manufactured by pretensioned normal wire ropes (or strands), or by cable manufactured by wires. Said hanger is the direct load bearing structure of the bridge surface and is manufactured by wires of low stress relaxation in preference. Normal wires although pretensioned but never reach the effect of low stress relaxation.

Besides wires used for the cable manufacture of stay cables, the using of 1×7 strand is also one of popular methods in the world. After France NFA35-035 declared the technical indexes of galvanizing strands, the (know-how) was also declared. Said standard is worked out by Freyssinet of France.

Table 1 Classification of wires and strands used for bridge cables

Bridge type	Variety selected	Specs (φ mm)	Strength class (MPa)
Cable stay bridge	Wires	5 ~ 7	1570 ~ 1670 [1]
	Strands	12.5 ~ 15.7	1770 ~ 1860
Suspension bridge	Wires for main cable	5 ~ 7	1570 ~ 1670 [2]
	Wires for hanger	5 ~ 7	1570 ~ 1670

3. Comparative analysis of standards from different countries

Product standard of galvanizing wires used for bridge cables is also regarded as the "know-how" by manufacturers. Until the declaration of the French standard, other corresponding standards are declared continuously including our nation's product standard based on our company's enterprise standard. Now the relevant standard and main parameters in the technical articles specially used in some engineering are listed on table 2.

It is known form Table 2:

3.1 For the technical index of galvanizing wires used for φ 7.0 mm stay cables, that of Thysson Company is the highest. Shanghai Shenjia Company developed wires with the strength class of 1670 MPa in responding the requirement from Tongji University "Ling-Li Company". It is used for Yellow Mountain Taipin Lake Bridge and is spread quickly in application.

3.2 The technical index of galvanizing wires used for main cables of φ 5.0 mm suspension cable of Japan Tokyo Steel Works and Akashikaikyo Bridge (supplied by Nippon Steel Corp.) are the highest. Shanghai Shenjia Company provided wires for Humen Bridge and Yichang Bridge, besides pursuant to said condition, increased also the examination of stress relaxation properties, among them the torsion index is the examination to the wire comprehensive properties, and has a considerable difficulty.

3.3 The standard level of said two wires produced by Shanghai Shenjia Company reached the world's advanced level.

4. Comparative analysis of the level of real product galvanizing wires from different countries

Our company collected data of galvanizing wires imported in recent years detected by Shanghai Commodity Inspection bureau or data of samples detected by Tongi University for the comparative analysis with the real product of Shanghai Shenjia Company. Results are listed on Table 3 and Tale 4 respectively according to the classification of stay cables and main cables.

It is known for Table 3 and table 4:

4.1 The stability of the tensile strength of Shanghai Shenjia Company's real product is better, except the strength value difference of Spanish galvanizing wires is the least, the fluctuation range of other countries are all larger.

4.2 For the diameter torelence and out of round of wires, those of Shanghai Shenjia Company are the least.

4.3 For the examination index of uniformity (copper sulfate times) of galvanizing wires used for main cable , except Shanghai Shenjia Company reaches the target, other wires do not meet the requirement.

4.4 The real product level of galvanizing wires from Shanghai Shenjia Company is at the line of the world's advanced level.

5. Brief analysis of features of production technology

The process is as follows:

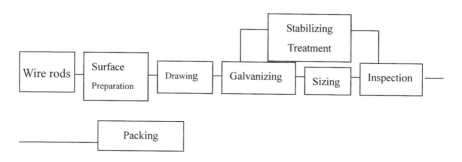

Although the technology is used as its own "know-how" in the world, but the basic process is similar.

5.1 wire rods

Before the putting into production of this company, the optimal selection of wire rods is an important work of the company, types of wire rods selected concern with many countries such as European countries, USA, Japan, Republic of Korea and so on. After the screening selection, 2 ~ 3 varieties from 2 ~ 3 countries, which mainly is Japan, are selected finally.. According to our production experience, the company determines to use salt patenting rod of DLP (direct lead patenting) supplied by Nippon Steel Corp which is the best of rod subcontractor. The chemical composition, mechanical properties and surface defects shall be strictly controlled and receiving inspection shall be conducted as per JISG 3502.

In order to explain conveniently, standards of wire rods used for main cables of Japan Akashikaikyo Bridge, wire rods of Japan Kawaichi (PAC wire rod) and wires rods used for Japan piano wires (JIS3502) are listed on Table 5:

Table 5 Comparison table of special requirements of wire rods for galvanizing bridge cables

No.		1	2	3	4
Standard code		JISG3502 SWRS82B	HBSG3507 (Honshu-Shikoku) HWRC82B	Kawaichi PACwire rod SWPR7BL	Nippon Steel Corp. SWRS82B-DLP real product
Item					
Chemical Contents (smelting) (%)	C	0.85~0.85	0.80~0.85	0.80~0.83	0.81~0.84
	Mn	0.60~0.90	0.60~0.90	0.76~0.86	0.70~0.77
	Si	0.12~0.32	0.80~1.00	0.15~0.32	0.17~0.25
	S	≤0.025	≤0.025	≤0.010	0.005~0.015
	P	≤0.025	≤0.025	≤0.015	0.007~.-13
	Cr	-	≤0.06	-	≤0.20
	Ni	-	≤0.06	≤0.10	-
	Cu	≤0.20	≤0.06	≤0.05	≤0.01
Impurity contents (%)		Agreement	≤0.07	≤0.04	≤0.10
Surface defect depth (mm)		≤0.10	≤0.07	≤0.05	≤0.10

| Decarbonizing depth (mm) | ≤0.07 | ≤0.07 | ≤0.03 | ≤0.07 |

It is known from table 5:

5.1.1 It must have high purity

In Akashikaikyo Bridge, the element Si is used for strengthening ferrite and Cr, Ni, Cu are controlled strictly to ensure that at the same time of wires have a high strength, also have a high ductility index. The impurity content is controlled at ≤ 0.07 %, (generally 0.10 % is required), Kawaichi PAC wire rod increases further the sulfur, phosphorous requirements, the copper content is controlled strictly to ≤ 0.05 %, the impurity content is ≤ 0.04 % to ensure the quality of bridge cables with the high purity of steels.

5.1.2 It must have high and uniform physical and chemical properties

On the control of chemical contents, because C, Mn, Si are main elements that determine mechanics properties of steels, so their fluctuation range must be controlled small. Except Akashikaikyo Bridge uses the steel grade that contains a higher silicon content, Kawaichi uses the reducing of the supplying range of C, Mn, Si to ensure the uniformity of contents, Japan DLP has a better uniformity.

In Japan, the production of said kind of wires has similar manufacturing technology requirements to P.C wires and strands, that is, wire rods must be drawn after the sorbitic treatment. Therefore, Japanese wire rods standards all haven't mechanics property requirements, the specifications of its selection all used as the know-how of manufacturers and shouldn't be transferred externally. In view of that at present, in Europe, USA, including Japan, wire rods manufacturers required the sorbite as their own operation, subsequently, Japan DLP, KKP wire rods appear in the world, metal products Works all use it as the direct raw material to conduct the processing. Therefore the uniformity of mechanics properties is required, especially the requirement of the control range of the tensile strength and the reduction of area, then it ensures that the discretion degree of mechanics properties of finished product wires could be reduced.

5.1.3 It must have a good surface quality

Galvanizing wires used for stay cable and main cable has the examination index of its ductility respectively. Besides the long gauge length (250 mm) must be used to examine the elongation of the plasticity for ensuring that at the same time of the high strength limit there is a good uniform elongation, non-cracking must be also ensured under the condition of 3D wrapping. Especially on the use performance, the stay cable sometimes uses heading to anchor securely. After the highest strength of BA type (Button Anchorage) wires in USA ASTMA421 is increased from 1620 MPa to 1670 MPa, the trend of the heading cracking is more larger, so as the requirement of the surface quality of wires is more strict. The main cable uses the twisting index ≥ 14 times to examine its surface condition; at the same time it is also an examination to the uniformity.

5.2 Surface preparation : Preparation of the steel surface is carried out for the purpose of removing scale and having wiredrawing coating. Hydrochloric acid pickling which is most effective cleaning method is used for scale removal. Phosphate coating and borax is used as a wiredrawing coating to improve the following wiredrawing operations.

5.3 Wiredrawing

The wiredrawing can be described as plastic deformation process of the steels. After the rod passes through 8 dies, The drawn wire not only has exact dia conforming to the requirement of final wire size but also meets requirements such as mechanical properties and surface quality prior to galvanizing. For the wiredrawing process it is necessary to create good conditions for conforming to finished product requirements and the following galvanizing process.

The wire drawing must satisfy requirements of mechanics properties.To design dimensions of wire rods and drawing dimensions of each draught will be conducted with the principle of satisfying mechanics properties of finished product wires. Therefore specs of wires rods and dimensions of wires before the galvanizing are important technical parameters. Our company could select it according to the need of customer's design for ensuring that at the same time of having sufficient mechanics properties, wires also have good plasticity and ductility.

5.4 Galvanizing

The galvanizing is an important measure of the anti-corrosion

According to conclusions of FIP (Federation Internationale de la precontrainte) commission reports, Hot dip galvanizing is known not to cause hydrogen embrittlement and utilize zinc in it's forms as a coating material for protecting steel from corrosion. The company possess sophisticated hot dip galvanizing technology and equipment supplied by FIB Belgium, refer to the following sequence of operation for details.

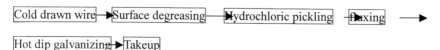

5.4.1 Surface degreasing : When the wires pass through molten lead contained in an elongated pan, Grease on wires surface is brunt out. Heat treatment can also be accomplished by passing the wires through molten lead pan.

5.4.2 Hydrochloric pickling : Hydrochloric pickling is commonly used to remove oxide films and activate the surface of the wire in order to have good adherence.

5.4.3 Fluxing : Fluxing is the form of chloride salt is applied directly on the surface of activated wire to assure that heat diffusion is carried out between the surface of cleaning steel and zinc.

5.4.4 Hot dip galvanizing can be described as heat diffusion process between iron and zinc (Fe-Zn). The company possess a zinc bath holding 100 MT of molten zinc of purity not less than 99.99 percent. Zinc bath is much longer with 9 m long, So it assures that zinc coating is much more uniform in thickness on smoother wire surface and time of immersion is enough to meet requirement of Fe-Zn diffusion. After galvanizing, The wire quality including zinc mass, uniformity and adhesion can fully meet requirements of ASTM designation: A90, A239 and A586.

5.5 Sizing (galvanized wire for main cable)

UK and Japan consider that wires couldn't be mechanically straightened or machined work again after the galvanizing, otherwise a great loss of its ductility and plasticity indexes may be caused. But our company has been established the sizing process with Shenjia " know how" basing on more than 40 years experience of galvanizing wire. Sizing process is different from machine straightness or any other form of cold working of wire. Wire sizing after the galvanizing is the important measure of ensuring the final technical indexes. With reference to PPWS method, for the sizing process it is necessary to assure that galvanized wire has good plasticity, excellent ductility and exact size. Also, it is necessary to improve corrosion resistance and wire straightness.

5.6 Stabilizing (galvanized wire for cable stay)

Stabilizing process is described as heat treatment for stress-relieved. After stabilizing process, the galvanized wire can meet max. 2.5% of low relaxation requirement.

5.7 Inspection of finished product : The company possess sophisticated inspection equipment for testing chemical and mechanical properties and controlling quality of the finished wire with the addition of qualified inspectors. The main inspector equipment used in the company is respectively introduced from Germany, Britain, Italy and Japan.

5.8 Packing : All finished products shall be stacked on rubber/plastic plates and handled with nylon rope prior to packing. In this way it is necessary to assure that the zinc coating is free from damage. Also, the finished products shall be packed by moisture-resistant paper and plastic knitting cloths. The finished products shall be wrapped to prevent damage during shipping and storage.

6.Conclusion:

Through the control of above technology, the quality of our company's products enters into the world's advanced line. So far, Examples of engineering that use our products reach 55.

Table 2 Level contrast of standards of hot dip galvanizing wires used for bridge cables in home and abroad

	Index of properties	Italy Redaelli co.	German Thyssen Co.	Japan Akasaikaikyo bridge		Japan---industrial Co.			France NFA35-035		National Standard of China				
Dimension	Diameter/mm	φ5	φ7	φ5	φ7	φ5	φ7	φ7	φ7		φ5			φ7	
	Tolerance/mm	≤0.06	+0.08~-0.02	±0.06	±0.08	±0.06	±0.08	±0.08	±0.07		±0.06			±0.07	
	Roundness	-	≤0.04	-		≤0.06	≤0.08	≤0.08	-		≤0.06			≤0.07	
	Tensile strength/Mpa	1600~1800	≥1600	1570~1770		1570~1770			1670	1770	1570	1670	1770	1570	1670
	Yield strength/Mpa	≥1300	≥1300	≥1160		≥1160			1480	1570	1180	1250	1330	1410	1500
	Elongation%≥(L=250mm)	≥4	≥4	≥4		≥4			3.5		4.0			4.0	
Physical properties	Elastic modulus/Gpa	≥200	195~210	-		-			-		190~210				
	Torsion(100d) ≥	14	-	≥14		14	12		-		-			-	
	Repeat bending	≥4	≥3(r=17.5mm)	-		-			≥5(r=22.5mm)		≥4(r=15mm)			≥4(r=20mm)	
	Wrapping property(3d)	8 turns	8 turns	8 turns		8 turns			-		8 turns			8 turns	
	Relaxation (1000h)	≤7.5%	≤2.5%	-		-			≤2.5%		≤8.0%			≤2.5%	
Zinc	Zinc mass/g/m2	≥300	≥300	≥300		≥300			190~350		≥300			≥300	
	Zinc Adhesive force(5d)	8 turns	2 turns	2 turns		-			6		8 turns			8 turns	
	Copper sulfate Dip	4 times	≥5 times	-		-			2		≥4 times			≥4 times	

Straig	Free cast/m	≥4	Chord	≤30mm/m	height	≥4	≥4	-	≥5	≤30mm/m
	Lift/cm	≤15		-		≤15	≤15	-	≤15	-

Table 3 Contrast of real product properties of galvanizing wires used for stay cables in home and abroad

No.	Item	Technical requirements of wires	Shenjia (Jinan bridge)	Shenjia (Taipin Lake Bridge)	Tysson (Nanpu Bridge)	Arbed (Yangpu Bridge)	ROK (Nanpu Bridge)	Disson (Tonglin Bridge)	Disson (Qiou Bridge)
1	Diameter and tolerance	$\phi 7$ +0.08 mm / -0.02 mm	+0.02 / -0.02	+0.03 / -0.01	+0.04 / +0.02	+0.07 / +0.03	+0.05 / -0.01	-	-
2	Roundness	≤0.04 mm	0.02	0.02	0.02	0.04	-	-	-
3	Tensile strength	$\sigma_b \geq 1600$MPa (or 1670MPa)	1710~1730	1710	1741~1808	1651~1712	1710~1780	1680~1700	1690~1720
4	Yield strength	$\sigma_{0.2} \geq 1300$MPa (or 1470Mpa)	1520~1560	1560~1570	1520~1600	1460~1540	1560~1630	1490~1520	1520~1750
5	Elongation	≥4% (250mm L. Distance)	5.0~5.8	4.5	4.1~4.3	4.1~5.2	5.0~6.5	4.3~5.2	4.5~6.0
6	Reverse bending	≥3 times (r=17.5mm)	4~5	4~6	>3	>3	4~8	6~8	4~8
7	Wrapping	8 turns (3D mandrel) not broken	Conforming	Conforming	Conforming	Conforming	Conforming	Conforming	Conforming
8	Relaxation	≤2.5%(0.7G.U.T.S.10	1.12	1.8	2.3~2.5	1.34~2.15	-	-	1.29~1.60

80 DESIGN AND CONSTRUCTION

No	Item	UK Bridon Co.							
9	Stress fatigue	360MPa(0.45GU.T.S)2x10^6 times max.	Conforming	Conforming	Conforming	Conforming	Conforming	Conforming	Conforming
10	Elastic modulus	(195~210) Gpa	200~206	200~204	195~205	199~204	205~208	195~197	200~203
11	C	0.75~0.85	0.82~0.83	-	0.78~0.81	0.75~0.78	0.78~0.80	0.79~0.82	0.76~0.78
	Si	0.12~0.32	0.22~0.24	-	0.20~0.24	0.26~0.29	0.22~0.28	0.22~0.26	0.18~0.21
	Mn	0.60~0.90	0.72~0.74	-	0.72~0.83	0.72~0.74	0.79~0.84	0.72~0.75	0.74~0.76
	S	≤0.025	0.008~0.009		0.010~0.012	0.010~0.013	0.003~0.013	0.006~0.011	0.007
	P	≤0.025	0.010~0.011		0.011~0.021	0.010~0.013	0.006~0.018	0.011~0.012	0.009~0.010
	Cu	≤0.20	0.01	-	0.024~0.044	0.04	0.02	0.005	0.03
12	Zinc mass	≥300 g/m	375~400	319~336	352~444	377~458	303~358	325~381	315~390
13	Adhesiveness	5D, 2 turns, zinc not peeled off	Conforming	Conforming	Conforming	Conforming	Conforming	Conforming	Conforming
14	Copper sulfate test	≥5 times	6~7	6~8	6~10	6~8	Conforming	6	5~6
	Date source		Tongji	National Detection Center	Commodity Inspection Bureau	Commodity Inspection Bureau	Commodity Inspection Bureau	Commodity Inspection Bureau	Commodity Inspection Bureau

Table 4 Contrast of real product properties of galvanizing wires used for main cables in home and abroad

No.	Item	Products supplying conditions of UK Bridon Co.	Shenjia bridge	(Humen bridge)	Spain (Sample)	Italy (Santao bridge)	UK (Xiling bridge)
1	Diameter and tolerance	φ5 +0.06 mm -0.06 mm	+0.03 -0.02		+0.17 -0.02	-	+0.04 +0.01

2	Roundness	≤0.04 mm	0.01	0.09	-	0.03
3	Straightness	≤30mm/m	(5~11)mm/m	(7~10)mm/m	-	28~50mm/m
4	Tensile strength	σ_b≥1600MPa	1730~1780	1780~1790	168~1780	1680~1750
5	Yield strength	σ_{0.2} ≥1180MPa	1380~1420	1660~1670	1180~1220	1400~1620
6	Elongation	≥4.0% (250mm L. Distance)	4.9~5.2	4.8~4.9	5.0~6.0	5.0~5.5
7	Reverse bending	≥3 times (r=15.0 mm)	7~9	7~10	≥4	6~8
8	Torsion	≥14 times (L=100d)	17~21	13~29	≥14	15~19
9	Wrapping	3d 8 turns (without any damage)	Conforming	Conforming	Conforming	Conforming
10	Relaxation	≤7.5%(0.7GU.T.S.1000h) .	4.0%	4.97%	6.58%	-
11	Elastic modulus	(195~210) Gpa	202~205	194~202	200~205	195~200
12	C	0.75~0.85	0.81~0.83	0.80~0.82	0.75~0.79	-
	Si	0.12~0.32	0.20~0.22	0.24~0.26	0.18~0.29	-
	Mn	0.60~0.90	0.71~0.74	0.69~0.72	0.69~0.80	-
	S	≤0.025	0.004~0.007	0.015~0.016	0.005~0.008	-
	P	≤0.025	0.009~0.015	0.027~0.029	0.015~0.018	-
	Cu	≤0.20	0.015~0.017	0.018~0.021	-	-
13	Zinc mass	≥300 g/m	322~345	439~490	326~418	391~412
14	Zinc Adhesiveness	5D, 8 turns, zinc not peeled off	Conforming	Conforming	Conforming	Conforming
15	Copper sulfate test	≥5 times (60s/time)	6~6	4~4	≥4	4~6
	Date source		Tongji	Commodity Inspection Bureau	Commodity Inspection Bureau	Commodity Inspection Bureau

Genoa Harbour Crossing

FAROOQ, A.,
High-Point Rendel, London, UK.

DAVIES, GW,
Earth Tech Engineering, London, UK. (formerly with Symonds Group)

GRASSI, M.,
Comune di Genova, Genoa, Italy.

1. Introduction

Over the centuries the city and port of Genoa have evolved almost as two separate entities. The City, with its rich historical heritage of palaces and churches, clings to the mountainside overlooking the historic harbour basin (figure 1.1). It has an almost picture-post-card juxtaposition with the medieval port which has now expanded to become one of the largest and most commercially active terminals for the import and export of goods for central and southern Europe. Genoa's economic success, built on a rich history of mercantile tradition, continues to grow and expand.

Figure 1.1-Aerial View of Genoa Harbour

In the city centre, traffic volume also continues to expand to the point where the existing infrastructure can no longer cope. It was partly to respond to this growth that a long elevated viaduct was built around the historic harbour during the 1960's. This viaduct links the Sampierdarena area to the west of the City with the Foce area to the east. The viaduct provides a link of sorts, but apart from having a weight restriction of only 2 tonnes, it is seen as an aesthetic aberration that severs the link between the medieval town and its harbour. Radical solutions to address the traffic problems are now considered essential, particularly if they involve the demolition of the Sopra Elevata and the restoration of the seafront area to maximise Genoa's tourism potential.

In January 2000 consultants were appointed by the Comune di Genova to undertake a feasibility study. The objectives of the study were as follows:
- To provide a link for vehicular traffic between Sampierdarena in the west of the city, with the Foce area to the east, thereby relieving the local road network of traffic congestion.
- To evaluate three alternative schemes, consisting of a bridge scheme over the port, a tunnel scheme under the port and an inland tunnel route around the north of the city under the high ground.
- To assess the practicality of demolishing the existing elevated highway, known as the Soprael Elevata.

The criteria for evaluation included the following:

> Relief of traffic congestion
> Environmental impact
> Effect on existing property, land use, and future development potential
> Navigational requirements for shipping
> Aviation requirements for aircraft using Genoa Airport.
> Capital construction and maintenance cost.
> Economic worthiness
> Potential for attracting private finance

2. Route Determination

Following a desk study of the existing local and regional road network, eight potential routes for investigation were identified. These comprised three bridge routes across the harbour, two tunnel routes under the harbour and three inland tunnel routes (figure 2.1).

Given the congested nature of the City and the Port, each possible route was assessed in terms of its physical constructability and where a specific alignment could not be found, sufficient engineering work was undertaken to identify a corridor in which a feasible engineering solution was thought to lie. This was particularly difficult in the case of the inland tunnel routes due to the complex topography with very high ground intersected by deep valleys and virtually the whole hillside occupied with intensive residential development. After several weeks of work, the only options which emerged involved long lengths of bored tunnel under the mountainside. In the case of the harbour tunnel routes, both bored and immersed tunnels were considered but due mainly to the ground conditions, only relatively shallow immersed tunnel options were considered economical. In all instances, key considerations were not to interfere with important historical buildings and not to impact on Genoa's strong cultural heritage.

A traffic model was built using a simplified representation of the road network and this was calibrated using survey data provided by the Comune. The effectiveness of each route was then assessed to determine the impact on traffic patterns and assess the relief afforded to the City Centre. A simple cost-benefit analysis was carried out and all eight options were evaluated in this way. Other important criteria including potential environmental impacts were assessed before a short list of three schemes was identified. This comprised an inner harbour bridge route, an inner harbour immersed tunnel route, and an inland bored tunnel route (routes 1, 4 and 8 in figure 2.1).

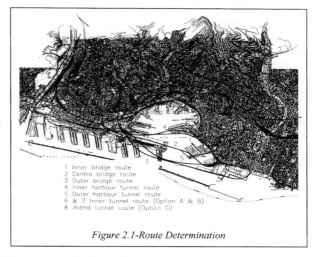

Figure 2.1-Route Determination

The traffic model was also used to investigate the effect of demolishing the existing Sopraelevata and a summary of the traffic findings is given in the table below. The figures show the flows with and without the Sopra Elevata. However in the case of the inland route, the relief to traffic in the harbour area is so poor that the removal of the Sopra Elevata is not realistic. Flows are 2-way vehicles per morning peak hour and are based on 1996 traffic data.

Table 2.1 Traffic Flows

	Base	Bridge Route	Tunnel Route	Inland Route	Bridge Route	Tunnel Route
Sopra Elevata		Retained			Removed	
A12 Motorway	7810	6926	6903	6195	6923	6974
City roads						
Corso Carbonara	1117	1233	1357	152	1295	1360
Galeria Garibaldi	1906	1767	1488	512	1751	1553
Via Gramsci	1473	1536	1684	878	1341	1685
Sopra Elevata	6984	1219	1458	4679	0	0
New Route	-	7412	7095	5894	8231	8337

3 Bridge Scheme

3.1 Layout Constraints

There has been a steady growth in cruise liner traffic to/from the Port of Genoa and the cargo handling facilities at the piers located around the 'arch' of the inner harbour have been largely replaced by those needed for passenger traffic. The piers now accommodate terminals for cruise liners and ferries. Increases in the volume of passenger traffic have been accompanied by increases in vessel size and those using the inner harbour now require an air draught of up to 62m. To allow for future growth in ship size, it was a requirement of the Port Authority that a navigational clearance of 75m under a bridge crossing must be provided.

Genoa Airport lies almost immediately to the west of the harbour and the Airport Authority has stringent landing and take-off clearance envelopes which must be complied with. These criteria, coupled with the navigational requirement, mean that only a narrow band is available in which any bridge superstructure can be accommodated.

There is no doubt that these restrictions severely limit the design of any bridge scheme and the route eventually selected for detailed investigation was route 1, Figure 2.1. It passes along the boundary between the inner and outer harbour and links the San Benigno area to the west of the harbour with the southern part of the Foce area to the east. A complex interchange is required in the San Benigno area and there is a further interchange at the eastern end where the scheme meets the existing road network near the Piazza Kennedy.

3.2 Bridge Arrangement

A suspension bridge with a main span of 610 metres was selected for the chosen inner bridge route and the engineering layout is illustrated in figure 3.1.

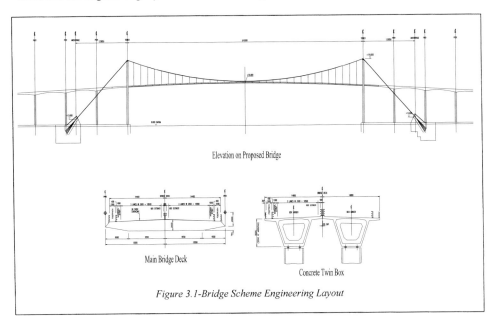

Figure 3.1-Bridge Scheme Engineering Layout

This was considered as the most appropriate structural form to achieve an acceptably economical bridge while complying with the opposing requirements of the navigation headroom and airport clearance constraints. A conventional three-span configuration was selected although only the main span is suspended from the cable. This improves the cable geometry, as the cable is taken down directly to the anchorage. The arrangement also has additional benefits for deck construction. For the 31m wide deck, supporting a dual 3-lane carriageway, a steel orthotropic box girder was selected and this is divided into 18m long segments coincident with the hanger spacing. The box girder depth is reduced to the minimum considered practical with this arrangement and this is 2.0 metres.

The selected layout, also means that excessively long side span cables are avoided (which a very high deck would otherwise require) and this has additional advantages in that the erection of the suspended deck is greatly simplified. All lifts are directly from the delivery barge. A more economical overall structure is also possible using cheaper balanced-cantilever construction for the side spans. An artist's impression is illustrated in figure 3.2.

Figure 3.2-An Artist's impression of the Bridge

3.3 Cable Design
In order to achieve the lowest practicable overall height of structure, a sag/span ratio of 1:15 was selected for the main span cables. This is somewhat greater than the 1:9 to 1:12 ratios normally expected for a suspension bridge of this type. It is however similar to that proposed for the new Tsing Lung bridge in Hong Kong which is also severely constrained by its proximity to an airport.

The cables are of conventional construction, made up from multiple small (5mm) diameter parallel high tensile steel wires with a minimum tensile strength of $1570N/mm^2$. These are compacted into a close packed circular shape and the required cable diameter in the main span is of the order of 600mm.

3.4 Environmental Loading
Genova has a relatively benign climate resulting from its north Mediterranean location, and is reasonably free from extreme temperature fluctuations or windstorms. This means that with this relatively short span bridge arrangement, the achievement of an adequate margin of aerodynamic stability is possible using the above shallow form of construction.

3.5 Towers

Except in areas subject to severe earthquakes, a reinforced concrete structure, stiffened laterally by a number of cross beams, is normally the most economical form of suspension bridge tower. As Genova is not in a highly active seismic area, an outline design was developed based on the use of concrete towers. The tower legs have a tapered hollow rectangular section, with three cross beams providing resistance to transverse forces. Foundations consist of a thick reinforced concrete pile cap supported by a rectangular array of large diameter bored cast-in-place concrete piles.

3.6 Anchorages

The anchorages must resist large horizontal loads from the main cables, and can be either gravity anchorages or tunnelled anchorages. In a gravity type anchorage, the cable pull is resisted principally by base friction generated by the mass of a large concrete structure. By contrast tunnelled anchorages mobilise a sufficient mass of rock along a sloping tunnel.

For the site conditions at Genoa, a gravity type anchorage was considered more suitable and the outline design was based on this. As a large above-ground concrete structure would be visually intrusive, this was avoided by placing as much as possible of the anchorage below ground level. Only the splay saddle housing projects above surface level.

3.7 Approach Viaducts

The approach viaducts and associated slip road geometry together with the congested site conditions were such that a form of construction comprising a prestressed concrete box girder built insitu by the balanced cantilever method was considered the most appropriate. This avoids the need for a precasting yard and associated handling facilities.

4 Immersed Tunnel Alternative

4.1 Route Alignment

The selected alignment for the immersed tunnel option is the inner harbour tunnel route shown as route 4 in figure 2.1 and follows very closely the alignment of the inner bridge. The routes considered to the south of route 4 entailed much longer tunnels and these are potentially much more disruptive to the Port in that they severely compromise future development plans. The selected route offers the shortest possible immersed tunnel length and the portal locations are almost completely hidden from the main residential and visitor areas in the town. An artist impression of the west portal is shown in figure 4.1.

Figure 4.1-An Artist's Impression of the Tunnel Portal

4.2 West Portal/Cut and Cover Tunnel

The western portal is located on the west side of the present dock access railway and the tracks are carried on the top slab of a short cut and cover tunnel forming part of the western approach. Various alternative alignments were examined in this area but the selected layout allows the tunnel to pass under the waterway in a location which minimises the likelihood of ships berthing immediately above it while they load and unload.

4.3 Immersed Tunnel

Various studies were undertaken and the scheme eventually selected is of the rectangular concrete reinforced concrete type. It carries two three-lane carriageways each of which is in a separate tube. Ventilation is by means of longitudinal jet fans placed above the traffic space and there is also a central duct for utilities and for safety/evacuation purposes (figure 4.2). The horizontal alignment is to 80kph standards and involves superelevated carriageways with an unavoidable "roll over" in the eastern half of the immersed tube.

The tunnel vertical alignment requires a trench of about 29 metres maximum depth (below mean tide level) and this is near to the maximum which can be tackled by a cutter-suction dredger. Preliminary indications are that the whole of the

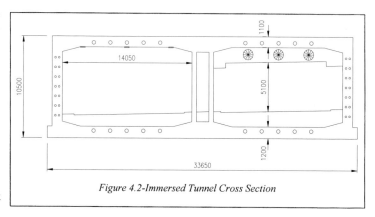

Figure 4.2-Immersed Tunnel Cross Section

trench will be above bedrock level but this will need to be confirmed. Should bedrock be encountered then underwater drilling and blasting will be needed or alternatively, (and more likely), the vertical and horizontal alignment will need to be re-appraised. It is also worth noting that the depth of the bed of the harbour varies considerably in the tunnel location suggesting that scour and accretion effects have taken place over the long term. These are almost certainly as a consequence of the Molo Vecchio reclamation plus the other reclaimed areas built on either side.

The optimum location of the junction between the immersed tunnel and the eastern cut and cover tunnel is heavily dependent on the practicality of dewatering deep temporary excavations for the cut and cover tunnel in this area, and there are additional engineering complications in that the excavations are close to property. The junction is close to the end of the Molo Vecchio and a ventilation building is incorporated into the cut and cover tunnel at this point. The overall length of the immersed tunnel is 930 metres, based on eight 116 metre long tunnel elements each approximately 10.5 metres deep and 33.5 metres wide.

4.4 Eastern Portal/Cut & Cover Tunnel

The eastern cut and cover tunnel extends through to a point near the Mercato Del Pesce (Fish Market) where the eastern portal is located and where there is an intermediate junction with the existing roads. The whole length of the route alongside the Molo Vecchio is thus in tunnel, with the space above being used for landscaping or development. The alternative is to construct part of this length in a retained open cutting. This has advantages from a tunnel ventilation viewpoint and is certainly cheaper, but the loss of useable land area was considered unacceptable in this key location.

In horizontal alignment terms the tunnel is located as far north as possible in the Molo Vecchio to leave space on the south side for a new reclamation. It is accepted however that a full landscaping and development plan for the area as a whole will be needed before a precise alignment could be confirmed.

4.5 The Eastern Foreshore

From the east portal of the tunnel, the route follows the eastern foreshore and is more or less at harbour level. A twin 810 metre long bored rock tunnel is required to take the route through a high cliff area where space at harbour level is not available. The route then terminates at the same point as the bridge crossing where it rejoins the existing road.

4.6 The Casting Basin

Reinforced concrete immersed tunnel units are normally built in a casting basin and if the geological conditions are suitable, and space is available, the most economical method is usually to excavate a large open earthwork excavation big enough to accommodate all of the tunnel units at once. In this instance an area of approximately 10 hectares is required excluding any additional work space required around the basin for working purposes.

At first sight there is only one area within the harbour area sufficiently large for such a facility and this is the reclamation area currently proposed by the Port Authority for an extension to their container handling terminal on the west side. A casting basin here would require almost the whole of this area to be used and would also envelope the area currently occupied by fuel storage tanks (which are due for relocation). This area could be used by the tunnel contractor and the basin backfilled afterwards as part of the tunnel contract. The idea would be to leave it in a condition suitable for the Port. However, the Port Authority advised that they have well developed plans for this area and that another site will need to be found.

At one time, it was suggested that one or more of the existing dry docks on the east side of the harbour might be used and might be large enough to accommodate one tunnel element at a time. While this may be physically possible, there are serious programme implications with such a proposal as well as severe access and working area constraints. Since the Port Authority use the whole of this area for ship repairs, this whole concept had to be abandoned.

A third possibility is to use part of the Foce foreshore lying immediately to the east of the Exhibition Centre. This is a prominent and environmentally sensitive part of the foreshore but there are proposals to develop this zone into a protected harbour area for yachts and other small craft. If it is used temporarily as a casting basin then it might be converted afterwards for this type of use. This concept was considered sufficiently attractive to the Comune for it to be taken forward into the costing studies and now forms part of the scheme.

90 DESIGN AND CONSTRUCTION

5 Inland Bored Tunnel

5.1 Route Alignment

The inland bored tunnel is much longer than the other routes and follows a curved horizontal alignment behind the town (route 8, figure 2.1). There are three intermediate junctions plus terminal junctions at the A7 motorway toll area in the west and at the Bisagno River in the east. Over 80% of the route is in tunnel. While the inland route provides good access to the higher residential areas of Genoa, it is not effective in reducing congestion in the harbour area and does little to enable the Sopra Elevata to be replaced. The route is also very costly and has substantial property and development impacts.

For these reason it was agreed with the Comune that this route would not be taken into the detailed comparative evaluations which were undertaken for the other two schemes. These evaluations compared, in detailed terms, all of the temporary and permanent impacts likely from each of the two schemes and included quantification of all of the differing relative effects including those which cannot be measured in monetary terms.

6 Bridge v Tunnel

6.1 Traffic Effects

There is little to choose between the two schemes in traffic terms (Table 2.1). Both carry a similar amount of traffic and both provide effective relief to the existing harbour side roads. Traffic relief in the City Centre is slightly greater with the tunnel and journey time savings (10 minutes or so) are approximately the same. Both provide sufficient traffic relief to make the removal of the Sopra Elevata realistically possible and to enable sufficient landscaping and local improvements to be made to improve the whole area of the foreshore around the inner harbour.

6.2 Visual Impact

For many people, the advantage of a bridge scheme lies in its visual attractiveness and there is no doubt that the suspension bridge developed in the study will be a dramatic structure in its own right. It will provide important new views of the City for residents and visitors passing across it from side to side. Its principal disadvantage however also lies in its visual impact. The study showed that it may not be a landmark structure due to the constraints imposed on the tower heights and the associated shallow and yet very high superstructure. The extent to which this is important is a subjective matter however which no doubt will generate different and probably opposing views. Irrespective of these, the study found that the scale of the bridge is out of keeping with Genoa`s medieval infrastructure around the inner harbour and also that there are important cultural heritage considerations to be borne in mind.

By contrast the advantage of the tunnel scheme is its lack of visual impact and the fact that most of it is completely out of sight. Its main impact is from the San Benigno junction. The tunnel is largely neutral from a visual point of view since its construction neither enhances nor detracts from the setting of the inner harbour.

6.3 Construction Disruption

Both schemes will involve extensive disruption to existing activities and both schemes involve extensive property demolition mainly on the east side of the harbour. The effect is greater with the tunnel but by contrast there are greater opportunities here for redevelopment using the space above it. The principal disadvantage with the tunnel is the disruption and disturbance to harbour users during the construction period and this will be much greater than with the bridge. There will also be a sizeable impact from the construction and use of the casting basin opposite the Piazza Kennedy.

7 Conclusions and Recommendations

A comparison between the inner bridge route and the inner port tunnel route indicated that both schemes were equally effective in terms of affording relief to traffic congestion. The capital construction cost was also similar. However the bridge solution, it was considered would be a dominant structure and not in keeping with the architectural heritage of Genoa. In contrast the tunnel scheme would be largely out of sight. The balance of arguments favoured the tunnel scheme and was recommended for development.

Full details of the Consultants findings and their conclusions and recommendations are contained in their study report which has been published and is available from the Comune.

8 Acknowledgements

The Consultants wish to express their thanks to all the participants for their co-operation in connection with this Study and in particular the Comune di Genova for its assistance and kind permission to publish this paper.

Our thanks to AIATI srl for its assistance in relation to organisation and management of operations in Genoa as well as his technical input and to,

Professor Del Grosso (University Of Genova), Specialist Advisor on the project, for his advice on a whole range of subjects.

The Planning and Design of Viaduct Construction in Route 9 between Tsing Yi and Cheung Sha Wan

NAEEM HUSSAIN BSc (Eng) DIC MSc CEng FHKIE FIStructE FICE
Ove Arup & Partners Hong Kong

1. Introduction

Route 9 is a trunk road linking Lantau and Sha Tin via Tsing Yi Island and West Kowloon (Figure 1). The North Lantau Highway and Lantau Link completed in 1997 form part of this route. Route 9 between Tsing Yi and Cheung Sha Wan (R9T) will connect Lantau Link with the West Kowloon Highway at Cheung Sha Wan. Another section of Route 9 between Cheung Sha Wan and Sha Tin (R9S) will extend the route to Tai Wai in Sha Tin. R9T will provide a route for container traffic from various part of the Territories to have an access to Container Terminal No. 9 currently under construction and other existing container terminals such as Container Terminal No. 8 (CT8). Together with R9S, Route 9 will provide a direct route for traffic from North East New Territories to North West New Territories and to the Airport at Chek Lap Kok.

Since its commissioning in 1997, the section of Route 3 comprising Cheung Tsing Highway, Cheung Tsing Tunnel and Tsing Kwai Highway has become a fast link between North West Tsing Yi and West Kowloon. The subsequent opening of the Ting Kau Bridge and the Route 3 Country Park Section in 1998 has attracted further traffic between the North West New Territories and West Kowloon to use these three highways. The traffic impact assessment study for R9T confirmed that these three highways will reach saturation by 2007, and will further operate much beyond capacity by 2011. The completion of R9T will form an alternative route to these three highways. This paper describes the planning and design of the phase 1 high level viaducts that lead upto the proposed Stonecutters Bridge.

2. Alignment

Phase 1 of the project is situated between Stonecutters Bridge across Rambler Charmel and the interface with West Kowloon Expressway at Lai Wan Interchange near Cheung Sha Wan.

The main carriageway is dual 3-lane with hard-shoulder and there are links to and from Container Terminal No. 8 (CT8). These links are designated as Ramps E, E1, F and F1 (Figure 2). Ramps E and F are dual 2-lane with hard shoulder whilst Ramps E1 and F1 are single lane with hard shoulder. At Lai Wan Interchange the main carriageway splits into two 2-lane with hardshoulder northbound and southbound links that connect with

Route 9 Cheung Sha Wan to Shatin. The main carriageway also connects with West Kowloon Expressway via Ramps G and H which are both 2-lane with hardshoulder.

The alignment is virtually straight in a east-west orientation between Lai Wan Interchange and CP3/End roundabout and will traverse over the Container Port Road South which will be re-aligned to run under the viaduct. There are two other major ground level roundabouts CP3/CP4 and CP3/D16 which connect with ground level distribution roads. From CP3/End roundabout the alignment swings north to connect with the proposed Stonecutters Bridge across the Rambler Channel.

The main carriageway height varies from approximately 70m at Stonecutters Bridge to about a low of 30m near CP3/CP4 roundabout and then rises to about 45m as it passes over West Kowloon Expressway.

3. Environmental Considerations

The viaducts runs alongside the CT8 container port area with its port scape of high derrick cranes and stacked containers, with generally high ambient noise level from container traffic and port activities.

The viaduct is at a high level, almost matching the level of the top of the derrick cranes and will be a very visible element, able to be seen from near and distant viewpoints (Figure 3). The viaduct leads up to the Stonecutters Bridge and is approximately 70m high at the common pier with Stonecutters Bridge backspan.

4. Design Principles

The basic design principle for the viaduct is to have a design the satisfies and integrates the disparate requirements of Environment, Aesthetics, Construction, Costs, Availability Reliability and Maintainability.

The viaduct will be viewed from a variety of viewpoints. From below by pedestrians and road users, from the side and above by the public in developments in CT8, Lai Chi Kok, Cheung Sha Wan and in distant views from marine traffic and Hong Kong Island.

The pier columns and soffit of the main carriageway and ramps will be the most visible elements at ground level, whilst the sides and top will be most visible from adjacent CT8 offices. The viaduct needs to have a uniformity of appearance for both the main carriageway and the ramps.

The viaduct furniture such as sign gantries, noise barriers etc need to be neatly integrated in the overall design. The finishes to the visible elements need to work at both the macro and large scale level.

The superstructure needs to be of a form suitable for construction at the relatively high height and in an economic manner along with quality construction.

5. Superstructure

The location of the piers is governed by the ground level roads and roundabouts, the transport corridor and multi-level roads at Lai Wan Interchange, and by utility equipment and drainage requirements. Considering the height of the piers, the optimum span between CP3/End roundabout and Lai Wan Interchange is in the order of 60m and between CP3/End roundabout and Stonecutters Bridge in the order of 80m. Typical span arrangement is shown in Figures 4 & 5.

The variable span lengths and the juxtaposition of down and up ramps leads to the choice of a constant depth superstructure between CP3/End roundabout and Lai Wan Interchange, from both structural and visual considerations. The superstructure will comprise precast post-tensioned concrete boxes (Figures 6, 7). Precasting also ensures quality fabrication under factory condition and is a well tried and tested method in Hong Kong. Prestressing will be a mixture of external and interal prestress.

Between CP3/End roundabout and Stonecutters Bridge, the spans will be increased to 80m with inset haunches provided over the piers.

6. Substructure

Between CP3/End roundabout and Stonecutters Bridge, the superstructure comprises three boxes which is supported by portal type piers comprising of two columns and a cross-beam (Figure 8).

Between CP3/End roundabout and CP3/CP4 roundabout the main carriageways are supported by single columns with a T-head whilst the ramps are supported on single columns (Figure 9).

Between CP3/CP4 roundabout and Lai Wan Interchange the main carriageways and ramps are generally supported on single columns.

With the variable heights and type of substructure it is important that there is a visual harmony at ground level. The shape of the columns for all types of substructure are similar and vary only in cross-section size, thus achieving a consistent visual appearance along the whole length of the viaduct. The portal cross-heads and single column T-heads spring from the sides of the columns in a consistent manner, thus also achieving a consistent visual appearance.

Figures 10 shows how the various piers grow out of a common theme.

At Lai Wan Interchange there are a multitude of existing structures at various levels with different pier column shapes. The new piers which will rise above the level of the West Kowloon Expressway will be the most visible columns and will be visually read with this new viaduct.

7. Surface Treatment

The predominantly visible elements of the viaduct are the concrete surfaces of the substructure and superstructure and the cladding of the noise barriers.

Concrete by its very nature does not weather well, especially the surfaces that are exposed to the wetting and drying action of rain. The rain affected surface stain, and this is particularly pronounced in Hong Kong. It is therefore essential that the flow of water on visible surfaces is controlled, thus obtaining a controlled weathering pattern.

The deck soffits will generally be protected from the rain and will normally weather uniformly. A smooth surface helps in this uniform weathering and this has been specified for the superstructure boxes.

The parapets will be the concrete elements most exposed to the rain. In order to minimise the flow of water on the external face, a generous fall towards the inner face will be provided on top of the parapets. The parapets will have a thin precast external element with insitu concrete on the inner face (Figure 11).

The tall piers have a sculpted cross-section. Vertical grooves are provided on all faces to control the flow of water and provide an element of self-cleansing and uniform weathering (Figure 12).

8. Drainage

The carriageways will be drained via catchpits contained within the depth of the parapet overhanges. The drainage pipe from the catchpit to the void within the superstructure box will be encased in concrete (Figure 13). The longitudinal run of the drainage pipes within the box will connect to drainage downpipes contained within the column voids, or cast into the solid columns. Rodding eyes will be provided as appropriate (Figure 13).

9. Acknowledgement

The author wishes to thank Mr Y C Lo, Director of the Highways Department Hong Kong, for giving his kind permission to publish this paper, and acknowledge that the views expressed in this paper are those of the author and not necessarily those of the Highways Department.

Figure 1   Location Plan

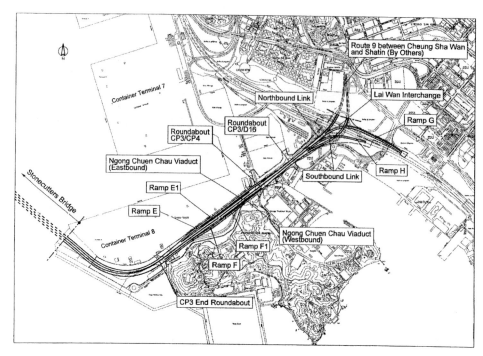

Figure 2   Plan of Phase 1 Works

Figure 3   Perspective View

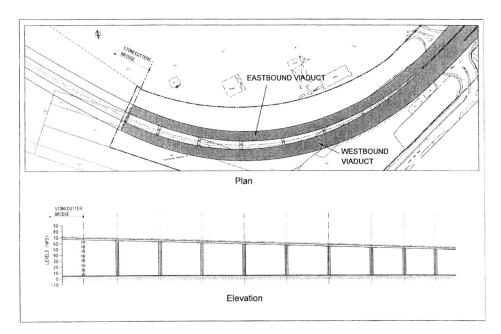

Figure 4 Plan & Elevation – Stonecutters Bridge to CP/End Roundabout

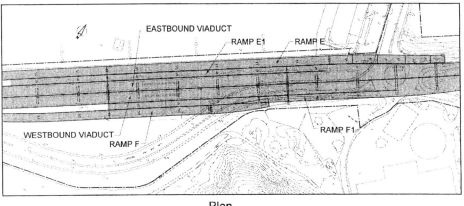
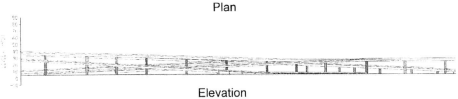

Figure 5 Plan & Elevation – At Container Port Road South

98 · DESIGN AND CONSTRUCTION

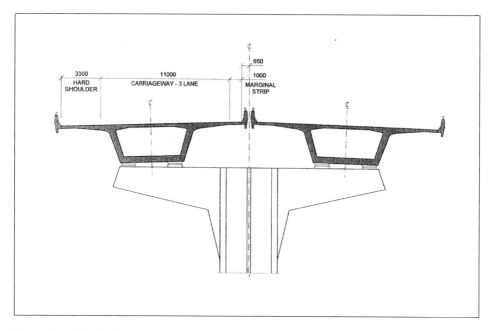

Figure 6　Typical Deck Section for 3-Lane Dual-2 Carriageway

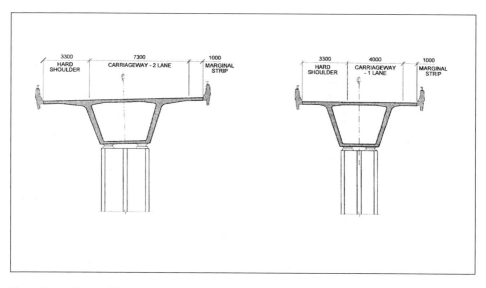

Figure 7　Typical Deck Section for 2-Lane and 1-Lane Carriageway

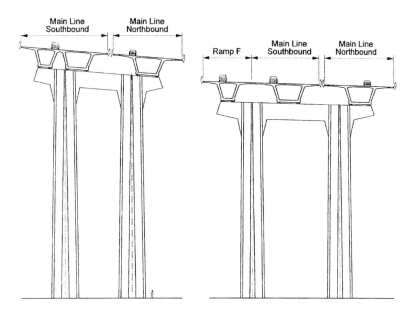

Figure 8   Typical Substructure between Stonecutters Bridge and CP3/End Roundabout

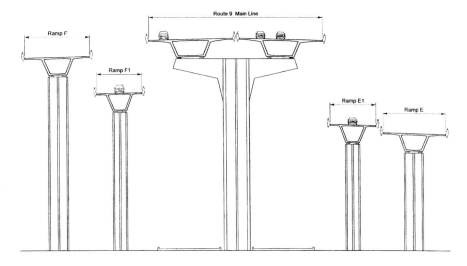

Figure 9   Typical Substructure between CP3/End and CP3/CP4 Roundabout

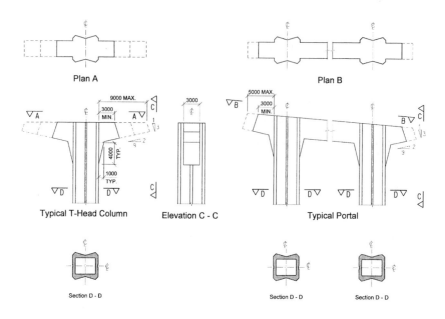

Figure 10   Pier Concept

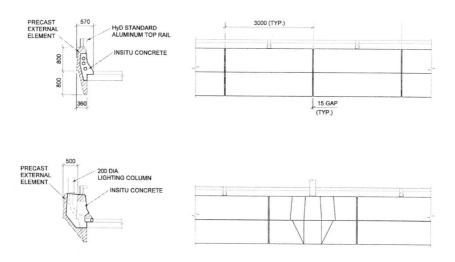

Figure 11   Typical Parapet Details with Lighting Column

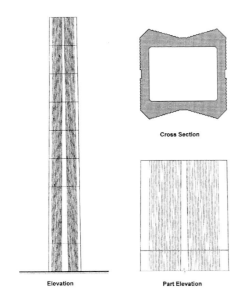

Figure 12   Typical Pier Finishes

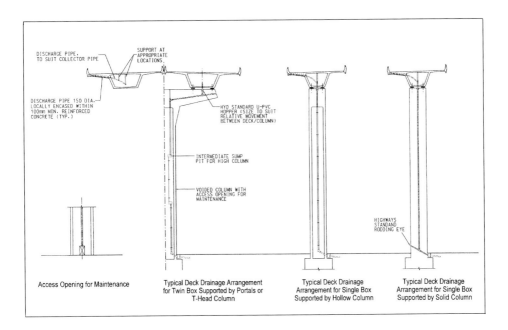

Figure 13   Typical Drainage Details

Recent achievements in Polish bridge engineering

WOJCIECH RADOMSKI
Warsaw University of Technology
Warsaw, Poland

Introduction

Bridge engineering in Poland has its long and interesting tradition relating – as in other countries of the world – to the Polish national history, especially in the Twentieth Century. In the first decade after the World War II many destroyed bridges have been reconstructed. During the next decades up to 1989 the development of the bridge engineering have been rather limited mainly due to economical and political reasons. Some more modern structural and technological solutions, such as cantilever methods of construction of concrete bridges, have been then implemented in several cases only, while the use of precast (mostly pretensioned) concrete elements has been dominated (about 90 % of the total number of concrete bridges in some years). An evident growth in the Polish economy resulting from the change in the political system and introduction of the market system in the national economy can be noticed from the beginning of the 90s. The economical growth is always accompanied by an increase in transportation needs and, therefore, the road network in Poland requires to be developed and many new bridges have to be constructed as well as many existing bridges require to be rehabilitated and modernized to be in accordance in the current or predicted traffic conditions.

The number of inhabitants in Poland is about 38 millions. The area of the country is in general flat with a chain of mountains in the South where two the biggest rivers, Vistula (Wisła) and Oder (Odra) begin their flow to the Baltic Sea. These natural conditions result in a great number of bridges crossing a lot of small streams and channels and a relatively small number of big rivers.

At present, there are about 29,000 highway bridges in Poland with the total length of about 550 km. About 76 % of them (70 % of their total length) there are bridges with reinforced or prestressed concrete superstructures, while 14 % (22 % of their total length) there are the bridges with steel or composite (i.e. with steel girders and reinforced concrete deck) superstructures. About 10 % of the bridges (8 % of their total length) there are temporary structures, i.e. the wooden bridges or the bridges with wooden deck and steel girders. Above mentioned natural conditions result in a great number of bridges crossing a lot of small rivers and channels and a relatively small number of large bridges. The average length of a highway bridge is about 18 m and the average distance between bridges on the road network is about 12 km as yet. The number of large bridges with the total length more than 200 m is about 120. However, some of them are more than 1 km long, e.g. the longest bridge structure in Poland is 2593 m long.

Total number of railway bridges in Poland is about 8,800. About 65 % of them there are concrete and masonry bridges and about 35 % there are steel bridges.

The bridge inventory in Poland is relatively old. About 50 % of highway and about 80 % of railway bridges have been more than 50 years in service. Nearly 20 % of the bridges are classified as structurally deficient or functionally obsolete and, therefore, require to be repaired, strengthened or modernized. It should be emphasized, however, that the situation described above may be considered similar to many other countries in the world, including highly developed ones. In some of them, the situation is even more critical. For instance, over 40 % of the more than 577,000 highway bridges in the United States have been classified as either structurally deficient of functionally obsolete (1).

Above presented situation shows two main fields of tasks for bridge engineering in Poland. One of them is construction of many new bridge structures resulting from the development of the road network, while the second one is rehabilitation and modernization of the existing bridges. Therefore, this situation can be considered as typical for many other countries. However, the needs to construct the new bridges is particularly pronounced.

As far as construction of new bridges, it is necessary, for instance, to construct by the year 2005, fourteen relatively big bridges with a total length of about 6,600 m, excluding the approach structures. However, the most important task in the near future is construction of a number of bridges in connection with the Project of Motorway Construction. This Project also includes three new transeuropean motorways A1 (North-South, Helsinki-Łódź-Budapest), A2 (West-East, Berlin-Warsaw-Moscow), and A4 (also West-East, Berlin-Cracow-Lvov) with a total length about 2,000 km, which should be constructed by the year 2010. It also requires more than 500 bridge structures with a total length of more than 35 km.

As far as bridge rehabilitation and modernization, the needs require application of the most modern and the most effective methods, especially for bridge strengthening. Transfer of the relevant newest technologies to Poland is very quick and more and more common during the last years.

The paper presents some selected and outstanding examples illustrating recent achievements in Polish bridge engineering and concerning both construction of the new bridges and the most modern techniques applied for bridge strengthening.

Main fields of development of Polish bridge engineering

The considerable development of Polish bridge engineering during the last years can be especially noticed in the following fields:
- applications of high-strength concrete for the new bridge structures;
- more common use of the modern erection methods such as incremental launching and cast-in-place cantilever balance;
- construction of relatively big bridges of modern structural system, such as cable-stayed bridges;
- the use of composite materials, such as carbon fibre reinforced plastic (CFRP) strips or fabrics, for bridge strengthening.

Above mentioned fields are discussed and exemplified below. They seem to be in accordance with the current world's trends in bridge engineering concerning both bridge construction and rehabilitation.

Application of high-strength concrete

Problems concerning applications of high-strength concrete (HSC) for bridge construction in Poland have been already presented by the author elsewhere, e.g. (2), (3), (4). Therefore, it should be only generally mentioned that research on high-strength/high-performance (HSC/HPC) concretes is in Poland relatively well developed and there are no major obstacles in obtaining these concretes. Economical considerations indicate that the use of B60 concrete instead of B40 concrete with the same volume, increases the total construction cost by only about 1.5% to 2.0%, while the use of B90 concrete increases the total construction cost by about 3.0% to 5,0% compared to the use of B40 concrete with the same volume. The above given information shows that an evident improvement in the bridge durability resulting from the use of HSC/HPC can be not too costly. Taking into account the other profits, first of all the reduction of the total concrete volume even by 20% to 30% depending on the bridge type, the use of HSC/HPC in bridge structures is economically justifiable.

However, research on HSC/HPC is more advanced in Poland than its structural applications. Similarly to the situation in other countries, a main barrier to the more common use of HSC/HPC in bridge engineering is lack of an adequate official design methodology. Unfortunately, the concrete classes, beginning from 60 MPa concrete, are beyond the scope of Polish and many other national or international standards or codes. Additional data for design are provided by the relevant tests as yet.

The tendency to apply HSC/HPC in Poland is demonstrated among others by the bridge in Chabówka (south of Poland), shown in Fig. 1, 267 m long, constructed by incremental launching and completed in October 1996. The superstructure of this bridge is made of concrete B60 (i.e. the conventional lower limit of HPC) with a total volume about 1,600 m^3 (5).

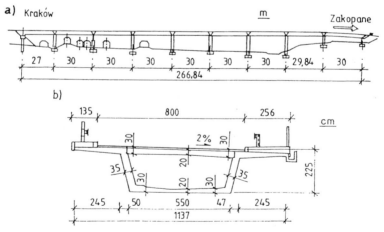

Fig. 1. Bridge in Chabówka: a) elevation, b) cross section

The trend to use HPC is also endorsed among others by the projects submitted in December 1996 for competition concerning a new bridge over Vistula River in Płock (central Poland). From fourteen projects, six concrete bridges of various structural system, including cable-stayed one, have been proposed. In five of them, B60 concrete as a minimum has been proposed.

Modern erection methods

Progress in the use of modern erection methods is somewhat more evident in the construction of concrete bridges than steel ones From among several methods applied for construction of relatively long concrete bridges, the most developed in Poland are two of them, namely: incremental launching method and cast-in-place cantilever balance method.

The first application of incremental launching method for construction of the concrete bridge took place in Poland till in the 80s (bridge over the Soła river in Oświęcim, south of Poland $37.8 + 3 \cdot 45.6 + 37.8 = 212.40$ m long, completed in 1987). Several further bridges have been constructed using this method during the last decade. However, the most advance example seems to be the Border Bridge between Poland and Czech Republic, 750.3 m long, completed in 1991, because of its complicated geometry shown in Fig. 2. This bridge is curved in plan and with a longitudinal gradient up to 4 %. Incremental launching is much more easy to use it for construction of straight bridges and, therefore, this is a major domain of its applications. In case of more complicated bridge geometry, the application of this method requires to be preceded by an adequate and very advanced theoretical analysis. Moreover, a longitudinal gradient of the bridge superstructure requires the use of some additional accessories, shown schematically in Fig. 3, to prevent the structure against a self-acting sliding down (brake system). Cross section of the bridge is presented in Fig. 4. Some other basic information on the bridge is given in the relevant captions.

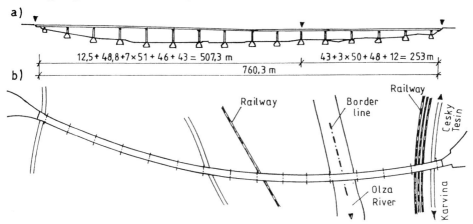

Fig. 2. Border Bridge in Cieszyn: a) elevation, b) plan. Height of the piers: 4.0 m to 26.4 m. Expansion joints are marked by black triangles.

The first application of cast-in-place cantilever balance method took place in Poland in 1963 (Bernardyński Bridge over the Brda river in Bydgoszcz, central Poland, 64 m long, with the main span of 40 m). However, during the next three decades this method has not been applied, mainly because of rather limited economical development of the country and predominant role of various prefabricated bridge systems. Beginning from the early 90s, the method has been used for construction of several bridges. Two relevant examples concerning the biggest concrete bridges in Poland are shown in Figs. 5 and 6.

Fig. 5 presents the bridge over the Vistula river near Toruń (central Poland), 955.4 m long, which is a part of the above mentioned motorway A1. The bridge was completed in 1998.

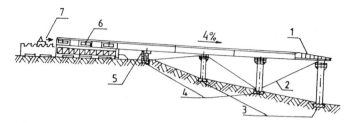

Fig. 3. The Border Bridge in Cieszyn during construction. Notations: 1 – steel nose, 2 – diagonal tie bracing, 3 – temporary supports, 4 – bridge piers, 5 – brake system, 6 – casting section, 7- driving mechanism (hydraulic jacks with capacity 2·1,500 kN to 2·3.000 kN).

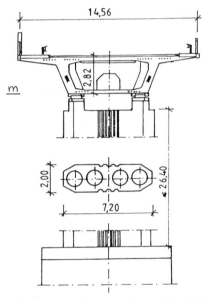

Fig. 4. The Border Bridge in Cieszyn – Cross sections of the bridge superstructure and substructure. The cable ducts for central (assembly) prestressing (cables 6L15.5 - black dots) and final prestressing (cables 27L15.5 - black rectangles) are shown.

The superstructure has been constructed using two different methods – incremental launching for the side parts, denoted in Fig. 5 by A1 and A2, each 219 m long, and cast-in-place cantilever balance for the central part composed by three spans, each 130 m long. and two spans, each 73 m long. Various cables of VSL type, imported from Switzerland, have been used for prestressing of the bridge superstructure. Concrete of the minimum standard strength of 50 MPa has been used for bridge superstructure. The average strength more than 40 MPa has been reached after 3 to 5 days after casting. The bridge has been entirely designed by the Polish engineers (6). One lane of superstructure is in service at present, while all the piers are completed for the second lane of the motorway.

Fig. 6 presents the bridge over the Oder river in Opole (south of Poland) with the total length of 385 m and with the central span 100 m long, completed in 1999. The side parts of the bridge, denoted in Fig. 6 by A1 (49.2 m long) and A2 (134.20 m long) have been constructed using stationary scaffolding, while the main part of the bridge, denoted by B (201.6 m long), with the use of cast-in-place cantilever balance method. The bridge superstructure has been designed by the Polish and Swiss group of engineers. Concrete with the minimum standard strength of 40 MPa has been used for bridge superstructure. BBR CONA Compact 1906 cables have been used for prestressing.

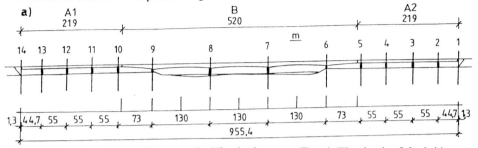

Fig. 5. Elevation of the bridge over the Vistula river near Toruń. The depth of the bridge superstructure is varied from 4.0 to 8,0 m. The total width of the bridge deck – 14.8 m (for one lane of the motorway). Two expansion joints only (over the abutment supports No 1 and 14). Piled foundations for majority of the piers (with diameter from 1200 mm to 1620 mm).

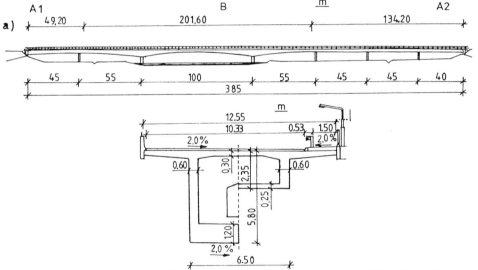

Fig. 6. Bridge over the Oder river in Opole: a) elevation, b) cross section. Piled foundations for all the piers (piles with diameter of 1500 mm and length from 18.25 m to 10.50 m).

Comparison of the basic technical characteristics of the bridge superstructures constructed by means of the cast-in-situ cantilever method (Figs 5 and 6) is given in Table 1.

Table 1. Basic technical characteristics of the bridges over Vistula and over Oder (for 1 m² of the bridge deck).

Specification	Bridge over Vistula	Bridge over Oder
Concrete	0.82 m³/m² (B50)	0.94 m³/m² (B40)
Reinforcing steel	190 kg/m²	108 kg/m²
Prestressing steel	37.5 kg/m²	30.9 kg/m²

Taking into account that construction of the bridge over Vistula river has been the first Polish application of the cast-in-situ cantilever balance method for erection of relatively big bridge, the characteristics given in Table 1 seem to be justifiable. The second application of the method (bridge over Oder river) has allowed to reach relatively better technical characteristics corresponding in general to those observed in the other countries with much longer tradition in the use of the method.

Modern structural systems

The use of modern structural systems in the Polish bridge engineering can be well exemplified by the cable-stayed bridges. Cable-stayed system has been applied rather exceptionally up to the second half of 90s and limited to relatively small structures, manly the pedestrian bridges. Beginning from the last few years, the cable-stayed bridges has became more frequently designed and constructed in Poland. The newest examples are shown in Fig. 7.

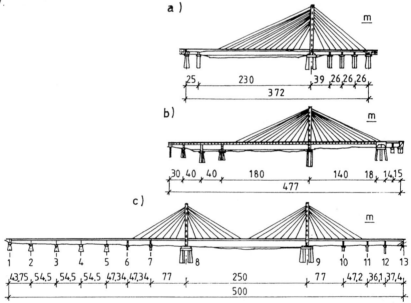

Fig. 7. Newest examples of the cable-stayed bridges in Poland; a) bridge along the Sucharski route in Gdańsk (u.c., to be completed in 2002), Świetokrzyski Bridge in Warsaw (completed in 1999), Siekierkowski Bridge in Warsaw (u.c., to be completed in 2002).

All the bridges shown in Fig. 7 are located in the towns and are constructed over the Vistula river. They are the first Polish experience concerning the construction of relatively big bridges with cable-stayed structural system. The pylons of the bridges, shown in Fig 8, are made of reinforced concrete. They have been constructed section by section using the climbing shuttering. The bridges have basically the same type of the superstructures, composed of steel plate girders and reinforced concrete deck (i.e. composite structure). However, the different methods of construction have been applied (7).

The superstructure of the Świętokrzyski Bridge (Fig. 7b) has been constructed with the use of temporary supports. Left part of the bridge (over the river, 290 m long) has been constructed

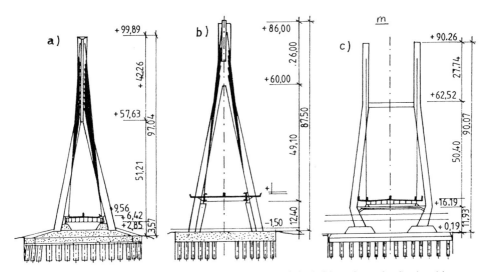

Fig. 8. Pylons of the bridges shown in Fig. 7: a) pylon of the bridge along the Sucharski route in Gdańsk, total width of the deck – 20.31 m, piles with diameter 1.80 m, 30 m long, b) pylon of the Świętokrzyski Bridge, piles with diameter 1.50 m, 25 m long, c) pylon of Siekierkowski Bridge, piles with diameter of 1.50 m, 24 m long.

with the use of longitudinal pulling of the steel structure, while the right part (over the land, 140 m long) has been assembled directly using the mobile cranes operating at ground level. The concreting of deck slab has been performed after the whole steel part of the superstructure was completed. The cables (BBR HiAm) have been assembled and tensioned with the prescribed forces as the last operation. The relevant control of the bridge geometry has been performed in every construction stages and after the bridge was completed.

The main span of the bridge along the Sucharski route (Fig. 7a), under construction as yet, is constructed using cantilever method (i.e. without any temporary supports), while the side spans (with the length of 61 m, left section, and 117 m, right section) are constructed by means of longitudinal pulling.

The main part of the Siekierkowski Bridge (Fig. 7c), under construction as yet, will be constructed by longitudinal pulling (the section between the supports No 6 and nr 11) with the use of temporary supports, while the side spans (between the supports No 1 and No 6, 251 m long and between the supports No 11 and No 13, 75.5 m long) – by means of the mobile cranes operating at ground level. The reinforced concrete deck slab will be made just after a given steel section is completed. Therefore, contrary to the Świętokrzyski Bridge, the sections of the superstructure will be pulled as a composite ("steel-concrete") structure.

All the erection methods used in the above presented cable-stayed bridges require the pylons to be completed before the assembly of the superstructures is finished. Therefore, the methods seem to be somewhat conservative compared to those applied in the other countries. It should be emphasized, however, that constructions of the presented structures are the first Polish experience as yet and some local conditions should be also taken into consideration in selecting of an adequate erection technique.

The use of composite materials for bridge strengthening

The use of composite materials and products, mostly CFRP strips and fabrics, for bridge strengthening is well developed in Poland. This technique belongs to the most modern ones. Beginnig from the first application of the CFRP strips for bridge strengthening in 1997 (bridge over the Wiar river in Przemyśl, south-east part of Poland, (8)), nearly 20 further bridge applications of this technique have been in Poland. The first Polish application is exemplified in Fig. 9.

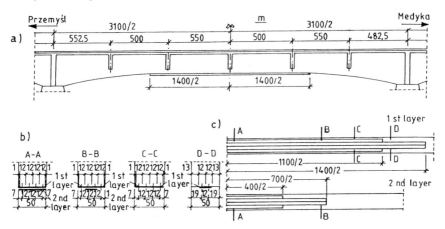

Fig. 9. Lay-out of the CFRP strips for strengthening of the concrete bridge in Przemyśl.

The first and many further applications of CFRP strips have been verified by the field tests and a high effectiveness of this technique has been confirmed.

Apart from the CFRP strips, the CFRP fabrics are also successfully used in Poland for bridge strengthening, especially in case when the shear structural capacity requires to be improved. For instance, the first world's application of the Swiss SIKAWrap system for strengthening of the shear zones in the bridge structure took place in Poland in 1998 (the bridge over Bystry Channel in Augustów, north-east part of Poland).

Moreover, the external post-tensioning is also used for bridge strengthening. More than 20 bridges have been strengthened in Poland using this modern technique. This problem has been already presented elsewhere (9).

Concluding remarks

Based on the above given information, the following concluding remarks can be formulated:

1. The scale of needs concerning the construction of the new bridges as well as rehabilitation and strengthening of the existing ones is very large in Poland. Beginning from the early 90s, a remarkable progress in the Polish bridge engineering can be noticed as a result of the political and economical changes.

2. Transfer of modern technologies as well as implementation of the original Polish structural

and material solutions are recently remarkable in Poland. The international cooperation in the bridge engineering is very wide and developed in many fields.

3. It is obvious that some selected problems are only presented in this paper.

References

1. Friedland, J.M., The Practice of Bridge Management in the United States, Proceedings of the International Bridge Conference Warsaw '94, Road and Bridge research Institute Warsaw, June 1994, Vol. 1, pp. 215-223.

2. Radomski, W., Development of Research and Application of High-Performance Concrete in Bridge Engineering in Poland, Proceedings of Second CANMET/ACI International Conference, ACI SP-186, Gramado, Brazil, 1999, pp. 537-554.

3. Radomski, W., Recent Investigations and Applications of High-Performance Concrete in Bridge Engineering in Poland, Proceedings of the 5th International Symposium on Utilization of High Strength/High Performance Concrete, Sandefjord, Norway 1999, pp. 881-890.

4. Radomski, W, Durability of Concrete Bridges – Polish Experience, Fifth CABMET/ACI International Conference on Durability of Concrete, Supplementary Papers, Barcelona, Spain 2000, pp. 693-707.

5. Gliszczyński, J. and Radwan, D., Construction of a Viaduct in Chabówka", Drogownictwo, No 2, 1997, pp. 39-41 (in Polish).

6. "On the Motorway Bridge over the Vistula River near Toruń", AKCES Publishers, Toruń, 1999, 210 pp. (in Polish).

7. Stańczyk A., Bridges over the Vistula River under Construction, Drogownictwo, No 10., 2000, pp. 308-316 (in Polish).

8. Siwowski, T., and Radomski, W., First Polish Application of CFRP Laminates to Strengthening of RC Bridge, Inżynieria i Budownictwo, No 7, 1998, pp. 382-388 (in Polish).

9. Radomski, W., Application of External Prestressing for Strengthening of bridge Structures in Poland, Recent Advances in Bridge Engineering, Ed. By J. R. Casas, F.W. Klaiber and A. R. Mari, CIMNE, Barcelona, 1996, pp. 597-612.

The Bridge on the Wadi Kuf valley – Libya
Rehabilitation and maintenance project

DR. ENG. E. CODACCI-PISANELLI – Design coordinator

Introduction
The cables stayed bridge in concrete on the Wadi Kuf valley, was designed by Professor Riccardo Morandi and was built during the period 1968 – 1971.
The structure represents the maximum expression of the technique concerning the prestressed concrete; in fact all the constructive elements as piers, deck and tie-rods were built by means of this system.
Considering the works constructed by the reinforced concrete technologies, the bridge on the Wadi Kuf valley is the crossing which has the greatest width between the pier axles (282 m).
This bridge was subject to an intervention of extraordinary maintenance recently concluded with the final static tests carried out on October 2000.
The purpose of this maintenance was the restoration of the structural limits of the bridge considering that at the time of the construction many static problems as creep phenomena were partially unknown.

General view of the bridge

It was also carried out the maintenance of the tie-rod superficial concrete and the substitution of the damaged bearing devices.
The design was studied to avoid all the possible alteration of the bridge shape allowing maintaining intact the beauty of this structure universally known and mentioned in the most important specialized books.
The intervention has foreseen the applying of an additional prestress inside the main box girder, the introduction of special anti-torsion rings and the construction of stiffening elements to conform the bridge for special loads. After that an anti-seismic adaptation of the structure was carried out, conforming to the recent methodologies of calculation, using special bearings and anti-seismic devices purposely designed and executed for this bridge.
The medium earthquake shock recorded in Cirenaica area in June 1999 has perfectly confirmed the validity of the design choices allowing to overcome the event without any problems.

Bridge description
The main road from Bengasi to El Beida through the plateau of Cirenaica, at about 30 km from El Beida, crosses a deep valley on the bottom of which flows the river Wadi Kuf. The crossing is done by means of a bridge of considerable size that has the following main characteristics:
- Mean elevation of the road-plan above sea level: 441.50 m

- Elevation of the road-plan above the bottom of the valley: 180.00 m
- Theoretical span scanning: 97.50 - 282.00 - 97.50
- Whole length of the work: 524.75 m
- Deck width: 13.00 m
- Road plan level above foundations origin: Bengasi-side abutment: 25.40 m
  Bengasi-side pier: 80.70 m
  El Beida-side pier: 63.70
  Benghazi-side abutment: 15.80
  Maximum slope toward the center: 0.5 %

The static system

The static complex comprising the whole bridge has been built employing two independent balanced systems connected by a central beam that is laying on its extremities; the two systems are balanced on the opposite ends (abutments) by a bilateral restraint.

General view

The static behavior of each system can be synthesized in the following way: - The truss must be considered as a continuous beam with variable section on three spans with a terminal cantilever; the supports are internal (elastic) with the exception of the fixed external support on the abutment. The masts and the oblique piers that support the deck are mostly compressed elements of high stiffness. The tie-rods, that represent the elements supporting the stronger stress in the system, have been placed as naked steel with gradual foretension in order to be able to control the distortion of the free- bouncing ends. After both main systems had been built, the tie-rods have been covered with partially precasted concrete elements precompressed by means of post-tensed auxiliary cables. In this way, the lowering of the bouncing end caused by variations in the cable tensions, due to the live load is smaller than what would result if the steel of the tie-beams had been naked. Even if the steel of the auxiliary cables is more strongly stressed a part of the action of the cables is only directed toward the precompression of the beam and the concrete makes the entire system heavier.

The external actions

The cable-stayed bridge has been designed considering the following actions:
Its own weight and permanent loads, live load according to the British Standard Specifications – Girder Bridges: Part 3/section A; longitudinal and transversal wind action of 200 kg/m2 with unloaded structure and 100 kg/m^2 on the loaded structure including the outline of the live load; variations of temperature in the range of +/- 30 °C; differences in the temperature of the extrados and of the intrados of the deck of 10 °C; longitudinal or transversal seismic actions verified by means of dynamic analysis.

Piers

The basis slab, 2.50 m thick, supports the system of sub-vertical elements with external design in pyramid shaped built as a hollow section with stiffening ribs.
These structures are 39 m (Bengasi side) and 22 m (El Beida side) high and end with a slab with ribs placed at 34.50 m below the deck plan. Four elements that support the truss are

114 DESIGN AND CONSTRUCTION

arranged in couples that diverge toward the exterior, with a slope of 34% from the vertical. These elements, built with a hollow section, vary in their external dimension from 5.87 x 3.50 m to 3.50 x 3.50 m and have a larger section at the origin.

Tower masts

The masts are formed by four concrete elements that originate from the external vertices of the sub-vertical system and converge in couples to join at a level of 53.90 above road plan. The section of each element varies from 6.20 x 3.00 m at the origin to 3.11 x 3.00 m at the top. Each couple of converging elements is connected with its homologous in correspondence of the box-girder intrados and at the top of the mast so that the system is independent from the truss. The total height of each system of pylon-pier is 95.50 m and it is identical for both systems since the differences are corrected for by the sub-vertical systems already described.

Deck

The whole deck, except the laying beam, has been built as mono-cellular box-girder with a variable slab from 20 to 45 cm, vertical elements of 40 cm and counterslab 20 to 50 cm thick. The total height varies from 7.00 m to 4.50 m in correspondence of the tie-rod juncture; it decreases to 3.63 m where the central beam is laying. The box-girder is stiffened by transversal elements in correspondence of the supports and of the tie-rods junctures and by nine internal crosses. A longitudinal reinforcement formed by cables with 8 to 16 (0.5") strands is also present.

The slab is transversally pre-stressed using 8 (0.5M) strand cables with a 1 m interaxis. The central truss has a theoretical length of 55.00 m and consists of a slab varying from 20 to 24 cm and of 3 beams with section varying from 3.40 m on the laying points to 4.00 in the midline section. All the structure was blocked on one end by a hinge made of neoprene and steel plates with steel pins to avoid sliding (El Beida side); on the other side the support was built using molded steel rollers 40 cm in diameter with a blocking system, in case of excessive movements due to seismic actions, also in steel (Bengasi side). On the abutment the deck is joined to the underlying structure by reinforced concrete rods with a section of 1.50 x 7.40 m and 4.00 *m* high; the upper and lower ends finish with steel plates that are in contact with the abutment and intrados counterslab counter-plates.

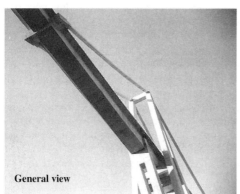

General view

The vertical movement of the truss rising from the abutment is prevented by a row of cables formed by 8 (0.5") strands anchored in the body of the abutment. These post-tensed cables are used to prevent any vertical movement of the truss on the abutment. The bending of the plates at the end of the pendulum, with a radium of 2 m, allows the horizontal movements of the deck structure.

Tie-rods

The tie-rods are formed by a parallel bundle of 90 high-resistance steel strands (R = 16,500 kg/cm^2), each with a nominal diameter of 1.125" (2.86 cm), and by a complex of 18 cables each formed by 4 strands of the same steel but with a diameter of 0.5 ". The entire complex of the strands forming the tie-beam is involved in a concrete element 154 cm wide and 104 cm high. The inclination angle of the tie-rods is approximately 30° on the horizontal line.

The main suspension system, formed by 90 strands, consists of 9 vertical alignments with a horizontal interaxis of 16 cm; these elements have been placed uncovered during the various steps of the construction of the work until the launching of the central truss was complete. In a later phase the tension of every strand has been verified and small variations of the geometry have been implemented in order to conform them to the project prevision.

At the end of these steps the precasted elements that form the external protection of the tie-rods have been added. These shells have an internal shape of combs with eight pins that are placed from below between the nine alignments of the main strands. At the end of every alignment it is present a cable formed by four 0.5" strands anchored with the Morandi M5 system. The precasted covers have been added after the positioning of the precasted elements, supported by twenty metal pins 16 mm in diameter.  At this time the sealing of the joints, the anchoring on the saddle and the maintenance of free cavities along the nine main ducts have been completed. The covers, free to slide along the tie-beam bundle, have been prestressed by stretching the eighteen external cables.

At the end the external concrete covers have been connected to the truss and the eight ducts containing each 10 main stay cables and 2 prestressed cables have been injected.

Bridge condition assessment

Various tests were carried out to verify the condition of the concrete and the reinforcing, and to determine the structural behavior of the bridge. At this purpose the following specific tests were conducted:

Detailed bridge inspection, topographic surveys, ultrasonic tests, sclerometric tests, Windsor probe tests, carbonate depth measurement, reinforcing steel corrosion analysis, compressive

strength tests on core samples drilled in situ, endoscopic visual inspection on prestressing cables, vacuum measurement on prestressing cables, expansion joint monitoring, determination of the stress on the tie-rods by means of dynamic tests, static load tests.

Rehabilitation works

On the basis of the obtained results the following rehabilitation works have been designed:

Additional prestressing

For each bridge section connected to a mast (that is, limited by two opposite pairs of tie-rods) the integrative post-tensioning system is composed of:
1) N. 14 high-strength alloy-steel Macalloy bars, 26.5 mm in diameter, installed close to the top slab soffit, approximately for the whole length of the bridge section (from the abutment to the tip of the internal cantilever that supports the central span).
2) N. 20 tendons, each one composed of no. 4 greased HDPE-coated 0.6" strands. Tendons are installed close to the upper surface of the bottom slab. They are limited to the mid-span bridge sections comprised between the stay anchorage and the mast supports.

Integrative prestressing is aimed at recovering the losses of prestress generated by the time-dependent phenomena in the concrete and the relaxation of prestressing steel during the past life of the bridge.

The work procedures for installing the integrative prestressing are described separately for the two post-tensioning systems (bars and strands) due to the different installation methods and equipment.

Prestressing bars

The procedure regarding duct placement, bar insertion and assembly, bar stressing, and grouting injection, is described hereinafter.

Placement of ducts and bar insertion

Service scaffolding was assembled underneath the whole surface of the top slab in order to allow safe access to the working area.

Before bar insertion, the anchorage was individually inspected in order to check the jack clearance for stressing operation. If clearance problems should occur, concrete was locally chipped and demolished until obtaining proper clearance. In this phase, planarity of the contact surfaces for the anchor plates was checked. In case of need, the concrete surface was rectified by means of epoxy concrete.

Bar assembly was carried out by inserting the single, loose bars through the holes already present in the original structure as well as through the holes created into the new S beams and torsion stiffening rings.

Because of the clearance required, all bars were inserted in the central zone of the deck (the zone comprised between the two supports at the mast).

From this zone, the bar elements were pushed toward the abutment in one direction, and toward the cantilever tip in the other direction.

Bar elements were inserted and transferred each-by-each into their final position, since bar-couplers do not pass through the holes in the transverse ribs of the top slab. As a consequence, each bar was assembled in its entire length by starting at the opposite anchorage and proceeding toward the mast section, by progressively adding and coupling new bars singularly fed from this central zone.

Assembly of the bar elements and the PVC duct elements was proceeded contemporaneously. For each span comprised between two transverse ribs of the top slab, two PVC tubes and a central PVC coupling was positioned first (one tube on each side of the bar coupling point, spanning from this point to the relevant concrete rib, and the coupling inserted along one of these tubes). Then, the new bar was coupled to the bar already installed by means of the specific Macalloy coupler.

Special rubber shutters were adopted in order to divide the grout injection of each duct into several subsequent segments. This was shorten the average duration of each grouting operation and was reduced the number of duct-joints affecting each grout operation. Rubber shutters were located at a mutual distance of about 25 meters.

Bars stressing

After the completion of the bar and duct assembly, the anchor and special anchor bolts were installed on the concrete surfaces at the anchorage.

Bars were tensioned according with the stressing program issued by the designer. This program was defined the stresses at the jack, the tensioning stages, sequences, theoretical bar elongation, etc. Bar tensioning stresses were defined by taking the transverse dissimetry of the bar layout into account. Bars and strand-tendons were tensioned together, according with a common procedure.

Bars were tensioned at both anchorages in order to reduce the average elastic elongation at the anchorage. Upon placement of the stressing jack, bar tensioning was produced by increasing the hydraulic pressure at the jack, and by contemporaneously screwing the nut along the bar. Once the jack end-of-stroke has been reached, the bolt is locked, oil pressure is released, and the jack is closed idle and anchored again to the bar to start a new stressing stroke. This sequence is repeated until the required stressing load has been reached. Bar elongation was calculated by subtracting the elongation read in the presence of an initial pre-load necessary for the bar alignment to the elongation read under the final stressing load.

Sealing of the pvc ducts

PVC ducts were sealed by means of two different products:

The first product was used to scal the 73-mm PVC tubes to both the galvanized steel pipes containing the rubber shutters for the grouting fractionation and the concrete surface of the top slab transverse ribs. For this last operation, the concrete surface was engraved with special cores in order to deepen and extend the sealing zone. The second product was used to glue the 73-mm PVC tubes to the 80-mm PVC coupling tubes.

The PVC tubes were scaled only upon completion of the stressing operations in order to case inspection. During the sealing operations, the flanges of the rubber shutters were tightened so as to press the rubber ring against the bar. This produces the grout-tightness of these sections.

Grout injection in the pvc tubes

Cement grout was vacuum-injected in the approximately 25-m-long tube sections limited by the rubber shutters described above.

Grouting procedure was carried out as follows:
1) Cleaning of the duct segment to be grouted by compressed air blowing. Then 2-bar air pressure was kept for 5 minutes in order to verify the air-tightness of the duct. If air pressure was appreciably decrease during this time, air-leaks was looked for. Once the defective joints have been found, they were treated with a second sealing.

2) Cement grout preparation. Grout is composed of a special injection cement and water, compounded by means of a turbo-mixer until obtaining a colloidal cement suspension.
3) Air aspiration from the duct section to be injected until an air-pressure of about 0,2 bar (80% of vacuum) is stabilized
4) Cement grout injection from the rubber-shutter at one end of the duct section up to its complete filling. Filling is signaled by grout spilling from the air-outlet of the rubber-shutter at the opposite end.
5) Cement grout compression to 2 bars for 1 minute and final valve locking.
6) Removal of valves once grout has hardened.

Strands
Procedures for duct placement, strand insertion, stressing, and grouting, for the lower tendons of the integrative prestressing are described hereinafter.

Duct placement and strand insertion
The external PVC tubes was joined to the PVC tubes already embedded into the anchorage and deviation concrete blocks by means of PVC coupling tubes sealed with the same glue used for the bar ducts.
Strands were cut in length prior to their insertion. The HDPE strand coating will be removed from the strands at each tendon end, for the length needed to insert the strands into the anchorage.

Strand stressing
After duct placement and strand insertion, the anchor bushes and their wedges were inserted onto the strands and transferred into contact with the anchorage.

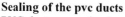
Ducts detail

Tendons were tensioned according to the stressing sequence issued by the Designer.
Strands were tensioned from one end only. Upon placement of four mono-strand jacks at the pulling anchorage, tensioning was occurring by increasing oil pressure. Once the end of stroke of the jacks has been reached, anchor wedges will be locked off. After releasing of the oil pressure, the jacks were placed again in contact with the anchorage to carry out a new stressing stage, and so on until the design stress has been reached in the strands. For details on anchorage, jacks and stressing pump.
Strands elongation was calculated by subtracting, the elongation read under an initial pre-load necessary for the strand alignment (100 bar at the hydraulic pump) to the elongation read under the final load.

Sealing of the pvc ducts
PVC ducts were glued as previously mentioned. Gluing was occurred only upon completion of the stressing operation in order to case inspection.

Grout injection in the pvc tubes
Cement grout vacuum-injected for the full length of the PVC tubes.

Anti-torsion rings

Following static load tests results it was necessary to insert some integration to the original schemes because of the apparent limited torsional stiffness of the deck due to the shape loss of the box girder. Because of this fact, it was designed specific anti-torsion rings that allow the increasing of the global transversal stiffness of the deck and restrain the values connected to the stresses about the 10 – 15% on the basis of the load position.

General view of the anti-torsion ring

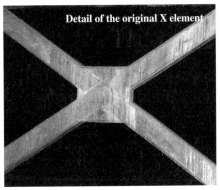
Detail of the original X element

Lifting of the central span for substitution of the damaged bearing devices

At the end of November '98 the central span of the bridge has been lifted and the existing old damaged bearings have been replaced with new mechanical ones responding to the seismic actions resistance required. Hereafter are resumed the carried out operation phases.

Detail of the original bearing

Detail of the original bearing

- **Preliminary operations**
  Preliminary operations, in order to prepare the top stiffening concrete beam and all the other devices necessary for the lifting, started.
- **Load transfer**
  Central beam dead load had been transferred on the lifting jacks. The operation went on up to 2.5 mm in order to discharge completely the existing bearing pads, that showed only a little residual elasticity, confirming the need of their substitution. Once the load transfer had been achieved, valves were closed and the safety ring nuts were definitely tightened. Transducers were removed and jacks protected with wooden boxes.
- **Old bearings removal**
  Bearing seats have been dismantled with mechanical means. It has been used first drilling tools to weaken the concrete then the pneumatic hammers.

- **Lifting phases**

After the transversal and longitudinal anchorage of the central span, the lifting hydraulic circuits were installed and the displacement transducers positioned and set to zero. After two hours the lifting phase was successfully ended in the full respect of the design prescriptions. The maximum lifting value was equal to 12 mm.

- **New special bearings installation**
  After the lifting phase, the insertion procedures for the new bearings have been started. In order to guarantee a perfect diffusion of the forces on the bearing seats, resin has been injected in between the two bottom counter-plates of the bearings.
- **Deck repositioning**
  The central has been repositioned on the new bearings discharging the lifting device slowly by means of micrometric valves. During the operation, pressure was controlled on the manometers so as to decrement it continuously and uniformly on all circuits.

Repair of the stay cables concrete procedures
- Demolishing of the damaged concrete surface by means of compressed air light hammer.
- The thickness of the demolishing has been decided, time by time, according to the condition of the reinforcing steel.
- Removal of the rust on the reinforcing steel by means of sandblasting.
- Cleaning of the surface by means of air jet.
- Protection of the reinforcing steel by means of anti-oxidation painting.
- Preparation of the concrete surface for the reconstruction by means of water spray until saturation.
- Concrete reconstruction phases.
- Surface curing.
- Protective final painting cycle.

Reinforcing steel protection   Concrete reconstruction

Final static load testing

The purpose of the static load testing carried out was to compare the behavior of the structure with the results obtained using the same test at the end of the construction. For this reason the same procedure proposed by the designer had to be followed and extended with special asymmetric load schemes.

After the rehabilitation works a final static load testing has been carried out. In order to calibrate the model, the environmental survey has been executed before starting the test. All activities are resumed as follows:

- **Temperature and relative humidity survey**
  The measures have been executed each hour for 24 hours/day.
- **Topographic levelling**
  During the load testing the surveys have been performed before and at the end of each load phase and after every unload phase for all the load schemes
- **Monitoring of the joint longitudinal excursion**
  As well as topographic levelling, the longitudinal excursion measures have been carried out every day at the same time of the topographic survey.
- **Monitoring of the stress condition of the tie-rods**
  All tie-rods have been checked by means of two strain-gauges fit up on each one.
- **Elastic rebound survey**
  After 12 hours from the complete removal of the load, the residual deflection in the indicator section has been less than the maximum in admissible value

Test results

The maximum allowable displacement for each loading scheme has been surveyed inside the maximum admissible values conforming to the theoretical calculations. So after the completion of the loading test the static behavior of the bridge has been verified in conformity to the expected results.

Acknowledgements

The author gratefully acknowledge Prof. M. P. Petrangeli consultant for the structural solutions, Prof. F. Brancaleoni consultant for the seismic and visco-elastic analysis, Mr. G. Cicchetti, Engineer, for calculation and plans, Mr. A. D'Adderio who supervised the field tests and Delma Maltauro Company who managing the works.

Special thanks to the members of the Libyan Road and Land Transport Department for their kind permission to publish this paper.

Metsovitikos Bridge – A Towerless Suspension Bridge

ALLEN PAUL & IAN WILSON
Ove Arup & Partners International Ltd., UK

Metsovitikos Bridge is part of the Egnatia Motorway which crosses northern Greece for 687km from Igoumenitsa in the west to the Turkish border in the east. In the western section the road runs through the Pindos Mountains with the consequence that there are many tunnels and viaducts. Egnatia Odos AE, the Government appointed company charged with delivering the road, and Haliburton Brown & Root, their project managers, saw the opportunity to create a landmark structure within the dramatic scenery of these mountains. Ove Arup & Partners, with Wilkinson Eyre Architects, won the resulting design competition for the Metsovitikos Bridge.

The bridge is located to the south of the town of Metsovo about 150km east of Igoumenitsa. It spans 565m between steep-sided mountains on either side of the Metsovitikos River 150m above water level. The spectacular design is a rock anchored towerless suspension bridge in which the cables are anchored directly into the mountainsides (Fig.1). At both ends of the bridge, the dual 2 lane road emerges from tunnel directly onto the bridge deck from the tunnel portal. In elevation the bridge is viewed from the village of Metsovo with the cables providing a fine visual cradle to the high mountains beyond. As the driver crosses the bridge, he will be struck by the drama of the mountain scenery, the simplicity of the bridge and the sweep of the suspension cables (Fig 2). The drama of the cables is enhanced by inclining them to the vertical. The resulting composition is a powerful and balanced statement for driver and local resident alike whilst making minimal intervention in the local scenery.

The bridge is located within a region of moderately high seismicity. However, the natural flexibility that the suspension bridge solution gives has meant that seismic considerations do not govern the sizing of any of the main structural elements. Seismic considerations are paramount, however, for consideration of the global stability of the slopes around and below the anchorages.

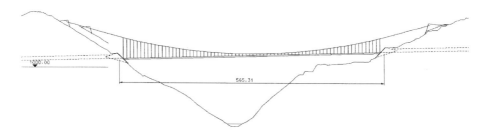

Figure 1 Metsovitikos Bridge Elevation

Current and future trends in bridge design, construction and maintenance 2, Thomas Telford, London, 2001, 122–130

Figure 2 Longitudinal View

Eurocodes have generally been adopted as the design standards for this bridge supplemented by DIN, British and American standards where appropriate.

There are four main elements to the bridge: the anchorages, the abutments, the suspension system and the deck structure. This paper considers these key elements of the structure and some parts of the analysis as described below.

Geology

The rocks in the area of the bridge comprise sandstones, siltstones and shales of the Pindos Flysch, which have been subjected to both thrusting and folding. On the western side of the Metsovitikos River, thick bedded sandstone dipping steeply towards the east outcrops at the anchorage and portal locations. On the eastern side of the Metsovitikos, the portal is also located on the thick bedded sandstone. However the eastern anchorages lie above a thrust fault which runs between the anchorages and portal, and are underlain by a complex tectonic unit. This is composed of stacked slices of lithic sandstone and debris flow units, blocks of thinly bedded sandstone/siltstone and of red shale (Fig 3). The ground investigations have included rotary cored drillholes, geophysical logging, borehole imaging with an acoustic televiewer for structural measurements, crosshole tomography and high pressure dilatometer testing.

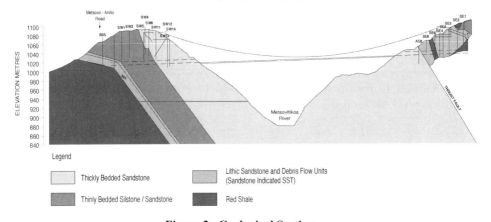

Figure 3 Geological Section

Anchorages

The four anchorages are located high on the mountainsides above the deck. It was originally proposed that these should all be tunnelled socket anchorages whereby the cable force is taken over the saddle and back into the mountain through a tunnel to an enlarged end section where the force is anchored off (Fig 4). During the site investigation however, a landslip was found behind the southwest anchorage leading to a change of type and of location of the anchorage. Instead of the tunnelled socket which would have extended into the landslip material, a gravity solution was adopted in which the mass of the anchorage is used as restraint against the cable force. The northwest anchorage was also changed to a gravity type as the topography did not lend itself to an economical tunnelled socket solution.

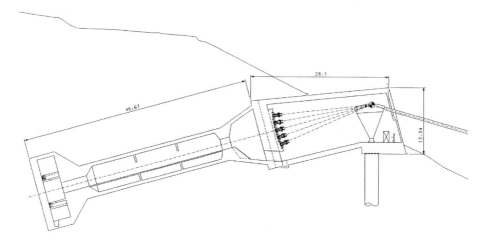

Figure 4 Tunnelled Socket Anchorage

In the tunnelled solution, the anchorage is approximately 75m long, 15m wide and 15m high. The cable passes over the 7m high saddle to be anchored off in the conventional manner by strand shoes on the back face bulkhead wall of the splay chamber. Prestressed strand then carries the load from the bulkhead wall through the 6.8m diameter tunnel to the enlarged end which allows load transfer into the mountain by direct bearing.

The gravity anchorage is much more substantial than the tunnelled solution, 60m long, x 30m wide x 20m high. This is because it relies upon the dead load of the anchorage and on the shear keys to resist the forward pull of the cable by friction on the base and a shear key (Fig 5). However, much of the anchorage is buried so that the visible part is similar to the tunnelled solutions giving an overall balanced appearance at the four anchorages. A large part of the self weight is provided by the bucket at the rear of the gravity anchorages which is backfilled with excavated material. The stability of these gravity anchorages and the surrounding mountain slopes has been the subject of detailed 3D FE analysis as reported below.

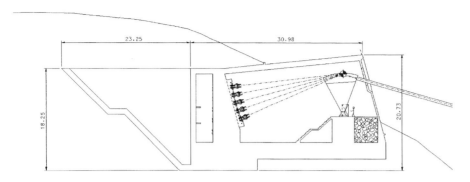

Figure 5   Gravity Anchorage

Tunnel Portals and Abutments

The abutment at each end is integral with the tunnel portal. The portal is formed from an elliptical shaped reinforced concrete cone that widens as the driver leaves the tunnel and approaches the bridge. The mouth of the cone receives the bridge deck and the twin bore tunnels intersect it at its rear. The front of the cone is shaped to follow the existing slope of the hillside. The form of the portal graduates the environmental transition from tunnel onto the bridge deck with the bridge and scenery opening out splendidly before the driver.

The Superstructure

The 560mm diameter suspension cables span approximately 775m between the anchorages and have a maximum design force at ULS of approximately 150kN. They are unique in that they will be slewed over to give a cable and hanger plane inclined to the vertical. This will be achieved by spinning the cables in a free hanging (vertical) catenary using the aerial spinning technique. Temporary saddles (or similar) will be required in front of each of the permanent saddles to deviate the cable from its final plan direction to its temporary direction, which is a straight line in plan between anchorages. The completed suspension cables will then be slewed towards each other and held in position by temporary ties.

The cable saddles, splay chamber and strand shoes follow conventional practice. The strand shoe crosshead slabs are stressed to the anchorage structures using galvanised, wax coated, sheathed prestressing strands in grouted tubes to give a robustly corrosion protected, and replaceable, anchorage system.

Conventional locked coil hangers suspend the 25m wide road deck from the suspension cables. The orthotropic steel deck has a curved soffit and comprises two box girder 'hulls' with crossbeams connecting them at 15m centres (Fig 6). The gap between the two carriageways adds excitement to the design from above and below as well as for users of the highway. The deck has been developed on a 15m module with parapet posts at 2.5m, cross diaphragms of the box at 3.75m, hangers at 7.5m and construction joints at 15m.

126 DESIGN AND CONSTRUCTION

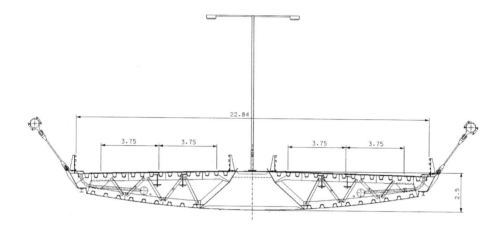

Figure 6 Section Through Twin Hull Deck

Articulation

The bridge is unusual in that, due to the 'V' shape of the valley, the deck does not span the full distance between the suspension cable saddles. To avoid the end-hangers de-tensioning under live and associated loads two measures were taken:

- Firstly the end hanger on the deck was moved 4.6m into the span. This means that the effects of geometric deformation of the suspension cable on the end hanger are reduced by the associated deflection of the deck.
- Secondly the four deck corner end hangers were prestressed. This provides a 'virtual support' to the cable and constrains its local geometric displacement to follow the axial stiffness of the end hanger.

Like all suspension bridges the applied vertical loading is resisted by the geometric alteration of the cable geometry combined with the increased stiffness resulting from the additional load. When combined with temperature effects, deflections of the deck under live load are considerable and patch live loading gives substantial rotations at the ends of the deck. 'Flexing panels' (Fig. 7) have been provided at the ends of the deck to smooth the transition between the fixed gradient of the road in the tunnel and the variable gradient of the deck. These panels comprise the box girder deck plate and modified deck trough stiffeners. The 6.75m long flexing panel commences before the pendel support point to facilitate the smooth transition.

The bridge is supported and restrained by a pair of pendels at each end of the main deck. The main deck is guided transversely at the pendel locations but has no longitudinal restraint bearings. Longitudinal restraint is achieved by restoring forces from the suspension system which occur when the deck is displaced.

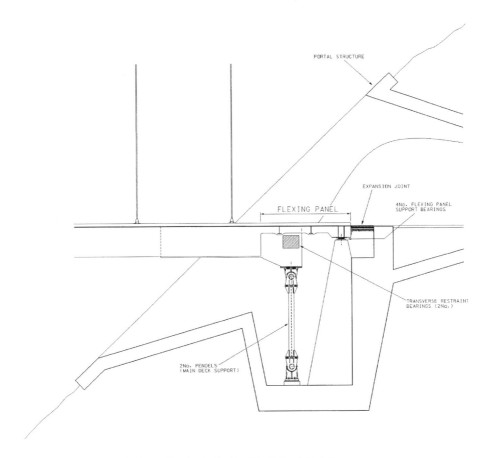

Figure 7 Articulation Deck End Detail

Stability Analysis of Anchorages

Verification of anchorage stability under both static and seismic loading has been a major aspect of the design. As part of this work a programme of 2D and 3D numerical analyses of the southwest anchorage and adjacent hillside has been carried out. For numerical analysis of jointed rocks either a discrete block or ubiquitous joint model may be used. In the discrete block approach, the rock is modelled as an assemblage of rigid or deformable blocks that can rotate and slide relative to one another under the action of disturbing forces. In a ubiquitous joint model the rock is modelled as an elastic - plastic continuum with lower strength limits specified on the planes orientated in the direction of discontinuities and no tension allowed perpendicular to these planes. In view of the fact that the discontinuity spacing was relatively small in relation to the size of the anchorage a ubiquitous joint model was adopted for the majority of the anchorage analyses.

Three different types of analysis were carried out as follows

- An analysis in which the cable force was progressively increased from the characteristic value. This type of analysis was used to investigate sliding failure mechanisms

- An analysis in which the rock strengths were progressively reduced from the characteristic values with the cable force maintained at the characteristic value. This type of analysis was used to investigate bearing capacity and slope failure mechanisms

- An analysis in which horizontal and vertical accelerations were applied in combination with the characteristic cable force. This type of analysis was used to investigate behaviour under seismic loading

Initially a range of parametric studies was carried out using 2D models. These were mostly carried out using *Oasys* SAFE, but a comparative analysis was carried out for one case with VISAGE, also using a ubiquitous joint model, and UDEC using a discrete block model. Good agreement was obtained with each of the models. Subsequently 3D analyses taking full account of the geometry of the hillside, anchorage and discontinuities in the rock, were carried out. These were mostly carried out using *Oasys* LS-DYNA but again a comparative analysis was also carried out using VISAGE. The 3D model of the anchorage and hillside that was used is illustrated in Figure 8.

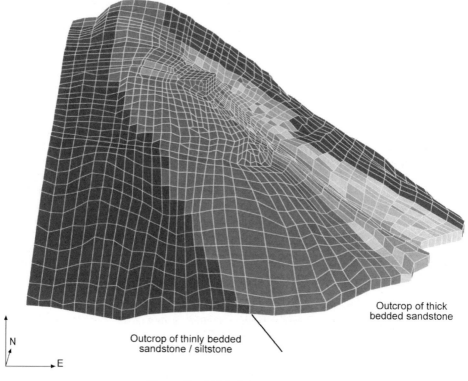

Figure 8 FE Model for SW Anchorage

Sophisticated 3D analyses with jointed rock models such as those used for the southwest anchorage are rarely undertaken, but are now becoming practical with developments in software and the continuing increases in computer power.

Seismic Design

The bridge site is located in an area of moderately high seismicity and the performance of the bridge under seismic loading is an important design consideration. The seismic design has adopted a performance based approach following the principles embodied in ATC-32. The seismic performance objectives, in terms of the required serviceability, permitted degree of damage and the associated probability of exceedance in the design life of the bridge are summarised in the table below:

Design Ground Motions			Seismic Performance Criteria (From ATC-32)			
Evaluation Level	Chance of Exceedance in 120 years	Return Period	Service Level	Damage Level	Ductility Level	Ductility Value
Functional	40%	235 years	Immediate	Minimal	Elastic	1.0
Safety	10%	1139 years	Immediate	Repairable	Elastic	1.0

Although ATC-32 recommends that moderate structural damage that is repairable be acceptable for the safety evaluation of important bridges, the objective adopted for the Metsovitikos bridge for the safety evaluation event is that no structural damage should occur to primary structural components. This takes account of the fact that the major structural components, such as the main cables and the hangers are made of high strength steel and thus possess little ductility capacity.

The design seismic parameters for the functional and safety evaluation levels have been obtained from a detailed site specific probabilistic seismic hazard assessment together with a study of the activity of faults in the vicinity of the bridge site. The spatial variation of earthquake ground motions at the bridge site due to loss of coherency and wave passage have been taken into account and the effects of topographic amplification have also been included.

Superstructure Analysis

The global analysis takes account of the geometrical non-linearity of the suspension system since the deflections of the main cable will alter the structural action of the deck. The *Oasys* program GSA Fablon has been used to perform a force equilibrium analysis determining the geometric profile and axial forces of the main suspension cables and hangers under SLS permanent loads. The basic structural model is a three-dimensional skeletal space frame consisting of one-dimensional elements with pin ended nodes. The twin hull deck is represented as two sets of longitudinal beam elements located at the shear centre of each hull. Beam elements are capable of resisting axial force, biaxial shear, biaxial bending moments and torsional load effects. Transverse outrigger beam elements notionally span from the extreme edge of the deck to the longitudinal centroidal members to allow modelling of load interaction with the hanger elements and the transverse boxes which span between the twin decks. Local wheel load effects are modelled using linear elastic models of representative portions of the steel deck.

KCRC West Rail Viaducts – Design Development

NAEEM HUSSAIN BSc (Eng) DIC MSc CEng FHKIE FIStructE FICE
Ove Arup & Partners Hong Kong

ALAN CROCKETT B.S. M.S.
Wilson Ihrig & Associates Hong Kong

1. Introduction

West Rail Phase 1 is an undertaking by KCRC to build a new twin track railway on a western alignment from West Kowloon to Tuen Mun (Figure 1). From Kam Sheung Road Station to Tuen Mun Station the railway runs on approximately 13.4km of viaducts through existing built-up areas and future close proximity urban development along the viaduct route. This paper describes the evolution, development and design of the viaducts, which was undertaken by Ove Arup & Partners as the Lead Designer for all the West Rail Viaducts.

2. Design Criteria

The Hong Kong Noise Control Ordinance (NCO) sets statutory noise limits on the operation of railways and other sources. During the night-time (2300 to 0700 hrs), operational noise levels along sections of the West Rail alignment cannot exceed an $L_{eq,(30min)}$ of 55 dB(A) at the facades of noise sensitive receivers. To meet this limit at all properties outside the boundary of the railway, with the proposed peak night-time headway of 4 minutes, the maximum noise level (L_{max}) cannot exceed 64 dB(A) at 25m from track centre for 9 car transit trains travelling at 130 km/hr. With an assumed reference wayside noise level of 88 dB(A) for an unmitigated viaduct, a 24 dB(A) noise reduction is required for compliance with the legal limit.

As shown in Figure 2, the total wayside noise from a train passby consists primarily of 1) direct (airborne) noise from the train undercar, generated by wheel/rail interaction and the propulsion system, 2) structure radiated noise from vibration transmitted through the trackform and 3) air conditioning system noise. If air conditioning noise is adequately limited, achieving compliance, as regards total wayside noise, requires significant reductions in both the airborne and structure radiated noise to a maximum level for each of 61 dB(A).

In addition, to satisfy the ride quality of the viaducts in accordance with ISO standard 2631 (1985) the following structural deflection requirements were stipulated:

- For non-repeating spans or spans ≥ 45m, deflection under live load had to be in accordance with UIC 776-3R 1989 - Deformation of Bridges.

- For repeating spans (≥5) and spans upto 40m, deflection under live load ≤ $l/5000$, where l = length of span.

3. Development of the Viaduct Superstructure

3.1 Technical Studies Stage

The elevated stations along the viaduct route have island platforms which require the twin track viaducts to diverge into single track viaducts at the station approaches. For attenuating air-borne noise high parapets and noise covers are required as described in para 4.0.

In order to avoid resonance between the vibration of the superstructure cross-section elements and the global superstructure vibration, the following was specified:

- The frequency of the fundamental longitudinal flexural mode of a typical span, assumed to be a beam simply supported at the piers, shall be in the range of 2.6Hz to 5Hz.
- The resonance frequency of the rigid viaduct in vertical motion acting against the bridge bearings atop the piers shall be in the range of 25Hz to 30Hz.
- It shall be demonstrated that the fundamental resonance frequencies of the rigid viaduct on the bridge bearings, the lateral and longitudinal flexural bending modes and the torsional bending mode of a typical span are sufficiently separated from the resonance frequencies of the Floating Slab Trackform, FST (14 - 18Hz), the primary suspension (5 - 10Hz) and the secondary suspension (1 - 1.5Hz) of the train vehicle.
- It shall be demonstrated that the ride quality of trains to be operating on the viaducts at planned operational speeds and consist size shall satisfy the 4 hour reduced comfort boundary given in ISO Standard 2631 (1985).

The above requirements alongwith the necessity to cater for the overturning effects of wind on the high noise barriers led to the choice of a concrete box girder superstructure. The concrete box girder is well suited to reducing low frequency structure radiated noise. In order to further control structure radiated noise a FST is used to reduce the transmission of vibration from its origination at the wheel/rail interface to the viaduct. Ideally the webs of the box girder should be placed directly under the FST rubber bearings in order to increase the mechanical impedance beneath the FST.

At the Technical Studies stage the single track viaducts were supported on a single cell box girder, and the twin track viaducts were supported on a three-celled box girder, (Figure 3). For the three-celled box girder the webs were located directly under the FST bearings.

3.2 Detailed Design Stage

The Technical Stage design did not adequately address the construction aspects of the design. For a viaduct of this long length, it was evident that quality and economy of construction could best be met by use of precast segmental construction. However the transition from a three-celled box to two single-celled boxes could not be easily achieved with the Technical Studies Stage cross-section. For simplicity and ease of construction each track could be supported on a single-cell box girder and the two single cell box girders could be brought together and the top slabs connected to form the twin track viaducts (Figures 4).

As intimated above the webs of the box girder should be placed directly under the FST rubber bearings in order to increase the mechanical impedance beneath the FST. However placing the webs directly under the FST results in narrow spacing of the webs and bearings leading to overturning instability. Two solutions were therefore possible.

- Wider spacing of webs to increase the spacing between the bearings. However this required use of large fillets between the deck and girder webs to reduce the vibration of the top slab (Figures 4).
- Narrow spacing of webs with a diaphragm at the pier supports to increase the spacing between the bearings (Figures 5).

The box girder section with wide spacing of webs, as opposed to a box girder section with narrow spacing of webs and end diaphragms, was chosen as the section for tender design. It was considered that the narrow box with diaphragms would make the choice of an under slung launching gantry difficult and at the same time not be visually attractive. The sides of the box girder were kept vertical in order to have as large a length of single track box girders as possible before they merge to form the twin track viaduct.

In order to reduce the extent of structure surfaces which radiate noise, the width of the top slab needs to be kept as small as possible. This results in an outer slanted curved parapet to accommodate the trackside evacuation walkway and yet provide the desired volume for the noise plena (Figures 6).

4. **Attenuation of Air-borne Noise**

    The air-borne noise is primarily generated by wheel/rail interaction, and is best arrested by devices close to the rail.

    It is not possible for a simple edge wall barrier, with or without sound absorption, by itself to reduce airborne train noise by 27 dB(A); thus, additional mitigation is necessary. The proposed noise reduction scheme, called the multi plenum noise reduction system, consists of three components: an undercar sound absorbing plenum; "under walkway" sound absorbing plena on either side of the vehicle; and edge walls with sound absorption applied, as shown in Figure 6.

    The under walkway plenum on the viaduct wayside is bounded by the parapet, the deck, the walkway and the vehicle. Sound absorption is placed on the edge wall and the underside of the walkway. The outlet of the plenum is the gap between the walkway and the vehicle which is limited by the vehicle kinematic and curvature envelopes. For KCRC West Rail, the minimum gap size is 250 mm on tangent track and 350-400 mm on curves. Derailment safety requires that the vehicle can move laterally by roughly 600 mm during derailment, implying that part of the walkway must be friable to prevent damage to or detachment of the parapet.

The under walkway plenum at the centre of the viaduct is bounded by a median wall, the deck, the top plate and the vehicle. Because of viaduct width limitations, the volume of this plenum is not as large as those beneath the edge walkways, and therefore, not as effective in attenuating noise. The median wall must be strong enough so that a contained derailment will not send debris onto the other track.

A comprehensive noise model of the multi plenum system, combining plenum and sound wall attenuation equations, was developed. Before predictions were made as regards the West Rail multi plenum system, this model was validated against measurement data taken on skirted trains, absorptive parapet walls on transit viaduct structures and close-in sound barriers placed adjacent to a transit train undercar. The prediction results are summarised in Table 1, wherein it can be seen that the multi plenum system satisfies the design maximum of 61 dB(A) on both the wayside and trackside for parapet walls 2.9 m high above the deck.

Table 1 Airborne Noise Levels (L_{max}) for Edge Barriers with and without the Plenum System at 25m Setback and Level with Top of Rail

		Edge Barrier Height	
	Mitigation	0.0 m	2.9 m
		- dBA -	- dBA -
Wayside	Parapet Wall Only	88	72
Wayside	Parapet Wall With Plena	--	56
Trackside	Parapet Wall Only	88	75
Trackside	Parapet Wall With Plena	--	61

5. Attenuation of Structure-Borne Noise

Structure radiated noise from train passbys (130 kph, at 25 m setback) on a concrete viaduct with stiff rail fixation and no particular attention paid to the noise aspects of the viaduct section is roughly 80 dB(A). A structurally similar section constructed of steel is about 10 dB(A) higher than a concrete section. Preliminary considerations (para 3.0) resolved that the West Rail viaducts would be designed as a simply supported structure with a twin box section constructed of concrete and with the deck stitched together along the viaduct centre line. This means a reduction of about 19 dB(A) is required for the West Rail concrete viaducts to comply with the NCO. Additionally, rumble noise is limited to 72 dB in any low frequency 1/3 octave band.

A structure borne noise model was developed to predict the wayside noise levels and aid in the design development. It is described in detail in References 1 and 2. Starting with a suitable wheel/rail roughness spectrum, vibration levels are calculated in the structure using a finite element analysis and vibration levels are converted to noise using analytical formulae. Before making predictions for West Rail, the model was validated against vibration and noise measurement data taken on the Kwai Fong covered viaduct structure on the MTRC and on the MARTA system in Atlanta. Validations were also made as regards rail vibration levels measured on the A-13 viaducts on the WMATA system in Washington and measured vibration to noise conversions obtained from the Tsing Ma Bridge.

In the study of structure radiated noise, the following design variations were considered:

- Type of trackform.
- Mass and stiffness of the section.
- Deck thickness.
- Distribution of mass and stiffness as regards number size and location of fillets and webs; and
- Noise radiating area.

The two major findings of the study are the following:

- Floating slab trackform (FST) with soft baseplates is required, regardless of the extent to which the viaduct cross section is optimised with respect to noise.
- In addition to the FST with soft baseplates, the viaduct section must be optimised with respect to noise.

A Finite Element Analysis was performed on a complete span to evaluate the vibration modes of the elements of the box girder. A Boundary Element Analysis was then performed to calculate the noise levels from the vibration, and the results compared with the specified criteria. The input parameters for the vibration analysis was the KCRC reference wheel/rail roughness spectrum.

A Finite Element Analysis followed with a Boundary Element Analysis was also carried out on the full noise cover described in para 7.0 to ensure that generated radiated and rumble noise were within specified limits.

In Table 2, predictions of airborne, structure radiated and total wayside noise are given for the optimised viaduct section (wide spaced webs plus fillets) for three different trackforms: resiliently booted sleepers (LVT), soft resilient baseplates only and FST with soft resilient baseplates. Note that the structure radiated noise from the top deck and the FST slab is included in the airborne noise, as it is attenuated by the multi plenum system. It can be seen that only the optimised viaduct section with the FST system satisfies the noise target of 64 dB(A).

Table 2 Noise Levels (L_{max}) Estimated on the Trackside for the Optimised Viaduct Section

Trackform	No Noise Mitigation - dB(A)-	Airborne Noise - dB(A)-	Structure Noise - dB(A)-	Total Noise - dB(A) -
1. Soft Baseplate	88	62	75	75
2. LVT	88	62	74	74
5. FST	88	63	58	64

6. Parapets and Walkway

6.1 Parapets

To prevent noise transmission through the parapets, mass is required and the minimum thickness to achieve the required specification was a 100mm thick concrete wall, which needs to continue 1.2m above the walkway level. The resultant parapet on the outer face is 3.3m high, leaning out from the superstructure box edge (Figures 6). The outer parapets must not only structurally cater for their own weight but also the weight of the high noise barriers and full noise enclosure.

The concrete stitch between the precast parapet unit and the top flange of the superstructure has to have sufficient strength not only to cater for the weight of the parapet unit, noise barriers, noise enclosures and walkway but also to resist the impact load generated by the walkway in the event of derailment of a train.

6.2 Walkway

The walkway has to satisfy the disparate requirements of :

- Sufficient mass to meet the noise transmission specification.
- Lightweight to reduce superimposed load.
- Sufficient strength and stiffness to cater for crowd loading from passengers disembarking from a stalled train.
- Be friable in the event of train derailment and not cause excessive damage to a derailed train nor impose excessive loading on the concrete parapets and cause their failure.
- Be durable with minimum maintenance.

These conflicting requirements led to the choice of a walkway being assembled from 6mm thick aluminium plate supported on an aluminium frame (Figure 7). In order to minimise the horizontal load on the concrete parapet, generated due to the in plane buckling of the aluminium plate caused by derailing train, real time dynamic analysis was carried out using the Arup program DYNA3. Initially a single aluminium plate was used and impact forces on the concrete parapet for various impact angles were evaluated. The analysis showed that peak forces in the range of 450 to 500kN were generated.

In order to reduce the impact forces, so that a practical and manageable concrete stitch could be provided between the precast parapet and the deck slab, overlapping aluminium plates with a gap between the plate and the inner face of the parapet were investigated. At 20° angle of impact the force generated was about 150kN. However a realistic angle of attack is 4°, which is the angle derived from one axle of a train carriage being derailed and arrested by the derailment kerb, whilst the second axle is still on the rails. This angle of attack generates a force of around 60kN to 70 kN and does not cause catastrophic failure of the concrete stitch connecting the parapet to the superstructure (Figures 8,9).

7. **Noise Barriers and Full Noise Enclosure**

In order to meet noise transmission specification of the barriers and noise enclosure, mass is required, and this can be achieved with use of 75mm thick concrete panels. Precast panels supported by steel frames were therefore proposed with an exterior cover in profiled aluminium sheeting (Figures 10). Noise absorption material is applied to the inner face of the concrete panels.

8. **Acknowledgement**

The authors wish to thank Kowloon-Canton Railway Corporation for giving permission to publish this paper, and acknowledge that the views expressed in this paper are those of the authors and not necessarily those of KCRC.

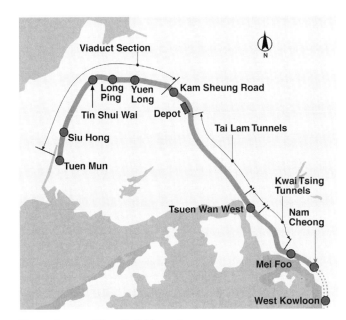

Figure 1   West Rail Phase 1

138 DESIGN AND CONSTRUCTION

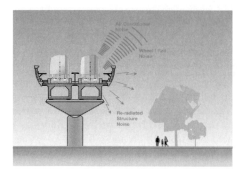

Figure 2   Constituents of the total wayside noise from a train passby on a viaduct

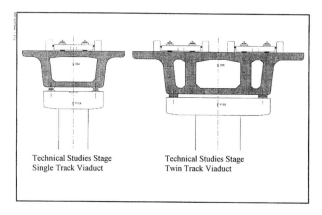

Figure 3

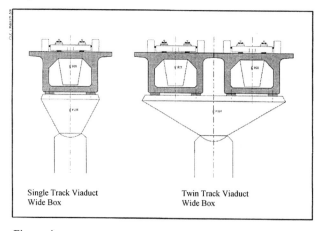

Figure 4

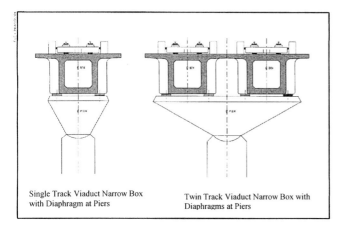

Figure 5

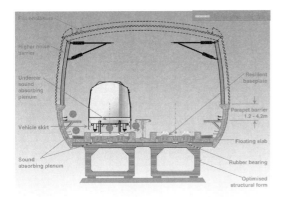

Figure 6   Multi plenum system, the floating slab track and viaduct cross section

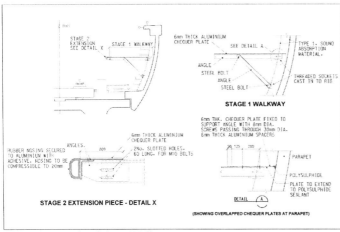

Figure 7   Aluminium Walkway

140 DESIGN AND CONSTRUCTION

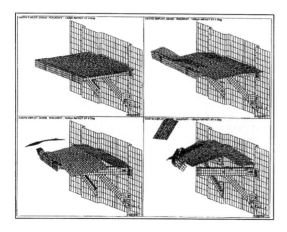

Figure 8   Analysis of Typical Walkway Section – Impact at 130kph, 4° - Three Plates

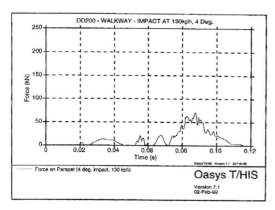

Figure 9   Impact Load Exerted by Walkway Structure on Parapet – Impact at 130kph, 4° - Three Plates

Figure 10   Twin Track Viaducts with Noise Covers

The Alternative Design of the West Rail Viaducts

N.J. SOUTHWARD, CEng MICE MHKIE, Director
J.H. COOPER, MA CEng MICE MHKIE MHKIVM, Director
Robert Benaim & Associates, Hong Kong

Introduction

In June 1998, the Kowloon-Canton Railway Corporation ("KCRC") invited contractors to pre-qualify for the forthcoming construction contracts of the West Rail Phase I project in Hong Kong. The West Rail Phase I project consists of a new 30.5km long railway with 9 stations, between Nam Cheong station in Kowloon and Tuen Mun station in North West New Territories. Of the overall route, approximately 13.4 km between Kam Tin and Tuen Mun is elevated, on viaduct and station structures. The 10 km of viaduct works are comprised in two contracts, CC-201, Kam Tin to Tin Shui Wai Viaducts and CC-211 Tin Shui Wai to Tuen Mun Viaducts.

The conforming design for these viaducts was prepared by Arup and Maunsell, as KCRC's appointed detailed design consultants. The final design solution developed by Arup (with an individual concrete box girder supporting each of the twin tracks) was quite different from the more conventional solution (with a single large box girder supporting both tracks) that had been put forward during the earlier Technical Studies and Preliminary Design stages. This change, and the detailed design consequences that followed, resulted from the onerous constraints imposed by the Hong Kong Noise Control Ordinance ("NCO"), the Environmental Permit granted by the Environmental Protection Department ("EPD") in respect of the West Rail project, and KCRC's own required performance criteria, respectively.

Within the length of contracts CC-201 and CC-211, West Rail runs through predominantly rural and less developed areas. However, in some places, the viaducts pass close to residential blocks, schools, and other noise sensitive receivers. EPD therefore required that noise levels created by trains running on the viaducts must be controlled within very stringent limits. These limits, and the design of the various mitigation measures found necessary to meet the EPD criteria, are described in more detail in the accompanying paper by Hussein and Crockett.

Joint venture contractor Maeda-Chun Wo Joint Venture (Viaducts), ("MCW"), were determined to submit a most competitive bid for contracts CC-201 and CC-211. Consequently, during the pre-qualification stage, they retained Robert Benaim & Associates in October 1998, to begin to investigate possible options for an alternative design of the West Rail viaducts, with the objective of finding a more economic design solution.

Prior to this, Benaim had already concluded that, for an alternative design to be considered seriously during the tender assessment process, notwithstanding any cost savings achieved, it was essential for the tender submission to demonstrate conclusively that the proposed alternative design would meet the requirements of the NCO, the Environmental Permit, and KCRC. It would therefore be essential to submit with the tender an authoritative report analysing the noise and vibration performance of the proposed alternative and comparing this with the performance of the conforming design.

Noise and Vibration Effects from Trains

There are three principal noise effects from a train running on an elevated railway:
- plant noise, emitted by the rolling stock traction motors, "choppers" and air-conditioners
- airborne noise, emanating directly from vibrations at the wheel-rail interface
- re-radiated structure-borne noise, caused by train-induced vibrations exciting the bridge deck

The first effect, plant noise, is controlled mainly by the performance specification for the rolling stock itself, and is therefore outside the scope of the alternative viaduct design.

The second effect, airborne noise, is controlled by the trackside installations, including side and central plenums, under-car plenum, parapets, and (where specified) high noise barriers or full noise enclosures. In principle, none of these items has been changed in concept or performance in the alternative viaduct design, although some minor modifications of details have been made.

The third effect, re-radiated structure-borne noise, is the component of greatest importance when considering the structural design, whether of the conforming design or an alternative design. As trains run along the tracks, vibrations at the wheel-rail interface are transmitted via the floating slab track ("FST") supporting the rails into the viaduct deck. The excitation of the deck that causes audible "rumble" noise to be radiated is generally that related to the transverse vibration of the panels forming the webs and flanges of the deck cross-section. Concrete viaducts, because of their greater mass and less flexible cross-sections, are inherently well suited to minimising the re-radiated low frequency "rumble" noise characteristic of elevated railways.

Noise and Vibration Studies for the Alternative

The originally announced tender period for the two West Rail viaduct contracts was three months, from February – April 1999. Benaim realised that a longer period would be required, to identify and develop an alternative tender design that was both economic in its use of materials and compliant with EPD and KCRC requirements, and to prepare an appropriately authoritative report on its noise and vibration performance to demonstrate such compliance. It was therefore necessary to begin work prior to confirmation of pre-qualification and issue of tender documents.

With MCW agreement, in December 1998, Benaim engaged Dr MK Harrison (now of Cranfield University School of Mechanical Engineering) and Drs DJ Thompson and CJC Jones, of the Southampton University Institute of Sound and Vibration Research Dynamics Group ("ISVR") to undertake a series of noise and vibration studies.

The ground rules adopted by Benaim and ISVR in developing an alternative were to maintain totally unchanged the railway trackwork and trackside systems and noise mitigation measures. Only the structural elements of the viaduct superstructure and substructure would be varied.

Preliminary studies were carried out by ISVR iteratively and interactively with the development by Benaim of a number of potential structural alternatives for the viaduct cross-section. The expected "noisiness" of each potential viaduct cross-section was compared with that of the conforming design, initially using simplified analytical methods. The effects of varying the concrete cross-section dimensions were also considered. It quickly became apparent that the thickness of the box girder elements in the conforming design was greater than was necessary purely for structural design reasons, and it was concluded that the extra thickness was provided to achieve the required noise and vibration performance.

Thus, to create an alternative design that would achieve both optimum economy and acceptable noise performance, some radical departure from the concepts of the conforming design would be necessary.

Development of the Alternative Design Concept

Benaim and ISVR established early that it is not only the mass of the viaduct deck itself which is important in controlling structure-borne re-radiated noise from the viaduct, but also the careful disposition of that mass in the deck cross-section. The studies showed that the magnitude of noise radiation from the deck depends not only on the thickness of the flange and web panels comprising the box girder, but also critically on the panel dimensions and their placement relative to the source of the vibration.

Vibration is caused by roughness at the wheel–rail interface, and is partially attenuated by the FST which supports the rails. However, residual vibrations are transmitted to the deck through the elastomeric bearings which support the FST. A line of FST bearings is located directly under each rail, at 1500 mm transverse centres. Studies identified that the closer the viaduct web panels were placed beneath the floating slab track bearings, the greater the mechanical impedance of the deck structure opposing the input vibration, and the less the resulting noise radiation from the deck.

In simple terms, this is because with the narrower box, the input vibrations are resisted primarily by the much greater in-plane stiffness of the box girder webs, rather than the much smaller transverse stiffness of the top flange. It was found that the beneficial effect of the narrower box girder was obtained, even when the panel thicknesses were quite significantly reduced.

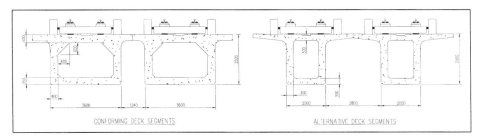

Figure 1 – Comparison of Conforming and Alternative Deck Cross Sections

Once it had been identified that a suitable alternative deck cross-section did exist, which had the potential to make substantial material savings, Benaim were commissioned by MCW to prepare the alternative tender design in a suitable format for their Tender Submission in April 1999.

In turn, Benaim instructed ISVR to prepare detailed analytical and numerical analyses of the chosen design, to predict absolute values of noise radiation and compare these with the conforming design, and to prepare an appropriate report for submission with MCW's tender.

Benaim's strategy, of making an early start to finding a viable alternative design, and engaging world-renowned specialists to provide a comprehensive report with numerical and analytical justification in support of the noise and vibration performance of the proposed design, was fully vindicated when KCRC's tender documents were issued. The Instructions to Tenderers and Particular Specification explicitly required any tenderer wishing to put forward an alternative design, to demonstrate its compliance with the Environmental Permit and KCRC's own requirements by carrying out just such analyses as had already been planned and commenced by Benaim and ISVR.

Tender Design

The adopted alternative deck cross section has a box width of only 2000mm, compared with 3600 mm for the conforming design. Thus, if the alternative boxes had been supported on bearings in a conventional manner, the spacing between the bearings would have been only 1000 mm, which would have been unstable under the high overturning moments created by typhoon wind forces. Various options were considered to provide stability, but the adopted solution was to make the bridge deck spans monolithic with their piers.

The only other alternatives to avoid the problem of overturning stability would have been either to provide "outrigger diaphragms" extending laterally from the box girder, to allow the bearings to be spaced further apart than the box girder webs, or to use tie-down bars. However, it was considered that "outrigger diaphragms" would be extremely unattractive, from an aesthetic point of view, while tie-down bars would be a potential maintenance problem. Thus, neither of these schemes was pursued further.

In fact, the monolithic piers of the alternative design provide an added advantage to KCRC, by removing all the bridge bearings, and thus avoiding a significant maintenance issue for the Corporation, since bridge bearings have a life considerably less than the 120 year design life of the structure itself.

Conventionally, however, bridge bearings are also provided to cater for relative longitudinal movements between bridge decks and piers, under longitudinal forces such as thermal expansion and contraction, seismic events, and live load traction and braking. Since it was proposed to make every pier monolithic with the viaduct deck, over a 10km length, some other provision had to be found to accommodate this movement.

KCRC required continuously welded rail to be used along the entire route length. In order for this to be possible, while accommodating longitudinal structural movements and keeping rail stresses within allowable limits, KCRC specified that structural movement joints in the viaduct should be spaced not more than 80 metres apart. This was one of the principal reasons why simply supported spans were adopted in the conforming design.

In the alternative design, Benaim retained the individually articulated spans, but made each end of each deck span monolithic with an individual column support. In this way, each span formed a portal frame with the piers at either end, and each pier comprised two independent columns framing into the spans on either side.

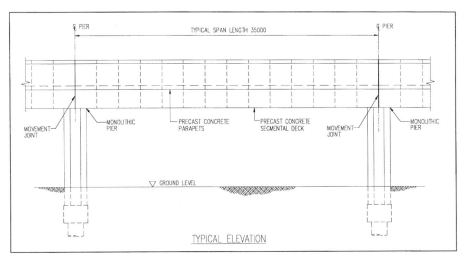

Figure 2 – Elevation of Alternative Design

Thus, at each typical pier location there are four leaf columns, each supporting its own deck span, resulting in a viaduct that consists of a series of back-to-back portal frames, with no bearings.

The dimensions of the reinforced concrete leaf columns were carefully proportioned to allow the piers to flex due to the flexural rotation of the deck together with its thermal, creep and shrinkage movements, while still providing sufficient rigidity and strength to sustain the applied loadings.

The four columns at each pier location are supported on a common reinforced concrete pilecap, which in turn is typically founded on 2 No 1.8m diameter cast in-situ bored piles.

The piles are typically founded in end bearing at 5000 kPa on Grade III rock, although in some locations where the rockhead is very deep, friction piles are designed to minimise pile length.

For the Tender submission, preliminary drawings and detailed quantities for the proposed alternative design were prepared, together with descriptive text, draft Design Statement, and submissions for the HK Advisory Committee on Aesthetics of

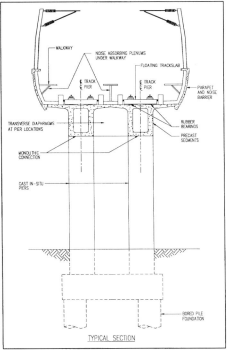

Figure 3 –Alternative Design Typical Section

Bridges and Structures (ACABAS), as well as the planned comprehensive report on the noise, vibration and ride quality performance characteristics of the proposed alternative viaduct.

After extensive tender assessment procedures, including some further work to corroborate the findings presented in the initial ISVR report, KCRC awarded MCW both CC-201 and CC-211 contracts, based on Benaim's alternative design, for the sum of HK$2,100 million.

Detailed Design Stage

In an alternative design scenario, detailed design is a fast track process that requires close coordination between all parties to ensure that the main contract programme is not compromised. The construction programme for the alternative design was essentially the same as would have been adopted if the conforming design was being built, and thus the design programme for the alternative had to be carefully planned to complement the early construction activities, and to provide certified drawings well in advance of construction.

The overall scope of work was divided into a large number of Work Breakdown Packages, each of which grouped similar items of design that shared similar site requirement dates. Each of these Work Breakdown Packages was separately programmed, to allow time for checking and certification by the Independent Checking Engineer, and incorporation of comments by KCRC and Arup, prior to the site requirement date. Careful monitoring of the progress of each Work Package was essential to ensure no delay to the works in the critical initial months.

During the initial months, as the detailed design of the alternative was being developed, MCW were selecting and appointing specialist subcontractors for piling, precasting and deck erection. Each of these subcontractors, once appointed, had certain particular requirements that impacted on the design, and which had to be taken into account. To manage this process effectively and efficiently, to ensure the maximum benefit to the progress and economy of the work, was a major task for MCW and Benaim management teams in the first few months of the design period.

Piles

The piled foundations of the alternative design are generally designed in a conventional manner, in accordance with Hong Kong Geoguide 1/96. The alternative design uses soil-structure interaction and flexure of the piles to resist the longitudinal forces and bending moments on the piers, rather than the more conventional but less economic multi-pile group using push-pull action between the vertical loads in the piles. As a consequence, it was found that no advantage was to be gained in the alternative design by adopting the less conservative bearing pressures contained in the Technical Memorandum published to report the results of pile load tests carried out by Arup and KCRC in the preliminary Technical Study TS-500.

KCRC require the longitudinal deflection of the viaduct deck to be limited to a maximum of 40mm under live load traction and braking forces. The diameter of piles that was found to be necessary to provide the necessary longitudinal bending stiffness to meet this specification requirement was also sufficient to allow use of conventional 5000 kPa bearing pressures on Grade III rock. This avoided the need for any special pile testing to be carried out for end-bearing piles, which comprise the majority of piles, founded typically at between 20 – 40 metres below ground on Grade III rock.

A smaller number of piles, where the depth to rockhead is excessive, or where no rock is found, are designed as friction piles. For the friction piles, the two main piling subcontractors were using different construction methods, and therefore different pile design parameters were appropriate.

For friction piles constructed by grab under water within temporary casings, in the conventional Hong Kong manner, the Geoguide design parameters were adopted, as follows:

Ultimate resistance is calculated based on shaft friction fs = 0.8N kPa, where SPT-N is taken as not more than 200, plus an end bearing component of 1000 kPa. Working load capacity is derived based on a Factor of Safety of 2.0.

For friction piles constructed by MCW subcontractor Bauer, using their proprietary rotary drilling equipment under bentonite, the TS-500 Technical Memorandum suggested that a higher shaft friction could be used, subject to confirmation by a pile load test. MCW and Bauer decided to carry out the pile load test, which confirmed the proposed design parameters, as follows:

Ultimate resistance is calculated based on shaft friction fs = 1.2N kPa, where SPT-N is taken as not more than 200, but with no end bearing component. Working load capacity is derived based on a Factor of Safety of 2.0.

This has allowed consistently shorter piles to be designed for construction using the Bauer rigs and method.

Other differences between the two piling methods have included, for example, a difference in the diameter of the pile reinforcing cages that could be accommodated by the casings used by each subcontractor. This also led to different designs being prepared by Benaim for each of the piling subcontractors, in the situation where maximum economy was required at each step.

A major review of the pile design was carried out once the initial issue of a significant batch of pile toe levels for construction had been achieved. This allowed greater savings to be made, especially in the lengths of rock sockets, where these were required to carry bending moments as well as axial loads.

Piers

Piers are of three basic types. Typical piers have a column beneath each end of each box girder, thus comprising four individual columns resting on a combined pilecap. "T-Piers" are used where constraints at ground level require use of a single pair of columns, beneath the deck centreline, each supporting its two box girders via a transverse crosshead beam. Portal frames are used where the obstructions at ground level are such that the columns cannot lie either beneath the box girders or beneath the deck centreline. For each type of pier, a reinforced concrete monolithic connection is made between the precast deck segment and the supporting column, crosshead beam, or portal frame beam.

Typical leaf columns are 2000 x 700mm in section, increasing to 2000 x 1000mm for the piers up to 25m tall. A cast in-situ transverse diaphragm is provided linking the boxes where the tracks are at standard spacing of 4.8m. Where the two deck boxes are separated by a greater distance, for example at the approaches to the island platform stations, the diaphragms are omitted and the columns are increased to 2400 x 700mm in section.

Pier-Deck Connections

For most bridges with monolithic piers, the superstructure is cast in-situ in order to create a reinforced concrete connection. However, with a precast concrete superstructure there is no immediately obvious way to create a monolithic connection with the piers. Benaim investigated several different options during the Tender stage, and the method that was presented in the Tender was to stress the deck onto the piers using typically 4 no. vertical 75mm diameter alloy high tensile prestressing bars at each pier. It was planned that this vertical stressing operation would occur immediately after the segments had been erected and the longitudinal prestressing had been applied.

Early in the detailed design phase, it was proposed by the specialist erection subcontractor VSL and agreed by MCW and Benaim, to switch from prestressing bars to prestressing strand, in order to provide greater flexibility in fine tuning the design for maximum economy by eliminating significant step changes in the quantity of prestressing when required. However, as the detailed design progressed, it was decided by MCW that a greater tolerance was required between the deck and the setting out of the piers than was afforded by the prestressing solution. Thus, at a relatively late stage in the programme, the design concept for the pier-deck connection was changed to that of a reinforced concrete joint.

To allow for construction of the connection, the diaphragm and the top and bottom slabs between the webs of the deck segment above the pier are "boxed out" during precasting. After the span is erected and supported on temporary packs under the webs, reinforcement is coupled to project from the central portion of the pier into the diaphragm area and is then concreted to provide the monolithic reinforced concrete connection.

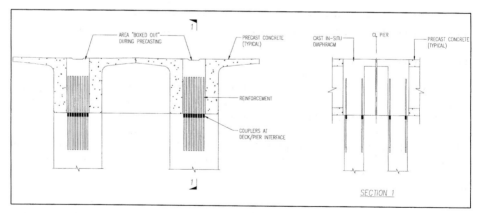

Figure 5 – Deck–Pier Connection Details

Box Girder Deck

In the Tender stage, the prestressing of the deck was designed for maximum economy of materials by carefully tailoring the layout to suit the bending moment diagrams in the bridge deck spans. Some of the tendons were extended the full length of the span and were anchored in the pier segments, while some tendons were only partial length, anchored in blisters in the internal void of the deck segments. However, early in the detailed design period, it was identified that an overall saving in construction cost would be achieved if the blisters inside the deck segments could be eliminated. This would simplify the precasting moulds and eliminate the requirement for stressing operations to be carried out in the narrow internal void in the deck.

The prestressing cables were therefore extended to the end of the spans, which increased the quantity of prestress by a factor of 1.45, over that anticipated in the tender design. This is a good example of where optimum overall economy of construction is not necessarily achieved by a design using minimum materials. It is essential that all aspects of buildability are taken into account when considering an alternative design.

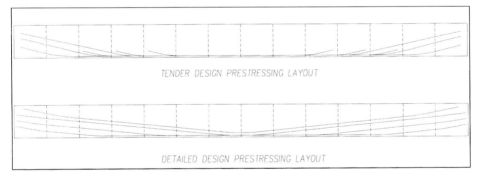

Figure 6 – Comparison of Tender and Detailed Design Prestressing Layouts

There are typically four tendons per web, of maximum size 31 no 12.9mm strands, and are stressed in two stages. Only six cables are stressed during segment erection, in order to limit the transfer stresses, while the remaining two tendons are stressed after the parapets have been installed. In order to provide access for stressing, the second stage tendon anchorages are located in boxouts at the ends of the pier segments.

At major obstacle crossings, the longer span continuous viaducts were designed to be erected with precast segments in balanced cantilever, with a conventional prestressing tendon layout.

Parapets

The unusually large parapets comprise 2500 mm long precast concrete units extending to 2070mm above rail level, and whose main purpose is that of a noise barrier. Although Benaim reviewed the design of the parapets during the tender period, and considered a number of different possibilities, the alternative parapets as finally designed are very similar to those of the conforming design. This decision was made to allow the design of all the various trackside installations, such as overhead line masts, walkways, cable brackets, etc, to remain as per the conforming design.

The main difference is that within each 2500 mm parapet unit, whereas the conforming design has two structural ribs at 1250 mm centres, the alternative design has a single main structural rib at the centreline of the parapet unit with a narrower width rib at each end. This arrangement was chosen in order to suit the increased 2500 mm spacing of the structural steel frames to the noise enclosure and noise barriers in the alternative design, while still providing ribs at 1250 mm spacings to accommodate the mountings of the walkways, cable brackets, etc.

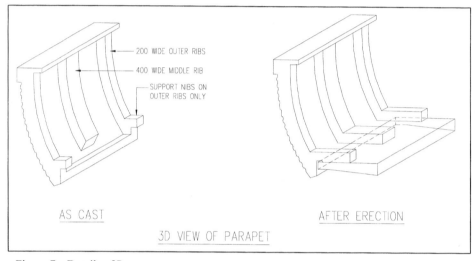

Figure 7 – Details of Parapets

The method of stitching the parapets to the viaduct deck was also modified in the alternative, to reduce the total superimposed load which had to be carried on the viaduct, by reducing the volume of concrete and reinforcement. In place of the continuous stitch of the conforming design, individual discrete stitches are provided at each structural rib in the alternative.

Noise Enclosure

During the tender period, Benaim also reviewed the design of the noise enclosure, and concluded that useful materials savings could be made in the total weight of structural steel framing, while not compromising the noise attenuation performance.

The conforming design comprised transverse structural steel frames at 1250 mm centres, with longitudinal purlins at 600 mm centres. The purlins supported 600 mm x 600 mm x 75 mm thick precast concrete slabs, which were designed to provide the mass necessary for the required degree of noise attenuation, and also carried the specified aluminium profiled sheet cladding.

In the alternative design, the structural steel frames were spaced at 2500 mm centres, with a frame located at the centre of each parapet panel. The concrete noise attenuating panels, instead of being supported by purlins, are design to be self supporting. To this end they are designed as prestressed planks, each 2500 mm x 1200 mm plank being reinforced with 4 No 7 mm pretensioning wires in the longitudinal direction. The pretensioning allows the planks to withstand typhoon wind loads and the "piston" pressures caused by passing trains, as well as spanning between the structural steel frames under their own weight. The concrete slabs also provide local buckling restraint to the compression flanges of the steel framing, while the overall stability of the frames is provided by a series of CHS and angle bracings. The aluminium profiled cladding is unchanged from the conforming design.

Subsequent to the award of the contracts, MCW have proposed to adopt a different system of noise absorption, to replace the precast concrete planks. The new system, which has now been approved by KCRC and is being adopted, has proprietary noise absorptive panels comprising a perforated aluminium casing on aluminium profiled framing, infilled with rockwool layers of graduated densities. The advantage of the new system is that the panels are lighter in weight than the concrete panels, simplifying and speeding up the erection process.

Whilst having no responsibility for the noise attenuation performance of these panels, Benaim have taken on board the responsibility for the structural design of the panels and their mountings on the structural steel frames of the noise enclosure and noise barriers. This has involved a full structural check of the panel design, and necessary modifications being made to suit the high wind pressures experienced in Hong Kong.

Construction Technology

Benaim's scope of works for the alternative design included not only the engineering design of the viaduct structures, but also provision of construction technology services.

Thus, Benaim provided MCW with:

- Setting out coordinates for all foundation and substructure elements.
- Fabrication ("shop") drawings, showing, for each of the 8000 precast segments individually, the casting geometry data to be used for the short-line matchcasting process, together with details of all the cast-in fixings and fittings that were required in each segment, both for the permanent works and for the temporary erection stage requirements.
- Calculation of the theoretical casting geometry data of each individual segment from the KCRC railway alignment data
- Computer software for day-to-day management of the casting geometry data, to track the actual as-cast geometry of each segment and to project the required geometry of each new segment to continually correct any previous casting errors and adjust back towards the theoretical alignment.
- Recalculation of casting geometry to suit spans which were cast in a different sequence to that originally planned.
- Checking of the permanent works design of all structural elements of the alternative for the temporary loading effects arising during erection by the adopted system.
- Dimensions to allow positioning within the precast segments of the starter bars required for the upstand plinths that locate the floating slab track elements and contain derailments, in the multi-track crossover and switch areas, where the tracks are not parallel to the segments centrelines.
- Length dimensions for precasting parapet units.
- Calculation of precambers for all viaduct deck spans, both for the simply supported spans and for the continuous viaducts.

- Theoretical elongations for all prestressing tendons.
- Reinforcement bar bending schedules for all structural elements.

Conclusion

This paper has described the principal aspects of the alternative design of the viaducts for the KCRC West Rail project in Hong Kong. This project has demonstrated that, with good co-operation between all parties, it is possible for the alternative design of a major project to be completed on a fast-track basis during the originally allocated construction period, yielding significant cost benefits to the employer.

An audit of the total materials quantities for the project recently carried out by Benaim has shown that, resulting from design changes subsequent to award of contract, the overall quantities have remained within the levels predicted at the time of tender. This, together with the achievement of predicted rates of progress in erection of the deck spans, confirm that the alternative design has met its principal objectives.

Acknowledgement and disclaimer

The authors would like to thank KCRC, Arup, Maunsell and MCW for their ongoing co-operation and the very many positive contributions which have contributed to making this alternative design a success story for West Rail.

The authors also wish to thank Kowloon-Canton Railway Corporation for giving permission to publish this paper, and acknowledge that the views expressed in this paper are those of the authors and not necessarily those of KCRC.

Construction of KCRC's West Rail Viaducts

HUGH BOYD, T. GREGORY, N. THORBURN
Hong Kong, China

Introduction

Previous papers have dealt with the Design Development of the West Rail viaducts and the Alternative Design adopted for Construction. This paper will describe the various elements of the construction process of the viaducts within Contracts CC201 & CC211. The paper shall be presented in 3 parts.

Part 1 shall provide an overview with particular attention given to the overall programme, construction constraints and those construction methods for the substructure which vary from those usually deployed in Hong Kong. Part 2 shall provide an overview of the precasting operation for the superstructure segments. Part 3 shall deal with the practical aspects involved in the superstructure erection.

PART 1 – CONSTRUCTION OF SUBSTRUCTURE

1.1 **Programme**

Contract CC201 connects Kam Sheung Road Station (near Kam Tin) to Tin Shui Wai Station via Yuen Long and Long Ping Stations and Contract CC211 connects Tin Shui Wai Station to Siu Hong Station (near Tuen Mun).

The two viaduct contracts were awarded on the basis of the Contractor's Alternative Design.

The critical lead in activities to commencement of superstructure erection were:-

- Detailed Design
- Substructure construction
- Precasting of superstructure segments
- Design, fabrication, delivery and assembly of deck erection equipment

The interdependencies between each of these required the implementation of a fast track technical coordination process during the initial period of the contracts. Meetings were attended by: the JV, the JV's Designer, the Specialist Superstructure Erection Contractor, and the Specialist Precast Company. During these meetings critical coordinated decisions were achieved to enable each of the parties to proceed with their responsibilities in a timely and certain manner.

Overall working sequences were dictated by the Access-Constraints and Key-Dates for completion.

In summary the major work elements for the combined contracts include:-

692	No. Large diameter bored piles;
317	No. Pile Caps;
1119	No. Columns;
124	No. T-Heads and Portals;
8728	No. Precast Segments; and
607	No. Spans. (Which equates to approx. 22km of viaduct erection)

1.2 **Construction Constraints**

Throughout the length of the combined contracts numerous construction constraints existed or became apparent which had to be overcome to maintain progress.

Extensive coordination and liaison between many parties, including KCRC, the Engineer, the Contractor, Statutory Bodies, Utility Companies, District Councils, Rural Committees, KCRC Light Rail Division, and Traffic Police, has been continuous to mitigate the effects of these constraints on construction progress.

1.2.1 Major Drainage Channels (belonging to Drainage Services Department [DSD])

The local area is susceptible to flooding during inclement weather and construction activities impacted on many surface stormwater drainage channels. Works interfacing with these drainage channels generally were required to be undertaken in the dry season (October to March). Drainage Impact Assessments were required as part of the submissions made to DSD under the Lands & Drainage Ordinance (L.D.O.) for major drainage channels to ensure that the works would not have any significant impact on the capacities of the various channels. Compensatory mitigation measures such as temporary widenings, diversions, complex temporary-works were put in place together with flood contingency plans. These ensured that in the circumstances of a significant rainfall event flow obstruction could be avoided and personnel safety would not be compromised.

1.2.2 Highways and Roads

The route of the viaduct traverses many highways and roads. For each of these interfaces detailed Temporary Traffic Management Schemes were required to be endorsed by the Site Liaison Group co-chaired by KCRC and RDO/Highways Department with representatives from relevant Government Departments. Extensive liaison and coordination was required with the appropriate Authorities and bodies to achieve agreement to the proposed schemes in advance of implementation. Interfaces at many locations exist for several construction phases eg. utility diversions, piling, substructure, deck segment erection, and parapet erection.

1.2.3 Light Rail Crossing

As the viaduct approaches Siu Hong Station, its route comes over the KCRC Light Rail system (LRT). Accordingly, an interface exists for approximately 600m wherein construction activities must be strictly undertaken within the conditions set out in a permit issued by the LRT.

1.2.4 Urban Area

The viaduct passes through a busy, congested area between Yuen Long and Long Ping Stations which also contains a major DSD stormwater channel. This area brings together a complicated matrix of major constraints in terms of utility diversions, Temporary Traffic Management Schemes, the DSD interface and environmental considerations. The overall sequence dictates that this is the final section of viaduct construction by which time the appropriate measures necessary to manage the constraints will have been well established through implementation in other areas.

1.2.5 Others

The viaduct route interfaces with many rural communities and much effort is required in liaising with the various rural committees to ensure they are kept informed of construction plans and investigating any complaints that may arise as a result of the works. Sensitive issues such as graves and trees of fung shui significance have arisen and have been dealt with to mitigate impact on the progress of the works.

Utility diversions have been numerous. Many of these were anticipated but a significant number of uncharted utilities have been encountered during the course of the substructure construction.

Over recent years legislation has been introduced in Hong Kong to minimize the environmental impact of construction projects. The hours available to Contractors to carry out their works have been greatly reduced. Generally working hours are restricted to between 7am and 7pm, Monday to Saturday. Work outside these hours is subject to the issue of the appropriate Construction Noise Permit by the Environmental Protection Department. Working extensive overtime to recover delays is no longer an option. Increased productivity during normal hours through increased resources or increased efficiency is the principal remedy.

1.3 Construction Methods for Substructure

This subsection deals with particular construction methods for the substructure, which are not usually deployed in Hong Kong.

1.3.1 Large Diameter Bored Piles

For Contract CC201 the large diameter bored piles have been constructed under bentonite suspension using BG Rotary Drilling Rigs. Initial concerns which related to outputs and ability of the equipment to cope with rock sockets and pile depths of up to 70m, were quickly overcome. Piles of diameters varying from 1.2m to 2.5m were successfully constructed in durations of approximately one third of those achieved using the conventional Hong Kong methods deployed on Contract CC211. Typically a pile of 40m depth with a 300mm nominal rock socket has taken 3 days to construct using the BG Rotary Drilling Rig.

In addition, the environmental management of the recycled bentonite system has proved much less problematic than the huge amounts of water discharge from the conventional methods.

1.3.2 Reinforced Concrete Monolithic Pier/Deck Connection

The technical reasons for the adoption of the monolithic pier/deck connection have been dealt with in the previous paper on the Alternative Design of the West Rail Viaducts.

Two options for the construction of this connection were considered. One involved using post tensioned tie bars or strand and the other involved a reinforced concrete connection. After detailed consideration, the latter option was adopted, principally for reasons of constructability.

Reinforcement couplers were extensively used in the pier head and pier segment to ensure that the deck erection operation was not disrupted.

Templates were used during the top concrete pour to ensure that the couplers were maintained within allowable positional tolerance. Initially rebar fixing to the connection was slow but once the operatives became familiar with the operation, completion of the connection could be made within three days of span handover from the Specialist Erection contractor.

Concrete with low shrinkage characteristics was used for the connection. A testing regime was carried out to verify that the shrinkage properties of the proposed mix were satisfactory. As an additional precaution, a waterproofing treatment is applied to the deck surface at the construction joint between the pier segment and the insitu infill.

1.3.3 REED Column Method
This method involves the use of durable precast concrete forms with a high quality external face finish, and steel H-Section reinforcement which results in rapid construction with minimal labour resources.

The Contractor proposed to demonstrate the method by utilizing it for the construction of the columns to Piers 244, 245, 246 and 247 within Contract CC-201. Initially, a mock-up pile cap and column were constructed in a site works area. On the basis of this mock up the Employer agreed to the Contractor's proposal.

During the construction of the columns to Piers 244 to 247 the following advantages over conventional insitu concrete were identified:-

- Rapid construction
- Superior concrete finish which negates the need for access scaffolding over a prolonged period.
- Less environmental impact

Subsequently, the method was also adopted for the columns to Piers 57 to 61 on Contract CC-211 in order to overcome physical site constraints and meet programme requirements.

PART 2 – MATCHCASTING OF CONCRETE SEGMENTS

2.1 **The Factory**
Redland Precast is one of Asia's largest precast concrete manufacturers. Over the last ten years of operation, it has supplied many of Hong Kong's largest civil engineering projects.

The factory is located in Dongguan, China and it extends to 300,000sq.m. It occupies part of an island, enabling transportation of products to Hong Kong and elsewhere in the region by sea.

2.2 The West Rail Casting Area

The factory is divided up into covered areas for the production of architectural components and a large outdoor work area for major civils works projects. The production lines for the West Rail Contracts 201, 211 occupy an area of 30m wide by 465m long. The production lines are flanked by component assembly areas, such as the rebar workshop, and temporary storage areas.

The entire production area is serviced by gantry cranes on guide rails, free roaming gantry cranes, tower cranes and sundry mobile lifting equipment. Compressed air, steam and power are reticulated throughout the area and flood-lighting enables 24 hour operation.

2.3 The Moulds

The moulds for these projects were designed and manufactured by SGB in China. The moulds were brought to the factory in kit form and assembled on site. There are 29 steel moulds in use and the quality of these moulds, having to attain a F5 finish, has been a key element in the successful manufacture of the segments.

2.4 Concrete

The concrete for the segments is supplied from an on site batching plant. The concrete is grade 50/20 for severe exposure and it is subjected to the following tests:

workability;
compressive strength;
creep test;
modulus of elasticity; and
chloride ion content.

Together with these, a strict regime of continual testing is carried out on the concrete constituents, monitored by the QA/QC department. The testing is carried out on site and in Hong Kong by the only Hong Kong Laboratory Accreditation Scheme (HOKLAS) approved laboratory in China for construction materials.

2.5 **Steel Reinforcement**
The concrete for the segments is reinforced with deformed steel bar sourced from ISO suppliers who comply with CS2 and the relevant British standards.

The reinforcement bars are cut and bent in a dedicated, covered workshop in the factory. The reinforcement is tested in a HOKLAS accredited laboratory in Yau Tong to CS2.

The reinforcement cages are fabricated using purpose-made jigs to ensure that dimensional tolerances for cover and spacing are maintained. When each cage is complete, it is thoroughly inspected as part of the QA/QC regime.

2.6 **The Match-casting Process**
The basic philosophy of "match-casting" is that one manufactures the viaduct segments in line with the sequence that they will be erected on site. The keyed meeting surfaces of the segments will match exactly on site since they were cast together at the factory.

2.7 **The Process in Detail**

2.7.1 The Positioning of the Conjugate
The conjugate is a segment that has already been cast utilizing fixed and temporary bulkheads and is located next to the segment now to be cast in the finished viaduct. In effect, the conjugate forms one end of the mould.

The conjugate rests on a hydraulic cart at the side of the mould. The hydraulic cart guide rails permit the advance of the conjugate to the mould and the later disengagement of same.

Once the conjugate is correctly positioned, utilizing the hydraulic cart, the keyed surface that is to be cast against is cleaned and coated with mould oil. The correct position is verified by the geometry control surveyors stationed on the survey towers at each side of the mould, any adjustments are made using the hydraulic jacking system.

2.7.2 Preparation of the Mould
The mould is thoroughly cleaned by the surveyors for geometry.

When the surveyors and the QA/QC staff are satisfied with regards to the position of the mould and the conjugate, the work may proceed.

2.7.3 Placing of the Reinforcement Cage
The steel reinforcement cage is inspected for compliance by the QA/QC team. After approval, the cage is lifted into the mould and is placed upon concrete spacers to ensure adequate cover. Custom built lifting frames are used to move the rebar cages and they utilize 4 lifting points on each cage to maintain the

geometry of the reinforcement and other items incorporated into it, such as post-tensioning ducts.

The cage is finally checked by the QA/QC team to ensure all areas conform.

2.7.4 Placing the Inner Core Formwork

The inner core form is now moved into position, turn buckles extended to their full length and any necessary adjustment made.

Any remaining items to be cast-in are placed and secured at this stage.

2.7.5 Final Pre-pour Inspection

The mould, the conjugate segment, the rebar cage and all cast-in items are re-checked by the QA/QC team prior to the pouring of concrete into the mould.

2.7.6 Concreting

Concrete is delivered to the mould by a conveyor belt or by concrete skip, after compliance testing has been carried out the concrete is added section by section as shown on this diagram. Each section is compacted using hand-held vibrators before the next section is poured.

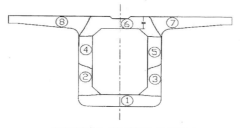

CONCRETE POURING STAGES

When the segment concreting is complete the top surface is treated, at required areas, with a retarding agent to enhance bonding with future insitu concrete.

The last activity at this stage is for the surveyor to install the geometry control pins and alignment plates.

2.8 Curing

The top surface of the concrete segment is treated with a sprayed application of curing compound and then the units covered with a tarpaulin.

After concrete has "taken up" approximately 1-2 hours after casting (dependent upon concrete temperature as monitored by thermocouples placed in the concrete and by ambient temperature monitored by hand held digital thermometers), steam is introduced through pipes situated at the bottom of the moulds. This is checked/recorded hourly and upon reaching $45°C \pm 5°C$ the steam is switched off. Any sudden drop in temperature recorded thereafter during the curing period is compensated for by application of further steam. Towards the end of the curing cycle, the steam is switched off and the concrete allowed to cool gradually until it is within 10°C of ambient. Steam curing is not required during summer due to both high temperatures and humidity. This humid atmosphere helps avoid concrete cracking during the curing process.

2.9 Test Cubes

Concrete cubes taken from the concrete destined for the mould are placed under the tarpaulin to ensure that their curing environment matches that of the segment.

After 12 hours, the test cubes are taken to the site laboratory for testing. If the cube strength has reached 12MPa, the steam supply is switched off, the tarpaulin is removed and segment demoulding may commence. Additional cubes will be tested for the handling and transportation of the segments and also for 28 day strength prior to stressing.

2.10 Demoulding

Firstly, the surveyor carries out an as-built dimensional survey of the segment. This information is then given to the JV to assist them in maintaining the viaduct alignment.

Next the inner form is removed, then the two side forms, then the bottom formwork is lowered.

The conjugate segment is withdrawn using its hydraulic cart, exposing the match-cast faces.

Segment identification marks and casting date are stenciled into the inside face of the segment, and segment joint notation at each end.

A coating of cement slurry is applied to exposed starter bars.

The conjugate segment is now removed to its temporary storage position.

2.11 Final Storage

The finished segments are removed to the final storage area once they have achieved 25MPa strength. Here they are generally cleaned up and any minor repairs made. The segments are stacked for storage and dispatched to Hong Kong by lighter in accordance with the site delivery requirements.

In general the standard span segments are typically 2.5m long by 2.5m deep. However the long spans which can extend to 80 m have a variable geometry with pier segments up to 5.5 m deep. The raking soffits and sides of the segments are created by the use of adjustable formwork for the bottom of the segment.

PART 3 – PRECAST SEGMENT ERECTION WORKS

A study of the alignment indicated that there were three principal structural arrangements that would need to be considered for the erection system.

i. Multiple parallel spans in the turn out areas at Kam Tin and Tin Shui Wai (105 spans total).

ii. Long runs of parallel twin spans which form the majority of the structure (474 spans total).

iii. A series of discrete parallel 3 to 5 span, Long Span Free Balanced Cantilever Structures (28 spans total).

The initial approach taken was to utilise overhead Launching Girders which could be designed to cater for all three structural systems. This choice was also driven by the relatively small wing cantilevers on the conforming design which virtually precluded the use of Underslung Launching Girders. These Overhead Girders were planned to be capable of a two to three day cycle and a total of four girders would have been necessary to complete the project on time.

With the advent of the alternative design the erection system was reconsidered as it was more easily possible to utilise an underslung system for the majority of the project. Given the requirements of the programme and the nature of the alignment it was decided that it would be more advantageous to utilise a greater number of Erection Systems which could be easily moved around the project. Thus the final decision was to adopt the following three different erection methods.

3.1 **Type 1 Erection Truss**

This Girder was developed specifically to cope with the multiple parallel decks at the station interfaces at Kam Sheung Road and Tin Shui Wai. The Erection Truss is made of a series of modules, which can be inserted or removed to allow a basic length variation from 21 to 40m. Final adjustment of the length is achieved by hydraulically operated telescopic arms at each end, which allow a range of up to +1.5m. As can be seen from the picture the Girder is designed to span between the piers or portal supports and to support the pre-cast segments below the soffit of the bottom slab.

The segments are lifted by crane and placed onto individual sledge supports which are adjustable in all 3 axis. Once all the segments have been loaded into the girder epoxy glue and temporary post-tensioning is applied to form one complete span. The permanent post-tensioning is then applied in stages and the dead load of the span is transferred from the truss to the span support jacks. Having achieved the load transfer the Erection Girder is slid out sideways and the span is lowered into its final position where it is left resting on grout packs until the final insitu diaphragm is cast. The support system at the piers is arranged such that after erection of the span the girder can be slid sideways into position to erect the next parallel span. When portal frames support the spans, the truss is in fact supported directly off the top of the portals and the span is built some 1.6m above its final position. Once the span has been assembled and become self-supporting the truss is slid out from below and the span is lowered into its final position by the span lowering jacks shown in the picture.

Once all the parallel spans between two gridlines have been completed the Erection Truss is slid further sideways into position such that the crawler crane can lift it and carry it forward to the next span to repeat the procedure.

The Girder, and its associated Crawler Crane, is capable of working at an average rate of 4 days per span.

3.2 **Type II Launching Truss**

Each Type II Truss comprises two 85m girders, composed of a central section with a fixed front and rotatable rear nose. These Trusses, of which there are currently seven, were designed to erect the standard span arrangement of two parallel spans supported either on individual piers, T-piers or portal frames. The system is based on the twin plate girder approach with 1 girder below each wing of the segments. The plate girders are supported either on brackets or on the top of the T Piers or Portals and are capable of individual self-launching.

Segments are loaded into the girders near the piers by cranes and are then taken along the girders by trolleys to their approximate position where they are left supported on three hydraulic jacks with ring nuts. The arrangement of the trolleys and support jacks is such that the two be completely independent and non-interfering (see picture).

Once all the segments have been loaded into the Truss the trolleys are then used to align and glue the segments together starting from one end. Once all the segments have been glued and temporarily stressed, the permanent post-tensioning is applied and the dead load transferred into hydraulic jacks at the piers. Having achieved load transfer the girders are then lowered to allow the span to be placed on grout packs on the piers and the girders are then launched forward to start the next span.

Where possible the Trusses are operated in pairs on parallel spans with one Truss about 3 spans ahead of the second. This allows the two Trusses to share cranage and also provides a more practical sequence of deck construction.

After the initial learning period the Trusses have settled into a comfortable 3½ day cycle per span with some spans being erected in as little as 2½ days.

3.3 **Type III Erection System**

Initially, it was decided to construct the Long Span Free Balanced Cantilevers as pre-cast segmental structures utilizing self-launching lifting frames or jib cranes mounted at the front of the cantilevers.

After much detailed study it was shown that it would be possible to carry out all of the segmental erection utilizing a combination of standard crawler and hydraulic cranes available in the local market.

The pier segments are delivered as two half units which are joined together at ground level before being lifted up to the top of the pier and set on brackets. The pier segments are then adjusted on the brackets to achieve the desired alignment and level, and the segment and is then temporarily nailed to the pier head.

Segments are then lifted up to alternate sides of the cantilevers where they are glued and temporarily stressed into position. After each pair of segments has been placed permanent cantilever post-tensioning tendons are installed and stressed prior to the next pair of segments being placed. After 3 pairs of segments have been erected the alignment is rechecked and the insitu connection to the pier is constructed.

Once the cantilever is completed the operation moves to the next pier. Sequential cantilevers are joined by a stitch joint sized to suit the specific span which consists of a pre-cast segment suspended between the two cantilevers with two 200mm un-reinforced insitu stitch joints to make the final connection.

Side spans are erected by crane onto temporary falsework.

Erection rates vary depending on the location of the span being constructed and the various constraints applied, however allowing standard working hours it is possible to erect 3 pairs of segments per day on one pier.

3.4 **Conclusion**

Considering the work carried out to date it may be concluded that the decision to change from the original plan to use complex new design Overhead Erection Girders to the relatively simple and well proven systems adopted was definitely beneficial for the following reasons:

i. The final systems are more simple to operate which allows for shorter learning curves and a safer working environment.

ii. The launching girders use less sophisticated equipment much of which can be bought off the shelf or rented and is easy to maintain and repair.

iii. The relatively small size and simplicity of the girders allow them to be more easily taken down and relocated than a complex overhead girder thus providing a more flexible system on the site.

Acknowledgement

The authors wish to thank Kowloon-Canton Railway Corporation for giving permission to publish this paper, and acknowledge that the views expressed in this paper are those of the authors and not necessarily those of KCRC.

Construction of Steel Girder Bridge Rigidly Connected to Concrete Piers with Perfobond Plates

HIROSHI HIKOSAKA[1], KATSUYOSHI AKEHASHI[2], YASUTAKA SASAKI[2], KIYOTAKA AGAWA[3] and LING HUANG[1]
1) Dept. Civil Engineering, Kyushu University, Fukuoka, Japan
2) Research Institute, Yokogawa Bridge Corp., Chiba, Japan
3) Kyushu Branch, Japan Highway Public Corp., Kagoshima, Japan

1. INTRODUCTION

A new cantilever-erection method has been developed recently in Japan for the construction of two parallel steel plate girders spaced widely apart between axes and rigidly connected to reinforced concrete (RC) piers. The Imabeppu River Bridge, located on the Higashi-Kyushu Motorway forming part of the national expressway network, was the first application of this erection method in Japan.

Steel bridges generally require more maintenance than concrete structures. However, the low dead load of steel girders results in the low construction cost of substructures when the potential for seismic activity is high as in Japan or where the foundation condition is poor. By rigidly connecting the steel girders to concrete piers, high maintenance cost items such as bearings and deck joints are eliminated, which will reduce the life-cycle cost of the bridge. The moment-resisting capacity of the connection creates the potential for additional redundancy in the seismic force resisting path. Wider spacing of girders means less steel fabrication cost although at the expense of deeper sections. Either a precast or cast-in-place prestressed concrete slab can be used for the deck, which may be designed as composite or non-composite. Use of the transverse prestressing increases the span length of the concrete slab with greatly reduced cracking. The large stiffness of the deck in the horizontal plane can also minimize the lateral bracing of plate girders.

The new cantilever-erection method for steel plate girders seems to be a better choice, especially in the mountainous regions where the use of other conventional erection method is difficult and the construction yard is limited to a rather small space around the pier. However, particular attention has to be paid to the stability of laterally unstiffened plate girders under cantilever-erection. Designing the rigid and durable steel plate girder-to-RC column connection is significantly important for the bridge constructed using the cantilever-erection method. The perfobond plate, that is a steel plate with large holes at close intervals, was used for the steel-concrete connections in the Imabeppu River Bridge.

2. BRIDGE DESCRIPTION

The general view of the Imabeppu River Bridge is shown in Fig.1. It has three continuous

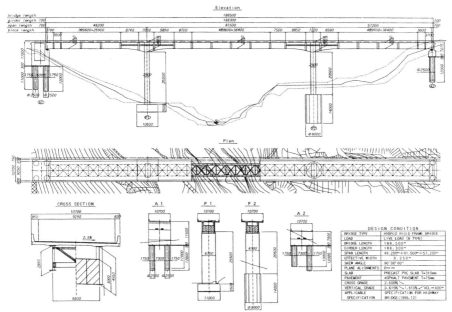

Fig. 1 General view of the Imabeppu River Bridge

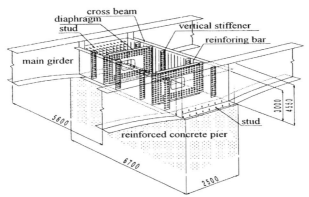

Fig. 2 Steel girder-to-RC column connection

spans, 48.2+81.5+57.2 m, for a total length of 188 m, with two RC piers, P1 and P2, which are rectangular columns 2.5x6.7 m. Only two lanes of the planned four-lane highway are built, in accordance with the prevailing traffic demands. The remaining two lanes will be constructed at a later date to meet the needs of future development along the route.

The main girders are divided into 23 steel blocks, which are made up of two parallel welded I-girders 5.6 m apart between axes. The girders are 2.9 m deep except for three blocks of varying depth (4.5 m maximum) adjacent to each pier. The longitudinally profiled steel plates are extensively used to achieve more efficient fabrication and material saving. The transverse beams are rolled steel profiles, spaced every 4.8 m.

168 DESIGN AND CONSTRUCTION

The girder-to-column connection (Fig.2) has two main I-girders made of 570 MPa steel (yield strength 450 MPa). The top and bottom flanges have a constant width of 800 mm and 950 mm to limit their thickness to 59 mm and 50 mm, respectively, and the web thickness is 32 mm. Two cross beams have a web of 3,000x28 mm and are expected to transfer various forces between main girders and RC piers by means of four perforated vertical stiffeners and two perforated diaphragms of 22 mm thick steel plate with 70 mm diameter holes. The perforated stiffeners and diaphragms become embedded in the pier when concrete is placed up to the upper flange height of the cross beams (Photo 1). Stud connectors are also provided on the webs and underneath the bottom flanges of main girders to improve the bonding at the steel-concrete interface.

Photo 1 Concrete placement in the girder-to-pier connection

The perforated steel plate was originally proposed and named "perfobond strip" by Leonhardt et al.[1] as an efficient bonding means for steel-concrete composite structures. In their push-out type experiments on the perforated steel plate embedded in concrete, it was observed that shear-slip relations depend on factors including: 1) concrete compressive strength, 2) hole diameter, 3) lateral confining pressure, and 4) amount of lateral reinforcement passing through the hole. A series of pull-out tests on perforated steel plates with a single hole was performed[2]. The test specimens had the same parameters as those used for the girder-to-column connection in the Imabeppu River Bridge, such as the concrete strength (30 MPa), the plate thickness (22mm) and the hole diameter (70mm). No lateral reinforcement was provided in both the specimen and the actual structure.

It is well-known that Japan is located on one of the most seismically active areas in the world. The current Japanese design code for highway bridges adopts the dual strategy of seismic design: 1) The structure should resist a moderate (Level 1) earthquake without damage to the basic system; 2) Limited, repairable structural damage is permitted during a strong (Level 2) earthquake which is very unlikely to occur within the life of the structure. The Imabeppu River Bridge was designed to the equivalent static force corresponding to a horizontal acceleration of 1.8 m/s^2 (Level 1) and to the seismic action characterized by a 6.6 m/s^2 maximum ground acceleration (Level 2).

3. ERECTION OF STEEL PLATE GIRDERS

Construction of Steel Girder-to-RC Pier Connection
The cantilever-erection method of the main steel girders as well as the construction sequence

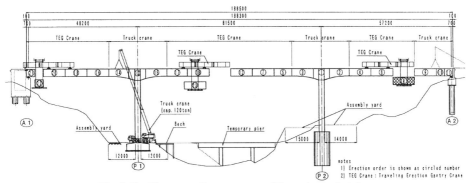

Fig. 3 Cantilever-erection sequence of the main steel girders

Photo 2 Cantilever-erection using TEG crane

is illustrated in Fig. 3. The first step for steel girder erection after pier construction was rigidly connecting the steel block to the pier P2. Two 7.22 m long main girders, two 5.6 m long cross beams and two perforated longitudinal diaphragms were bolted together on top of the pier to assemble the steel block No.1 using a truck crane. The steel block was set on four stay-in-place supporting devices and positioned 800 mm above the pier top. After providing local reinforcement underneath the bottom flanges of the girders and setting up casting forms, concrete was placed in the 2.5x5.6x3.8 m pier head.

Cyclic Cantilever-Erection of Main Steel Girders
The truck crane was also used to erect the steel blocks 2 and 3 on alternate sides of the pier, and then to erect the traveling erection gantry (TEG) crane on the main girders. The TEG crane traveling on rails over the top flanges of girders was used for the cyclic cantilever-erection of main girders and later for the laying of precast concrete deck slabs. The TEG crane weighed about 600 kN including two main beams, a traveling device, a hoisting device running on the main beams, and a supporting floor used as the work platform during erection.

Each steel block was assembled by bolting on the supporting floor at ground level, before it was hoisted and transported to the previously erected girders. The maximum block weight was 250 kN. The main beams of TEG, hanging the supporting floor with an assembled steel block as shown in Photo 2, traveled first to the cantilever tip. The hoisting device alone then moved to the correct position to connect the joints between two steel blocks. Combined welding and high-strength bolting was adopted for the field splice of main girders. The webs and bottom flanges were bolted together, whereas the welding of top flanges was required as the precast concrete deck slabs had to be laid on a smooth surface of the flanges.

After completing the field splice of the steel block, the TEG crane returned to its original position over the pier, and the cantilever-construction proceeded by erecting the steel blocks on alternate sides of the pier. The 4-day erection cycle consisted of the 2-day assembly of a steel block on the ground including trucking the steel members to the site and the 2-day erection including hoisting, traveling and field splicing.

The three end steel blocks (Nos. 9-11) at abutment A2 were erected without using the TEG crane, since they were within the reach of the truck crane and an accessible assembly yard could be secured behind the abutment. At abutment A1, however, all the steel blocks had to be erected using the TEG crane. The end block 21 was first fixed to the abutment after the TEG tip had been jacked up to the desired position on the abutment, and then the closure block 22 was spliced to the adjacent main girders.

Finally the closure block 23 in the center span was connected to the two cantilever tips (Fig. 4). Because only the cantilever from pier P1 was subjected to the weight of the TEG crane and segment 23 in this final stage of erection, the difference of vertical deflection between the two cantilever tips was about 100 mm and the horizontal clearance between them became too small for the closure block to be inserted. Thus, these alignment deviations were compensated by means of both temporarily lowering the support of main girders by 330 mm on abutment A1 and temporarily setting back the girders by 40 mm toward abutment A1, to splice first the connection between the blocks 19 and 23. The TEG crane then returned to pier P1 before the last connection between the blocks 12 and 23 was spliced without residual internal forces in the structure.

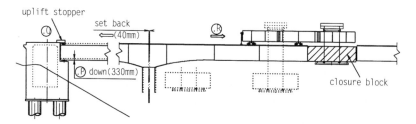

Fig. 4 Closure of center span

Safety Check for Steel Girders under Cantilever Erection

Since the lateral bracing was minimized in this two-I girder system and its cantilever during erection had relatively small torsional rigidity, the safety for lateral torsional buckling was carefully checked. The three-dimensional elasto-plastic finite displacement analysis was conducted for a simplified structural model, fixed at pier P2 cantilevering out critically $L=49.35$ m toward the side span as shown in Fig. 5. Both the typical residual stresses distributed in the shop-welded I girders and the lateral displacement of $L/1000$ at cantilever tip were assumed as the initial imperfections. The parametric study on the effect of lateral bracing was also performed to enhance the structural stability and safety during erection.

The loads considered were the structure's self weight, that of the TEG crane taking its eccentricity into account, and the wind load which was more critical than the seismic load. In

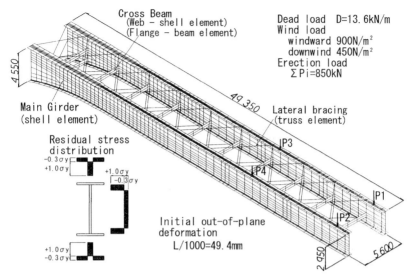

Fig. 5 Structural model of main girders under cantilever erection

the analysis, both the structure's self weight, D, to which a load factor 1.3 was applied and the horizontal wind load, W, were kept constant as the basic loads. The erection load, E_R, given by four vertical concentrated forces acting on the top flanges of main girders, was then incrementally increased until it exceeded $2.5E_R$ or the lateral displacement of the top flange became extremely large.

Fig. 6 compares the effect of lateral bracing for the main girders on lateral displacement versus vertical load during erection. The lateral bracing was placed in the horizontal plane of the top and/or bottom flange of the main girders. The analysis result clearly shows that the lateral bracing contributes significantly to the lateral stability of the girders under cantilever-erection. Only the lower lateral bracing was actually adopted in the structure, because the upper lateral bracing was revealed to be less effective than the lower one. Enhancing the aerodynamic stability of the two-I girder cantilever during erection was another favorable aspect of placing the lower lateral bracing[3].

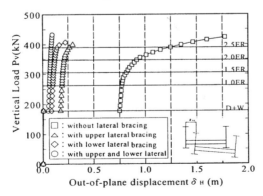

Fig. 6 Lateral displacement versus vertical load during erection

Monitoring of the Structure during Erection

Because a single TEG crane was used for the cantilever-erection of main girders, a large unbalanced pier moment took place during the erection. In addition to the survey control for the construction camber of each girder block, several parameters were monitored during erection in the steel-concrete connection at each pier head, including strains in the main steel girders, strains and cracks in the RC piers, as well as ambient and material temperatures.

Fig. 7 shows the variation of measured and calculated stresses in the longitudinal reinforcing bars at the positions 650 mm below top of the pier P2, as the main girders were cantilevered on alternate sides of the pier. The largest unbalanced pier moment occurs when the girder block G_{20} is being erected in the A2 side span before the block G_{12} in the center span is completed. The measured compressive stress in rebars shows close agreement with the theoretical calculation, whereas the measured tensile stress is lower than the theoretical since the latter was obtained neglecting the tension-zone in concrete. Fig. 8 illustrates the variation of axial stresses in the flanges and webs of the two parallel main girders, G1 and G2, measured at the cross-sections just out of the girder-to-pier connection. The differences between the measured and calculated values are relatively small, indicating that the analysis model used for the design is satisfactory.

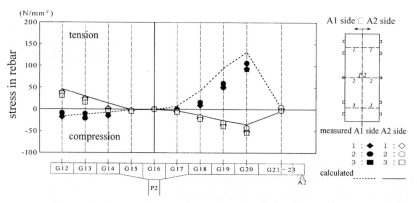

Fig. 7 Variation of rebar stresses in pier P2 during cantilever-erection

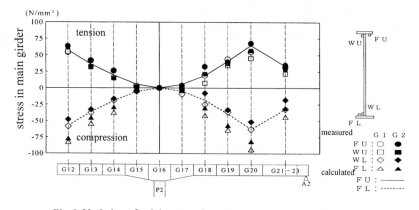

Fig. 8 Variation of axial stresses in main girders during cantilever-erection

4. PRECAST PRESTRESSED CONCRETE DECK SLAB

The non-composite bridge deck was composed of 46 precast prestressed concrete slabs, each of which was 10.4 m wide, 4 m long, 0.31 m thick and 333 kN in weight. Before the segmental slabs were laid on the main steel girders, each of them had been prestressed only transversely due to large spacing (5.6 m) of the girders. The longitudinal design of the deck slab was based on the conventional RC model without prestressing. The concrete mix was designed to reach 30 MPa in 3 days and 40 MPa in 28 days, and each precast slab was transversely post-tensioned after the concrete reached a strength of 30 MPa. Three segmental slabs per week were produced in the yard behind abutment A2. The 7-day cycle for producing a segmental slab consisted of setting up forms, installing reinforcing bars and prestressing tendons, casting, stressing and moving into the storage area.

After completion of the entire steel spans, the precast deck slabs were laid on the main girders simply from abutment A1 toward A2 by the TEG crane which had just previously been used for the girder erection (Photo 3). Each segmental slab had rectangular holes, 120 mm in length by 600 mm in width through the slab depth, provided at 1 m intervals along the top flange of main girders. Thus, the welded studs on the top flanges were embedded in the concrete as the holes were poured. The deck joints between each segmental slab were designed based on the RC model and completed with cast-in-place concrete (Photo 4). The end deck slabs adjacent to each abutment were cast in place to allow for any erection error and to install the expansion joints, and then transversely prestressed.

Photo 3 Laying of precast concrete slabs using TEG crane   Photo 4 Joint of precast concrete slab

5. CONCLUDING REMARKS

The newly developed cantilever-erection method for steel girders rigidly connected to RC piers was applied to the Imabeppu River Bridge in Japan. The traveling erection gantry crane was efficiently used for both the girder erection and the laying of precast prestressed concrete deck slabs, resulting in the completion of the bridge far ahead of schedule. Use of the perfobond steel plates in place of conventional stud connectors also led to simplification and rationalization in the construction of steel-concrete hybrid structures.

The inelastic stability and safety of the steel two-I girder system under the cantilever-erection were carefully checked using a three-dimensional elasto-plastic finite displacement analysis. Several parameters, including strains in the main girders and RC piers, were monitored during erection. The experience gained and the lessons learned from the design and construction of the Imabeppu River Bridge have enhanced the future of this type of erection method.

References

1) Leonhardt, F., Andrä, W., Andrä, H-P. and Harre, W., 'Neues, vorteilhaftes Verbundmittel für Stahlverbund-Tragwerke mit hoher Dauerfestigkeit', *Beton- und Stahlbetonbau*, (12) (1987) 325-331.
2) Watanabe, M. and Akehashi, K., 'Cyclic-Loading Test of Rigid Connection for Hybrid Frame Bridge Using Perfobond Strip as Shear Connector', *Proc. of the 54th Annual JSCE National Convention*, I-A148 (1999).
3) Nakamura, K. et al., 'Design and Construction of Imabeppu River Bridge', *Bridge and Foundation Engineering*, **34** (12) (2000) 2-9.

Twinning of Jindo Grand Bridge, Republic of Korea

MICHAEL JAN KING
Technical Director, High-Point Rendel, London, England

WOO JONG KIM
President, DM Engineering Ltd., Seoul, Korea

CHUNG YONG CHO
Managing Director Structures Group, Yooshin Corporation, Seoul, Korea

Abstract

During the early 1980's Rendel Palmer & Tritton (now practicing as High-Point Rendel - HPR) was responsible for the full detailed design and supervision of construction of a major cable-stayed bridge linking Jindo Island to the South West corner of the Korean Peninsular, for the Ministry of Construction & Transportation. This bridge, with a 344m main span and 70m side spans, was not only one of the world's longest cable stayed spans at the time, but more significantly, the slenderest, being just 11m wide. As a consequence, considerable effort was expended on ensuring the aerodynamic stability of the design, both during construction and in service. The bridge was opened to traffic in 1984.

Photomontage of the twinned bridge

Early in 2000, the same Client commenced the process of increasing the traffic capacity of the crossing and issued the documents for a Turnkey Competition to design and construct an additional crossing near the existing two-lane structure.

Two local design companies, Yooshin Corporation and DM Engineering, together with High-Point Rendel, were engaged by Hyundai Engineering & Construction, to carry out the necessary engineering services for this Turnkey Competition, and if successful to provide the further engineering services for completion of construction. For the tender submission the services included site investigation and route option studies, followed by wind tunnel testing, simulations, preliminary design and a full set of preliminary design drawings.

The dualling of this very slender bridge posed some challenges. Not only was the aesthetic treatment of the new crossing of paramount importance, but the option selected, the addition of a near identical and very slender bridge deck immediately adjacent to the existing bridge, required careful study in terms of its aerodynamic behaviour.

The Turnkey Competition was won by the Hyundai group submission. The chosen form of the new bridge is a 70 – 344 – 70m span configuration with steel 'A' frame towers. A particular feature of the twinning concept is the detailed consideration of aesthetics.

This paper describes some of the important features of the existing bridge and their impact on the form of the new bridge. It describes how the new bridge differs from the existing and the measures taken to harmonise the two together. The paper outlines the aerodynamic investigations and the proposals for ensuring aerodynamic stability of the twin bridges.

The site

Jindo Bridge is located on the south-west coast of the Korean Peninsular and connects the island of Jindo, the third largest of the Southern coast, to the mainland of Haenan-Gun in the province of South Cholla. The bridge crosses an historic stretch of water, the Straits of Uldolmok, where more than 300 years ago Admiral Lee of the Yi dynasty with his 'turtle' ships destroyed an invading fleet of the Japanese navy by using the fast tidal current of the straits. It is reported that, knowing the exceedingly swift currents, he lead the pursuing fleet into the straits, and as soon as his ships had passed, raised a heavy rope out of the water and dismasted the invaders.

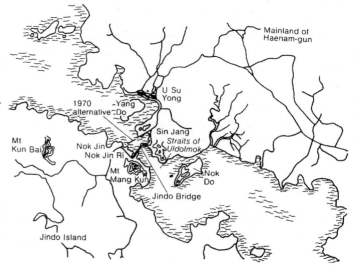

At it narrowest point, the straits are 320m wide between two promontories, with water depths up to 22m below datum, a spring tide range of 6m and maximum tidal currents of 12 knots. The coastline is very irregular with numerous islands, peninsulas, narrow valleys and mountains. The steep slopes are typical of an actively eroding landscape of resistant igneous and volcanic rocks. The geology at the site consists of a mixture of pyroclastic and tuffaceous rocks, heavily faulted and jointed.

This area is on the edge of the typhoon belt and a mean hourly wind speed of 48m/s was used corresponding to a gust speed of 80 m/s at deck level. The region lies outside the circum-Pacific earthquake belt, but there are records of limited seismic activity in the area. No specific seismic design criteria were adopted for the original bridge, it being considered that the wind loading would prove far more significant. This was proved later, when in the mid 1990's, HPR carried out a seismic analysis in accordance with the recently published Korean seismic design criteria, and established that the bridge could withstand the stipulated 0.1g (0.07g x 1.4 importance factor) without requiring any stiffening.

The existing bridge

Studies for the first crossing commenced over 30 years ago. Initially, in 1970, a suspension bridge was proposed at a site convenient for connecting into the exiting road systems on both island and mainland. In 1978 a further study rejected earlier proposals and proposed a shorter crossing of 320m span, but with less direct connections. After HPR joined the team, the span was slightly increased to 344m to ensure the main pier foundations could be built on land without the far more difficult prospect of working within the tidal streams. The local topography of the promontories at each side resulted in the selection of unusually short side spans; 70m each side.

The considerable volume of commercial shipping using the straits resulted in a requirement for a 20m clearance over a 200m wide navigation channel. The roadway design requirements were for a 2 lane width to HS20-44 (Korean DB18), a design speed of 60 kph and a maximum gradient of 6%.

The selection of a main span of 344m together with an overall deck width of 11.7m resulted in an extremely slender design, exceptionally so for cable stayed bridges both then and now. (The longest span cable-stayed bridge now under design, Stonecutters, with a 1018m span has proportionally a significantly wider deck and is proportionally less slender.) This, coupled with the particularly onerous wind conditions prevailing, dominated the design. The aerodynamic stability of the bridge superstructure was a major consideration, and for a deck less than 12m wide, the lateral strength was also a governing factor. Stability is best achieved by ensuring a wide separation of the first vertical and first torsional modes of vibration. This was attained by adopting a torsionally stiff deck and towers of an inverted 'V' form. Because of the unusually proportioned side span lengths, a fan cable arrangement was chosen as being more suitable than the alternative harp arrangement.

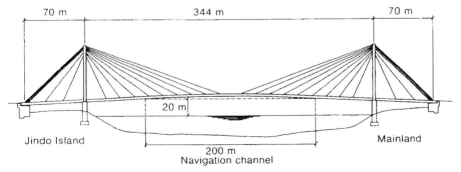

178 DESIGN AND CONSTRUCTION

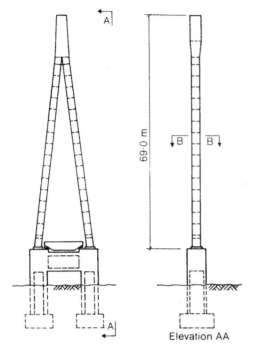

Elevation AA

The main bridge piers were located as close together as possible, consistent with construction in the dry. Ground conditions for both main piers permitted conventional spread footings, although the presence of significant faults filled with clay gouge on the mainland side resulted in a founding depth at −11m.

The piers were rectangular hollow concrete portals with solid 'knees' to support the stressing down of the towers with Macalloy bars. The abutments were proportioned to provide sufficient counterweight to the main span and were founded in a more weathered zone of the rock.

After consideration of alternatives, the torsionally stiff deck, was detailed as a trapezoidal steel box girder with trough stiffening throughout. The use of troughs for inclined webs and the bottom flange gave a less imperfection sensitive detail for a fabricator new to this type of bridge construction. The deck was detailed on the basis of modular erection and a cable spacing of 17m.

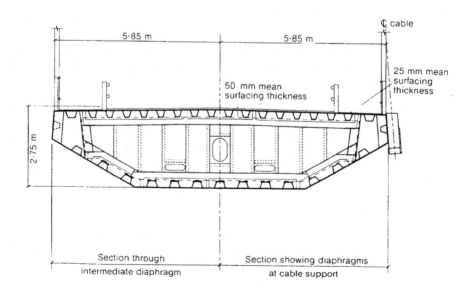

The inverted 'V' form of the towers with cable connections concentrated in the upper portion led to steel being chosen for the tower design. For the loads and cable sizes necessary, locked coil ropes were found at the time to be most suitable; rope diameters ranged from 56 to 87mm diameter.

The bridge was articulated with lateral restraint at all sub-structures and longitudinal restraint at one pier. Because of the span proportions, a particularly large tie-down force resulted. To permit the necessary longitudinal movements at the abutments a back-stay rocker detail was developed using fabricated links rotating on pins anchored within each side of the abutments.

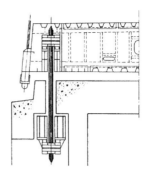

Link detail within abutments

In view of the long span of the bridge and other parameters of the structure it was decided to assess the aerodynamic characteristics by wind tunnel testing. As the bridge response can be highly dependent on comparatively small features, a 73m long section of the deck was modelled at a relatively large scale of 1:30. The testing proved the selection of box shape and cable configuration and resulted in a bridge with the required wide separation of first vertical (0.472 Hz) and first torsional mode (2.08Hz). However, it also determined that the critical speed for vertical motion due to vortex shedding was in the region of 16m/s with a predicted prototype response of 450mm, and the critical speed for torsional motion from vortex shedding was 45m/s with predicted amplitudes of 100mm. Neither of these would have been acceptable and accordingly turning vanes at the upper corners of the box were examined. These eliminated the vortex response from the practical range of consideration and reduced the vortex-flutter interaction to insignificant amplitudes. These were then introduced to the central part of the main span.

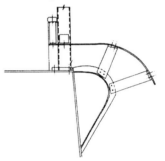

Vane detail on deck box corner

Cable vibration was considered, particularly of the closely spaced back-stays. This lead to high hysteresis rubber dampers specified to be fitted to the upper end of all bottom cable boxes.

The construction contract was awarded to Hyundai Engineering & Construction in December 1980. After an interrupted start because of budgetary problems, substructure construction began in March 1982. Steelwork erection commenced in March 1983 and was substantially completed by the end of that year. A total of 31 deck units were erected, using two rigs for cantilever erection of the main span. The average box erection cycle was less than 12 days with a minimum of 6 days.

Developments since opening.

As originally constructed, the mainland side approach to the bridge had poor alignment because of a significant re-entry in the promontory to the North. During the very late 1990's funding was provided to improve this alignment and the local sea inlet was blocked off with an embankment which now provides a reasonably straight northern approach. (Both the old and current northern approaches are shown on the topographic plan later.) In 1995, Hyundai was awarded a major engineering and maintenance contract on the bridge to include dynamic seismic analysis and assessment to the higher Korean DB 24 loading. This work was carried out by HPR. It was confirmed both that the bridge conformed to the recently introduced seismic design criteria and could carry the higher live loading. As well as resurfacing the bridge, together with extensive routine maintenance, dehumidifying equipment was installed to the deck box-girder. A major purpose built maintenance building was constructed to house the control and recording hardware for the extensive and comprehensive bridge condition monitoring system which was installed throughout the bridge deck, piers and towers.

The second crossing

In the spring of 2000, the bridge owners, the Ministry of Construction and Transportation, required the capacity of this key link to be increased. An outline design brief was prepared and prospective tenderers invited to register for inclusion in a short list to prepare fully engineered submissions.

The Client requirements were very sparse, but very specific in a few items. The scope of work for the submissions required:
- Consideration of all possible sites for a second crossing.
- Alignment and costing of all alternative connections to existing road approaches.
- Major improvement to the southern, Island side, approach to the existing bridge.
- Selection of preferred crossing type and location.
- Preliminary engineering design of new crossing and approaches.
- Necessary traffic, geotechnical, topographic and aerodynamic investigations.
- Outline of construction and erection proposals.
- Full set of outline and detail drawings.
- Full and fixed costings for the chosen scheme

Hyundai Engineering & Construction formed a team together with Yooshin Corporation, DM Engineering and High-Point Rendel for this project. For 6 months the studies continued and the necessary engineering work was undertaken. In September the final submission was made and a month later, the design and construct contract was announced, with the team lead by Hyundai being successful.

Selection of new bridge location

It can be seen from the area plan shown overleaf that the site permits a number of different crossing locations to be considered. Several alternatives were considered by the Hyundai team including suspension, cable stayed and multi-span box girder bridges with overall bridge lengths ranging from 484m to 850m.
These were optimised to a final 4 options that were then compared with the existing crossing on a comparative cost basis. This cost comparison was made on the basis of a series of unit rates for different forms of construction with appropriate weightings to include for location of main foundations (marine or on land), length of connecting approach roads and aerodynamic 'risk'. This last item was included to allow consideration of the possibility that considerable additional costs could be incurred with the addition of a further extremely slender deck in proximity to the existing bridge.

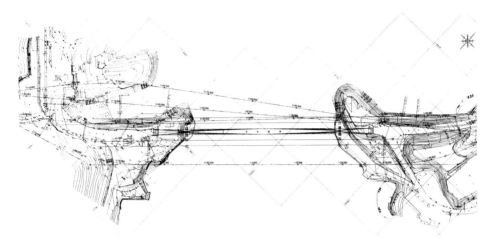

The finally selected alternative routes are shown on the plan. After the costing comparison was made the alternatives were ranked. Even with the inclusion of a hefty 10% additional cost for as yet 'unknown' aerodynamic problems, it was concluded that the option closest to the existing bridge would be the cheapest, by a margin of over 10%. However, this did not conclude the selection process. The two cheapest alternatives being an identical span configuration as close to the existing alignment as possible, or a longer, symmetrical cable stayed bridge some distance to the east.

It was appreciated that the existing bridge is in a remote and scenic location. Indeed, the existing bridge is considered by many, to be the most elegant bridge on the peninsular. Debates continued at length over the relative merits of 'twinning' alongside the existing bridge with a near identical form of new bridge or a separate and distinctively different bridge, positioned symmetrically relative to the existing bridge, but far enough away to avoid aerodynamic interference. Finally the adjacent "twinned" bridge solution was adopted.

Form of new bridge

It had already been concluded that the span configuration would exactly match the existing. However, in following a twinning concept it was felt important to be able to demonstrate some significant differences; hopefully improvements. In appreciation of the importance that aesthetics would be considered in the appraisal of design submissions, Hyundai added an Industrial Design Company to the team. Whilst not an architectural practice, this did have a major impact on the detailing of various aspects of the project and of the 'packaging' to the Client.

It was felt that the introduction of a near identical bridge adjacent to the existing, had to achieve two contrasting objectives. On the one hand the new bridge should have its own identity, whilst on the other, it should harmonise and add to the identity of the existing. It is considered that this was achieved in a very effective manner.

Initially the concept of brand identity was applied to the project. The locality, the local history and geographical features were all considered and the Jindo 'Twin' Bridge packaging resulted.

182 DESIGN AND CONSTRUCTION

The industrial design team considered the heavy angular pier portal shape of the existing bridge to be both aesthetically weak and masculine in appearance. As a result of the Twin Bridge idea, the form of the new pier was deliberately chosen to suggest a female 'twin'. This was achieved by selecting a flowing form to the new pier and adding a distinctive finish that was then repeated as an addition to the cross head of the existing piers.

Further, the detailed cross-sectional shape of the steel pylon legs, although structurally nearly identical, was enhanced to emphasise the differences. The effect of these two approaches is shown here.

A further item of detailed consideration was the historical connection with Admiral Lee and his Turtle Ship naval victory three centuries earlier. The combination of a prominent bridge feature in this scenic environment has resulted in the site becoming something of a local tourist attraction. Indeed there has been built an exemplary museum at the bridge site to house replicas, models and historical artefacts pertaining to the Admiral and his Turtle ships. In recognition of this, and to provide further interest for visitors who walk back and forth over the bridge(s), the footways on both new and existing bridge are to be taken around the pylon legs on a curved timber-decked platform shaped as the deck of a Turtle ship. An inexpensive but interesting addition; shown in the view overleaf.

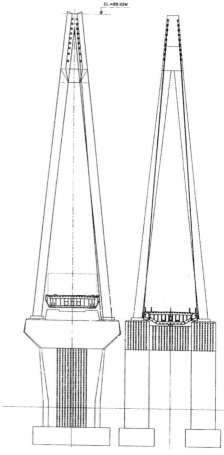

One of the problems encountered with the "twin" bridge was that of vertical alignment. As stated earlier, the brief in the early 1980's for the original bridge called for a design speed of 60kph with maximum gradients of 6%. Although, in fact, the approach slopes on the existing bridge are only 5%, the vertical curve length adopted over the navigation opening was selected from the required sight distance at 60kph. In drawing up the design criteria for the second bridge, the Client appeared not to have taken cognisance of this and required any new crossing to have an 80kph design speed. The increase in vertical curve length for this higher speed resulted in the need to introduce vertical separation between the decks; in the limit as much as 4.7m. This has either to be at mid-span, with complications for the aerodynamic modelling, or at the abutments with complications for tying-in the approach roads. During the tender period there was little scope to get a relaxation in this design requirement, although it is hoped, once final design is underway, that the Client will accept that there is little advantage, even a hazard, with an 80kph design speed Southbound and a 60kph limit Northbound.

Differences in detailed design

Despite determined attempts to show that the existing bridge form, articulation and basic arrangement could be significantly improved, it proved very difficult to make any significant improvements. The salient differences that were adopted included:

i) The use of partial concrete in-fill in the side spans – of marginal benefit in increasing the bending stiffness of the side spans.
ii) The use of flat rather than trough stiffening to inclined webs and bottom flange of the deck boxes – a more usual choice with the availability of good fabrication.
iii) The use of pre-formed parallel-wire strand cables – a natural successor to the earlier use of locked coil ropes.
iv) The adoption of cable rather than pinned-link tie-downs within the abutments.
v) The use of triangulated bracing to form intermediate diaphragms – certainly more economical to fabricate than plate diaphragms.

In the following aspects there are no significant changes to the details of the existing bridge:

- Deck box cross-section
- Cable and deck box module spacing
- Steel deck stiffening
- Material and stiffening to towers.
- Geometry of cable anchorages
- Method of substructure construction
- Method of pylons and superstructure erection.

Aerodynamic modelling

It was considered that the most adverse effects of interference between the two bridges would occur with a separation between the decks of between 3 and 5 times the depth of section. For these decks that is a central gap of between 8 and 14 m. To ensure no interference between the pier foundation bases the actual separation could not be reduced below 10m. Some aerodynamic interference effects were therefore to be expected.

Large scale wind tunnel testing was carried out using sections of both new and existing decks. Initially a single deck was tested to replicate the results of the testing carried out for the original bridge. The tests were then repeated with the second deck alongside and the two extreme conditions of one deck being either lower or higher than the adjacent. Additional vortex shedding was observed but was almost completely eliminated by adding turning vanes, as used on the existing bridge, to both decks.

The horizontal approach angle of wind was varied between +/- 30° in 10° increments. Buffeting of up to 30cm full scale was recorded at the maximum simulated wind speed of 63m/s. The direction of principal displacement being reversed with reversal of relative deck elevations. This testing suggested that the vortices from the windward deck would only have very small effect on the leeward deck and required no further control measures. Of more significance however, will be the effects of the existing bridge on the new one during construction.

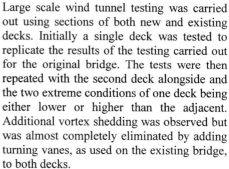

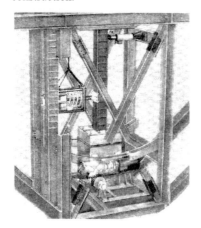

During the construction of the original bridge it was found necessary to add considerable additional temporary steel to the edges of the deck in the tower region. This was done by bolting heavy temporary steel box-beams down onto the deck which were extended outwards as erection proceeded. The wind testing so far, suggests that the temporary condition will prove problematic. Current proposals involve the use of mass tuned dampers in both the tower top and on the deck cantilevers to reduce the amplitude of dynamic deflections by 50%. The required tower-top mass damper has been calculated as 1.92 tonnes with a stroke of 1m. On the deck cantilever a pendulum type damper as illustrated here of 13.55 tonnes is required.

Reference and acknowledgement to:
RGR Tappin & PJ Clarke, **Jindo and Dolsan Bridges: design**, *Proc. Instn. Civ. Engrs. Part 1, 1985.*

The Design of the Stonecutters Bridge, Hong Kong

Stuart Withycombe
Technical Director
Halcrow Group Ltd.,
Swindon, UK

Ian Firth
Partner
Flint & Neill Partnership,
London, UK

Chris Barker
Associate
Flint & Neill Partnership,
London, UK

Introduction

This paper describes the design of the Stonecutters bridge - the winning scheme in the design competition held by the Hong Kong Highways Department. The competition attracted 27 designs from 16 pre-qualified international teams in stage one, and 5 schemes were shortlisted for stage two. The winning design team comprised Halcrow, Flint & Neill Partnership, the Shanghai Municipal Engineering Design Institute (SMEDI) and the Danish architects Dissing+Weitling. (Figure 1)

The bridge crosses the Rambler Channel between Tsing Yi in the north and the reclamation around Stonecutter's Island in the south. It has a world record cable stayed main span of 1018m and the 290m single leg towers are the highest in the world. It has been acclaimed as a "world class design" and includes a number of unique features. It has also received the seal of approval from the Feng Shui experts who believe the design resembles a dragon and will bring prosperity to Hong Kong and security to the harbour!

Aesthetic considerations were critical in the development and judging of the design, not least because the bridge will be a landmark clearly visible from Central and Western Kowloon, and will form a gateway to visitors arriving in Hong Kong by road from the north. However, this paper does not specifically address aesthetics which is covered elsewhere in this conference [1] but concentrates instead on the key technical design issues.

Figure 1 View of bridge from the north

Image: Dissing+Weitling Architects

186 DESIGN AND CONSTRUCTION

Principal Constraints

The detailed competition brief defined the following key requirements:
- A dual 3-lane highway bridge, with an interchange close to the Tsing Yi end.
- The need to connect to conventional elevated twin box approach viaducts at each end.
- A clear navigation channel width of 900m and height clearance (airdraft) of 73.5m.
- A clear distance between the faces of the towers of at least 983m and an overall bridge length not more than 1640m.
- The planned reclamation on the Tsing Yi side for Container Terminal 9 would not be finished until November 2004, and construction on that side could not commence until after that date.
- Construction had to be complete by March 2008. (This has subsequently been changed to December 2007.)
- Severe restrictions on the lifting of deck units from barges in the water due to the 1000 or so shipping movements per day under the bridge including some of the world's largest container vessels.
- Ship impact loading due to collision from an errant 190,000 DWT vessel.
- A requirement for aerodynamic stability in sustained wind speeds up to 95 m/s.
- Construction cost had to be within the budget of HK$3,500 million.

The technical design issues gave rise to further stringent aerodynamic performance criteria which dominated much of the conceptual design thinking. Typhoon conditions may occur during erection of the long cantilevers, and the use of temporary restraints into the navigation envelope was discouraged.

The geotechnical conditions also imposed tight constraints. The reclamations on both sides of the channel consist of 40 to 50 metres of fill and alluvium before reaching moderately decomposed granite, so deep foundations were essential. Not identified in the competition brief but evident from our own investigations, is the presence of the Tolo Channel fault on the Stonecutters side, running under the side spans.

Design Concept

Considerations of elegance, construction speed and economy, aerodynamic stability and low maintenance were among the main factors driving the evolution of the concept. The twin deck and single tower leg solution was fixed early because of the clear potential advantages for aerodynamic performance and construction speed, as discussed below. (Figure 2)

The restricted overall length and large main span result in unusually short side spans. As a result, the side spans are in concrete to balance the lighter but longer steel main span, maintaining a constant girder depth for visual consistency throughout. The concrete deck is integral with the intermediate piers and tower shaft, eliminating the need for any bearings. The effect of the stiff integral concrete girder and side span piers is to effectively stiffen the end parts of the main span resulting in a shorter effective span length.

The smooth tapered towers are designed for efficiency, elegance and economy and are the major features of the bridge visible from a distance. They also change from concrete to steel part of the way up, enabling the top steel sections to be pre-fabricated off site and achieve a faster overall erection time. The transition occurs just below the stay anchor zone at a height of about 175m, which is convenient for delivering large quantities of concrete by pumping during casting. This change in material also makes visual reference to the change from concrete to steel in the deck for consistency.

Foundations formed using barrettes were proposed to cope with construction through the soft overlying materials and drilling to depths of 60m, with arrangements driven by requirements to resist high lateral loads from seismic effects, ship impact and typhoon winds.

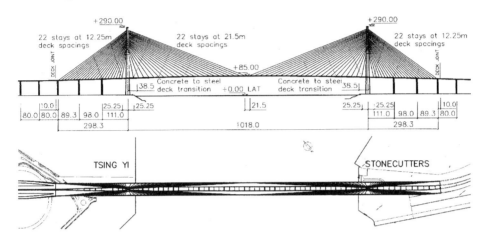

Figure 2   Bridge Elevation and Plan

The stay arrangement forms two inclined planes, meeting at the tower, and thereby provides high torsional and lateral stiffness to the span. Pre-fabricated locked coil or parallel wire cables are proposed as a key feature of the concept in preference to other types for the reasons given below. In the competition design, the main span stays were at 18m centres, but subsequent refinement and optimisation in the design has stretched this to 21.5m centres. Among other benefits, this creates a more logical rhythm for the steel deck diaphragms and shortens construction by reducing the number of individual deck unit lifts.

Towers

The lower part of the tower is in concrete offering high mass and strength for stability and resistance to loads including typhoon winds, seismic and ship impact effects. The upper part incorporating the cable anchorage zone is in steel. The single leg towers avoid the difficult and expensive portals or cross beams otherwise needed in the more common H or A frame solutions, with a single shaft allowing work to be focused on one rather than two legs. The shaft has a 14m diameter at deck level tapering uniformly to 7m at the top in the competition design. Below deck level the shaft flares out to be 23 x 17m at the base. Reducing the diameter by up to 2m from these dimensions is technically feasible, but there is a need to balance a desire for slenderness against the appearance of stability and at the same time maximise economy in construction. The tower proportions were carefully determined with these factors in mind. (Figure 3)

The lower shaft is designed as a reinforced concrete section throughout, constructed using either jump or slip form techniques. There is now considerable experience in constructing large towers using this approach, and the ability to achieve a dense durable high quality finish to the concrete is well proven. With the exception of the tower to deck connection zone, the section is made sufficiently rigid to avoid the need for internal diaphragms, simplifying construction and allowing the casting operation to progress the entire height of the shaft without interruption. Below deck level, the walls are up to 2m thick requiring cooling measures to control heat of hydration temperatures. A key feature of the design is the integral tower to deck connection formed by a special 11m wide crossbeam the same depth as the other crossbeams. As would be expected with a built-in deck, shear and torsion effects in the crossbeam are relatively high, but the section is designed as a reinforced concrete deep beam, and the overall dimensions have now been confirmed by finite element analysis.

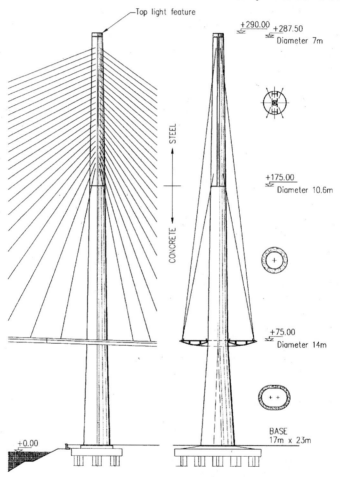

Figure 3   Tower elevations

The steel upper tower section is a simple thick-walled cylinder, with horizontal diaphragms at intervals corresponding to stay anchorage working platform levels. Two vertical planes of internal cross bracing help to distribute local horizontal shears along the line of the bridge and also assist in controlling geometry during erection. Horizontal forces from the stays are generally carried through the platform diaphragms at each anchorage level, but towards the top of the towers where these horizontal forces are greatest the vertical planes of bracing become two vertical plates for added strength. (Figure 4) Vertical stiffeners are not needed, except to distribute local stay anchorage forces into the tower walls, and to frame the vertical "light slot" detail on the front and back faces. The steel section is designed to be erected in pre-fabricated cylinder or half-cylinder units for maximum control of quality and stay anchor alignment, and rapid erection using a special climbing derrick mounted on top of the tower.

Side Spans

The twin concrete side span deck boxes are longitudinally pre-stressed using external unbonded tendons within the girders. Internal webs are needed to carry the high forces at the piers, and deep pre-stressed cross heads form the integral connection between the boxes and the tall slender piers. (Figure 5) The aerodynamic edge nosing on the main span is replaced with a gentle outwardly inclined edge face leaving a clean finish to the deck edge designed to catch light in a similar manner to the steel deck. The same simple pinned socket stay anchorages used in the steel deck are repeated in the concrete deck with the steel anchor plate embedded in the concrete edge sections to transfer the cable load through shear and bearing.

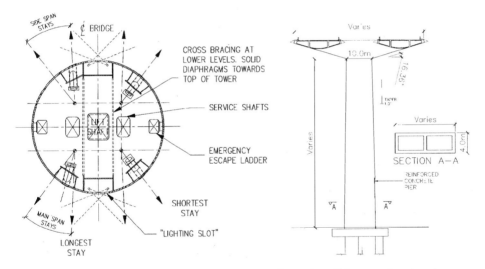

Figure 4 Steel tower section

Figure 5 Side span pier elevation

A key consideration was the method of construction and the desire to complete the side spans in advance of main span erection to provide an elevated working platform and support during cantilevering. After considering several alternatives, taking into account the fact that the heavy concrete spans needed temporary stiffening before erection of the cable stays, the decision was taken to cast the boxes on the ground and lift them up span by span. The piers were configured to enable the boxes to be built on line requiring a simple straight vertical lift before casting the in-situ integral cross head connection to the pier tops. The varying width of the deck due the presence of slip road ramps close to the bridge ends complicates the back span construction, making in-situ rather than pre-cast construction preferable. Significant benefits are gained by carrying out the box concrete works on the ground rather than 70m in the air. Temporary support is provided either by the lifting truss, a kingpost or a system of props prior to installation and stressing of the stay cables. Constructing the boxes on the ground in this way induces much lower loads in the piers than would arise if the decks were built by balanced cantilever methods.

The end pier cross head also provides a convenient point to form the transition between the aerodynamically shaped side span deck and the conventional box girder section proposed by others for the approaches. The side span decks are deliberately the same depth as the adjacent viaducts with a form that could readily be adopted in the approaches. If this option were to be accepted, the integration of the main bridge with the approaches would be improved, enhancing the visual balance and quality of the overall scheme.

Main Span

The main span boxes adopt conventional well proven stiffened steel panel design, but with the added development of the twin box arrangement and partially curved soffit. (Figure 6) Many major steel bridge fabricators have invested in specialist equipment for the rapid assembly and welding of trough-stiffened steel plates. This technology also enables curved stiffened panels, like the centre section of the bottom flanges, to be fabricated almost as simply as flat panels. Curved panels have an advantage over flat panels in that they are able to resist out-of-plane buckling more easily due to shell behaviour and that more of the steel is effective in the cross section. Towards the ends of the span close to the towers, two vertical internal webs are introduced to assist in carrying the high compressive forces.

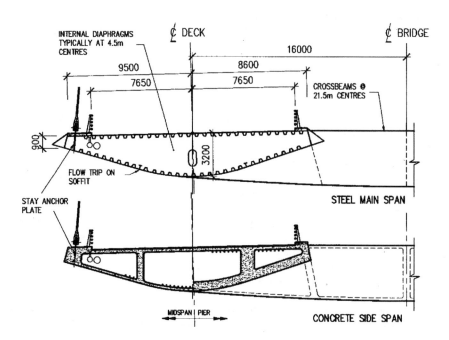

Figure 6   Typical deck cross sections

Steel box crossbeams connect the two girders together at every stay position. These act as deep beams between the stays to transmit the vertical loads, and also as horizontal beams in vierendeel action under transverse wind loading. Transverse diaphragms are typically at 4.5m centres, which is optimum for the orthotropic steel deck behaviour, with a 3.5m span every 5 bays corresponding to the crossbeam webs and stay anchor zones. The stay anchorages are simple flat plates protruding above the deck in plane with the stays. These transmit the local stay forces into the top and bottom flanges and the two adjacent internal diaphragms which form the webs of the crossbeams between the boxes.

The steel-to-concrete splice occurs 38.5m from the tower centreline and consists of an in-situ concrete connection, with pre-stressing bars attaching the first steel box unit to the concrete section. These bars are needed for the temporary erection conditions but are only lightly loaded in service because of the high net compression in the deck at this point.

The erection method assumes delivery of individual prefabricated steel box units to the work face along the deck, completely avoiding the need to lift sections up from the Rambler Channel. Sections would be carried along the deck on a purpose built transporter to the end of the cantilever to be picked up by the erection gantry and manoeuvred into place. Crossbeam units would be erected to connect the twin boxes using the same rig, and then the stays would be fitted and stressed.

Stay Cables

The cable stays are pre-fabricated locked coil or parallel wire strands with pinned sockets at the deck anchorages and cylinder block sockets within the towers where stressing takes place. There are many advantages of this type of stay over site-assembled bundles of pre-stressing strands within an HDPE tube. These advantages include the following:

- Their diameter is approximately half of the equivalent parallel strand cable duct, and this substantially reduces the wind load on the bridge, which is a significant consideration for this bridge at this site. The calculated consequent savings in the towers and foundations alone more than compensate for the higher stay supply costs.
- Off-site fabrication improves quality control.
- Quick erection and stressing time. Each cycle only involves handling a single unit.
- Easy to detect broken or damaged wires, and to check the corrosion protection system.
- Less susceptibile to wind-rain induced vibrations.
- Future stay replacement possible in a single operation.
- They also enable the use of neat and tidy anchorages at deck level, avoiding the heavy fabrications above deck level which often accompany the parallel strand type of cable.

The stays vary in diameter from 80mm to 170mm, which is within the range of currently available sizes. The largest stay, which weighs 98 tonnes, can be fabricated, transported and erected using currently available equipment. By comparison, the 987m long 127mm diameter locked coil suspension cables for the Kvisti suspension bridge in Norway weighed 110 tonnes which was approaching the limit in 1995 for transport by road.

Small diameter cross-stays or "aiguiles" are proposed to limit cross-wind response and control any potential vibrations in the very long stays.

Aerodynamic Design

The aerodynamic performance of the bridge was always going to be a key issue in this scheme, particularly considering the possibility of constructing the world's longest cantilevers under typhoon conditions. For this reason this matter was given the highest priority from the outset, with computational fluid dynamics (CFD) analysis using discrete vortex methods being used to examine the behaviour of different sections before testing the preferred arrangement in the wind tunnel. The twin deck configuration has been shown to offer several advantages for long spans, and has been proposed for the Tsing Lung, Messina and Java-Bali bridges among others.[2, 3, 4] The CFD analysis has been used to derive the static force coefficients C_L, C_D and C_M for input to the global analysis, and to estimate the flutter derivatives that are needed to predict the possibility of divergent response. (Figure 7)

The pleasing curved lines of the deck soffit produce a streamlined aerodynamic design, but is not without its own specific problems. Such curved sections can be prone to responses associated with low speed vortex shedding. In the twin deck arrangement, fluctuating down-wash currents between the decks can also exacerbate this problem and result in fluctuating early separation and reattachment of the flow past the downstream deck. These serviceability conditions have been countered by the addition of flow trip plates on the soffit to ensure that the underside has a stable, turbulent boundary layer at all wind speeds and a well defined rear separation point. This also has the considerable additional benefit that it enables the section to be tested reliably in the wind tunnel. Curved surfaces can be difficult to deal with because of the difficulty of matching Reynolds number in the wind tunnel, yet their vortex responses are known to be sensitive to Reynolds scaling. The addition of trip plates provides a section for which wind tunnel tests can give safe reliable predictions by ensuring the flow separation point for both model and prototype is the same. Internal girder edge guide vanes have also been proposed to maintain attached flow over the front surfaces of the downstream girder in the presence of fluctuating currents between the decks. However, the CFD analysis showed that these may not in fact be required.

Image: Halcrow Wind Group

Figure 7   Plot from CFD analysis.

Although in-line wind loading on long span bridges with shallow decks may be relatively easy to cope with, it is very important to deal properly with the significant cross wind loads that occur on these thin but wide sections. Cross wind loading can be a critical design load effect and has therefore been carefully considered. The calculated responses demonstrate that the design is satisfactory for the specified range of wind speeds with the levels of damping likely to be present in the structure.

The towers may appear at first sight to be prone to problems of vortex shedding, but the proportions and stiffnesses are such that with a Scruton number of about 20 they are not calculated to experience any significant responses either during construction or in service.

Construction Programme

While it is possible to gain access to the Stonecutters side as early as June 2003, in advance of the opposite shore becoming available, it was not considered desirable to extend the overall programme unnecessarily. By delaying commencement, the Contractor's site running costs, together with those of the Client and the Engineer, which are all directly related to programme length, can be kept to a minimum. With this is mind, a construction start date in March 2004 was proposed for the works on the Stonecutters side. These could then be progressed sufficiently ahead of the opposite shore to enable some later re-use of plant and equipment on the Tsing Yi side once construction works start there immediately on handover of the CT9 land in November 2004. The original completion date of March 2008 resulted in an overall programme duration of 49 months. The subsequent shortening to 45 months has led to a need to accelerate certain key activities but analysis of the options has demonstrated that this is feasible while still allowing contractors the opportunity to balance cost and duration in determining the total price. Allowance was made for contingencies to cover weather-sensitive operations, taking into account the typhoon seasons in particular.

Durability and Maintenance

Fundamental to our thinking from the outset was to design for maximum durability and minimum maintenance as far as possible. The selection of materials and structural form has been driven by these considerations as well as those of economy, efficiency and elegance. All bridge components are designed to be robust, and items that are likely to need more frequent inspection and maintenance have been made easily accessible. Principal features of the design which specifically reflect this approach include:

- No bearings at towers or piers. The bridge is fully integral throughout, so all thermal movements and other lateral effects from wind and earthquake are absorbed within the structural elements.
- Concrete substructures and side spans over the busy container areas. Concrete mix design and workmanship will be tightly specified to ensure maximum durability.
- Smooth external steel box and tower surfaces, free from dirt and water traps, prolong the life of protective treatments and facilitate future re-coating.
- Interior spaces of the steel boxes may be protected by dehumidification as an alternative to painting. Requirements to protect the steel during construction and running costs of the dehumidification system need to be balanced against the costs of repainting.
- Pre-fabricated stay cables composed of galvanised wires laid up in Metalcoat and coated on the outside, all in accordance with well-proven international best practice. These permit inspection of the outer wires giving immediate warning of any damage.
- The lift, service shafts and platforms at each stay level within each tower provide easy access and safe working areas for installation, inspection, maintenance and replacement of the stays. Emergency escape shafts served by ladders are also provided on both sides.
- Two travelling gantries slung under the bridge girder provide access to the bridge soffit over the full length of the main span.
- An external climbing tower platform, specially designed to pass the stays and driven from a winch at the tower top, provides access for external painting of the steel top section.

- The sign gantries and under-deck travelling gantries are designed in glass fibre reinforced polymer composites for low weight and maximum durability.

Concluding Remarks

The design is a unique and majestic structure with several innovative features all devised to maximise the long term economy and performance of the bridge. Because of the nature of the design competition and the requirements of the bidding process for the follow-on detailed design commission, a large amount of work has already been undertaken on the scheme proving the design to a high degree of detail.

The sinuous shape of the approach viaduct, particularly on the Stonecutters side, means that drivers will enjoy the changing perspective view as they approach and cross the bridge. (Figure 7) When completed, the bridge is set to become one of the well-known milestones in the history of long spans, complementing the already world-famous Hong Kong skyline.

Acknowledgements

We are grateful for the close collaboration and teamwork enjoyed by all members of the team, and for the enormous contribution to the design by a large number of people in Halcrow, Flint & Neill Partnership, Dissing+Weitling and SMEDI. We also wish to acknowledge the helpful advice received from Cleveland Bridge, UK, who independently verified the construction programme and cost estimates.

We also acknowledge the role played by the Hong Kong Highways Department Major Works Project Management Office, and are grateful for their support.

Image: Dissing+Weitling Architects

Figure 7   View from Tsing Yi

References

1. "Stonecutters Bridge Design Competition – Aesthetic Consideration". *M. Hui & C. Wong.* Current and Future Trends in Bridge Design, Construction and Maintenance, Hong Kong, April 2001.
2. "The Design of the Java-Bali Bridge." *I.P.T. Firth & P.O. Jensen.* Cable-Stayed Bridges – Past, Present and Future. IABSE Conference, Malmo, Sweden, June 1999.
3. "Aerodynamic Characteristics of Slotted Box Girders". *H. Sato, R. Toriumi, T. Kusakabe* International Conference: Bridges into the 21st Century, Hong Kong, October 1995.
4. "Investigation of Twin Box Suspension Bridge". *Y. Morita, Y. Yamasaki, T. Yamazaki* International Conference: Bridges into the 21st Century, Hong Kong, October 1995.

Maintenance practice

Safety and Performance of an Active Load Control System for Bridges

ATKINSON, T[1], BROWN, P[2], DARBY, J[3], EALEY, T A[4], LANE, J S[5], SMITH, J W[6], ZHENG, Y[6]
[1]PAT-GB, Wolverhampton, UK; [2]Oxfordshire County Council, Oxford, UK; [3]Consultant, Oxford, UK; [4]Buro Happold, Bath, UK; [5]Transport Research Laboratory, Crowthorne, UK; [6]University of Bristol, Bristol, UK

SYNOPSIS

It has been estimated that about 20% of the 155,000 bridges in the UK have strength deficiencies in some form. These are partly due to various types of deterioration and partly because loading standards have increased over the years. If a bridge fails its assessment there are several alternative actions available to the bridge owner. These include: complete replacement; strengthening and repair; reduction of the number of loaded lanes; posting a load limit; or "do-nothing" on the basis of advanced assessment using bridge-specific loading, sophisticated analysis or probabilistic risk assessment. An additional alternative is to employ active control of the bridge loading by using live information from weigh-in-motion (WIM) equipment linked to signals on entry to the bridge. The objective of this paper is to assess the performance and safety implications of an active load control system.

INTRODUCTION

Under European Union legislation, heavy goods vehicles of up to 40 tons gross weight were allowed on UK highways from January 1st 1999. The Highways Agency was already implementing government policy to upgrade the UK bridge stock by a programme of assessment and strengthening. There are approximately 155,000 bridges in the UK of which about 13,000 on major trunk roads and motorways are the responsibility of the Highways Agency. The remainder are in the care of local authorities, Railtrack and private organisations (Leadbeater, 1997). It is estimated that more than 20% have strength deficiencies in some form (Williams, 1997).

Bridges that fail their assessments should be repaired, strengthened, demolished or subjected to load limits. Any form of structural rehabilitation or complete replacement is likely to be very expensive and a financial burden on the bridge owner. Often it is necessary to post load limits to ensure public safety for a bridge that is significantly below the required strength. This may be satisfactory as a temporary measure, especially for any bridge that is not heavily trafficked. For a bridge that carries a high volume of heavy vehicles, the cost penalty of having to make lengthy and time consuming detours may be substantial. Another cheap and immediate measure is to reduce the number of lanes on the bridge. Most bridges carry two lanes of traffic and therefore reduction to single lane operation will reduce the bridge loading by half. Unfortunately, this will also cut the capacity by half and the effect on traffic delays and queuing may be unacceptable at busy times.

A feasible alternative is to install a system of active load control. The technology for weighing vehicles in motion is currently well developed (COST 323, 1995; McCall and Vodrazka, 1996). Weigh in motion (WIM) has already been used as a means of monitoring traffic loads on a bridge (Opitz, 1993). In that project the WIM sensor data was displayed in real time so that the operators could control bridge load by changing the permitted lanes for heavy vehicle access. In the current project, the objective is to go one step further and control bridge loading actively by a combination of WIM and traffic signals (Atkinson et al, 2000). This concept is analogous to single lane operation. The major difference is that, in a two lane highway, both lanes will be open most of the time while one lane will be closed briefly on the rare occasions when a combination of heavy lorries is predicted to exceed the bridge load limit. In this paper the hardware and software of the control system is described and its performance is examined by computer simulation. Load and traffic data obtained at the site installation is also assessed.

THE LOAD CONTROL SYSTEM

The load control system proposed here consists of sets of bending plate weigh-beams and vehicle detector loop arrays at the approaches to the bridge in question. These are shown schematically in Fig 1. An electrical induction loop embedded in the road surface detects the presence of a vehicle. The bending plate weigh-pad measures each axle load in turn and the second induction loop detects when the vehicle has completely passed over the array. The vehicle axle load distribution, gross weight and class may be interpreted from the data. An algorithm calculates the expected load or load effect (bending moment, shear etc) that will be applied to the bridge. This is added to the effects of load detected by the weigh-pad array in the other lane at the far side of the bridge. If the expected combined load exceeds a pre-calculated threshold, then traffic signals on the bridge are set to stop vehicles on one of the approaching lanes. Note that there is an induction loop on the far side of the bridge in each lane. This can be used to count vehicles off the bridge so that the signals may be set to green again when it is calculated to be safe to do so.

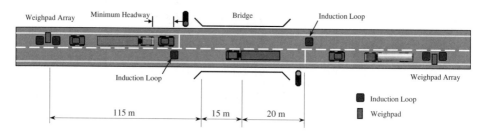

Figure 1. Schematic arrangement of weigh-pad and induction loops

ANALYSIS OF TRAFFIC AND LOAD DATA

The bending plate weigh-pads record axle load data continuously. The controlling program is able to evaluate the total load that will be on the bridge at all times. This is illustrated in Fig 2, which is a time history of total load on an assumed 75 m span bridge over a period of 14 hours. The highest loads may be observed as brief spikes. It is instructive to examine a part of this history in detail over a shorter period of time, as shown in Fig 3. This is a 60 second snapshot of loading on the bridge and demonstrates how vehicle loads on opposite lanes are added to give the total load on the bridge.

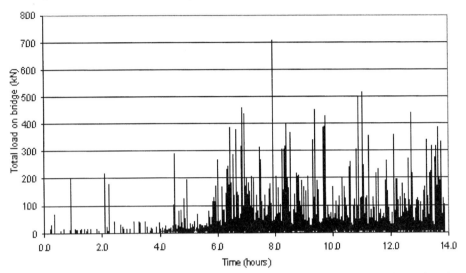

Fig 2 Time history of total load on bridge over 14 hours (75 m span; 6500 v/day; 2 lanes)

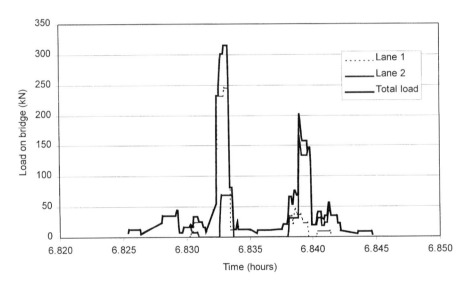

Figure 3. 60 second time history of load on the bridge (data from above)

Fig 4 shows the distribution of numbers of axles of goods vehicles throughout the week. The results show that, for the bridge site under observation, the numbers of vehicles having 5 or more axles represents about 9% of goods vehicles. Average two way weekday traffic flow is about 6500, of which about 10% are goods vehicles.

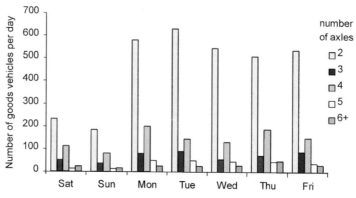

Fig 4 Axle load distribution through the week

An important objective of the study was to obtain site-specific data for the axle load distribution. The commercial vehicle load spectrum used for fatigue design of bridges is shown in Table 1 (BS5400, 1980). However, as can be seen from Fig 4, the number of five and six axle vehicles has grown significantly and these were not in the original load spectrum. Vehicles with seven or more axles need special permits and have been excluded. Therefore, the weigh-in-motion data was used to generate a site-specific load spectrum for the actual bridge site. These are also shown for comparison in Table 1. The vehicle designation indicates the number of axles, articulated or rigid, and heavy, medium or light.

Table 1. Commercial vehicle load spectrum

Vehicle Designation	Gross Weight (kN)	No. in each group per million commercial vehicles	
		Data from BS 5400: Part 10	Data obtained from WIM equipment
6A-H (new)	440	Not included	52500
5A-H (new)	380	Not included	85900
5A-H	630	280	4100
5A-M	360	14500	21400
5A-L	250	15000	7900
4A-H	335	90000	29000
4A-M	260	90000	29000
4A-L	145	90000	10450
4R-H	280	15000	36700
4R-M	240	15000	36000
4R-L	120	15000	21100
3A-H	215	30000	10200
3A-M	140	30000	9400
3A-L	90	30000	6900
3R-H	240	15000	31600
3R-M	195	15000	30800
3R-L	120	15000	18600
2R-H	135	170000	183500
2R-M	65	170000	182500
2R-L	30	180000	191500

The load distribution of the traffic at the site was examined for the frequency of maximum traffic loads on the actual bridge, which has two lanes and a span of 25 m. This was done by running 3000 groups of nose-to-tail vehicles, taken from the WIM data, over the span and producing a cumulative frequency or probability curve. This was compared with a similar curve produced by using a simulated stream of design traffic, as shown in Fig 5. It may be seen that the site-specific loading is much less because of the greater dilution by cars.

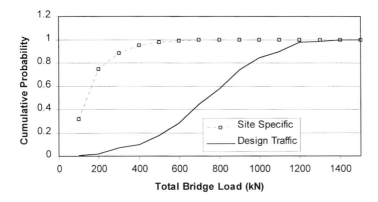

Fig 5 Cumulative probability of maximum bridge load being exceeded

OPERATION OF THE CONTROL SYSTEM

The load control system operates by interrogating the data output by the weigh-pads and setting signals to stop traffic in one lane as necessary. When a vehicle passes over a weigh-pad array, a computer program interprets the data, classifies the vehicle and records the vehicle speed, axle loads, axle spacing, overall weight and current time. This information is read line-by-line by the controller program. In addition, the induction loop at the exit of the bridge inputs a line of data. At each time step (10 milliseconds) the program predicts the total load on the bridge. This includes axle loads of vehicles in the opposing lane. If the total load exceeds a predetermined threshold, then a stop signal is set to halt traffic in one direction. As soon as the heavy vehicle causing this threshold to be exceeded has passed over the bridge, the signal is given to proceed.

Vehicles travel a considerable distance between the weigh-pad array and the bridge and are not detected again until exit from the bridge. Their positions are tracked by using their entry speeds (also output by the weigh-pad array) combined with car-following theory for closely spaced vehicles. In the event of the stop signal activating, the program analyses the slowing down and stopping behaviour, followed by acceleration as soon as the "go" signal is given. The induction loop at the exit of the bridge may be used to confirm the actual exit of the vehicle from the bridge.

The controlling program may also be run in simulation mode, either using traffic data recorded on a previous day, or by generating a synthetic traffic stream with volume and vehicle composition input by the user. For the latter it is assumed that vehicle arrivals at the weigh-pads in each lane conform to the Poisson distribution, thus generating a sequence of negative exponential headways. Vehicles are assigned a vehicle class by selecting them randomly from an appropriately proportioned spectrum of cars and goods vehicles. When running with a synthetic data stream, the program requires data for traffic flow, percentage of HGVs, bridge span, load spectrum and duration of simulated run.

The load control software runs in Visual Basic. It is able to display an animated image of the current state of traffic within the system as shown in Figure 6. The loads currently on the bridge in each lane are shown graphically and numerically on a load history screen. In the picture shown, a heavy petrol tanker has been stopped while two short wheel base HGVs are allowed to pass. The threshold has been set deliberately low for this illustration, although it can been seen in the load history that one of the vehicles exceeds 300 kN.

In the practical implementation of the system, permission to control live signals on the open highway has not yet been granted, so the system has only been able to demonstrate how traffic would behave provided that a stop signal is always obeyed. The controlling program outputs a voltage signal that can be used to trigger an external signal controller.

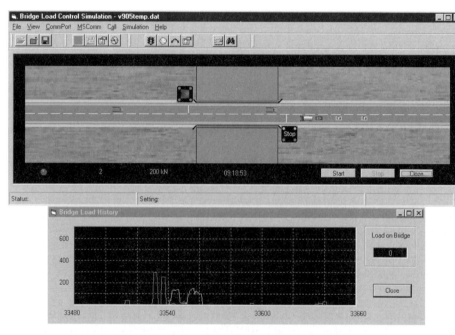

Figure 6. Animated traffic simulation window

The above description of the system refers to a very simple load control criterion, namely a total load threshold for the bridge. A more advanced procedure has been developed that makes use of influence lines for bending moment, shear or other stress resultant. An algorithm evaluates the positions of vehicles on the appropriate influence line and then calculates the expected bending moment or other stress resultant that will be applied to the bridge. This is added to the effects of load detected by the weigh-pad array in the other lane at the far side of the bridge. If the expected combined bending moment exceeds a pre-calculated threshold, then signals on the bridge are set to stop one of the approaching heavy vehicles.

The load control operates as follows: (i) an eastbound vehicle is predicted to create a bending moment, m_e ; (ii) a westbound vehicle is predicted to create a moment, m_w, such that $m_e + m_w$ exceeds the allowable moment; (iii) if these events occur within a time slot during which both vehicles could be on the bridge together, then the signals are activated to stop the first vehicle; (iv) when the second vehicle has been counted off the bridge, the first vehicle is released. It may seem more logical to stop the most recent vehicle that has caused the threshold to be exceeded. However, if the overload is caused by vehicles in only one lane, this can prevent that lane from ever being released. Therefore, traffic in the other lane is stopped to avoid combinations of heavy vehicles in both lanes being stopped simultaneously.

TYPE OF SIGNAL

An important consideration is the form of the traffic signal. Four possibilities have been considered as follows:

1. The normal 3 colour traffic signals, possibly with a simple explanatory notice, were considered to be rather similar to single lane operation, where two lanes are physically reduced to one. In the load control situation, however, the signals would remain green most of the time. This is not considered to be good practice for traffic signals since regular users of the bridge would infrequently or perhaps never see them turn red. It is important that drivers understand the purpose of the signals and it might be confusing to see an empty lane ahead on the rare occasions that they turn red.

2. An alternative is to adopt "wig-wag" signals as are used at level crossings or fire stations with flashing lights and a warning message. However, these are appropriate when there is a long warning period followed by a visibly significant event such as a train crossing. The lengthy warning period would require the weigh-pads to be located even further from the bridge. This is often difficult owing to the existence of side roads.

3. A third possibility is to use Variable Message Signs (VMS) with flashing lights, as illustrated in Fig 6. These would be more appropriate for occasional use. They would remain blank most of the time and would activate with four flashing lights and a message such as "Stop" or "Give Way to HGV", preceded by an early warning message such as "Weak Bridge - HGV Control - Delay Possible". This may be more understandable to drivers.

4. Finally, if it is possible to maintain complete control over traffic entering the bridge, as on a toll bridge, it would be possible to combine the signals with barrier arms to physically prevent vehicles from entering the span until safe to do so. This type of control could be contemplated for historic bridges (Mitchell-Baker and Cullimore, 1988).

PERFORMANCE OF THE SYSTEM

It is of interest to enquire how often overloading events will activate the control signals. In Figure 7 the frequency of simultaneous meetings of commercial vehicles on the bridge is shown. It is evident that the frequency increases with total traffic flow. Although it may be seen that two commercial vehicles may meet on the actual bridge about 25 times per day, a minority of these meetings are likely to be combinations that will exceed the allowable bridge load. The program was run in simulation mode using previously recorded day files for traffic data. The bridge load threshold was set to various percentages of HA load and the number of signal control events were observed for each day. The results are shown in Fig 8.

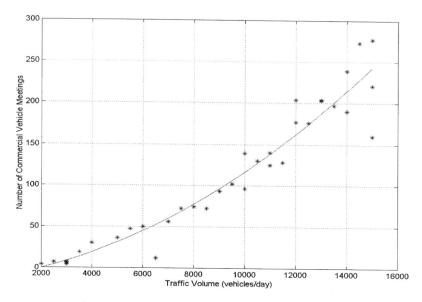

Figure 7. Number of HGV meetings versus traffic flow

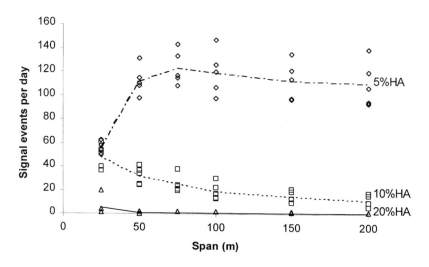

Fig 8 Frequency of signal control events for thresholds set as percentage of HA load

It may be seen in Fig 8 that signal control would be very rarely activated in normal traffic on a rural highway unless the bridge was significantly under strength. Even when the threshold is set at 20% of full HA load, the number of times per day that the signals would stop traffic could be counted on the fingers of one hand. This suggests that it would be possible to set the threshold deliberately low since the delays to traffic overall would be minimal. Figure 8 also shows the effect of bridge span on the frequency of signal activation. Apart from the very low case of 5%HA, it appears that shorter spans would experience the greatest frequency of high loads. This is because 20%HA load, for example, can be exceeded by a single vehicle, and certainly by two closely spaced vehicles. For longer spans, the total HA load is greater and it seems that the frequency of combinations of vehicles required to exceed it reduces with span.

An important consideration is the possibility of traffic signal violations. It must be emphasised that active bridge load control is primarily a tool for managing risk. If the signals on an under-strength bridge activate on average 3 times a day, it does not mean that without them the bridge would be overloaded and collapse. There are significant margins in the strength assessment calculations to ensure that there is low probability of a rare overload event coinciding with unusual lack of strength. The installation of a load control system would reduce the probability of failure. Therefore, violation of the control signals would result in the probability of failure increasing, not necessarily actual failure. It is very difficult to assess the probability of such violations. Some indication can be obtained from statistics of motoring offences. For example, in 1997 in the UK there were 480,000 recorded offences in the categories of "dangerous, careless or drunken driving" or "neglect of traffic signs and directions", amongst a population of 30 million drivers. This is less than 2% of offences per driver per year. Taking account of the brief time that heavy goods drivers spend crossing bridges, it will be appreciated that the risk is very small. A similar, or possibly greater, risk applies to bridges with posted load limits. The greatest risk applies to bridges for which the safe load can be exceeded by a single vehicle, usually short span bridges. There is an instance on record of an historic bridge in Ireland being destroyed by a heavy lorry that had violated the posted load limit.

It is difficult to be precise about cost comparisons of the various alternative management strategies. The costs would be significantly different for every bridge site. However, it is likely that a load control system would be a low cost option. The cost of installing a WIM weigh-pad array is about £25,000. A minimum of two would be required for a simple entrance geometry. Multiple lanes and more complex road layouts at a bridge approach would increase this number. But even with the addition of design and software costs, the total is likely to be much less than the costs of strengthening. Moreover, bridge repair and strengthening would entail delays to road users during civil engineering work. It has been pointed out that costs of road user delay, as calculated by the program QUADRO, can be as great or greater than the costs of the works themselves (Tilly, 1997). For example, even at a lightly trafficked site, such as in this project, an average delay of one minute per vehicle caused by single lane restriction on the bridge, would result in £2,000 per day delay costs. Road user delay costs do not fall on the authority responsible for the bridge, but can influence them indirectly as a result of hostile public opinion and political pressure.

CONCLUSIONS

1. Active control of traffic loads on bridges is a viable alternative to the posting of load limits or reducing the number of lanes for under-strength bridges.
2. Weigh-in-motion data may be used to monitor the total load on a bridge, or the magnitude of stress resultants such as bending moment or shear.
3. Variable message signals with flashing lights are likely to be the most readily understood form of control.
4. A simulation study has shown that overloading events are likely to be infrequent and the overall delay to traffic will be minimal.

5. Active load control is relatively inexpensive to install and is likely to be highly cost effective compared with other management options.

ACKNOWLEDGEMENTS

This project was funded by the DETR through the LINK Inland Surface Transport Research Programme.

REFERENCES

Atkinson, T, Brown, P, Darby, J, Ealey, T, Lane, J, Smith, J. W. and Zheng, Y (2000). Active Control of Bridge Loads. *Bridge Management 4*, eds. M J Ryall, G A R Parke and J E Harding, Thomas Telford, (475-482), April, 2000.

BS 5400 (1980). *Steel, Concrete and Composite Bridges*. Part 10: *Code of Practice for Fatigue*, British Standards Institution, London.

COST 323 (1995). *Post-proceedings of the First European Conference on Weigh-In-Motion of Road Vehicles*. ISBN 3-9521034-0-3.

Leadbeater, A (1997). Bridge assessment and strengthening - local authority perspective. *Safety of Bridges*, Parag C Das (ed), (12 - 19), Thomas Telford, London.

McCall, W and Vodrazka, W (1996). *States' Best Practices WIM Handbook*. Center for Transportation Research and Education, Iowa State University.

Mitchell-Baker D and Cullimore M S G (1988). Operation and Maintenance of the Clifton Suspension Bridge. *Proceedings of Institution of Civil Engineers*, Vol. 84, Part 1, pp291-308

Opitz, R (1993). BCS 200 Active Bridge Control System. *Bridge Management 2*, Thomas Telford, 989-997.

Tilly, G P (1997). Principles of whole life costing. *Safety of Bridges*, Parag C Das (ed), (138 - 144), Thomas Telford, London.

Williams, T K (1997). Bridge management problems and options. *Safety of Bridges*, Parag C Das (ed), (131 - 137), Thomas Telford, London.

Replacement of Steel and Composite Bridges under Traffic

REINER SAUL, SIEGFRIED HOPF
Leonhardt, Andrä und Partner GmbH
Stuttgart, Germany

Abstract
The increase of traffic requires more and more often to replace old bridges by new, wider bridges without interruption of the traffic. The low weight of steel and steel composite bridges permits to maintain old masonry piers. As steel is 100 % recyclable, the recycling costs are much smaller than those of concrete bridges. Different techniques for the replacement are presented.
Keywords: Replacement, steel bridges, steel composite bridges, lateral shifting, masonry piers, auxiliary piers.

1 Introduction

The replacement of steel and steel composite bridges is more often required due to an increase of the traffic volume and of the live loads than due to corrosion, fatigue and lack of maintenance. The - in comparison with concrete bridges - considerably smaller weight permits to erect and remove large elements in a short time and thereby limit the inconveniences to traffic as far as possible.
The following examples have in common that parts of the existing bridge are reused and/or that the axis of the old and the new bridge are coincident; they are arranged according to the construction procedure in the transverse direction
- lateral shifting of the complete cross-section
- lateral shifting of the superstructure of one roadway
- construction of the new superstructure for one roadway after partial dismantling of the existing bridge
- removal of the old bridge and construction of the new one in slices
- special procedures.

2 Lateral shifting of the complete cross-section

Example: „Oberkasseler Bridge" across river Rhine at Düsseldorf [1], Fig. 1

The main spans of the Oberkasseler Bridge at Düsseldorf were reconstructed after World War II as truss girders with 4 spans of 94,5 m each and a traffic width of 12,5 m. The insufficient traffic width and the danger caused by the small spans to the ship traffic required the construction of a new bridge in the early seventies.
The new bridge is a single plane cable-stayed bridge with a total length of 590,75 m, a main span of 257,75 m and a single tower on the left embankment of the river. The 35 m wide superstructure consists of a 20 m wide box girder with long cantilevers. The single cell steel tower has a height of 100 m above the bridge deck, and the 4 cables have a harp arrangement.

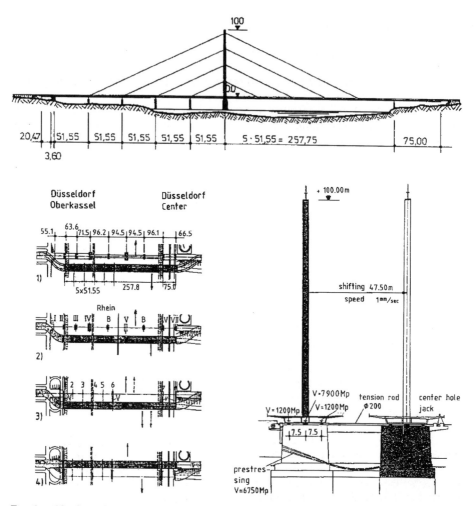

Fig. 1. „Oberkasseler Bridge" across river Rhine at Düsseldorf
a. New bridge view; b. Construction sequence; c. Cross-section at main pier

The new bridge was built 47,5 m upstream of the old one. After deviating the traffic to the new bridge, the old bridge - including the piers - was dismantled and new piers were built in the axis of the old bridge. After that, the provisional and final abutments, tower piers and embankment piers were linked by sliding tracks. Sliding tracks at the backstay cable anchorages (pendulums) were not required as they are without load under permanent loads. After setting free the pendulums, the bridge was pulled to its final position on 7 and 8 April 1976. A total weight of 12700 t had to be moved, 10.300 of them at the tower pier.

3 Lateral shifting of the superstructure for one roadway

Example: Highway bridge across the Werra Valley at Hedemünden [2], Fig. 2

The first highway bridge across the Werra Valley was built from 1935 to 1937 with spans of 80 - 96 - 96 - 80 - 64 = 416 m and a total width of 21,5 m. Each roadway had a truss girder with a system depth of 8 m. After its destruction during World War II the bridge was reconstructed in the early fifties with a slightly increased width of 22,3 m and with 6 m deep plate girders. The superstructure towards Hannover had an orthotropic plate and that towards Kassel a concrete slab.

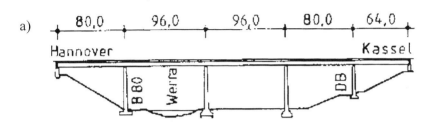

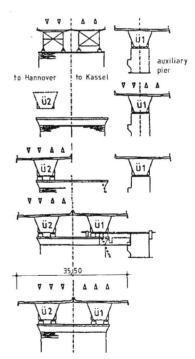

Fig. 2. Highway bridge across the Werra Valley at Hedemünden:
a. View; b. Construction procedure

The increased traffic required to widen the bridge to RQ 37,5 at the early eighties. Both roadways have now steel composite superstructures with trapezoidal box girders. The roadway slab is longitudinally reinforced for a crack width of 0,1 mm and transversely prestressed. The construction depth of 5,85 m corresponds to 1/16,4 of the biggest spans. As the existing piers were capable to carry the increased loads of the new bridge they had not to be replaced. In order to allow for the necessary distance of the bearings they were widened at their top like a hammer head.

The new superstructure for the traffic towards Kassel was built on auxiliary piers and abutments and later the total traffic diverted to this bridge. After the dismantling of the existing superstructures and the refurbishment of the piers and abutments, the second superstructure was built in its final axis. After diverting the total traffic to this bridge, the first superstructure was laterally shifted into its final axis.

4 Construction of the new superstructure for one roadway after partial dismantling of the existing bridge

Example: Highway bridge Wilkau-Haßlau [1], Fig. 3

The existing bridge, built in 1938/39 as part of the highway Chemnitz-Hof, crosses the valley of the „Zwickauer Mulde" with spans of 69,9 - 110,0 - 110,0 - 99,0 - 99,0 - 99,0 - 88,0 = 674,9 m. The bridge deck was a plate girder with 4 webs and had a width of 24,5 m and a construction depth of 6,2 m corresponding to a slenderness ratio of 1/17,7. As usual at the time of construction, the piers were constructed solid and received a cladding of masonry from natural stones. As the adjacent highway had not been completed yet, only the northern part of the bridge was under traffic.

a)

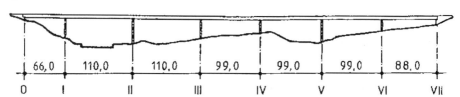

The new bridge has a width corresponding to RQ 30, but the central reserve a width of 3 m only instead of the usual 4 m. Both roadways have steel composite superstructures with trapezoidal box girders. The construction depth is 5,08 m corresponding to 1/21,6 of the biggest span, and the roadway is longitudinally reinforced and transversely prestressed. The old piers are maintained, but they have at their top a 2 m deep bearing block from reinforced concrete.

In a first step, the half bridge which was not under traffic was dismantled. In order to avoid strengthening of the existing structure, auxiliary piers were used. After that, one half of the abutments and the top of the piers were adapted to the new bridge and the first new superstructure was constructed by incremental launching. Due to the extremely reduced construction time and due to the auxiliary piers required for the dismantling of the existing bridge, the complete cross-section was launched, following a proposal of the contractor. After the dismantling of the second half of the existing bridge and the adaptation of the rest of the abutments and piers, the second new superstructure was built as the first one.

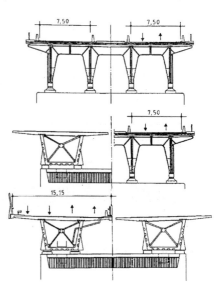

Fig. 3. Highway bridge Wilkau-Haßlau: a. View; b. Construction Procedure

5 Removal of the old bridge and construction of the new one in slices

Example: Heinrich-Erhardt-Steel Bridge at Düsseldorf [4], Fig. 4

The Heinrich-Erhardt-Steel Bridge carries the Federal Roads 1, 7 and 8 across the trunk railway line Düsseldorf-Duisburg and the shunting station Düsseldorf-Derendorf. The bridge was built in 1909 and had a truss girder with spans of 39,6 - 48,6 - 39,6 = 127,8 m, short side spans of 15,08 m and 11,56 m respectively and a 10,5 m wide roadway. The insufficient traffic width and load carrying capacity required the construction of a new bridge at the end of the seventies.

The new bridge is a single plane cable-stayed bridge with spans of 14,0 - 30,0 - 88,8 - 30,0 - 14,0 = 176,8 m. The total width of 39 m is built up by three 5,7 m box girders and by orthotropic plates in between and cantilevering. The steel towers raise 15,3 m above the bridge deck and the cables consist of 8 locked coiled ropes with a diameter of 78 mm.

Neither the railway traffic underneath the bridge nor the road traffic on the bridge could be restricted during the replacement. These requirements resulted in the following construction procedure

1. Dismantling of the cantilevers of the existing bridge.
2. Construction of the Southern third of the new bridge by launching from East to West. For this, auxiliary piers could be provided at the places of emergency piers of the old bridge, which correspond fairly well with the cable anchorage points.
3. Construction of the northern third of the new bridge as before.
4. Dismantling of the old bridge with a portal crane running on both parts of the new bridge.
5. Launching of the centre third of the new bridge, coupling of the three parts, construction of the towers, erection and stressing of the cables.
6. Final arrangement of traffic.

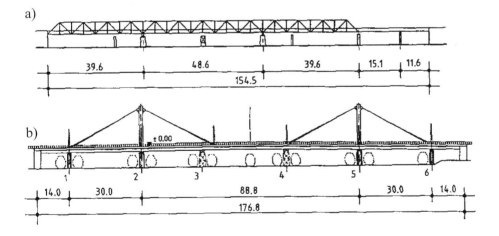

Fig. 4. Heinrich-Erhardt-Steel Bridge at Düsseldorf: a. Old bridge; b. New bridge, view

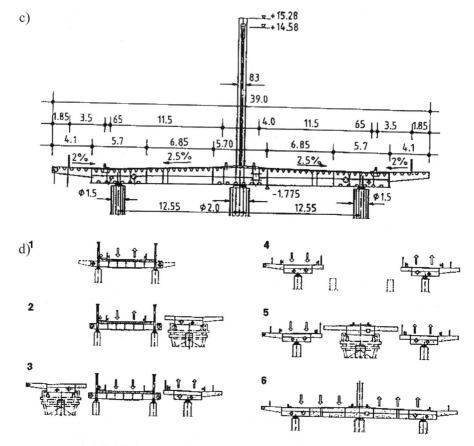

Fig. 4. Heinrich-Erhardt-Steel Bridge at Düsseldorf:
c. New bridge, cross-section; d. Construction

6 Special procedures

6.1 Schwarzbach Valley Bridge at Wuppertal [5], Fig. 5

The Schwarzbach Valley Bridge built in 1885 crosses two streets and a residential area with a maximum height of 23 m above the ground. The superstructures of the three main spans, of about 30 m each, had a fish belly shape and were from St 48. The track was ballasted. Considerable corrosion damages required a substantial refurbishment of the superstructures in 1980. Following an alternative proposal of a steel contractor, the superstructure were not refurbished, but replaced.

The new bridge has single span plate girders with a constant construction depth of 2,08 m = L/15 and has again a ballasted track.

In a first step, the ballasted track of the existing bridge was removed and the new superstructure rolled into its final position upside down. At their ends, both superstructures were attached to giro-wheels and later rotated by 180°. After that, the old superstructure was rolled out. Due to this sophisticated construction procedure the traffic was interrupted for 2 months only.

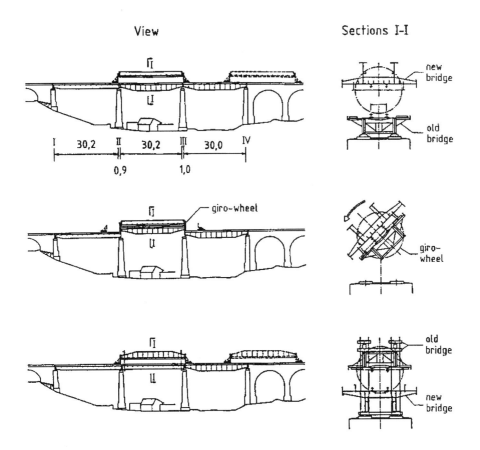

Fig. 5. Schwarzbach Valley Bridge at Wuppertal, Construction procedure

6.2 Neckar Viaduct at Marbach upon Neckar [6], Fig. 6

The single track railway bridge across the Neckar river, opened to traffic in 1879, had 5 single span truss girders of 67,0 m each with a construction depth of 7,0 m = L/9,6.

The age of the existing bridge required its replacement in the early seventies. The old piers with a masonry cladding of natural stones, nevertheless, had to be maintained. The new superstructure is a continuous beam with spans of 67,7 - 3 x 68,05 - 67,7 = 339,55 m. Its 3,2 m wide box girder has a construction depth of 4,8 m = L/14. The track is ballasted.

For its reconstruction, the bridge could be closed to traffic for 3 months only. In order to scope with this extremely tight schedule, the following construction procedure was chosen. The superstructure was transported in parts to the site by rail with half its depth and lengths corresponding to one third of the span and later assembled to full spans. After closing the old bridge to traffic, the five spans were transported on the old bridge to their final position and connected by welding. After that, the old bridge was hung to the new one and each span was cut into 3 pieces, lowered to the ground and scrapped. During this phase, the new bridge was supported by portal frames which allowed to lower it to its final position after the refurbishment of the top of the piers.

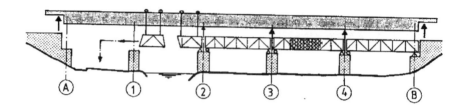

Fig. 6. Neckar Viaduct at Marbach upon Neckar, construction procedure

7 Summary

The replacement of bridges without interruption of the traffic - or with a very short one only - is one of the biggest challenges for designers and contractors. The rather reduced weight of steel and steel composite bridges is in many cases a decisive advantage as it permits a quick erection and removal in big sections. As steel is 100 % recyclable, the recycling costs are considerably smaller than those of concrete bridges.

The widening of some highways to RQ 37,5 during the last years has required the widening of big valley bridges. Also here the reduced weight of steel and steel composite bridges was a decisive factor as it permitted to maintain the old piers cladded with masonry from natural stones and in many cases protected as historic monument. As the concept for the replacement depends essentially upon the load bearing capacity of the piers, an independent checking of the piers prior to the tendering phase is strongly recommended.

8 References

[1] Beyer, E. et al.: Neubau und Querverschub der Rheinbrücke Düsseldorf-Oberkassel (Reconstruction and lateral shifting of the Oberkasseler Bridge across river Rhine at Düsseldorf). *Der Stahlbau* 46 (1977), pp. 65-80, 113-120, 148-154 and 176-188.
[2] Rabe, D.: Um- und Ausbau der Autobahnbrücke über das Werratal (Replacement of the highway bridge across the Werra Valley). *Baustatik - Baupraxis* 4, pp. 14.1 - 14.24.
[3] Autobahnamt Sachsen, Krebs und Kiefer GmbH: Ausschreibungsunterlagen für die Talbrücke Wilkau-Haßlau (Tender documents for the highway bridge Wilkau-Haßlau). Dresden 1993.
[4] Beyer, E. and v. Gottstein, F.: Ersatz von drei Straßenbrücken in Düsseldorf ohne Verkehrseinschränkung (Replacement of three road bridges at Düsseldorf without interruption of the traffic). *Bauingenieur* 59 (1984), pp. 27 - 37.
[5] Gerhards, K., Rademacher, C.-H. and Ramberger, G.: Die Erneuerung der Schwarzbachtalbrücke (Replacement of the Schwarzbach Valley Bridge). *Bauingenieur* 57 (1982), pp. 473 - 481.
[6] Bundesbahndirektion Stuttgart: Erneuerung des Neckar-Viadukts zwischen Marbach (Neckar) und Benningen (Replacement of the Neckar viaduct between Marbach upon Neckar and Benningen). Stuttgart 1979.

RECONSTRUCTION OF THE LIONS GATE BRIDGE

MICHAEL J. ABRAHAMS, P.E.
Senior Vice President
Parsons Brinckerhoff, Inc.
New York, New York, USA

JOSEPH K. TSE, P.E.
Assistant Vice President
Parsons Brinckerhoff, Inc.
New York, New York, USA

Introduction

The Lions Gate Bridge is a 1518 m long bridge with a 473 m long suspended main span over the Burrard Inlet in Vancouver, British Columbia, see Figures 1 and 2. It was opened to traffic in 1938 as a two-lane toll facility. The bridge was built at a cost of CAN $6 Million for the Guinness Family, of Guinness Stout fame, to connect the districts of North Vancouver and West Vancouver to downtown Vancouver. In 1952 the bridge was purchased by the Provincial government and converted to a three-lane facility to meet a growing traffic demand. Today, the Lions Gate Bridge serves as a vital link between Vancouver, the North Shore and communities beyond.

In 1993, the Province of British Columbia launched a public consultation process known as the Choices Program to address the future of the aging bridge. The program concluded that the existing corridor was "clearly superior" to other routes due to various reasons including cost effectiveness and compatibility with existing road networks. The options were therefore narrowed down to (i) building a new four-lane toll crossing in partnership with the private sector, or (ii) rehabilitating the existing three-lane crossing with the construction fully funded by the provincial government. The Province has selected the second option to rehabilitate the existing bridge.

The reconstruction of the Lions Gate Bridge, including its north approach viaduct, is a part of a CAN $100 Million rehabilitation of the Lions Gate Crossing. The major structural components are:

1. Replacement of the entire suspended span deck including hangers (This is to be done without interruption to bridge traffic during daytime hours; only nighttime closures lasting 10 hours are allowed for operations that interrupt traffic)
2. Widening of the sidewalk for the approach spans
3. Seismic retrofit of both the north approach viaduct and the bridge

In order to accomplish these modifications, the owner selected a rather unique combination of the traditional design-bid-build with newer design-build methods. A number of engineering challenges combined with the two methods of procurement make this project unique.

The seismic retrofit portion of this project was a design-build undertaking. The Contractor, American Bridge-Surespan, a joint venture, retained an engineering firm, Klohn-Crippen, to conduct a seismic analysis and retrofit design and, as required by the contract, an independent consultant firm, Parsons Brinckerhoff, to conduct an independent analysis and design check.

The sidewalk widening had been developed by the owner's consultant, N.D. Lea / Buckland & Taylor, and was detailed in the contract plans, so no contractor consultant was required during the construction phase.

The suspended span deck replacement had been designed by N.D. Lea / Buckland & Taylor and was included in the contract documents. However, other than a limited erection analysis, which considered only dead and live loads, it was left to the contractor and his consultants to develop the deck replacement methodology and erection equipment.

The deck replacement has presented many challenges and these are discussed below. As required by the contract documents, the erection analysis was conducted by one consultant, the Parsons Transportation Group Major Bridge Division (PTG-Steinman), with an independent analysis by a second consultant, Parsons Brinckerhoff (PB). Further, all erection equipment required an independent check, and in this case the erection equipment was designed by American Bridge-Surespan engineering staff with an independent design check by PB.

The Existing Structure

The existing bridge main span is 473 m in length, with supporting towers located in the foreshore zone on either side of the Burrard Inlet. The two side spans on either side of the main span are 187 in length. The total suspended deck is thus 847 m. The north approach viaduct is 671 m in length while the south end abutment is at the end of the south side span. The total length of the bridge from abutment to abutment is therefore 1518 m. The main shipping channel has 61 m horizontal clearance and, at mid-span, 61 m vertical clearance.

Each tower is 109.2 m high and comprised of two columns spaced 21.0 m apart at their base and 12.2 m at their top. The towers are constructed of steel plates and angles stiffened internally by a succession of diaphragms.

Each suspension cable is 330 mm in diameter with 61 strands of galvanized steel wire. Each cable is secured to the anchorage chambers by 21 anchors and supported by saddles, located on a cable bent and the towers. The cables, like the top of the tower columns, are 12.2 m apart. The hangers supporting the deck are typically spaced at 9.8 m on centers.

The two 4.6 m deep stiffening trusses of the bridge serve to distribute local live loads among the suspenders. Floorbeams connecting the trusses form a pony truss and carry the roadway and sidewalks. The roadway of the existing deck consists of three vehicle traffic lanes, each approximately 3 m wide, and two sidewalks, each 1.2 m wide. No barrier is provided between the sidewalks and the vehicle lanes, see Figure 3.

Each tower is supported on a concrete pedestal. Resistance to ship impact was improved in 1986 by the addition of a berm around the base of the north tower and in 1988 with placement of a concrete collar around the base of the south tower. Due to the age of both the Lions Gate Bridge and the north approach viaduct, neither meets the current seismic standards of the British Columbia Ministry of Transportation and Highways.

Seismic Retrofit

The seismic retrofit of the north approach viaduct is believed to be the first of a major structure to be carried out in a design-built environment. The structure is being retrofitted to a level of performance such that it can remain open to traffic following the 475-year design level event. A detailed engineering design and detailed independent design check was part of the construction for the seismic retrofit of the north approach viaduct. The north approach viaduct was evaluated for serviceability under the design level earthquake using performance based analysis and design procedures. As a part of the independent check, ADINA models were developed both for the global seismic demand assessment and for capacity evaluations using pushover models. The seismic analysis included evaluation of the effect of the interaction of the suspension bridge with the viaduct. The retrofit scheme involved strengthening steel bracing members, the existing foundations, and implementing remediation measures against potential ground settlement under liquefied soil conditions.

The retrofit of the viaduct structure primarily involved the strengthening of steel bracing members and the foundation at the north cable bent. Strengthening of the foundation was achieved by introducing tie-downs anchored in a new concrete collar placed around the existing footing.

Deck Replacement

The New Deck

The new deck carries three 3.6 m traffic lanes, and two 2.7 m sidewalks. The deck is of welded orthotropic steel plate construction with stiffening trusses utilizing welded rectangular hollow sections. The new stiffening trusses are located below the deck, see Figure 4, and are 2.08 m deep. The number and location of the hangers that support the new deck remain the same as the original deck. The new deck is 4.5% lighter than the existing deck in the main span, and 3.1% heavier on the side spans; a significant design consideration for the erection engineering.

Existing Deck Conditions and Challenges

The condition of the bridge prior to construction is poor for several reasons. First, traffic loads are substantial greater than originally intended. Then, load changes and geometric changes since construction completed in 1938 has resulted in a sag at mid span of over one meter. Corrosion of the floor system has further reduced the capacity of many of the stringers and floorbeams.

The challenge in this project was the very limited additional capacity of the bridge to accommodate erection loads and deformations. The original design of the bridge economized on every element to the maximum extent possible, and at the time of its design, was only for 2 lanes of traffic with no sidewalk. And at that time, there were no seismic forces, nor were analytical methods very accurate. Over time, the effects of section loss, cable and suspender stretch and cable slip have further reduced the bridge's capacities.

The following summarizes several aspects of the bridge that required special attention while performing our analyses –

- The preconstruction (present day) geometry of the bridge - the contract documents listed changes in the bridge geometry between the original design and field survey data collected through 1998. The Contractor was required to conduct an independent survey to verify the listed changes and provide updates as necessary. With all information gathered, interpreted and resolved, our computer models were adjusted to account for these primary geometric corrections, namely, the suspension cables had sagged around 1120 mm at mid-span, and the north and south towers had deflected 160 mm and 90 mm, respectively, towards mid-span.

- The difference in the vertical alignment of the existing deck and the new deck – the existing main span deck has an abrupt 8% grade change at the center (believed to be associated with the need to reduce construction costs). This break in grade will be taken out with the installation of the new deck. The computer models were developed to account for this difference by the systematic tracking of hanger fabrication lengths and intermediate hanger adjustments for proper alignment across the continuity link.

- Limited excess capacity in the existing structure – the existing structure has little reserve capacity for construction loading and geometric effects. As a result, fine-tuning of the hangers, especially around the work front, is necessary in each stage to keep hanger loads within allowable limits. Likewise, the existing truss diagonal members required reinforcement to avoid being overstressed. A hydraulic "spring" is being employed at the main span continuity link to prevent overstressing the lower chords in case of high winds when replacing the main span sections. This required the models to include multi-linear springs to simulate the hydraulic system. The flexibility introduced by the continuity link in turn required measures to mitigate overstressing of the chord members due to significant inclinations during construction in the existing and new hangers.

Need to Maintain Traffic

One of the requirements of this project is that interruption to the public's use of the bridge be kept to a minimum. The contract stipulates that deck replacement, with a few exceptions, can only be performed during ten hours of nighttime closures. The special operations for which weekend closures may be scheduled include primarily –

- Installing the erection equipment and replacing the first deck section at the north cable bent
- Replacing the deck sections adjacent to the north tower
- Replacing the deck sections adjacent to the south tower
- Replacing the deck section at the south abutment and dismantling the erection equipment

Contractor's Replacement Scheme

After evaluating alternatives, a scheme that is similar to one suggested on the contract documents was selected. The scheme involves replacing the existing deck in 19.66 m long sections for the north side span and the main span. The 19.66 m sections are raised and lower to, and from, the ground or barges in the water. At the south side span, the steep slope and environmental sensitive nature of the park grounds below required the deck to be removed from above; there 9.83 m long sections are to be removed / delivered through the south end of the bridge.

Special Equipment & Hardware

The basic scheme is illustrated in Figure 5. The more notable equipment that the Contractor custom designed and fabricated in conjunction with his replacement scheme are as follows –

Jacking Traveler –
Essentially a steel lifting frame, the jacking traveler advances with the work front by rolling along the deck on Hillman Rollers. In its operating mode, the frame is attached to the hangers with the supporting legs retracted. Strand jacks that are mounted on top of the frame are engaged to support the weight of either the existing deck or the new deck during the various stages / steps of the replacement operation. There are two versions of the traveler – one for the north side span and main span, and a shorter version to be used at the south side span, see Figure 6.

Continuity Link –
When the bridge is reopened to traffic at the end of each replacement stage, a continuity link is needed at the erection front to join the existing and new deck sections together. The continuity link has two steel members; each attached to one of the two bottom chords of the existing truss. Each steel member houses an extendable / retractable "key" that extends to engage its counterpart (called the "link nose") attached to the new deck. The key maintains vertical alignment across the work front providing shear continuity and thus ensuring ridability between the new and the existing deck sections, see Figure 7.

By design, the continuity link provides a variable degree of flexural continuity in the lateral direction. This is necessary when replacing deck sections at the main span, where the design wind would have otherwise cause significant overstress in the existing truss lower chords.

Temporary Deck Section –
The Contractor is required to design, furnish, and store on site, a temporary deck section that can be deployed rapidly to fill a gap in the deck in the event of unforeseen circumstances such as an failure of the erection equipment or misfabrication of the new deck section. The temporary deck was designed to support 3 lanes of reduced traffic load and full design wind load.

Hanger Extensions –
The existing hangers are to be replaced by new and longer hangers attached at the roadway level of the new deck. This replacement operation is scheduled to trail behind the deck replacement, thus removing it from the critical path. Before the new hangers

can be installed, temporary hanger extensions are required to provide a link between the connections at the new deck and the shorter existing hangers. The hanger extensions have been detailed with adjustment capabilities so that adjusting the hangers disperses localized overstresses in the hanger and trusses members near the work front.

Hinge Details at Hanger Connections –
The bridge under construction is much more flexible than either the existing structure, before construction began, or the new structure, when completed. This is due to the lack of continuity across the erection front. As a result of detailed analyses, a significant number of connections were found to require special details to accommodate the rotations of the suspenders associated with the increased flexibility.

Wind Restrictions on Replacement Operations

The bridge is vulnerable to high winds when the deck section is discontinuous. As a result of extensive wind studies, the Contractor has installed fairings at the main span to improve aerodynamic stability of the structure during reconstruction. It has also been established that the bridge can withstand a wind speed of 12 m/s in its open condition, i.e. with a deck section missing. Therefore, night closures for deck replacement are ordered only when the forecast wind gust speed is less than 10 m/s.

Typical Replacement Stage (North Side Span and Main Span)

The following are Stage / Steps that best characterize the major operations within a typical replacement stage -

Step One - Initial Condition. The jacking traveler has just been rolled into position for replacing a given deck section. The continuity link engages the new deck to the existing deck. Traffic is running.

Step Two. The jacking traveler is attached to the hangers and its weight is transferred to the hangers as its supporting legs are retracted.

Step Three. Traffic is stopped. The continuity link is disengaged and rolled southward, clearing the target deck section for replacement.

Step Four. The strand jacks take up most of the weight of the target deck section. The existing truss hanger socket plates are removed. The existing truss top chord, bottom chord and diagonal are cut.

Step Five. The section of existing truss is lowered to the ground or barge. The continuity link nose is removed from the bridge and attached to the next new deck section.

Step Seven. The next new deck section is hoisted into place and incorporated. Temporary hanger extensions are installed and the strand jacks released.

Step Nine. The continuity link is moved northward as required to engage the link nose. The continuity link traffic plates are installed. The new deck splice traffic ramp is moved ahead.

The hangers are adjusted to predetermined forces / extensions that will accommodate all the loading conditions (DL / LL / Temperature etc.) expected until the next stage.

Step Eleven. The bridge is reopened to traffic. The jacking traveler rail is moved ahead and the jacking traveler legs are extended to engage Hillman Rollers. The jacking traveler is winched forward on the Hillman Rollers. Various pieces of construction equipment are relocated ahead for the next stage.

South Side Span

As mentioned above, the south side span sections will be replaced in 9.83 m segments, which is half the length of those for the north side and main spans. Removal of the existing truss involves cutting the section free, lifting and rotating it 90-degrees, and lowering the section onto the back of a specially design transporter that will be waiting on the deck at the work front. To accommodate the heavier construction loading for these operations, both existing and new hangers will be adjusted at each stage to more favorably re-distribute the loads near the work front. (On the north and main spans only the new hangers need to be adjusted.)

Analyses of the Deck Replacement Procedures

Accurate computer models are essential to this project because every stage / step needs to be accounted for to ensure success in removing and replacing a deck section within the 10 hours of nighttime closures. The analytical process sorts out the required procedures and hanger adjustments while the analytical results are the basis for generating field control data such as those required for setting up the jacking traveler, relieving loads at the cut sections, and aligning the new deck at installation.

PTG-Steinman chose the software (LARSA98, Verion 5) that can solve nonlinear, large deformation problems that include cable elements. Their computer model takes into account the geometric change due to various sources such as the stretching of the main cable. Then a combination of 2-D and 3-D nonlinear analyses were used to generate all the required loading cases. 2-D analyses were used to compute the effects from dead load, construction live load and traffic live load. 3-D analyses have been used primarily to calculate the effects of wind. Natural frequencies and mode shapes from various replacement stages were generated and combined with results from wind tunnel testing, performed by West Wind Laboratories, the wind specialist. Equivalent static wind loads were then developed to simulate dynamic wind for each replacement stage.

For the independent analysis, Parsons Brinckerhoff chose the computer program GTSTRUDL with a series of pre and post-processors to analyze critical stage/steps of the erection scheme. On average, as with PTG-Steinman's analysis, 15 erection steps were analyzed for the replacement of each deck section. These erection steps are required to capture critical stresses in the existing and new structure, demands on equipment, or to provide essential construction data such as strand jack loads and geometric controls. A projected total of well over 3000 load cases are expected to complete the analyses according to Service Limit State and four different Ultimate Limit States as required by contract. And as each analysis is completed, PTG-Steinman and Parsons Brinckerhoff compare their results to develop concurrence.

Conclusion –

The Lions Gate Bridge is believed to be the first major suspension bridge that will be seismically retrofitted under a design / build format. It is also believed to be the first major suspension bridge that has its deck completely replaced while maintaining traffic. In the case of the deck replacement, wind issues combined with the need to maintain traffic is perhaps more complex than the construction of a brand new suspension bridge. Compounding the wind issue is the fact that the existing bridge lacks reserve capacities for construction loading and deformations. Moreover, solutions to these engineering challenges are being provided under significant time constraints. The construction engineers solved the engineering challenges by using sophisticated computer software, old fashion hard work, and engineering judgment. At the time of this paper, replacement of the north side span is complete. The bridge is behaving close to the construction engineering teams' prediction.

Acknowledgement

Owner
The Government of British Columbia

Contractor
American Bridge / Surespan, Joint Venture

Rehabilitation Design
N. D. Lea / Buckland & Taylor

Seismic Retrofit Design / Independent Checking
Klohn-Crippen / Parsons Brinckerhoff

Erection Analysis / Independent Check
PTG-Steinman / Parsons Brinckerhoff

Equipment Design / Independent Checking
American Bridge/Surespan / Parsons Brinckerhoff

Wind Consultants
West Wind Laboratories / Rowan Williams Davies & Irwin

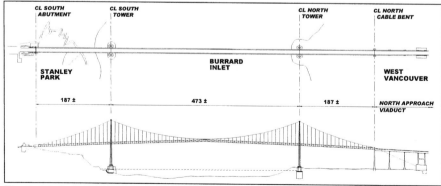

Fig. 1: Main Bridge - Plan & Elevation

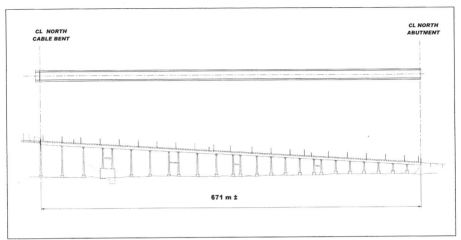

Fig. 2: North Approach Viaduct - Plan & Elevation

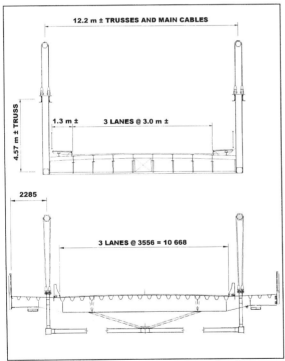

Fig. 3 & 4:   Existing and Final Cross Sections

224 MAINTENANCE PRACTICE

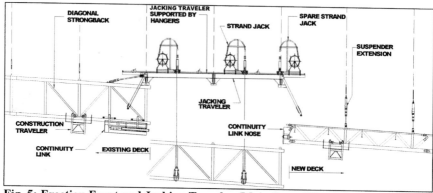

Fig. 5: Erection Front and Jacking Traveler (Main Span / North Side Span)

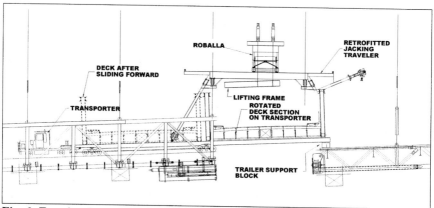

Fig. 6: Erection Front and Jacking Traveler (South Side Span)

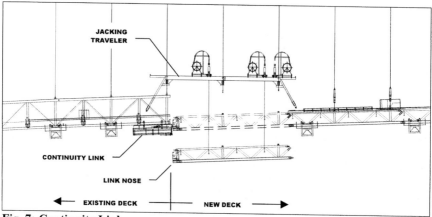

Fig. 7: Continuity Link

KINGSTON BRIDGE PHASE 1 STRENGTHENING

MATTHEW COLLINGS BSc CEng MICE MIStructE
Technical Director
Gifford and Partners, Southampton, England.

IAN TELFORD BSc CEng MICE
Project Manger
Glasgow City Council, Glasgow, Scotland.

Presented with the permission of Neil Mackenzie, Director, Road Network Management & Maintenance Division, Development Department, Scottish Executive and Alastair Young, Director of Land Services, Glasgow City Council.

ABSTRACT
Kingston Bridge carries the M8 motorway across the river Clyde to the west of Glasgow City centre. It is a strategically important bridge providing the primary crossing of the Clyde within both the regional and local traffic networks. This paper describes the development and implementation of the £32m strengthening contract undertaken by Balfour Beatty as main contractor, Glasgow City Council as employer and agent for the Scottish Executive (previously the Scottish Office) between 1996 and 2001 to rectify significant deficiencies. The paper includes a description of the defects, the bridge assessment findings and the strengthening and repair techniques employed in the works which included a complete jacking operation to lift and temporarily support the 52,000 tonne superstructure. The management of the bridge to maintain its operational condition throughout the project is also addressed.

1. INTRODUCTION

Kingston Bridge was constructed of in situ prestressed concrete and built in balanced cantilever. It was opened by the Queen Mother in 1971 and with a main span of 143m remains amongst the largest examples of this form of construction (Refer to Photograph in Figure A). The bridge was designed to carry five lanes of traffic in either direction. It now carries in excess of 170,000 vehicles per day and forms a strategically important crossing of the Clyde within the regional and local traffic networks. There is no suitable alternative for the majority of the traffic using this route and therefore any disruption would have severe economic and environmental impact on the region.

Figure A   Kingston Bridge Prior to Phase 1 Strengthening

The bridge comprises a linked pair of triple cell, variable depth, post-tensioned concrete superstructures having a 143m main span over the river and two side spans of 62 metres carrying the motorway some 18 metres above the street level. (Refer to Figure B). The relatively short side spans contain ballast concrete placed within the deck cells at the outer ends to generate positive reactions at the end supports under permanent loads. The original post-tensioning consists of single and four bar Macalloy tendons longitudinally in the top and bottom slabs of the main and side spans and single bar vertical tendons in the webs.

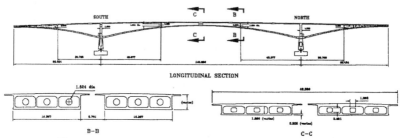

Figure B   Kingston Bridge Structural Arrangement

In its original articulation the bridge acted as a 'three pin' portal with knuckle pin bearings at the base of both north and south piers and rocker bearings at the top of the north pier. The end supports have cylindrical bearings with sliding surfaces together with sliding 'hold-down' bearings. The bridge connects at the end of both north and south side spans with significant lengths of elevated approach viaducts so that the articulation of the main bridge must be considered along with that of the approach viaducts.

2.   DEFECTS, DEVELOPMENT OF PROJECT TEAM AND INITIAL WORKS

In the late 1980's inspections suggested that the bridge was not behaving as the designers had intended;

- The Quay Wall in front of the northwest footing was bulging and spalling concrete was recorded at the rotation joint at the base of the north piers.

- The north pier was leaning permanently to the north and was 165mm out of plumb.

- The north pier upper rocker bearing had slipped from its installed position.

- The expansion joint at the south end of the bridge, where we would expect to see least movement, was excessively open. The teeth of the expansion joint were designed to have a minimum overlap of 75mm but in midwinter the tips of the teeth had separated by 25mm.

- The profile of the deck had sagged by 300mm at mid-span.

Strathclyde Regional Council (SRC) were the Roads Authority for the M8 at that time and set up an 'in-house' Project Team to carry out an Assessment of the structure. It was recognised that the apparent instability of the Quay Wall posed a threat to the bridge and following consideration of various options the ground between the northwest foundation and Quay Walls was stabilised by jet grouting and rock bunds were formed in front of all the Quay

Walls. These complex works were carried out in 1991/1992 and have been reported elsewhere.

Assessment of the structure began in 1991 with the 'in-house' team working in parallel with a team from consultants Scott Wilson Kirkpatrick (SWK) who also undertook a Principal Inspection on behalf of SRC.

Initial Assessment identified inadequacy of the prestress in the deck and it was recognised that a system of external post-tensioning would be required. As a result, Balvac Whitley Moran (BWM), as UK agents for VSL (an international manufacturer of prestressing systems) at that time, in association with Gifford and Partners (who had recently designed the externally pre-stressed Camel Viaduct in Cornwall) were appointed to assist the in-house team.

The enlarged team completed assessment of the structure and began to develop a refurbishment strategy to address the following defects;

- Insufficient 'continuity' pre-stress in the main span with small tensile stresses under permanent load alone and hypothetical tensions of up to $10N/mm^2$ under full live loading. More seriously, the ULS flexural capacity of the deck at the shallow mid span region was found to be (theoretically) capable of carrying only a small proportion of 40 tonne Assessment Live Loading (ALL) when appropriate partial load factors were included.

- Movement of the north pier upper bearings in excess of capacity as a result of the combined effects of creep/shrinkage/temperature, etc.

- The end support bearings had insufficient movement capacity and resistance to live load uplift. Furthermore, the north expansion joints were locked in a closed position for much of the year inhibiting free articulation between Kingston Bridge and its north approaches and giving rise to significant longitudinal 'push forces'

- Pier shafts understrength.

- Pile caps understrength.

- South pier foundations understrength for longitudinal loads arising from closure of north hall joint.

- Deck understrength in shear. (This 'defect' was subsequently overcome by agreement of a 'Departure from Standard').

- It was clear that the scale of the works involved to remedy such a catalogue of defects would take some time to develop and implement, particularly in view of the overriding need to carry out the works with minimum disruption to traffic. Thus, when the inadequacy of the mid span region was identified, various short term measures were implemented and these have been reported elsewhere.

3. SCOPE OF PHASE 1 STRENGTHENING WORKS

The scope of the Phase 1 Strengthening Works are summarised as follows;

- Vertical and horizontal jacking operation to lift the 52,000 tonne superstructure off its original piers, slide it south and temporarily support it during reconstruction of the substructure.

- Strengthening of the superstructure by installation of additional external prestress within the decks.

- Reconstruction of the main piers, including more uniform distribution of load onto the existing pile caps to overcome their original inadequacy, and construction of additional piles at the south piers.

- Rearticulation of the bridge on new bearings at the main piers.

4. METHODOLOGY

It became necessary to develop a complex and integrated methodology for the required strengthening works that took cognisance of the interrelated nature of the defects and the various components of the remedial works. The adopted Methodology is summarised in Figure C. It was impossible to consider the various aspects separately and a single unified process that embraced all the objectives and constraints was developed.

Two guiding principles were identified early on:

- **Progressive Strengthening** – The structure was to be progressively strengthened. Intermediate construction stages should not reduce the overall safety of any element or the bridge as a whole.

- **Security at Every Stage** – The security of the structure at every stage should not be dependent on an assumption regarding subsequent stages, ie the security of the structure should not be jeopardised if a subsequent stage was delayed. This principle is of particular significance in view of the interaction with the north approaches which produces longitudinal loads which are temperature and seasonally dependent. This precluded certain options and required contingency plans at some stages, as will be discussed later.

In addition the following principal technical criteria were identified:

- The bridge and roads under should remain open to traffic for the duration of the works with the exception of only limited weekend and nighttime closures.

- Externally imposed movement of the decks were not to be permitted until the north pier upper bearings were 'secured'. This originally precluded stressing of the deck post-tensioning system until the decks had been lifted. However, on the basis of observation during the works, and the desire to accelerate the programme, this constraint was in the end relaxed.

- Riverward load on the north pier foundations should be avoided in the permanent works in view of the history of riverward movement at this location.

- The east and west decks had to be lifted simultaneously due to the substantial transverse connections between them.
- Absolute lift and the differential lift between the north and south piers was severely restricted by the limited 'hold down' capacity at the half-joints and the inadequate mid span bending capacity.

The longitudinal 'push force' arising from interaction of Kingston Bridge with its north approaches had a significant influence on the development of the Methodology and imposed a number of constraints.

The magnitude of the 'push forces' to be accommodated was predicted by correlating computer modelling with loads observed during the summer of 1995. If 'additional gap' could be created in the half-joint by, for example, global translation of the decks, shortening due to prestress or other measures at the joint, then the predicted 'push force' after rearticulation would be reduced. However some 150mm of 'additional gap' would have been necessary to eliminate the force.

On lifting the decks the 'push force' had to be resisted by the longitudinal jacking systems that react against the supplementary piers and it was impractical to design these to accommodate the maximum predicted loads. A 'seasonal restriction' on the initial lifting operation was therefore adopted in order that a lower design temperature and corresponding lower 'push force' could be defined for this critical period.

In the permanent works some 'gap creation' is guaranteed due to global translation of the decks and shortening due to the additional prestress. However the scope for global translation of the deck was limited by the need to maintain the half-joint bearings at either end of the bridge within their travel and at the design stage a conservative lower bound figure for gap creation had to be adopted. A significant 'residual push force' was therefore accommodated in the permanent design and it was for this reason that additional piles were provided at the south piers.

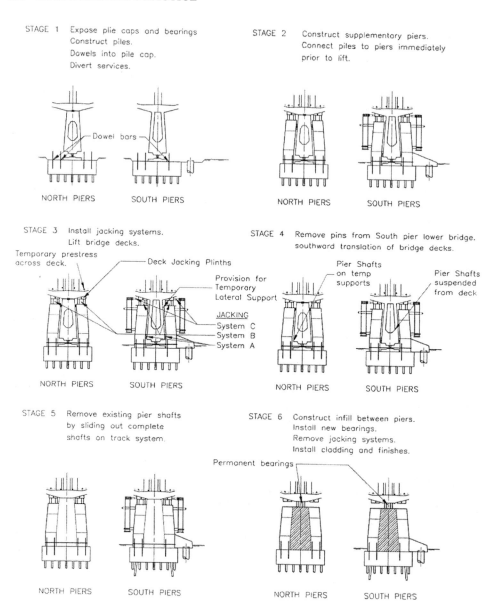

Figure C Methodology for Phase 1 Strengthening Works

5. CONTRACT AWARD

The contract for the Phase 1 Strengthening Works was awarded in 1996 to Balfour Beatty in association with VSL as specialist jacking subcontractor and Balvac Whitley Moran as specialist prestressing subcontractor by SRC. On 1April 1996 following Local Government re-organisation the M8 through Glasgow was trunked and responsibility for the Kingston Bridge transferred to the Scottish Office (now the Scottish Executive).

6. JACKING OPERATIONS

A key feature of the project was a sophisticated, state of the art, computerised jacking control system (the system) required to lift, support and move the 52,000 tonne deck to facilitate the works. Responsibility for the design of the system, in accordance with the employer's requirements, rested with VSL. The main vertical jacks were located in linked groups either side of the existing pier centrelines and in order to maintain free articulation, the jacking system could not be 'locked off' and the decks were supported on 'live' hydraulics for a period of ten months while the bridge remained opened to traffic. Tolerances of 0.1mm had to be achieved in order to ensure the integrity of the bridge during the jacking operations and complex failsafe provisions were also required.

The jacks that were used to lift the bridge and unload the existing substructure comprised four separate jack types A, B, C and D, each of which was operated and monitored by the system.

The B jacks were used to lift the bridge and support it on hydraulics. It comprised 128 No. 1000 tonne jacks which were supported on new piers constructed either side of the original piers on the existing foundations and reacted against downstand plinths connected to the deck. The B jacks (Refer to Figure D) consisted of 32 active jacks at the north and south piers respectively which were capable of lifting the bridge. Because the bridge would be supported on hydraulics for such a long time, 32 reserve jacks at each pier provided back up in case the active jacks failed. Both the primary and reserve jacks were controlled by the system and the reserve jacks were programmed to 'follow' the primary ones during lifting, 'maintenance' and lowering routines with a nominal 90/10 per cent load split between the active & reserve jacks. This required careful consideration of the control mechanisms and, in particular, sophisticated 'triggers' within the system which ensured that the reserve jacks followed the primary during normal operation but in the event of a failure would 'catch' the deck. Each jack was equipped with a locking collar as an ultimate rigid support in case of failure of both the active and reserve systems. All B jacks were capped by sliding and rotating devices to permit horizontal and rotational movement.

232 MAINTENANCE PRACTICE

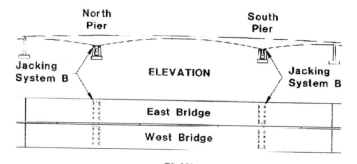

Figure D   **Location of Jacking System B**

The final modifications to the employer's requirements for the system were developed in liaison with the specialist jacking subcontractor VSL. It had been the original employer's requirements intent that the four lines of B jacks under each diaphragm should be independently controlled (as illustrated in Figure Ei). This was because the diaphragm was extremely lightly reinforced and it could not be established with certainty that it had the necessary strength to distribute uniform support reactions provided by a linked jacking system. VSL however wished to link the four groups into two pairs (as illustrated in Figure Eii). This was because the diaphragm is extremely stiff and control of four separate lines of jacks might present problems (ie very small movements on an individual line would result in significant change in load). However, such a support provides no restraint if the diaphragm started to fail in bending. Subsequent to additional structural assessment to increase confidence in both the likely load distribution across the width of the deck and the capacity of the diaphragm, a compromise was reached. This permitted VSL to link the lines of jacks as they preferred but incorporated additional safeguards in the system which ensured that the separate lines would be instantly isolated, at any stage while the decks were hydraulically supported, if the diaphragm began to deflect beyond a very small tolerance (of 0.5mm distortion of the diaphragm at its mid point). Hydraulic multipliers were also incorporated into the system to provide the option to proceed with the jacking operation with a fixed ratio of pressure between each pair of hydraulically linked lines, should this prove necessary.

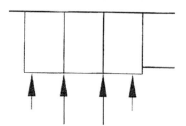 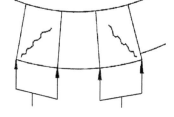

Figure Ei   **Originally Intended Support Condition**   **Figure Eii**   **VSL's Preferred Support Arrangement**

The cost of including these contingency features within the system was not insignificant. However their inclusion was justified on the basis that lifting of the bridge decks off the existing piers was considered an irreversible operation and any risk that problems could arise which might result in damage to the bridge and delay the reopening of the motorway had to be avoided. Closure of Kingston Bridge for any significant period would have had very serious economic and political implication for the region.

The A, C and D systems were horizontal systems that facilitated transfer of the longitudinal and transverse loading and provided fixity of the bridge decks when they were supported by the B jacks. The C jacks were also used to move the bridge deck to the south to relieve the load on the north approaches and to partially redress the excess travel on the half-joint bearings.

Prior to bringing the system to site a very sophisticated sequence of factory acceptance tests were carried out to demonstrate that the system would comply (under all circumstances) with the specification. The individual jacks were all tested prior to installation and the assembled system was also load tested in situ prior to the lifting operation.

During a 17 hours total closure of the motorway, which took place in October 1999 the 52,000 tonne, bridge deck was lifted 20mm. One week later during another overnight closure the deck was pushed south 30mm.

The system was required to provide extremely tight controls and checks on both position and pressure. After the bridge had been lifted and moved the system was then placed in 'maintenance' mode. In this mode, the bridge position and pressure status of all the hydraulics was continuously monitored by the system while demolition and reconstruction of the new piers took place.

7. DECK STRENGTHENING WORKS

The deck strengthening works comprised the installation of approximately 85MN of unbonded, external post-tensioning in each of the two bridges. The tendons are anchored in the side spans adjacent to the internal diaphragm beyond which the deck cells are filled with 'ballast' concrete.

In order to avoid excessive local strengthening of the deck it was necessary to distribute anchorage of the additional prestress around the section. The solution adopted, following extensive analysis, was to provide 65 % of the required prestress in the form of six 43 strand tendons, per bridge, anchored via in situ concrete blocks secured to the bottom slab. The remaining additional prestressing is provided by 20 smaller 11 strand tendons per bridge, anchored in balanced pairs on each face of the two internal webs of each deck.

Detailed discussion of the deck strengthening works is beyond the scope of this paper and will be published separately in due course.

8. DEMOLITION AND RECONSTRUCTION OF THE PIERS

New 'supplementary piers' were constructed on the existing pile caps either side of the original piers. These were connected into the pile caps by 40mm Ø reinforcing bars which were grouted into cored and reamed holes (percussive drilling of the pile caps was precluded due to its delicate nature). The 'supplementary piers' were inclined inwards to follow the profile of the existing piers and ensure that the vertical jacks B could be located as close as possible either side of the deck diaphragm. In order to provide the required robustness against longitudinal loads the separate sides of the supplementary piers where joined together by shear walls on the bridge centreline and by portal frames at their outer ends.

The 'supplementary piers' were in fact a frame structure comprising columns under each of the jack groups and transverse spreader and capping beams. Such an arrangement minimised additional weight on the foundation and facilitated access to the original pier shafts. The structure was, however, extremely heavily reinforced to accommodate the onerous loading arising from its complex behaviour under vertical and longitudinal loads.

Upon lifting the decks the four original pier shafts each weighing up to approximately 800 tonnes were pulled out from between the supplementary piers in their entirety and demolished remote from the bridge.

The void between the supplementary piers was then infilled with reinforced concrete to form a new pier structure that is almost the full width of the pile cap and thereby resolved the substantial inadequacy in shear and bending that previously existed.

9. REPLACEMENT BEARINGS

The replacement bearings are unusual in a number of respects. At the north pier there is a special requirement, as noted earlier, that no riverward load should be applied by the decks to the foundations. Longitudinal fixity is provided at the south piers, and thus at the north piers, longitudinal freedom is required. However a conventional sliding bearing would produce friction in either direction which was not acceptable. A number of options were considered including inclined sliding bearings and visco elastic type springs but the solution adopted is an elastomeric bearing with provision to preshear.

The elastomeric components are large (vertical capacity 40MN (SLS), strain 180mm (SLS) approximate dimensions 1250 x 1250 x 450mm) but confidence was gained from the use of similar sized bearings in the Second Severn Crossing. Provision for preshearing is made by incorporation of sliding plates, below the elastomeric component, which can be jacked relative to one another and locked in position. The initial preshear is applied after the bearing has been installed and loaded vertically. It can be adjusted relatively easily at any point in the future.

At the south pier, facilities for adjustment of the longitudinal fixity are provide in order that the longitudinal position at the bridge decks can be 'fine tuned' in the future if necessary. For this reason the functions of fixity and vertical load capacity have been separated into different bearings since to combine them was considered too complex. Due to the 'residual push force' from the north approaches a longitudinal capacity of 16 MN/bearing (at ULS) is required.

The design of the bearings in accordance with the employers requirements was the responsibility of VSL.

10. AESTHETIC CONSIDERATIONS

Kingston Bridge is a significant landmark and the aesthetics of the proposed strengthening works were given careful consideration. The 'cutwaters' at the piers, although cosmetic, are a distinct feature of the bridge and an early decision was made that they should be reconstructed in a similar form.

As a consequence of the phased construction and the risk of damage to and contamination of the concrete surfaces during the process of the complex works, some form of skin to the structural form was deemed necessary to provide a reasonable finish. This could have been achieved with an in situ concrete skin but the additional cost of cladding with reconstituted stone to achieve a high quality finish compatible with the existing structure was considered justified. Sketches illustrating the existing and proposed piers are provided in Figures F and G.

 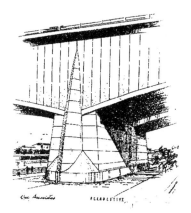

Figure F   Original Piers   **Figure G   New Piers**

11.   CONCLUSIONS

Successful completion of the Kingston Bridge Phase 1 Strengthening work has been a significant and truly international achievement. Balfour Beatty, Balvac, VSL, Glasgow City Council, Scott Wilson, Halcrow, JMP, the Scottish Executive, Frank Rowley of Tony Gee & Partners and ourselves Gifford & Partners have all been involved and had an input to the final design and construction solutions. Whilst there have been many problems along the way, these have been overcome and the prime objective of maintaining and securing the operational status of this strategically important transport link has been achieved throughout.

Acoustic Emission – A Tool for Bridge Assessment and Monitoring

J.R.Watson[†], P.T.Cole[†], S.Yuyama[‡] and D.Johnson[*]
Physical Acoustics Ltd, Cambridge, UK[†]
Nippon Physical Acoustics Ltd, Tokyo, Japan[‡]
Physical Acoustics Corporation, Princeton, New Jersey, USA[*]

Abstract

Structural assessment of bridges using computational techniques can effectively identify areas potentially vulnerable to damage in structures. This may be due to changes in use, structural alterations or most commonly degradation. Physical inspection of bridge infrastructure at regular intervals serves to increase the confidence of engineers with regard to the condition of the bridges and their reliability. Presently inspection of the bridge structures greatly relies upon visual identification of damage which can prove ineffective in identifying flaws until they are of such as size that integrity has been compromised. In addition, defects may be hidden or located in inaccessible areas of the structure preventing visual assessment or detection using traditional non-destructive testing.

All materials used in bridge structures (e.g. steel, concrete and composites) emit transient elastic stress waves, Acoustic Emission (AE), from initial degradation through to cracking and failure. These stress waves radiate from defects or damaged areas through the structure as material-borne sound. The stress waves can be detected as waveforms at the surface of the material. Each waveform holds information about the type and severity of defect. In addition, knowledge of the rate and characteristics of AE can allow prioritisation of defects/structures for further assessment and repair. One of the greatest strengths of AE is that it is capable of locating the position of active defects. Signals from initiating, hidden and remote defects can travel many metres along plated steel structures, steel cables and through concrete masses, can be detected by surface mounted sensors. If two or more sensors detect the waveform, the location of the origin of the signal can be identified using the time of arrival, similar to that used in seismology.

The unique abilities of AE can be used to identify and assess active damage in the component parts of bridge structures. This paper summarises the practical use of AE in the assessment of CFRP strengthened structures, steel plate structures and concrete bridges, using case studies from around the world.

Introduction

Acoustic Emission has been known about for over a 100 years and its use in monitoring and testing has steadily increased for a number of reasons. Advances in electronic equipment, starting with the development of the first oscilloscopes, aided the pioneers, followed by the almost exponential grow in computing power in recent decades. Today systems have the processing speed to allow measurement of more than 1500 waveforms per second per sensor.

These advances coupled with the associated cost reductions have allowed a vast increase in applications. Furthermore, the enhanced power and speed of computers allows a multitude of features to be extracted from the emission allowing a detailed signature of different types of damage to be identified and stored. Example of such signatures are fatigue crack growth in steels, damage in fibre reinforced composites, tensile fracture of concrete, micro fracture of corrosion product (rust) and wire fracture in ropes.

Acoustic Emission will always be an experience based technology. In order to understand the AE signals you receive from the material in real structures it is important to be able to correlate the data to experiences where the damage process has been studied in greater depth. All successful applications like aircraft fatigue monitoring, corrosion monitoring of petrochemical tank floors and crack detection in metals have been due to a thorough understanding of the relationships between AE and damage growth. It is this knowledge that allows AE to be used confidently as a non-invasive monitoring tool identifying different types of damage using sensors mounted on the external surface to assess the structure.

Today AE is widely used in the oil and chemical industry. The results of monitoring, even for short periods of time, can quickly identify defects and prioritise structures for maintenance, even giving indications of when further action is required. This form of maintenance is termed Risk Based Inspection (RBI) and its periodic use is shown in Figure 1.

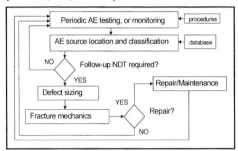

Figure 1: RBI cycle of maintenance using AE

RBI utilises AE to determine if a structure is undergoing actual active damage and, if so, to locate its position and severity. Using additional information such as strain or load the conditions under which damage occurs can be identified. RBI acknowledges the fact that AE should be used as a tool to identify structures or areas of concern which then can allow cost effective focusing of further non-destructive testing techniques and analysis (such as fracture mechanics) to determine the severity of the defect. It is at this point that the decision is made whether to repair the defect or to re-monitor at a later date or to monitor continuously. This strategy of short term monitoring structures both allows early detection of defects and aids the development of a cost effective priority based maintenance strategy depending on the actual damage and its significance in the structure.

With the increasing applications and awareness of AE in the world-wide civil engineering industry this paper goes on to profile the use of AE in different construction materials, structural elements and strategies, in three case studies.

Local Area Monitoring of Steel Corbels for Fatigue Cracking, Queenhill Viaduct, M50, UK

The Queenhill Viaduct consist of two independent concrete box girder viaducts carrying traffic to and from South Wales from Herefordshire. Each carriageway consists of approximately 30 metre spans joined by a half joint, Figure 2. Within these half joints are two steel corbels on which the bearings sit. The corbels underwent numerical fatigue analysis and were found to be nearing the stage when fatigue cracks might initiate. In particular the fillet welds on the plate identified by bold black lines, Figure 3, were particularly prone.

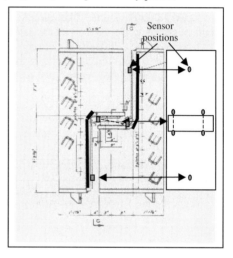

Figure 2 (above): Corbel site with scaffold access into the joint. Mobile laboratory can be seen at the base of the right hand column.
Figure 3 (right): Corbel design. Dashed line denotes area accessible, bold line denotes area with suspected fatigue cracking.

The corbels are cast into the half joint leaving only very small amount of their structure exposed, in addition to this the top corbel sits tightly on the bearing, leaving a gap of 40mm between the two concrete spans. The combination of these two factors greatly limited the inspection of the corbel, with the majority impossible to assess for fatigue cracking even with non-destructive testing techniques such as radiography or ultrasonics. Acoustic Emission was proposed as the technique that could identify and locate any active fatigue cracking.

AE Monitoring of the Corbels

Piezoelectric sensors were mounted in a local array on the two upper and lower corbels of the west bound (pairs "A" and "B") and east bound (pairs "C" and "D"), to provide planar location of active cracking in the hidden welds shown in red in Figure 3. The two triangular arrays were linked so that any emission from bearing movement was distinguishable by both its location and from the signal features, which are very different to emissions from crack growth. Location checks using a Hsu-Neilsen source (simulating an AE source using the fracture of a pencil lead) determined that location accuracy was very precise, both on the bearing and the vertical plate. High frequency sensors were used to limit environmental noise (which is present at lower frequencies) from passing vehicles from corrupting the data acquired.

From previous experience of monitoring motorway bridges for cracks [1] it had been established that one day's traffic provided sufficient fatigue cycles to conclusively identify active defects, so monitoring was carried over a 17 hour week day period, from 05:00 to 22:00 hours.

Results of AE Corbel Monitoring

AE data was collected from all four of pairs corbels, with some of the results presented in Figure 4. A small number of emissions were detected from the corbels, however they were judged to be insignificant. Focused AE sources were detected in the region of the bearing block. The significance of sources was determined from an extensive database accumulated during 8 years of AE monitoring fatigue tests at Cardiff University [2].

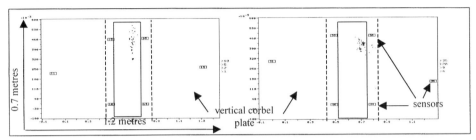

Figure 4: Planar location of AE sources on corbel pairs and bearing, B (lhs) and D (rhs). Each location plot shows two corbels separated by the bearing.

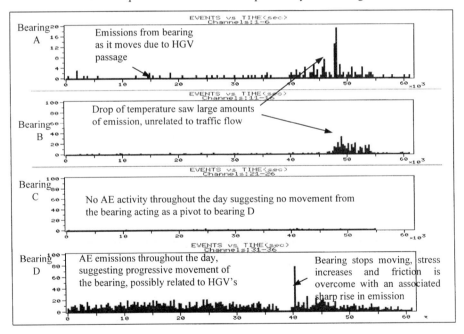

Figure 5: Located AE source activity over the day of monitoring from the bearings.

The activity of the source over the course of the day, Figure 5, provides information on the bearing behaviour.

Conclusions

AE monitoring of the corbels was successful, allowing assessment of the concrete encased corbel with only very limited access to its surface. Hsu - Neilsen source location was found to be accurate both on the corbel surface and the bearing. Monitoring showed no indication of fatigue crack activity during the test, the only located activity originated from the bearing block.

AE Monitoring of Concrete Bridges (NPA, Japan)

RC Slabs of a Highway Bridge in Service [3]

AE was monitored under live loads in RC slabs of a highway bridge, which had been in service for more than 27 years. Figure 6 schematically illustrates the structure and the location of AE sensors. The dimensions of the slabs are 4.8m in length, 3.4m in width and 180mm in thickness. Asphalt pavement of 60mm in thickness is placed on the surface of the slabs. A median strip covers almost half part of the upper side of the slabs. AE signals were monitored continuously for two hours under live fatigue loads in service.

Number of AE hit increased almost linearly as a function of time during the continuous monitoring in CH1. The same tendency was observed in CH2, CH3 and CH4 sensors. Other AE features such as energy and count gave the same kind of history. Therefore, it was thought that there was no significant change in the live loads due to traffic during the monitoring and AE measurement was carried out under a stable condition.

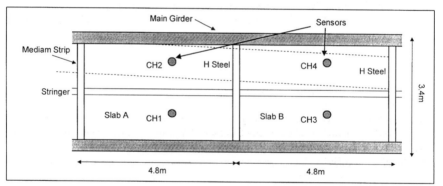

Figure 6: Schematic illustration of the bridge in service. Note that two PAC R6 (60kHz resonant AE sensors) are attached to each slab.

Figure 7 indicates cumulative number of AE hits and AE energy. The number of detected hit is almost the same in CH1, 3 and 4 (about 8×10^4) but it is about 50 percent greater in CH2. In contrast, AE energy exhibits about 10^6 in CH2 and CH4, while 2.6×10^7 and 3.0×10^7 are detected in CH1 and CH3, respectively. Thus the detected AE energy is more than one order of magnitude greater in CH1 and CH3 than in CH2 and CH4. This is because CH1 and CH3 sensors detected the signals generated by live loads due to the traffic since the sensors were attached to the underside of the traffic lane.

However, AE signals due to indirect causes such as vibration of the structure were thought to be detected in CH2 and CH4 because the median strip was on the area where the sensors were placed. Comparison of AE energy in CH1 attached to Slab A and that in CH3 attached to Slab B indicates that it is about 15% greater in CH3 than in CH1 although AE measurement was made under the same traffic condition.

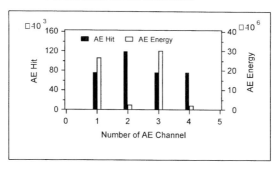

Figure 7: Cumulative number of AE hits (a) and AE energy(b).

Visual observations of crack density found that the damage level of Slab B is greater than Slab A. Accordingly, the difference of the detected AE energy between CH1 and CH3 was attributed to the difference of damage levels between the slabs.

RC Beam of a High Speed Railway Bridge in Service [4]

AE was monitored in an RC beam of a high speed railway bridge in service. Figure 8 schematically illustrates how the tests were performed. Six PAC R6 (60kHz resonant) sensors were placed on two sides of the beam. A strain gauge was attached to a main bar to monitor loading process. The beam was shown to be repeatedly loaded and unloaded due to train passage. Accordingly, the beam has been exposed to fatigue loads since its construction.

Given in Figure 9 are histories of AE hit rate and strain during a passage of one train carriage. It is clearly observed that AE activities are detected during loading and unloading. It should be noted that no activity is seen at the maximum loading phase. Visual observation found that the maximum width of surface cracks reached 0.3mm. As has been reported elsewhere [3], AE activity is strongly dependent on loading phase under fatigue loading. It has been shown that the AE activity due to crack extension is detected near the maximum load.

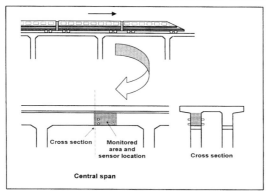

Figure 8: Schematic illustration of AE monitoring in RC beam of a high speed railway bridge in service.

Since no AE activity was detected at the maximum loading phase, it was concluded that no crack extension took place in this beam. The AE activities observed during loading and unloading were attributed to frictions of the existing crack faces.

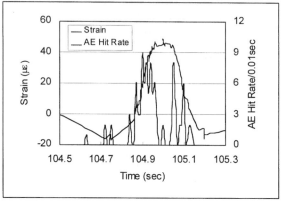

Figure 9: Histories of AE hit rate and strain during a pass of one car.

CFRP Laminate testing on RC Beams (PAC, USA)

New York State DOT decommissioned several reinforced concrete members from a bridge near Albany, New York. These members showed premature signs of decay and were chosen to be sent to the Non Destructive Evaluation Center (NDEVC) of the Federal Highway Administration (FHwA) in McLean, VA. Several methods of testing are being performed on these members, including the evaluation of CFRP laminates as repair methods. Several evaluation techniques were prepared for these tests, including the collection of Acoustic Emission as the bridge member is loaded to failure.

Test Procedures

Upon the bridge's decommissioning, NYSDOT shipped the beam, with others, to the NDEVC for this testing. The beam is a 68'10" box beam with multiple layers of CFRP laminates applied to the bottom of the beam before testing, as shown in Figures 10 and 11. A PAC LAM acoustic emission system was used to record AE data from 8 sensors.

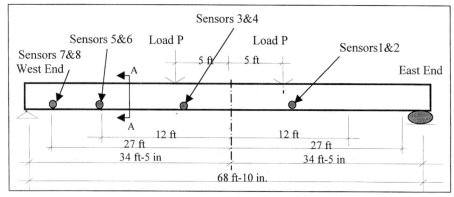

Figure 10 – Side view of beam

For this test resonant sensors were selected and were located as shown in Figure 10. The array of sensors utilised allowed linear source location of damage between the four sensors on both the front and back of the beam. The sensor configuration also allowed for a full 2D planar damage location across the bottom of the concrete beam [5].

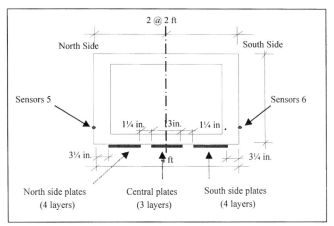

Figure 11 – Cross Section A-A.

The instrument was calibrated in the field by using a H-N source, location checks determined that lead breaks could be detected between any two sensors at a distance up to 15'.

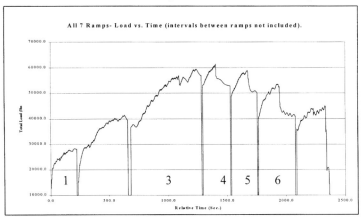

Figure 12 – Loading cycles.

Loading was applied to this beam using a four-point loading scheme. This load was applied in seven incremental loadings until the failure of the carbon laminates took place. Each loading of the beam culminated with either a visual crack location or large audible emissions from the beam. Figure 12 shows the combined load from the four load cells. Time increments between loadings have been shortened for graphing purposes.

One goal of the AE testing here was to locate AE events as they take place during the testing. The following graphs of Figure 13 represent three different ways the damage was located.

244   MAINTENANCE PRACTICE

The first graph shows the linear location of damage along the back of the beam, where sensors 2, 4, 6 and 8 are located. The second graph shows the detected by the sensor at the front of the beam. The third graph shows planar two dimensional location of damage over the base of the beam.

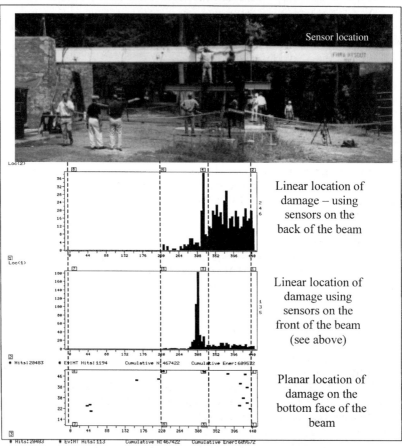

Figure 13 – Location graphs for Loading 1 of 7 (start of test).

Analysis

In reviewing the AE location data, it was found that most of the initial AE activity occurred between the two loading points at the center of the beam, with most AE activity coming from the leftmost load cells corresponding to AE sensors 3 and 4. No AE events were detected from the end of the concrete beam until the third loading, and minimal AE activity was detected from that location until load cycle number 6, well after the concrete beam had reached maximum loading in cycle number 4. This AE activity at the end of the concrete beam during loading cycles 6 and 7 is likely caused by failures at the CFRP laminate level, either from the CFRP itself, at the bond level or in the concrete adjacent to the bond. When ultimate failure occurred during loading cycle 7, the CFRP laminates gave way from the concrete beam, releasing any stresses that had transferred from the concrete beam to the

CFRP laminates. The AE events detected from that area of the beam are likely indicative of the failure at the CFRP laminate level.

Conclusions

AE was found to be an effective tool to monitor the failure process and pinpoint locations that generated AE stress waves, both from within the concrete and at the CFRP interface. In this test it was found that the initial failure process began at centre span, but ultimate failure did not take place until the failure occurred along the CFRP laminate at the left end of the beam. High levels of energy were recorded with AE during all phases of this test, including the initial loading of the beam. These failures were effectively located, providing valuable information about the failure process in action, before ultimate failure.

Paper Conclusions

This paper demonstrates the practical use of acoustic emission in a variety of applications, in assessment and detection of damage in a variety of materials. Acoustic emission has been shown to detect and locate active damage within materials during service conditions. As the age of our bridges increases there is a need to accurately identify the extent of damage and its location. Acoustic emission is well placed as a tool to facilitate this and can greatly aid civil engineers, especially when combined with the philosophy of risk based inspection and used to assist a priority based maintenance strategy.

Acknowledgements

The authors would like to acknowledge and thank the Highways Agency (UK), WSP Consultants (UK), New York State DOT, FHwA, Turner Fairbanks NDE Validation Center, McLean, VA Wiss, Janey, Elstner Associates, Inc.

References

1. Watson, J.R., Holford, K.M., Cole.P.T., Davies, A.W., "BOXMAP –Non-Invasive Detection of Cracks in Steel Box Girders" University of Surreys 4th International Bridge Management Conference, 17th- 19th April 2000, Guilford, Surrey, UK, pp 80-87, ISBN 0727728547.
2. Holford, K.M., Cole.P.T., Carter, D.C., Davies, A.W., "The Non-Destructive Testing of Steel Girder Bridges by Acoustic Emission". 14th World Conference on Non-Destructive Testing, Vol.4, pp 2509-2512, ISBN 8120411269, 1996.
3. Yuyama, S., Li. Z.W., Yoshizawa. M., Tomokiyo. T., and Uomoto, T. (2000) Evaluation of Fatigue Damage in Reinforced Concrete Slab by Acoustic Emission, *Non-Destructive* Testing in Civil Engineering-2000, Uomoto T. ed., Elsevier, 25-27 April 2000, Tokyo, Japan, pp.283-292.
4. Yuyama and Li, Z. W. (2000) Acoustic Emission Method, Technical Report, Research Committee on Evaluation of Degradation in Concrete Structures, Institute of Industrial Science, The University of Tokyo, pp.51-57.
5. Dr. Adrian A. Pollock , Physical Acoustics Corporation, "Acoustic Emission for Bridge Inspection, Applications Guidelines", June 1995.

The Strengthening and Refurbishment of Westfield Pill Bridge Pembrokeshire

D T GULLICK
Parsons Brinckerhoff Ltd, Cardiff, Wales, U.K.

ABSTRACT

Westfield Pill Bridge was built in 1969 and is a post tensioned seven span continuous bridge with an overall length of approximately 260-m. Parsons Brinckerhoff Ltd was commissioned in 1997 to carry out a Special Inspection of the Post Tensioning System in accordance with UK Highways Agency Standard BA 50/93. As a result of this inspection, and a subsequent load assessment, and due to matters unrelated to the condition of the post tensioning system, the bridge was closed to all traffic because of serious structural inadequacies. Following temporary strengthening, the bridge was reopened to limited traffic while permanent strengthening was designed and procured. Innovative Design and Contract Procurement methods have allowed the bridge to be strengthened quickly to allow the bridge to be opened to unrestricted traffic. This paper describes the lifecycle of the project.

Plate 1 - Westfield Pill Bridge

INTRODUCTION

Description of the bridge

Westfield Pill Bridge is a seven span structure carrying the A477 road over the Westfield Pill, a tidal inlet in Pembrokeshire, UK. The bridge has a width of 11.6-m carrying a single two-lane 7.3-m wide carriageway with two 1.8-m wide footpaths. The overall length is approximately 260-m comprising five main spans of 40 m and two end spans of 30 m. The overall height of the six piers varies from 13.6 m to 37.7 m with a maximum pier height of approximately 27 m above high water level. The superstructure comprises two precast prestressed post-tensioned box beams of trapezoidal cross section 2.4 m deep, supporting a 230 mm thick in-situ reinforced concrete slab. Transverse diaphragms at the piers and abutments connect each pair of box beams. Reinforced concrete abutments and hammerhead piers support the superstructure. (Refer to Figure 1)

When the deck was poured, the gaps between the beam-ends were also filled with concrete to form a 'stitch joint'. At the same time a concrete hinge was formed connecting the infill concrete to the top of the crosshead. This hinge became the main (and only) support for the superstructure at the piers. (Refer to Figure 2)

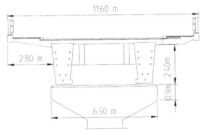
Figure 1 - Typical cross section

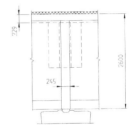
Figure 2 - Typical stitch joint detail

DESK STUDY (PHASE 1)

The Phase 1 Desk Study highlighted that Pembrokeshire County Council inherited the bridge in April 1996 following Local Authority Reorganisation in Wales (UK). At the time the desk study was carried out no inspection or assessment records were made available. As-built drawings were provided together with some limited construction photographs.

PRINCIPAL INSPECTION (PHASE 2)

When carrying out a Special Inspection in accordance with BA 50/93, a Preliminary Site Inspection (Visual Inspection) is all that is normally required. However, as there were no Inspection Reports available, the Client requested that a Principal Inspection to within touching distance be carried out. The main results of the inspection are given below:

- Severe cracking to the pier crossheads (some cracks as wide as 0.7 mm (refer to Plate 2),
- Severe cracking to the diaphragms and the stitch joints at the beam ends (some >10mm),
- Regular transverse cracking to the soffit of the deck cantilevers with stalactites forming,
- Poor provision of deck drainage through the cantilevers,
- Poor quality concrete to the main beams with widespread poor compaction and spalling,
- Severely corroded steel bearings and parapet mesh infill.

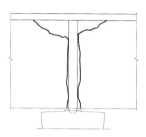

Figure 3 - Cracking to stitch

Plate 2 - Cracking to crosshead

SPECIAL INSPECTION OF POST TENSIONING SYSTEM (PHASES 3&4)

Due to the relatively difficult access to the bridge, the Special Inspection was carried out consecutively with the Principal Inspection. This was more cost effective for the Client and ensured that the underbridge unit was easier to obtain. Prior to carrying out the site works PB devised a preliminary investigation regime that could easily be adapted to suit the findings of the Principal Inspection. Following completion of the Phase 3 works it was found necessary to carry out an additional Phase 4 Investigation. The main findings of the Phase 3 and 4 works are as follows:

- The post tensioning system was in good/fair condition,
- Some fine cracks were reflective of the duct positions and in some cases reached the ducts,
- Leachate deposits indicated that water was migrating through the concrete,
- Corrosion testing and sampling indicated that pitting corrosion of reinforcement would generally not be expected but that general corrosion was possible.
- Alkali contents were above the upper limit recommended for the avoidance of alkali silica reaction,
- Cement contents were medium / high and sulphate contents were below those required to initiate sulphate attack of concrete.

ASSESSMENT

Stitch Joints

The exact design philosophy followed by the bridge designers is unknown, although it is reasonable to assume they intended that the stitch joint would be in compression under all loading scenarios. The assessment highlighted that the cracks probably opened as a result of the combined effects of pier flexure, traction, live load sagging moment and creep deflection of the prestressed beams.

However, the cracks noted in the Principal Inspection were so extensive that it was apparent that an alternative load path, probably via the diaphragms was transferring the loads.

The assessment found that this alternative load path was not adequate to transmit loads and it was therefore concluded that the shear forces were being transmitted across the cracks at internal points of contact between the beam-ends and the insitu stitch. The extent of these points of contact could not be quantified and neither could their capacity. All that could be said was that the joints were capable of carrying dead load with an unknown factor of safety.

Crossheads

The crossheads were found to be capable of carrying approximately 93% of the dead load and no live load in the cracked state . Additionally, a preliminary analysis using state-of-the-art upper bound plastic methods indicated that the assessment of the crossheads was conservative. However, the application of current UK assessment philosophy resulted in a zero live load carrying capacity.

Other Bridge Elements

A summary of the capacities of the rest of the bridge elements is given below:

- The bridge deck and beams were adequate in flexure to carry full Assessment Loading and full abnormal vehicle loading. In shear the beams were limited to 17 Tonne Gross Vehicle Weight,
- The columns were found to be inadequate to resist the effects of design transverse wind loading, which was a gust speed of (118mph).

Conclusions of the Assessment

It was clear that the bridge had been carrying the applied loads for some time in its poor condition. However, the mechanism by which it was carrying the loads was indeterminate and the period for which the bridge would be able to carry load safely was unknown.

The Principal Inspection highlighted some diagonal cracks emanating from the stitch cracks and running out to the intersection between the soffit cantilever and the beams. The assessment highlighted that this was a possible collapse mechanism caused by shear, causing great concern to the Client. (Refer to Figure 3)

BRIDGE CLOSURE

Closure

Following verification of the assessment results by the Independent Checkers, and in the interests of public safety, the bridge was closed to all traffic immediately on October 1st 1999.

Risk Assessment

The result of the bridge closure was to divert traffic around the bridge via an alternative route. Although the route was a public highway it could be considered unsuitable for pedestrians and cyclists because there is poor visibility, no provision of footways and no adequate means of escape for those caught on a narrow road with large vehicles running nearby. These factors were highlighted by the occurrence of a nearly fatal accident at this time (Refer to Figure 4).

It was also clear that in the interests of public safety the bridge could not be reopened to traffic until some remedial work was carried out. Additionally, the strengthening works would take time to procure and install given the nature of the problems and the difficult means of access. The Client was thus faced with the risk to pedestrians, cyclists and motorists using the diversion route remaining for upwards of 6 to 8 months, a situation that needed to be improved if possible.

PB was commissioned to look at temporary solutions that would allow the bridge to be reopened and that would reduce the risks associated with the bridge collapse and the use of the diversion route to acceptable levels.

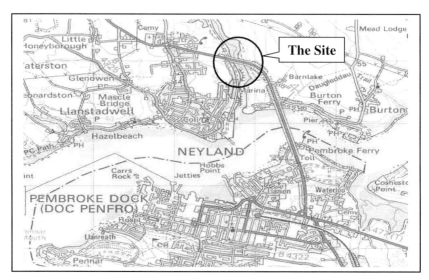

Figure 4 – Site Plan

Temporary Strengthening

There were a number of problems associated with devising a temporary strengthening solution:

- Without the installation of full strengthening the risk of failure of the bridge under load could not be completely eradicated,
- Access was particularly difficult and any system would thus have to be easy to install and comprise lightweight components, installed by men supported by a very lightweight access platform.
- The system would probably need to be in place for a period of 6 to 8 months and would thus have to endure the rigours of the winter months.
- Time was also important to the Client. Every day that the traffic had to use the diversion route there was a risk to public safety that was greater than if they could travel by using the bridge. The bridge closure had an impact upon local businesses close to the bridge in that their source of passing trade was removed. Additionally traffic from the south crosses the tolled Cleddau Bridge and the closure resulted in a reduction of toll revenue from the bridge.

PB devised a temporary solution that involved the insertion of temporary packs beneath each beam comprising neoprene pads supported on cementitious mortar pads.

The purpose of the packs was to ensure that, in the event of failure of the stitch, the beams would settle onto the temporary supports rather than impact upon the tops of the crossheads causing catastrophic failure of the bridge. The assessment using current standards was conservative and as such the crossheads probably had a higher load carrying capacity than the value given by the assessment. Therefore, provided the impact of the beams following failure of the stitch joint could be prevented, the risk of failure of the crossheads was reduced to acceptable limits. Additionally in the event of failure of the stitch joint the vertical movement of the beams would be negligible with no effect upon bridge users immediately above.

The temporary packs were in contact with the soffit of the beams but did not support them, allowing the bridge to continue to rotate and articulate as necessary. However, in the event that failure of the stitch joint occurred, the beams would become simply supported changing the articulation of the structure as a whole. If this occurred the bridge would need to be closed to traffic until full strengthening could be procured and installed.

If failure of the stitch joints occurred, the cost of strengthening of the bridge would increase significantly over and above those envisaged with the bridge in its un-strengthened state. It was therefore agreed that the bridge should be opened to the lowest level of traffic that could be specified, 3 Tonnes GVW. This limit ensured that the bulk of the traffic using the diversion route was transferred onto the bridge together with pedestrians and cyclists who were at greatest risk when using the diversion route

Monitoring System

It was necessary to ensure that any change in the status of the crossheads or the stitch joints would be notified immediately and PB therefore needed to devise a monitoring system. The problems associated with the installation and durability of the temporary strengthening also applied to the monitoring system.

The monitoring comprised electronic vertical movement transducers, coupled with visual crack tell-tales that would give clear indications of vertical movements of the beams, and widening of the cracks in the stitches and the crossheads. The transducers were linked to a data logger situated at Pier 6 that had a semi-permanent scaffold around it to provide easy access. This system was monitored daily with weekly visual inspections of the crack tell tales.

Installation of Temporary Packs and Monitoring

In order to minimise the loading exerted on the bridge during installation, roped access was used and an innovative lightweight aluminium CANSPAN System to provide a safe and stable working platform.

The neoprene packs were glued to the soffit of the beams and the cementitious mortar pads were poured to fill the void between the pad and the top of the pier crosshead.

The bridge was opened to restricted 3 Tonne GVW traffic on 15th November 1999, approximately 6 weeks after it was closed.

Plate 3 - Installation of Temporary Packs Pier

Plate 4 – Monitoring System at

PERMANENT STRENTHENING

Once the bridge was opened to 3T traffic the Client was faced with the problem of the design, procurement and construction of the permanent strengthening work. If possible the Client required the bridge to be opened to unrestricted traffic by the start of the busy tourist season beginning at Easter (20th April 2000).

Procurement

The option of a conventional Design, Procure and Construct method using competitive tender was required by the Authority's procedures, but given the fact that the bridge had to be opened within 5 months, adopting this method of procurement was unacceptable. Additionally the option of competitive Design and Build that could have saved some time was unrealistic due to the need for a tender process. The Client decided to opt for a single tender action, negotiated contract using the Engineering and Construction Contract Option E (Cost Reimbursable Contract) saving approximately 3 months in procurement time.

The strengthening works was split into separate Phases 1 and 2. Phase 1 comprised the strengthening of the pier crossheads and stitch joints to allow the opening of the bridge to unrestricted traffic. Phase 2 comprised the completion of the routine maintenance and refurbishment of the bridge.

Design Options

The Contractors input to the design phase was critical. At a single meeting attended by all parties involved all the available options for the design and installation of the strengthening works were assessed on the basis of efficiency, economy, safety, buildability, durability, aesthetics and programme. The method of strengthening of the pier crossheads was critical because this was going to be the most time consuming and costly exercise. (Refer to Table 1).

Assessment Criteria	Stressed Options								Unstressed Options				
	Reinforced Concrete	Steel bar/strand	Carbon Plates	Kevlar rope	Steel Plates	Carbon Wrap	Kevlar Wrap	Internal Anchors	Reinforced Concrete	Carbon Plate Bonding	Carbon Wrap	Kevlar Wrap	Steel Plate Bonding
Life Expectancy	3	3	2	2	3	2	2	3	3	2	2	2	3
Programme achievable	2	3	3	3	3	3	3	2	3	3	3	3	2
Eng. Concept/Materials suitability	3	2	3	3	2	2	2	3	1	1	1	1	1
Tech. Confidence/Track record	3	3	2	1	2	1	1	2	3	3	3	3	3
Safety/Risk Management/CDM inc. demolition?	1	2	2	3	2	3	3	2	1	3	3	3	1
Ease of Installation	1	2	3	3	2	3	3	1	1	2	2	2	2
Aesthetics	2	1	3	2	1	3	3	3	2	2	2	2	2
Installation Cost	2	3	2	2	2	3	3	2	3	2	2	2	1
Service Cost	2	2	3	3	2	3	3	2	3	3	3	3	2

Table 1 – Strengthening Options

The issue of access was also critical and the use of innovative lightweight materials such as carbon fibre composites for wrapping and plate bonding, together with the use of Kevlar Ropes for stressing the crossheads was explored. However it emerged that although these methods were attractive due to the weight of components required, the magnitude of the forces that were to be resisted, together with some of the dimensional limitations of the structure, meant that it would not have been an efficient use of the materials.

Crossheads

The crossheads were strengthened using 4 No. 40mm diameter stainless steel Macalloy bars each carrying 750 kN. Several methods available to stress the Macalloy bars but the one chosen involved the erection of a steel jacking frame collar around each of the crossheads. The frame was then jacked off the ends of the crossheads using a system of flat jacks that were all linked hydraulically. However the analysis of the jacking frame and proposed jacking sequence showed that it was very sensitive to distortion. In consultation with the Contractor's jacking specialist it was agreed that a method of operating each jack separately or a number of jacks in sequence needed to be incorporated into the system. This method was very successful. Prior to installation of the strengthening system, all cracks were pressure grouted. For aesthetic purposes a partial composite enclosure system was installed at each pier position (refer to Plate 7)

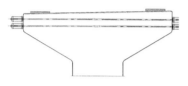

Figure 5 - Crosshead prestressing

Plate 5 - Crosshead prestressing

Stitch Joints

The stitch joints were strengthened using conventional bolted steel plates bonded to the faces of the beams across the stitch. Acrylic templates were used to transfer the hole locations to the steel plates that were then pre-drilled before they were delivered to the scaffolding at each pier. In order to prevent the possible slippage of the plates around the holes caused by the conventional tolerance necessary between the holes and the bolt shank, the Contractor was required to place a 32 mm diameter bolt in a 32 mm diameter hole. This was a very onerous requirement as there were 480 bolts to be installed. The Contractor achieved this feat on all but 7 of the holes that needed to be filled and re-drilled. Prior to installation of the plates and bolts, all cracks were pressure grouted.

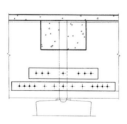

Figure 6 - Stitch joint strengthening

Plate 6 - Stitch joint strengthening

REFURBISHMENT AND GENERAL MAINTENANCE

The following works were required to complete the refurbishment and general maintenance of the structure:

- Extensive concrete repairs,
- Installation of new deck drainage,
- Grit-blasting and refurbishment of the bridge bearings, including bolt replacement,
- Replacement of the parapet mesh infill along the whole length of the bridge,

The following works are necessary but have been deferred due to budgetary constraints;
- Replacement of the bridge deck expansion joints,
- Waterproofing and resurfacing of the carriageway and footways,
- Surface impregnation of exposed concrete surfaces using Silane

CONCLUSION

Pembrokeshire County Council inherited a structure with almost unique structural problems. The project has comprised many interesting phases most notably:

- The load carrying assessment,
- The closure of the bridge
- The use of risk assessment to assess the problem posed by the closure,
- The installation of the temporary packs and the remote monitoring system,
- The use of innovative forms of Contract and fast-track procurement methods resulting in substantial time and cost savings over more conventional approaches.

In future, regular targeted inspection, completed with cost effective maintenance should ensure that the bridge remains serviceable for the future.

Plate 7 – Completed Strengthening

MONITORING SYSTEM FOR FATIGUE CRACK PROPAGATION BY IMAGE ANALYSIS

Kazuo TATEISHI[1], Takeshi HANJI[1] and Makoto ABE[2]
[1] Department of Civil Engineering, Nagoya University, Nagoya, Japan
[2] BMC Co.Ltd, Chiba, Japan

INTRODUCTION

Recently, a reasonable and effective maintenance strategy for steel bridges suffering from fatigue damage is required[1]. Fatigue damage in actual bridge has various characteristics in severity and deterioration rate, which should be carefully considered when maintenance plan is determined. However, because information obtained from fatigue damage, particularly, quantitative data, is usually limited, these important damage characteristics are estimated by a knowledge and experience of engineers or experts. In order to develop more quantitative maintenance system and health monitoring system for steel bridges, mechanical approaches should be introduced.

One of effective tools for considering fatigue problem in steel material is fracture mechanics. Particularly, stress intensity factor, which is a parameter used in fracture mechanics, can be related to the crack propagation rate and the criteria for the occurrence of brittle fracture. Therefore, if stress intensity factor can be quantitatively estimated for a crack in actual bridge members, the progress of the crack and the residual life of the member can be predicted.

In former studies, some methods to estimate stress intensity factor were proposed. For example, a stress analysis method, such as FEM, is available to estimate stress intensity factor. However, this approach is difficult to be performed because fine and large number of elements consuming enormous computer resources and computing time are required for calculating stress intensity factor so that local stress field around a crack in large scale structural member can be analyzed. Experimental approach was also proposed[2,3]. This method was based on local strain field measured by strain gauges placed around crack. However, the feasibility of this method is also limited because almost fatigue cracks are formed in welded joints where the geometrical shape is so complicated due to the existence of attached plate, welded bead, etc. that strain gauges can be hardly placed.

In this study, for obtaining quantitative information on the severity of fatigue damage in actual bridge members, a monitoring system for fatigue crack propagation is proposed. It is based on the combination of fracture mechanics and image analysis technique. In order to investigate the accuracy and the limitation of the proposed method, fatigue test was carried out on welded joint specimen.

TESTING METHOD

A welded joint specimen made from mild steel (SS400) with a longitudinal attachment was used in this study. The dimensions of the specimen are shown in Fig.1. Cyclic stresses with the constant amplitude from 0 to 200MPa with 10Hz were applied to the specimen so that a fatigue crack was developed along the welded toe as shown in Fig2. Then, cyclic loading was paused, and static load was applied to the specimen. During static loading, crack opening behaviors were recorded as digital images by using a digital microscope of which specifications are shown in Table 1. The lens unit was installed at the position about 20mm away from the specimen surface. Main unit as well as lens unit of the microscope was compact and portable. After recording images, cyclic loading was resumed and continued in 1000 cycles. Then, the same procedures were repeated until the specimen was broken. Based on the obtained digital images of the crack, stress intensity factor and crack propagation rate were estimated by digital image analysis and fracture mechanics theory.

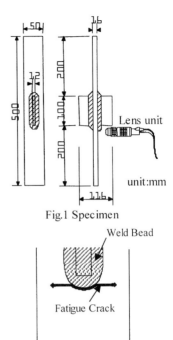

Fig.1 Specimen

Fig.2 Image of Fatigue Crack

IMAGE PROCESSING

Fig.3 shows an example of the digital image taken by the microscope. The image consisted of 752 x 480 pixels, and one pixel covered 0.0026mm x 0.0026mm area in the real space. A ruler in the upper part of the image was graduated in 0.5mm, and used for transferring the image scale to the real scale.

The original image shown in the figure was processed by using a retouch software, and converted to 'bit map image' in which each pixel had grayscale data between 0 and 100. In order to measure the crack opening width from the image, an measuring line was set up transversally to the crack direction as shown in Fig.4, and the grayscale data along the line were picked up. Fig.5 shows an example of grayscale data along the measuring line. In the figure, cracked part can be clearly recognized because cracked area has larger grayscale values than other parts. However, in the boundary region, the grayscale data change gradually from the cracked part to the metal body. This may be

Table 1. Specifications of digital microscope

CCD size	1/2 inch
Pixel format	752x 480
Lens magnification	x175 (for 14in. monitor)
Weight	8.3 kg
Size of lens unit	diameter: 35mm length: 208mm

Fig.3 Example of Image

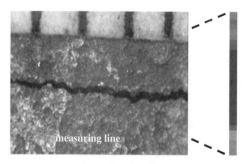

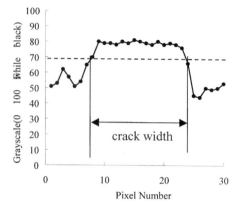

Fig.4 Set up of Measuring Line

Fig.5 Distribution of Grayscale Data

caused by the limitation of lens resolution and the slight loss of the lens focus. In this study, a threshold value was determined at the 90% level of the maximum intensity as shown in Fig.5, and the region with the intensity more than the threshold level was taken as the cracked region.

ESTIMATION OF STRESS INTENSITY FACTOR

In this study, the following relationship was assumed between crack opening width and stress intensity factor, which was obtained for semi-infinite crack in infinite plate shown in Fig.6.

$$K_I = G\sqrt{\frac{2\pi}{r}} \frac{\phi}{\kappa+1} \qquad (1)$$

ϕ : crack opening width
K_I : stress intensity factor
G : shear stiffness
r : distance from crack tip
$\kappa = 4 - 3v$,
v : poissons ratio(= 0.3)

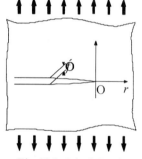

Fig.6 Model of Crack

This general relationship can be applied to the vicinity of the crack tip for any types of cracks. It has both advantages and disadvantages to use this relationship. The advantage is to be able to apply the proposed method to any cracks regardless of length, direction, shape, etc. However, for using this relationship, measurement must be carried out at the very near position to the crack tip where the crack opening width is very small. This may produce large errors. In this study, after some trials, r was set for 3mm.

In the specimen, the crack was clearly opened even when no load was applied. Therefore, the crack opening width without load was measured and determined as the initial crack opening width. The crack opening width under loaded condition was determined as the difference between the measured width and the initial width.

Fig.7 shows the relationships between nominal stress and stress intensity factor calculated by Eq1. Theoretically, they should have linear relationship. Direct verification on stress intensity factor can't be performed, because it is very difficult to calculate the theoretical stress intensity factor for this type of crack, However, linear relationships shown in the figure can be considered as a proof that the proposed method estimates stress intensity factors properly.

According to the curves shown in Fig.7, regression lines were obtained by a least square method. Then, regressed stress intensity factors were calculated. Fig.8 shows the ratio of estimated stress intensity factor to regressed one. In the region less than 10MPa√m, a large scatter can be seen. Stress intensity factor of 10MPa√m corresponds to the crack opening width of 0.0043mm. As mentioned before, the size of each pixel in the real space is 0.0026mm. Therefore, more than three pixels should be used to capture a crack. This is the limitation of the proposed method with the current system. The limitation can be eased if a device with higher resolution was used.

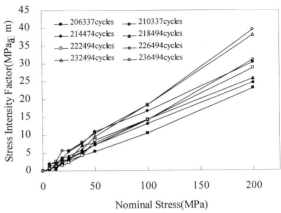

Fig.7 Estimated Stress Intensity Factor

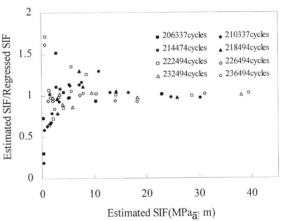

Fig.8 Comparison of Estimated SIF and Regressed SIF

ESTIMATION OF CRACK PROPAGATION RATE

Experimentally, the following relationship has been confirmed between stress intensity factor and crack propagation rate[4].

$$da/dN = C(\Delta K^m - \Delta K_{th}^m) \quad (2)$$

a : crack length(m)
N: number of cycles
C: material constant: 1.5×10^{-11}
m: material constant: 2.75
ΔK_{th} : stress intensity factor threshold : 2.9MPa√m

By substituting the stress intensity factor estimated from crack opening width by Eq.1, crack propagation rate, da/dN, can be calculated. Constant values in Eq.2 were derived from many experimental results to express the average relationship between crack propagation rate and stress intensity factor.

On the other hand, the crack extension at every 1000cycles was measured by visual observation. The average crack propagation rate during 1000 cycle loadings can be easily calculated as,

$$da/dN = da/1000 \quad (3)$$

Fig.9 shows the comparison of crack propagation rate estimated by Eq.2 and measured

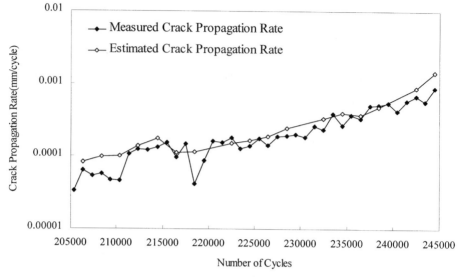

Fig.9 Verification of the Proposed Method

by Eq.3. Estimated values have good agreements with measured values, which shows the proposed method can properly estimate stress intensity factor and crack propagation rate, though it requires only small device and simple measurement works.

CONCLUSIONS

A new method to estimate stress intensity factor was proposed. The method was based on the crack opening width measured by image analysis. The results of fatigue test carried out on welded joint specimen showed this method could be applied to estimate stress intensity factor and crack propagation rate with enough accuracy for the engineering use.

The proposed method has many advantages for filed use because, 1) necessary device is only a portable and compact microscope, 2) non-touch measurement can be performed, and 3) accessibility to complicated structural details is high.

ACKNOWLEGEMENT

This research was supported by grants from the Japanese Ministry of Education, Culture, Sports, Science and Technology (Project No. 12650471).

REFERENCES

[1] Miki,C., Sakano,M., Tateishi,K. and Fukuoka, Y.: Database for fatigue experiences in steel bridges, J. of Structural Eng./Earthquake Eng., No.392/I-9, 1988.
[2] Dally,J.W. and Sanford,R.J.:□Strain-Gage Methods for Measuring the Opening-Mode Stress-Intensity Factor, KI,□Experimental Mechanics, Vol.27, No.4, 1987.
[3] Dally,J.W. and Sanford,R.J.: An Overdeterministic Approach for Measuring KI Using Strain Gages, Experimental Mechanics, Vol.28, No.2, 1988.
[4] Japanese Society of Steel Construction: Fatigue Design Recommendations for Steel Structures, Gihodo, 1993

Health Monitoring System for Bridge Structures Based on Continuous Stress Measurement

NORIYO HORIKAWA, New Energy and Industrial Technology Development Organization, Kusatsu, Japan
HIRONORI NAMIKI, Kyobashi Construction Corp., Osaka, Japan
TAKAYUKI KUSAKA, Ritsumeikan University, Kusatsu, Japan

INTRODUCTION

Fifty years have passed since many bridge structures of small or middle scale were constructed after the World War II. Though these bridge structures were designed for the use of a hundred year, their life has been decreased with increasing the volume of traffic from year to year or by an earthquake loading. It is also predicted that these old bridge structures increase rapidly in number after ten years. Recently, for making full use of such existing bridge structures, they are frequently repaired or reinforced to extend the life of the bridge structures.

Recently, various behaviors of the bridge structures are investigated to predict their life and check the structural health of them [1-4]. For determining optimum repair schedules for extending the service life of the bridge structures, first of all, it is necessary to measure and monitor stress acted to the bridge structures. In the investigation, stress (strain) is often measured directly using a strain gauge. Most of the measurements are carried out continuously for a long term. The measuring devices are set up in the open air, where the difference in temperature between daytime and the night is large. However, under these conditions, there are some cases that the strain meters incorporated into the measuring devices are unstable to temperature changes for a long term. Therefore, the measuring devices that have excellent property for temperature changes are necessary for long-term measurement.

In the present work, a strain meter, which was stable to temperature changes for a long term, was developed. Health monitoring system suited for long-term measurement was also developed. For confirming availability of this system, the stress of steel bridge was measured for three months continuously. A parameter related to the life of steel bridge was also estimated by a two-dimensional rain-flow method.

HEALTH MONITORING SYSTEM

Configuration
The health monitoring system consists of a strain meter, an analog-digital converter, wireless devices to send and receive the stress data, a personal computer for storing the stress data and a solar cell as power source as shown in Fig. 1. Data obtained were processed as the following. The stresses were detected by a strain gauge and were amplified by the strain meter. They were converted from the analog data to the digital data by the analog-digital

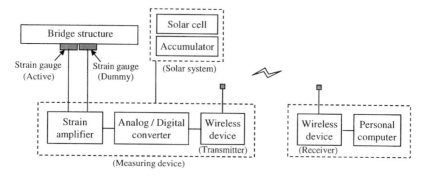

Fig. 1 Health monitoring system

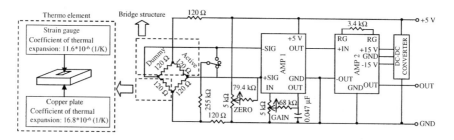

Fig. 2 Circuit diagram of the strain meter

converter and were sent by wireless from the point of measurement to the personal computer in the office and were compressed and stored.

Strain Meter
The primary advantages of the health monitoring system are that the strain meter itself is stable to temperature changes for a long term. Figure 2 shows circuit diagram of the strain meter. Active-dummy method in the bridge circuit was employed. In the present work, the reliability of the strain meter for temperature changes was guaranteed by thermo element as shown in Fig. 2. This thermo element consists of a strain gauge and a metal plate. For example, pasting strain gauge (11.6×10^{-6} (1/deg.)) to a copper plate (16.8×10^{-6} (1/deg.)), the thermo element can detect the apparent strain (5.2×10^{-6} (1/deg.)). In this work, by incorporating the thermo element into the bridge circuit, the apparent strain of the strain meter caused by temperature changes eliminated.

Figure 3 shows temperature characteristic of strain meter. Three types of metal plates as the thermo element were used as a thermo element. The ordinate shows output voltage from the strain meter. The abscissa shows ambient temperature of the strain meter.

For the steel plate, output voltage decreased with increasing ambient temperature. The value of apparent strain was $1.89*10^{-6}$ (1/deg). For the aluminum plate, it decreased with increasing ambient temperature. The value of apparent strain was $-0.63*10^{-6}$ (1/deg). However, for the copper plate, it was constant and the strain meter did not show temperature dependence. The value of apparent strain was $-0.10*10^{-6}$ (1/deg). Therefore, the copper plate was used as the

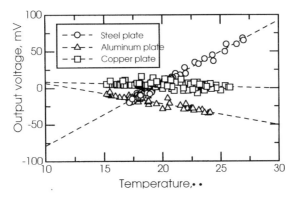

Fig. 3 Temperature characteristic of strain meter

thermo element in our strain meter.

Details of the strain meter developed were the following. The calibration voltage is 2000 (mV). The calibration strain is $219.7*10^{-6}$. The apparent strain of the strain meter caused by temperature changes is $-0.10*10^{-6}$ (1/deg) when the calibration voltage is 1000 (mV).

Wireless Device and Solar Cell

The output data from the strain meter were converted from the analog data to the digital data by the analog-digital converter and compressed as stress or strain data and stored in the personal computer. The data can be measured at 18-millisecond intervals.

The solar-cell power system was employed to operate the strain meter, an analog-digital converter and wireless devices to send the data. The solar-cell power system consists of a solar panel and an accumulator. The wireless device and the solar cell enable unattended monitoring system.

MONITERING FOR BRIDGE STRUCTURE (FIELD TEST)

In this present work, the stress of the bridge structure under the actual loading was measured by the health monitoring system. The bridge is located in Osaka. Over 100 trains per day cross the bridge. This bridge has a sponson beam of 14.3 m in length. Its overhang length is 2.3 m. The point of measurement was at a center of railway bridge across the Shirokita River as shown in Fig. 4, 5. The stress of a main beam was measured. The measurement was carried out for five months continuously.

RESULTS AND DISCUSSION

Waveform

The value of stress calculated from dead load, which was about 23 MPa, was added to the measured stress. Figure 6 shows a typical stress-time curve obtained by the passage of a seven-car train over the bridge. Each peak of the waveform corresponds to two or four axles of the cars.

264 MAINTENANCE PRACTICE

Fig. 4 View of the railway bridge across the Shirokita River

Fig. 5 View of the point of measurement

The electrical noises, which were caused by the electromagnetic waves from the transmission line and the train, was not observed. The maximum value of stress in the passage of the train was about 45 MPa. The increase in stress from dead-load stress is about 22 MPa.

Figure 7 shows deflection of the railway bridge caused by the passage of the train. (a), (b), (c), (d) and (e) correspond to Region E, A', B, C and D in Fig. 6, respectively. For illustrating emphatically deflection of the bridge during the train passing, the deflection caused by the dead load was disregard in Fig. 7.

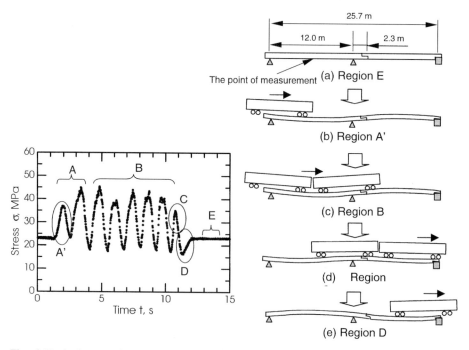

Fig. 6 Typical stress-time curves obtained by the passage of a seven-car train

Fig. 7 Deflection of the railway bridge caused by a train passing

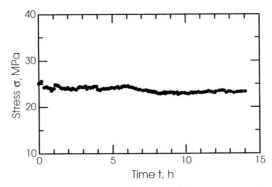

Fig. 8 Change of stress with temperature in a day

The first peak and the last peak of the waveform were lower than the other peaks in Fig. 6 because there are only two axles on the bridge as shown in Fig. 7 (b) and (d). The fourth peak of the waveform was low because the cars, which were equipped with motors, cross the bridge. The values of stress were partially lower than that of dead load stress in Fig. 6 because of a rebound phenomenon of the bridge caused by passing a train as shown in Fig. 7 (e).

Temperature Characteristic of Strain Meter
Figure 8 shows the change of stress with temperature in a day. Solid circles show the average value of the stresses in region E obtained by every train passing.

It was found that the change of stress was very small in a day though the measuring devices were set up in the open air, where the difference in temperature between daytime and the night was large. The stress did not change for three months. This result means that the strain meter is stable to temperature changes for a long term. Therefore, it was found that this system was

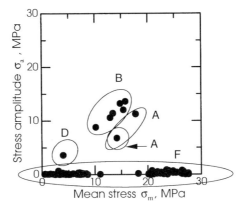

Fig. 9 Relationship between the stress amplitude and the mean stress obtained by the passage of a seven-car train

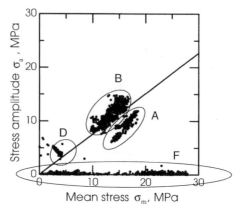

Fig. 10 Relationship between the stress amplitude and the mean stress obtained by the passage of 122 seven-car trains

suited to measurement for a long term continuously.

Data Analysis
Generally, stress amplitude and the ratio of stress amplitude to the mean stress relate to the life of steel bridges. In this work, this parameter was estimated by using tow-dimensional rain- flow method [5].

Figure 9 shows relationship between the stress amplitude and the mean stress obtained by the passage of a seven-car train. The data were divided roughly into three regions of A, B and D. The region A, A', B and D correspond to those in Fig. 6. The stress in region A were caused at the early stage of the train passing. The stress in region C were caused by the rebound phenomenon of the bridge as shown in Fig. 7 (e). The data in region F were noise or vibration of the bridge.

Figure 10 shows relationship between the stress amplitude and the mean stress obtained by the passage of the trains in a day. The numbers of trains were 122 in a day. Solid line was fitted to the data in region A, B and D by using the least squares method, to obtain the ratio of stress amplitude to the mean stress from its slope. The ratio of the stress amplitude to the mean stress obtained from the slope was 0.76. The maximum value of the stress amplitude was about 15 MPa.

CONCLUSIONS

In the present work, health monitoring system, which was suited for long-term measurement, was developed. For confirming availability of this system, the stress of steel bridge was measured for three months continuously. A parameter related to the life of steel bridge was also estimated. The following conclusions were reached.

1. The strain meter developed in the paper, which was stable to temperature changes for a long term, enable long-term measurement of the stress.

2. As a result of the field test using the health monitoring system, the availability of this system was confirmed. The health monitoring system was suited for long-term measurement.

3. The available parameters related to the life of steel bridge were obtained from the data measured in the field test.

REFERENCES

1. J. F. Elliot, "Continuous acoustic monitoring of bridges", IBC, 451-459, 1999.
2. T. Inada, Y. Shimamura, A. Todoriki, H. Kobasyshi and H. Nakamura, "Health monitoring system of composite laminated structures using vibration data", Proc. ACCM-2, Vol.1, 2000.
3. S. Kawamura, T. Okabayashi and S. Takagi, "Development of remote measurement system for bridge viblation by mobile communication, J. Structural Engineering, Vol. 46A, 539-546, 2000.
4. T. Okabayashi, T. Yoshimura, S. Kawamura and M. Hosokawa, "The telemetry system based on Internet technology with wireless LAN for execution management of bridge construction", J. Structural Engineering, Vol. 47A, 285-292, 2001.
5. Y. Murakami, T. Morita and K. Mineki, "Development and application of super-small size strain history recorder based on rainflow method", J. Soc. Mat. Sic. Japan, Vol.46, 1217-1221, 1997.

Bridge Inspection in Steel Road Bridge Based on real measurement

AKIHIRO KOSHIBA*, MAKOTO ABE*, TOSHIAKI SUNAGA**, HIDEKAZU ISHII***
*BMC Co.Ltd., Chiba Japan, **TEIKOKU Design Office Co.Ltd., Hokkaido Japan
*** KITAMI Industrial University, Hokkaido Japan

1. Introduction

Many steel road bridges were built in Japan at the advent of the high growth of Japanese economy in and after around 1960. More than 600,000 bridges of 2m or wider have been built since then. In recent years, it is believed that in order to keep those bridges functioning properly and maintain the safety thereof, the bridges, deteriorating as aging, have to be maintained and managed properly, and it is predicted that the maintenance and management costs will exceed building costs in 20 to 30 years, which will result in a situation in which renewal plans are difficult to be completed.

It is clear from this background that the prolongation of the lives of bridges plays an important role in reducing the maintenance and management costs while maintaining the safety thereof, and to make this happen, grasping the actual states of bridges is required in which the performances or retained performances of the bridges be evaluated in an objective fashion.

Taking the load-carrying capacity and fatigue of a bridge as the current retained performance thereof, this paper will describe an example of an evaluation method based on the measurement of an actual bridge.

2. Problems with Grasping Actual States of Bridges

A steel road bridge used for the survey this time was built in 1954 and the actual state of the 46-year old deck bridge plat truss was grasped. Since the width of this bridge is narrow, with the traffic volume of large-sized vehicles increasing, large-sized vehicles cannot pass each other on the bridge, and to cope with this situation, it was required to determine which is better; modifying the current bridge to get a wider road width or replacing the current bridge with a new one.

To determine which way to take properly, it was required to know the currently retained load-carrying capacity of the bridge and how long the bridge will be able to be in service.

To know them properly, the following problems needed to be solved:
(1) Designed to the load of 13tf class (TL-13), can the bridge carry the load of 25tf class (Live load B) which is the current design load?
(2) No method has been established in general for guaranteeing the current design load, although there has been partially disclosed a method for guaranteeing the traffic load for the time being.
(3) From the fact that fatigue cracks were identified in the main members of the bridge during the inspection thereof, the other similar portions of the bridge needs to be checked for fatigue cracks. In addition, the bridge needs to be checked for other problems which would occur during a long service in the future.

(4) No method has been established in general for predicting the occurrence of fatigue cracks.

In other words, to summarize the problems with grasping the actual state of the bridge, the methods for "evaluating the loading capacity of the bridge" and "predicting the occurrence of fatigue cracks" have not yet been established in general.

3. Method for Grasping Actual States of Bridges

To solve the problems with grasping the actual state of the bridge, the following items were generalized and systematized as well based on the actual measurement of the bridge.
  (1) Generalization of loading evaluation method,
  (2) Generalization of fatigue evaluation method, and
  (3) Standardization of actual bridge measurement

3.1 Method for Evaluating Loading Capacity

Old bridges were designed to a load which is smaller in magnitude than the current design load. Due to this, there are some organizations which disclose methods for guaranteeing the safety of bridges for the time being against the traffic loads.

However, the methods are not such that they guarantee the loading capacities for the current design load. To cope with this, a "method for evaluating the loading capacity of a bridge through a load test using a loading vehicle" was determined to be provided.

According to this method, a load as close to the current design load as possible is applied to a bridge, a structure analyzing model is prepared based on the results of the measurement of the actual bridge, and the loading capacity against the current design load is evaluated through a design calculation method. Fig. 3.1 shows the flow of the evaluation.

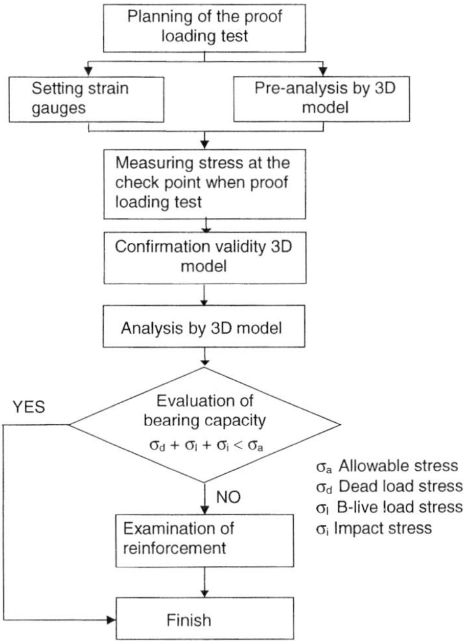

σ_a Allowable stress
σ_d Dead load stress
σ_l B-live load stress
σ_i Impact stress

Fig. 3.1 Flow of Loading Capacity Evaluation

Flowchart contents:
- Planning of the proof loading test
- Setting strain gauges
- Pre-analysis by 3D model
- Measuring stress at the check point when proof loading test
- Confirmation validity 3D model
- Analysis by 3D model
- Evaluation of bearing capacity: $\sigma_d + \sigma_l + \sigma_i < \sigma_a$
- YES → Finish
- NO → Examination of reinforcement → Finish

3.2 Method for Evaluating Fatigue

Fatigue damages emerge to the surface of a structure as cracks, and fatigue cracks are different from other damages in that they are difficult to be identified and quick to propagate once they occur. Due to this, since visual inspection is not good enough to cope with the fatigue cracks, with a view to complementing the visual inspection, there has been established in general for steel bridges a method for quantitatively grasping the prediction of occurrence of fatigue cracks in and the degree of accumulated fatigue of a bridge.

This paper also discloses a method for evaluating the fatigue of a bridge through a measurement method developed by taking into consideration the singularity of road bridges based on the aforesaid generally accepted idea.

The flow of the fatigue evaluation is shown in Fig. 3.2.

270 MAINTENANCE PRACTICE

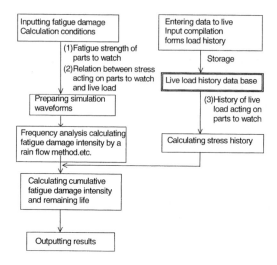

Fig.3.2 Flow chart of fatigue damage life evaluation and calculation

3.3 Standardization of Actual Bridge Measurement

Since the aforesaid methods for evaluating the load-carrying load and fatigue of a bridge are based on the measurement of an actual bridge, actual acting stress acting on respective members of the bridge needs to be measured.

Strain gauges are generally applied to the respective members of the bridge used to measure this actual acting stress acting thereon.

However, since there is caused a difference in actual acting stress so measured depending upon where to apply the strain gauges, locations where strain gauges are applied need to be standardized in order to make objective the results of diagnosing the bridge with respect to the loading capacity and fatigue thereof.

Fig. 3.3 shows an example of standardization of strain gauge application locations.

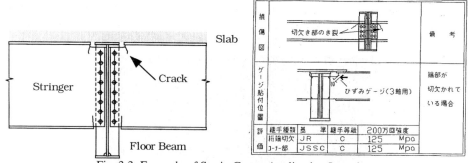

Fig. 3.3 Example of Strain Gauge Application Locations

3.4 Systematization

These evaluation and actual bridge measurement methods are being tried to be systematized as a bridge diagnosis system so that structure engineers can use them simple and easily. Below are main functions of the system:

  (1) Measuring function,
  (2) Analyzing function, and
  (3) Diagnosing function (for loading capacity, fatigue and the like)

4. Example of Grasping Actual State of Bridge
4.1 Summary of Bridge

Fig. 4.1 briefly shows a bridge whose actual state was grasped this time, and the specifications thereof are shown below.

Bridge type: steel deck bridge plat truss
Span: 14.0 + 110.0 + 14.0m
Design load: TL-13
Date of build: November, 1954

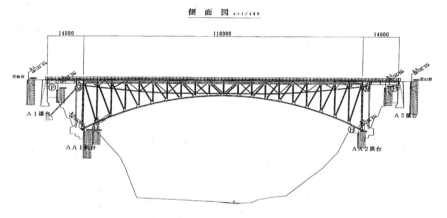

Fig. 4.1 Bridge Overview

4.2 Loading Capacity Diagnosing

The loading capacity of the bridge was diagnosed while following the aforesaid flow of the evaluation, and the following steps were taken of measuring the actual bridge using a load test with a load-applying vehicle, preparing a structure analyzing model and calculating the results of the measurement using a design calculation method for the current design load (Live load B).

Photograph 4.1 briefly shows the load test using the load-applying vehicle, and Fig. 4.2 shows the analyzing model used. In addition, Table 4.1 shows the results of the evaluation of the loading capacity according to the method.

Photo 4.1 Summary of load test using a loading vehicle

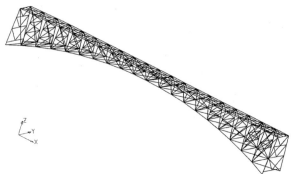

Fig. 4.2 Structure analyzing model

Table 4.1 Results of load-carrying capacity evaluation

Member	Panel Point	Measurement stresses of load test A (kgf/cm^2)	Calculate stress B (kgf/cm^2)	Ratio of real stress (α) A/B	Calculate stress of B live load C (kgf/cm^2)	Dead load stress (σ_d) E (kgf/cm^2)	Impact stress (σ_i) F (kgf/cm^2)	Total allowable stress (σ_a) G=D+E+F (kgf/cm^2)	Allowable stress (σ) (kgf/cm^2)	Judgment
L/8 Dia.mem.	U0-L1	92	95	0.97	392	402	58	852	1300	OK!
L/8 Ver.mem	U1-L1	-85	-92	0.92	-238	-373	-87	-698	-660	NG!!
L/8 Low.Chord	L1-L2	-80	-84	0.95	-251	-705	-32	-988	-1077	OK!
L/8 Upp.Chord	U1-U2	8	9	0.89	-65	-113	-8	-186	-1067	OK!
L/4 Upp.Chord	U4-U5	-32	-35	0.91	-172	-355	-22	-549	-1072	OK!
L/4 Dia.mem.	U4-L5	110	122	0.9	486	700	73	1259	1300	OK!
L/4 Low.chord	L4-L5	-97	-104	0.93	-283	-713	-36	-1035	-1072	OK!
L/4Ver.mem.	U5-L5	-103	-113	0.91	-343	-605	-123	-1071	-764	NG!!
L/2 Upp.chord	U9-U10	-45	-49	0.92	-209	-548	-25	-782	-1075	OK!
L/2 Dia. Mem.	U9-L10	-82	-85	0.96	354	188	53	595	1300	OK!
L/2 Low.chord	L9-L10	-20	-22	0.91	-156	-323	-19	-498	-1078	OK!
L/2 Ver.mem.	U1-L10	-113	-122	0.93	-191	-121	-68	-380	-1045	OK!
End Stringer	S1	203	225	0.9	783	225	281	1289	1300	OK!
Int.Stringer	S1	197	231	0.85	747	225	265	1237	1300	OK!
Int.Floor Beam	F1	-101	-111	0.91	604	375	214	1193	1300	OK!

The results show that the shortage of load-carrying capacity was found in vertical members at 1/8 and 1/4 points, and that the other members were found to be suffering from no problem.

4.3 Fatigue Diagnosis

A partial fatigue crack was identified as occurring in a notched portion in a stringer of this bridge. The bridge was actually measured at other similar portions of the bridge for fatigue cracks in order to study about a range to be coped with, and the evaluation of fatigue was carried out based on the results of the measurement of the actual bridge.

Figs. 4.3 and 4.4 show the results of a stress frequency analysis and prediction of occurrence of cracks based on the measurement of the actual bridge, respectively.

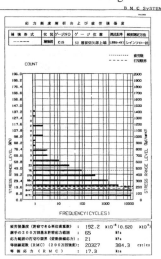

Fig. 4.3 Results of stress frequency analysis

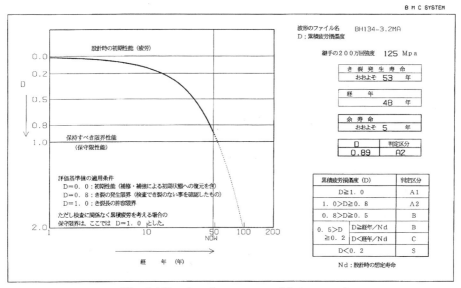

Fig. 4.4 Results of occurrence of fatigue cracks

The results show that it is highly possible that cracks will occur at similar locations of the notched portions in the other stringers in the near future.

5. Conclusion

The paper proposes the method for objectively evaluating the currently retained performance of a road steel bridge with respect to the load-carrying capacity and fatigue thereof, which were taken for example as factors for evaluation, based on the results of the measurement of the actual bridge.

In addition, the evaluation method is now being tried to be systematized for simple use by engineers who manage structures.

It was concluded from the results of the evaluation that the structure would continue to be used sufficiently from now on, and therefore, it was determined that the following countermeasures be taken on the vertical members which are short of load-carrying capacity and the strings which are concerned about fatigue, and the road width widening plan was determined to be effected by making needed modifications to the bridge, whereby the prolongation of the life of the bridge and reduction in costs involved in the plan could be achieved.
(1) Sway Bracing are provided so as to reduce the elongation ratio of the vertical members.
(2) The stringers are replaced during floor slab work when a road width widening program is carried out.

References
1. "Fatigue Design Guide" Japanese Society of Steel Construction (JSSC), 1993
2. "Road Bridge Design Manual" Japan Road Association, Dec. 1996

Installation of advance warning system at highway structures which are susceptible to flooding

P C WONG and C Y WONG, HIGHWAYS DEPARTMENT,
LEWIS H Y HO and Y H LEUNG, ELECTICAL & MECHANICAL SERVICES DEPARTMENT, HONG KONG SAR GOVERNMENT, HONG KONG

Introduction

During rainy seasons, flooding occurs in pedestrian subways and vehicular underpasses due to unexpected heavy rainfall, trapped rubbish, incoming power failure, tripping of electrical protective device or pumping system failure. In order to adopt a proactive approach to avoid flooding, an advance warning system is installed into the existing drainage system to detect the fault signals in association with the flooding and subsequently relay the warning message to the fault call centre for immediate follow-up actions by the Government.

Background

In accordance with the Hong Kong Observatory's record, the annual average rainfall in Hong Kong is about 2,200 mm. Situated in the tropical area, Hong Kong is one of the areas in the Pacific Rim having heavy rainfalls. It is not uncommon to have heavy rainstorms every year. During the rainstorms, serious flooding particularly occurs in the northern part of the territory where most of the flood plains in Hong Kong are found. The situation is heightened by heavy rainstorms and typhoons, often concentrated into short periods of time and leading to the cases of severe flooding.

On the other hand, to cope with the high degree of urbanization and increasing population in Hong Kong, the territory development as well as the construction of infrastructure has been expedited. As a result, the network and coverage of the highways system has been extended and there are nearly 400 subways and underpasses throughout the territory, facilitating both pedestrian and traffic flows and ensuring easy and unrestricted access to public areas in Hong Kong. It becomes increasingly important to ensure the free traffic flow on highways. In order to ensure that the drivers and passers-by can use the highways and footways as usual during the rainy seasons, drainage system with pumping facilities is already installed to avoid flooding at the pedestrian subways and vehicular underpasses.

However, for years, in particular during the heavy rainstorms, the pumping systems suffer mal-functions. It caused much inconvenience to the public as the roads and highways were flooded. Despite great effort in attending the faults all over the territory, the problem could not be satisfactorily resolved.

The main causes of the flooding, based on the past records, are due to a variety of reasons. These include:

- Interruptions to the mains power supply which is much more likely during bad weather
- Malfunction of the pumping system and its controls
- Heavy rain outweighing the system's capacity
- Drains blocked by debris

Needless to say these problems are exacerbated during very heavy rain. The public experiences considerable inconvenience when individual system breaks down. Whenever flooding happens, pedestrians and vehicle drivers would be forced to pass over flooding water. Moreover, vehicles may break down during flooding and then result into undesirable traffic jam. The flooding situation can only be rectified after the event. Measures can only be taken to clear the flooding after it has taken place. Should the main causes be detected in advance, appropriate action can be immediately taken to ensure that the pumping system function in the way it is purposed for. In order to achieve this purpose, a centralized advance warning system for the drainage system was therefore proposed as a preventive measure to avoid flooding by remotely and constantly monitoring the condition of the pumping system and the pre-flooding situation as well as raising an alarm if a problem is about to occur.

Advance Warning System by Auto-dialing

The advance warning system is mainly composed of sensor devices, remote monitoring units and a fault call receiver at the fault call centre. The communication medium between the respective remote monitoring units and the receiver is the mobile network commonly used in the territory. The wireless model of the advance warning system is shown in Fig. 1.

Fig. 1 – Wireless Model of the Advance Warning System

The tailor-made sensor devices are designed to detect those common flooding symptoms including incoming power failure, high water level in pump pit of the pump house, pumping facilities failure as well as flooding in barrel. The control output of these sensor devices will subsequently initiate the respective fault signals to the on-site remote monitoring unit for signal processing and reporting. Then, a warning message will automatically be relayed from the remote monitoring unit to the receiver in the fault call centre through mobile network. A typical subway layout is depicted in Fig. 2 to illustrate the general arrangements of the respective sensor devices and the remote monitoring unit.

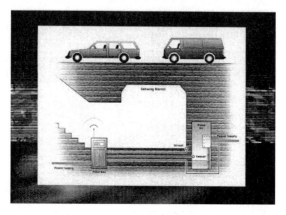

Fig. 2 - Typical Subway Layout

The simplified workflow for the operation of the advance warning system is shown in Fig. 3. The fault call centre is operated around the clock. On detection of any of the warning signals from the sensors, the remote monitoring unit will send the electronic signal to the fault call receiver. The fault call centre will then inform the maintenance personnel for immediate emergency repair so as to resolve or alleviate the flooding problem.

Fig. 3 - Simplified Workflow

There are four different types of tailor-made sensor devices. Three of them are used as preventive means to detect the common fault symptoms leading to flooding. The last one is to detect flooding in the barrel. Basically, the sensors will feed the fault signals to the remote monitoring unit for warning message referral. The schematic diagram showing the detailed control interface of the sensor devices and the remote monitoring unit is in Fig. 4.

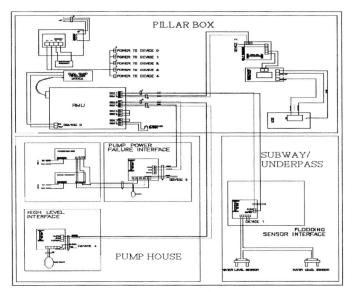

Fig. 4 - Schematic Diagram of Sensor Devices

The voltage output from the mains power source is continuously monitored by a voltage sensing unit. Once power failure occurs due to cable fault or power interruption from power company, substantial voltage drop will be detected and then incoming power failure alarm will be triggered immediately to alert the maintenance personnel who will in turn inform the power company to arrange site checking. The unavailability of the electrical power is always the potential cause to the serious interruption of the pumping operation leading to prolonged flooding.

Inside the pump house, there is a pump pit to collect the stormwater. When the water inside the pump pit reaches a designated high level, the pump will start automatically to drain off the water through the drainage pipework system. A logic detection unit is incorporated into the pump control circuit to check whether its control functions are normal under this circumstance. In case of failure of the control circuit due to poor relay contact or some other reasons, the pump will not be operated and a pump failure alarm will be initiated by the logic detection unit accordingly. The maintenance personnel will then check the control circuit of the pumping system.

Apart from the above two prevention measures, the pre-flooding situation is continuously detected by a level sensor which is placed at the pump pit inside the pump house. If the pump house gets flooding, the water will flow into the subway barrel soon. Therefore, this pre-flooding sensor will operate to give an advanced warning signal indicating that the subway will be flooded shortly. The maintenance personnel must be informed of this pre-flooding situation.

As a last resort, level sensors are installed at the low level of the side-wall of the barrel. In case flooding occurs in the barrel, the sensors will be immersed in water and then a flooding alarm will be forwarded to the remote monitoring unit. In this case, the maintenance personnel must arrive at site immediately to drain the water away using hand-held portable pumps.

Each sensor device is also incorporated with a time delay function to filter all the transient signals leading to false alarm. The filtering mechanism is intended for filtering the transient power interruption, short-lived flooding, etc.

Remote Monitoring Unit (RMU)

The major functions of the RMU are to monitor the status of the drainage system by gathering the respective fault signals and then transfer the warning messages to the fault call receivers by means of mobile network.

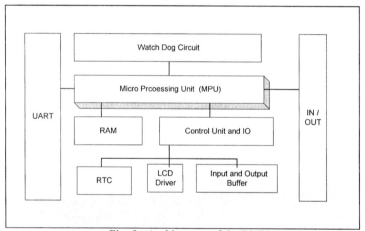

Fig. 5 - Architecture of the RMU

The architecture of the RMU is shown in Fig. 5. The RMU is a piece of hardware installed in the subways and underpasses. The major components of the RMU consist of a Micro Processing Unit (MPU), 256k bit Random Access Memory (RAM), self-contained Real Time Clock (RTC), self-contained Watch Dog Circuit (WDC) and Universal Asynchronous Receiver Transmitter (UART). The RMU interface is also equipped with eight external monitoring node and four internal system node. The external nodes can be connected directly to the alarm sensor devices.

The RMU contains system watch dog circuit. The MPU of the RMU is programmed to generate a pulse in every 100ms and then the signal is feed into the watch dog circuit. If the watch dog circuit fails to detect the 'live' pulse, it will trigger the reset pin of the MPU for a cool reboot. At the boot-up cycle, the MPU also checks the content of the RAM. If the content is found inconsistent, the MPU will clear the content and refresh a new set of data into it. The watch dog circuit is capable of monitoring the incoming DC power supply too. When the supply voltage drops to a certain unacceptable level, it will inform the MPU and raise an alarm flag.

In order to ensure that the RMU is still working under emergency situation, the RMU is equipped with built-in backup battery to sustain 6-hour operation once power supply is failed.

Receiver

The receiver consisting of the master terminal unit performs like a central host to gather all the field data in the subways and underpasses. The basic functionality of the receiver includes :

- Data display showing the fault status
- Event acknowledgment and logging
- Scheduled checking on remote sites
- Maintaining fault calls referral list
- Accepting incoming calls
- Database management and linkage with other supporting database
- Capturing data and report generation

The computer-based receiver is located in the fault call centre to centrally collect the fault messages from the respective remotely-located on-site RMUs. After receiving the fault message, the receiver will alarm the fault call centre and the fault message will be flashing until the operator acknowledges the fault on the screen. Then, the fault message will be placed in the waiting queue of Fault Message Window and waiting for the signal feedback for site attendance and completion of the repair. The relevant details including the date and time of receiving the fault, arriving the site and clearing the fault are all automatically logged by the receiver for record and analysis. The screen of the receiver is shown in Fig. 6.

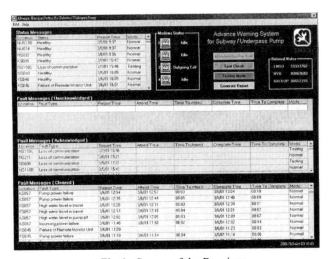

Fig.6 - Screen of the Receiver

Moreover, the receiver shall, based on the fault location and fault type, automatically and directly refer the pre-set voice message to the relevant maintenance personnel to inform them to correct the situation as soon as possible.

An interesting feature of the receiver is that it is incorporated with a feedback function for tracing the work status of the on-site repair work. There is an attendance switch in the control panel of the RMU. The maintenance personnel should press the attendance switch when arriving at site to confirm his attendance and a feedback signal will be sent back to the receiver by the RMU. The receiver programme will log the attendance time for record and analysis. Moreover, when all the monitoring parameters resume healthy after repair, a completion status will be automatically generated by the RMU and communicated to the receiver. These features will enable the two-way communication between the on-site RMU and the receiver in the fault call centre.

The software of the receiver also has an escalation function. If the maintenance personnel can not reach the site to rectify the situation within the designated time period, escalation alarming will be automatically initiated to alert the supervisory staff. He should either immediately tackle the problem personally or provide support, if needed, so that the flooding situation will not become worsen.

During the raining period, there may be several flooding areas around the same region and a number of flooding messages will be brought up to the receiver at the same time. In order to ensure that these fault messages are not missed, the receiver is supported with four modems for receiving the mobile signals. If the first modem is busy, the incoming call will be re-directed to the stand-by modem by the phone hunting system.

Everyday the receiver will also dial to the respective on-site RMUs to constantly check the equipment status to see whether they are healthy or not. This regular communication with the RMUs will ensure the subways and underpasses are under close monitoring and the warning messages will not be missed by this daily back-up checking arrangement.

Innovative Design

As there exists technical difficulty in installing data line at the subways or underpasses because of site constraints, complicated roadworks, high capital cost, etc., the system adopts the advanced wireless technology. Mobile modem is chosen as the data transmission media to automatically make the call. The digital data is transmitted to the fault call centre using the mobile phone network. With this new concept of design, we not only save the telephone cable laying cost but also greatly speed up the installation time. At the same time, no excavation or trench work was necessary for cable laying, thus minimising any disturbance or inconvenience to the public.

Conclusions

We are committed to providing an improved and more convenient traffic and pedestrian services throughout Hong Kong, even in the worst of weathers. Using the advance warning system, the fault attendance team, on receipt of the fault reports, will provide prompt and efficient services. The flooding situation can be rectified anywhere and anytime so that the inconvenience caused to motoring public and pedestrians will be scaled down to the minimum.

From an operational point of view, the maintenance staff are able to rectify the problems almost immediately, controlling the situations before they could grow in seriousness or complexity. Preventive action could now be taken as faulty pumps could be fixed before the rain arrives. This is in direct contrast to the past practice when staff were only able to respond to the situation after flooding had taken place, often relying on a member of the public to raise the alarm. Additionally, with the installation of these systems, the underpasses and subways should experience almost no flooding whatsoever.

Acknowledgements

The authors would like to express thanks to the Director of Highways and Director of Electrical and Mechanical Services of Hong Kong SAR Government for their permission to publish this paper and to all those people who have assisted in making this a successful project.

REHABILITATION OF TSING YI SOUTH BRIDGE, HONG KONG

PC WONG and CY WONG, Highways Department, Hong Kong SAR Government
F KUNG, Hyder Consulting Limited

INTRODUCTION
This paper first describes the recent investigations and studies carried out by Hyder Consulting Ltd. on behalf of the Hong Kong SAR Government Highway's Department for the rehabilitation of the Tsing Yi South Bridge in Hong Kong. The work included a detailed inspection of the structure, extensive materials and load testing, assessment of the load capacity, design of hinge replacement and proposals for rehabilitation.

The remaining sections of the paper discuss the construction of the rehabilitation works, particularly the hinge replacement, which was carried out and completed during a complete closure of the bridge for 7 months in last year 2000.

Fig 1 General View of Tsing Yi Bridge

Fig 2 Side View of Bridge

HISTORY
Tsing Yi Bridge is an important structure being the first major bridge sea crossing in Hong Kong. It was constructed between 1971 and 1974 and carries 2 lanes of road traffic and two footways between Kwai Chung and Tsing Yi Island over a busy shipping channel. The bridge also carries key utility services to the residents of Tsing Yi Island and from the power station on the island including water and gas mains, telephone cables and high voltage electric cables.

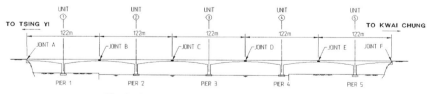
Fig 3 General arrangement showing notation

The deck is a six span prestressed concrete box girder constructed as five 122m long balanced cantilever units built symmetrically from each pier. The cantilevers are linked together with metal roller hinges at midspan. The ends of the bridge are supported on reinforced concrete bank seat abutments. The reinforced concrete piers are supported on a combination of spread footings and large diameter bored pile foundations. The water is up to 18m deep and the minimum headroom for the navigation channel is 17m.

The bridge has been the subject of various previous inspections, deflection monitoring, load assessment and materials testing work: In 1983 Highways Department (then Highways Office) appoints Mott, Hay and Anderson to carry out an extensive study of the bridge, particularly for deflections of the cantilevers. They recommended additional external prestressing work, which was implemented in 1990. In 1991 Ove Arup and partners were commissioned to carry out a survey on the concrete spalling and cracking of the bridge pier and deck. Rusting of reinforcement and spalling of concrete was located and made good in 1994. The previous studies of the bridge included materials testing, load testing and radiography investigation of the prestressing cables.

The structure has been subjected to a number of collisions from barges. Most of them are impacts on the bridge superstructure by derrick booms of the barges. The worst in a typhoon in 1992 caused damage to 52m length of catwalk and an 8m section of the main deck edge cantilever.

Serviceability Issues
Significant vibrations in the bridge deck are experienced under normal traffic loading particularly at midspan. This is mainly due to the flexibility design of the deck and is exacerbated by the movements across the damaged hinges. The ends of the cantilever units have deflected visibly due to long term creep of the concrete. Additional external prestressing tendons installed in 1990 raised the bridge levels at the joints by about 100 mm on application which reduce to 80 mm and appears to be fairly stable thereafter.

Fig 4 View of road alignment   Fig 5 View of internal tendons

INSPECTION AND ACCESS
In 1999 Highways Department appointed Hyder to carry out an investigation on the deteriorated bridge hinge joints and to conduct a comprehensive inspection on the bridge. The inspection was carried out in accordance with the UK Highways Agency guidance for inspections for Assessment[2]. A variety of access techniques and safety precautions were required to inspect the different parts of the bridge safely. These included: with high visibility jacket for all roadside inspections; with harnesses for the high level inspection; using the Highways Department's underbridge inspection vehicle and the abseiling method for the inspection of the deck soffit and the piers. Of particular note was the requirement of wearing full asbestos protection equipment to inspect inside the box. The China Light and

Power cables inside the box are coated with asbestos materials and safety precaution for working with asbestos must be taken. The China Light and Power is an electric company supplying electricity to Kowloon and the New Territories. A specialist diving contractor was appointed to carry out the inspection of the piers and substructure below water level.

Fig 6 View inside box

Fig 7 Rope Access

Inspection Findings

The bridge was generally found to be in good condition with isolated areas of concrete spalling. The midspan hinges are all in a serious state of distress with extensive corrosion. The hinge mechanism is a roller that is designed to bear on the upper and lower bearing surfaces of the joint. This design requires a small tolerance in the joint to allow the rolling movement. This tolerance gap has increased with wear and has been deteriorated to a state that significant vertical movement across the joint can be seen both at road level and at the hinge itself.

At the end of each T units, the concrete at the bottom slab of the box shows significant spalling which is aggravated by rain water leaking through the access manhole at the deck surface. The top of the bottom slab is in poor condition near the piers in areas where the cover is minimal and water has been standing. This does not cause immediate concern as the loss of concrete section is not a significant percentage of the deck cross section but it should be repaired. The abutment provides both reaction support and uplift restraint. The original bearings are located at both top and bottom of the web extensions of deck.

Fig 8 View of manhole access and exposed rebar in bottom slab

Fig 9 View of existing hinge from inside

MATERIAL TESTINGS
Concrete Testing

Concrete tests were carried out to determine the strength and condition of the concrete and investigate the level of chloride contamination and corrosion of reinforcement. These included: (i) concrete cores for crushing, aggregate cement ratio, original water content and modulus of elasticity laboratory tests; (ii) drilling dust samples for chloride content and sulphate content; (iii) carbonation depth in holes used for dust sampling; (iv) hammer tapping and pull off tests of mortar to identify delamination; (v) covermeter survey to check the depth of cover to the steel; (vi) half cell potential and resistivity testing to check for the presence of corrosion; and (vii) concrete breakout to expose the reinforcement for examination.

The results showed the deck concrete to be generally in good condition with little risk of durability problems except in isolated areas. The pier concrete results indicate that the repair mortar placed in 1994 is resistant to chemical ingress though it has uniform high levels of sulphate which are likely to be from a mix additive. The chloride profiles indicated a possible back migration from the contaminated older concrete. The average depth of carbonation measured was 3.3mm on the external face, which gives a rate of carbonation in the order of 0.55mm per year, indicating that the coating is effective in terms of reducing the rate of carbonation. The mortar pull off tests on the repair concrete around the piers showed the bond to be weak (average failure stress 0.78MPa), but no delamination was identified by the tapping survey.

Insitu Stress Measurement
The residual stress measurements can provide useful information on the structural behavior of a bridge. The insitu strains in the concrete were measured using the stress relief coring method and were used to calculate the actual stress in the concrete section at various positions along the deck. The measurement involved taking cores near the extreme fibres of the concrete section and measuring the change in strain around the hole.

The accuracy of the residual stresses determined using this method is highly sensitive to effects caused by the misalignment of the drilled hole, presence of reinforcement, local thermal effects and global temperature differences. Every precaution was taken during the tests to minimise these effects or to carry out additional measurements to correct these effects. However, due to congestion of reinforcement and variation of temperature, the residual stress results obtained from the site tests were difficult to compare with the analysis predictions.

Stress in External Tendons
A tendon pull-off test was carried out in order to measure the force in the external tendons. A jack was placed onto the tendons at the jacking end and a force applied until the end plate has seen to lift off the anchorage. This gives a measure of the actual force in the external tendons. An average of 3% loss of initial jacking force was measured indicating that the external tendons are performing well.

Special Investigation of Post-tensioning
A series of radiographic photographs of the internal tendons in the top slab were taken at various cross sections. The radiographs showed about 10% of the internal prestressing tendons contain voids but did not indicate any significant damage or corrosion. About 10% of the voids were investigated by drilling into the tendon duct and inspecting by boroscope. The tendons inspected were generally in good condition with minor surface corrosion. No water was encountered in the voids.

LOAD TESTINGS
Static Load Tests
A series of load tests were carried out on the bridge using the China Light and Power heavy vehicle loaded with concrete blocks. The objective of the load testing was to study the

behavior of the structure for comparison with the assessment. The static load test involved strain and deflection measurements caused by a large test vehicle positioned at various locations along the bridge. Strain gauges were located in the longitudinal direction near the extreme fibres inside the bridge and the readings were electronically recorded. Horizontal and vertical deflection measurements were taken using traditional surveying techniques.

The measured deflections were compared with the predicted deflections assuming a free or a partially restrained hinge using the assessment methods as described in the later sections. It was generally suggested that there were some jamming effects on all of the hinges, except for Hinge C. The jamming effects could be due to the condition of the horizontal hinges at top of box or due to the fact that some shim plates were temporarily installed to control the vertical deflections or even probably due to some foreign materials trapped in the expansion gap.

The measured and the predicted strain readings were also compared. Based on the section properties including the concrete upstands and footways, the predicted strains are in close agreement with the measured strain.

Dynamic Response Measurements

Dynamic response measurements were taken under traffic loading to estimate the modes of vibration, the mode shapes and the corresponding damping ratios. The dynamic testing was carried out using a series of accelerometers placed along the length of the bridge. The vibration monitoring was carried out under normal traffic flow to measure the mode shapes for the vibration of one cantilever unit and compare the general movements of the other units. The principle modes of vibration and a measure of the amplitude and frequencies of those vibrations were determined and the results showed eight distinct mode shapes. The first eight natural frequencies were within the range of 0 to 2.5 Hz. The lowest natural frequency was 0.703 Hz which corresponded to the pier torsion dominated mode. The highest natural frequency was 2.461 Hz which was an out of plane pier bending and deck bending mode. No deck torsional mode was identified in the first eight natural frequencies.

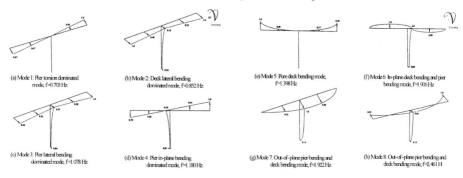

Fig 10 Vibration Modes and Relative Amplitudes

The measured damping ratios were between 1% and 5%. The lowest damping ratio was 1.18% which corresponded to the 3rd out-of-plane mode (Mode 7), and the highest damping ratio was 4.79% which corresponded to the 1st in-plane mode (Mode 1). It was concluded that the east end of unit 2 and both ends of unit 3 move more freely than the others. The measurements also confirmed that the lateral vibrations at the base of unit 3 were greater than those of unit 2.

The analytical assessment as described in the later section suggested that Mode1 to 4 of the site measurements were in close agreement with the predicted. The remaining 4 modes showed some variation, but as those modes had relatively lower amplitude it would not be considered any further.

ANALYTICAL ASSESSMENT

The assessment was generally based on the principles of BD 21/97 and BD 44/95 but to the current design loading given in the HK Structures Design Manual for Highways and Railways. The scope of the assessment was: (i) to determine, in terms of vehicle loading, the maximum load that the structure will carry without suffering serious damage so as to endanger any persons or property; (ii) to predict the residual life of the structure based on the results of the materials testing; and (iii) to investigate the feasibility of rehabilitation options, i.e. to predict the behavior of the bridge for proposals which change the form of the structure.

The structure was modeled using three different analytical models in order to ensure that the structural behavior of the bridge is correctly assessed. The models were verified against historical records and recent testing. The analysis included the following work:
- a basic check on maximum service limit state stress levels at critical points and a calculation of the ultimate limit state load carrying capacity.
- a sensitivity analysis to check for variations in prestress level, dead load, superimposed dead loads (SDL), temperature effects.
- a calculation of the natural frequencies of the structure to compare with the site testing and normal design limitations.
- a prediction of the long term deflection of the structure.
- a check of the structural behavior and effect of proposed remedial measures.

SuperSTRESS[3] *model*

A simple 2-D plane frame model of the entire Tsing Yi South Bridge was established using the SuperSTRESS program. For the model all hinge joints were free for longitudinal and rotational movements. The abutment joints were free in the horizontal direction and fixed in the vertical direction. The spread foundations for Pier Nos. 1 and 4 were modelled as rigid supports. The piled foundations for Pier No.'s 2, 3 and 5 were modelled as spring supports.
The model was used to calculate the ultimate load capacity, to assess the global effect of the rehabilitation options and to check the output from the ADAPT model.

ADAPT[4] *Model*

A 2-D plane frame model of the entire bridge was generated using the ADAPT program. The model was built up in stages to match the construction sequence and strengthening works. The hinges were idealised as members with released longitudinal movement, axial force and in-plane bending moment. The program can predict the time dependent deflections of the structure during each stage of construction. The analysis can be projected forward to predict the future behavior. The model was used to determine the serviceability limit state stresses during construction and the current situation. The deflection results were compared with the historic survey data.

SAP2000[5] *Models*

Two separate models were generated using the SAP2000 program. The first model of one cantilever section from the face of the pier to the hinge point was generated using solid elements. This was used to study the transverse effects on the box girders and the possible effect of shear lag. The second model of the entire bridge was generated as a line model.

The hinges were idealised as members with released longitudinal movement, axial force and in-plane bending moment. The model was used to determine the natural frequency of the bridge and to compare with the site tests.

Assessment Findings
The bridge satisfies the serviceability limit state stress limitations even with a 30% additional loss of internal prestress. The structure can still sustain 150% of the current ultimate design loading with this assumed further loss. The effect of this additional loss is a further deflection of the cantilever tips of 120mm at Hinges C & D and 70mm at Hinges B & E.

The site deflection measurements are more than those predicted from the ADAPT model and, as indicated from a sensitivity analysis, they are very sensitive to temperature gradient. The discrepancy could be due to differences in the elastic modulus or greater than predicted relaxation of the tendons.

The analysis of various rehabilitation options has shown that the possibility of modifying the articulation of the structure is significantly restricted. Providing concrete stitches across all the joints would cause overstress on the piers and foundations due to longitudinal temperature effects. This option would require major strengthening work to the substructures. If just Hinge joints B & E were stitched then the strengthening work would only be necessary at each of these stitched joints.

The analysis and testing has highlighted the following features:
- The downward deflection at Hinges B & E are up to 20% greater than at Hinges C & D. This is caused by the tilting of the end units as the downward creep of the end cantilevers is restrained by the abutment supports.
- The joints or hinges at B, D & E are experiencing some jamming.
- The footways and upstands not included in the structural model may be contributing to the live load capacity.
- The dynamic analysis and the dynamic response measurements both indicate that in accordance with SDM, the structure does not cause significant discomfort to drivers though very noticeable to pedestrians

REHABILITATION AND HINGE REPLACEMENT DESIGN
The options for rehabilitation included the following:
- do minimum
- carry out hinge replacement and remedial works to specific areas in poor condition.
- remove the hinges by replacing them with stitched joints.

The first option was not appropriate as the serviceable life in the current condition is too short. Hinge replacement in the second option was considered to be essential in the short to medium term as the hinges are in a very poor condition. The third option to stitch the bridge deck, as discussed in the analysis above, is limited by the effects on other parts of the structure. The possibility of removing the maintenance problems of two of the hinge joint areas is attractive but proves to be a costly alternative. Solutions to raise the midspan to level out the vertical alignment proved to be uneconomic. There is limited space for additional prestress. Improving the road alignment at road level by resurfacing imposes additional load that will deflect the cantilevers further.

A major issue for either the short term or long term works to Tsing Yi Bridge is that of traffic management and road diversions, including disruption to the navigation channel. Despite the fact that there are new bridges on either side, the area is highly sensitive to proposed developments and traffic growth being part of the link to the airport. The proposed rehabilitation works were designed to minimise any disruption to the existing traffic.

The existing hinge comprises a vertical roller mechanism in each web and a horizontal roller in the top slab. The replacement hinge has been designed to allow longitudinal and rotational movements across the joint thereby maintaining the original articulation. It was decided to replace the vertical web roller system with a purpose made pot bearing. The moving parts shall be easily inspected and will be replaceable without extensive concrete breakout. The top horizontal bearing does not require such a large rotational capacity so a separate simple guided bearing will be installed to prevent differential transverse movements across the joints. Combined with these works will be a complete rehabilitation of the hinge area and replacement movement joints.

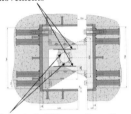

Horizontal sliding surfaces for translational movements

Spherical sliding surfaces for rotational movements

Fig 11 View of replacement hinge   Fig 12 Detail of replacement hinge

CONSTRUCTION WORKS
Following the investigation study, Highways Department decided to proceed with option 2 for rehabilitation immediately. Listed below are the major works items.
- ***Replacement of Hinges (12 nos.), Abutment Bearings (8 nos.) and Movement Joints (6 nos.)***: including supply/installation of bearings and M.J., and concrete breakage and reinstatement.
- ***Concrete Repairs***: particularly the spalled concrete at webs, top of bottom slab and local areas near assess manholes.
- ***Surfacing***: removal of existing surfacing; re-laid of a new waterproofing system and finally reinstate new surfacing including re-profiling near the mid-span hinges.
- ***Temporary Traffic Diversion***: including road markings, traffic signs, road widening works and pavement modification/reinstatement

The construction works were included as part of the Highways Department's Maintenance Term Contract No. 19/HY/1998. Chiu Hing Construction and Transportation Co. Ltd. was the main Contractor, with specialist works on hinges and joints replacement carried out by Freyssinet Hong Kong Limited and CCL respectively.

ANTICIPATED SITE CONSTRAINTS/DIFFICULTIES
The extent and type of work has not been done previously in Hong Kong. The Project required high technical skills and supports of all parties involved, and a careful planning of the works. The bridge was constructed nearly 30 years. While a good record of the as built information was kept in Highways Department, there were inconsistencies found in the reinforcement details, which required on site adjustment including addition of extra

reinforcement around the hinge. The time for breaking out concrete and removal of the hinges took longer than originally planned. The setting out of the hinge to its final position also imposed a challenge to the specialist sub-contractor. A special steel frame was developed to control and adjust the final levels and setting out of the cantilever tip for the installation of the hinges.

Inside the box, there were a number of high voltage cables and other utilities. They need to be protected, in particular during the hinge replacement works. Safety requirements over Rambler Channel and Container Terminal were strictly followed.

Majority of works was carried out in daytime during the complete closure of the bridge. Resurfacing was however carried out during night time in order to minimise the impact on the other works on site.

TEMPORARY TRAFFIC DIVERSION ARRANGEMENT

In order to carry out the proposed hinge bearing and abutment bearing replacement works, Tsing Yi South Bridge was closed for 7 months from 18 December 1999 to 22 July 2000. The other rehabilitation works were also carried out during the full road closure. The closure of the existing bridge did not affect the Tsing Yi bound traffic. The Kwai Chung bound traffic was however affected. Traffic from both the Tsing Yi Heung Sze Wui Road and the Tsing Yi Road were diverted to the newly constructed CT9 access road via the existing double roundabouts and then accessing the new Duplicated Tsing Yi Bridge No. 2 using the new underpass. Figure 13 presents the outline traffic diversion arrangement. It can be seen that the affected traffic was required to perform a U-turn at the CT9 access road before joining the new underpass to the new Duplicated Tsing Yi Bridge No. 2. Two U-turn facilities were provided, the first U-turn was located at the upper end of the CT9 access road and provided a U-turn facility for vehicles less than 6.0 metres in length. The second U-turn was located at the lower end of the CT9 access road and provided a U-turn facility for vehicles greater than 6.0 metres in length. This traffic arrangement has proved to be very successful, with no major problem/accident during the period.

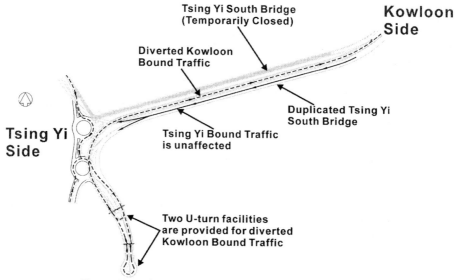

Figure 13 Outline Temporary Traffic Management arrangement

HINGE BEARING TESTS

As a contract requirement, the replacement bearings were to be load tested to verify their structural performance before they were installed on the bridge. According to the works programme, the vertical and horizontal hinge bearings were to be tested first. Due to the size of the bearings and the need to manufacture a purpose made jig, the contractor proposed to conduct the tests at Ruhr University in Germany. The duty visit by Ir. P. C. Wong and Ir. C. Y. Wong of Structures Division, was carried out to witness the testing of the special bearings. One vertical and one horizontal hinge bearings were load tested to the design maximum SLS and ULS loads. The maximum load applied was 996kN (static). The laboratory adopted a Universal Testing Machine, which has capacities of 1,000 kN in compression and 1,000 kN in tension. A special jig, using 75mm thick steel plate, was manufactured by the laboratory for this purpose. The criterion for acceptance is the load deflection for all major components should be within acceptable limits and elastic with acceptable recovery. After the load tests, a dye penetration test was conducted on the critical elements of the bearings to check for any damage. The test results were considered acceptable and are of high reference values for future bridge designs where horizontal thrust bearings are needed.

CONCLUSIONS

The inspection, testing and construction of the rehabilitation works were successfully completed without any incident. The investigation revealed that Tsing Yi Bridge is generally in good condition though there are localised areas in very poor condition particularly the midspan hinges. The completed rehabilitation works rectified these defects. The bridge is considered to be structurally sound and does not have any areas of structural inadequacy.

This project however highlights the importance of considering the durability and maintenance of a structure at the design stage. Items subject to wear such as bearings, joints and hinges need to be easily inspected and replaced. The life of these items is much less than the design life of a major structure and failure should be anticipated. In recent years the guidance to designers has been to avoid joints and bearings altogether where possible and design integral bridges thereby avoiding the problems that have occurred with the hinges on this structure.

References:
1 The Hong Kong Structural Design Manual for Highways and Railways.
2 UK Highways Agency guidance for Inspections for Assessment.
3 SuperSTRESS: an elastic analysis computer program. By Integer date 1989 – 1997.
4 ADAPT: a computer program to model time dependant effects such as creep. By Adapt Structural Concrete Software System Date August 1997.
5 SAP2000: a finite element computer program By Computers and Structures Date July 1997.
6 Rehabilitation of Tsing Yi South Bridge: 4th International Conference on Bridge Management, Surrey, UK, April 2000.

Acknowledgments:
The authors would like to express thanks to the Director of Highways, Highways Department of Hong Kong SAR Government for permission to publish this paper and to all those people who have assisted in making this a successful project. Particular thanks are extended to Hong Kong University for advice on the dynamic testing, Testconsult for advice on materials testing and Taywood Engineering the testing contractor.

Post Tentioning of Steel Beam Using High Strength Steel Plate

SAKANO, MASAHIRO
Kansai University, Osaka, Japan

NAMIKI, HIRONORI
Kyobashi Construction Corp., Osaka, Japan

Abstract

A reinforcing method using outer cables is popularly used, but it has very little effect on reducing live load stress. However, by using pre-stressed reinforcing steel plates instead of cables, we can reduce both dead load stress and live load stress in steel beams. In this study, we investigated the behaviour of a reinforcing steel plate and a reinforced steel beam under prestressing and live-loading. As a result, we confirmed that a new reinforcing method for steel beams using pre-stressed steel plates is very effective in reducing both dead load stress and live load stress.

1. Introduction

A reinforcing method using outer cables is popularly used, but the structural detail of the anchor area is complicated, and the effect of reducing live load stress is very little because of small stiffness of cables. On the contrary, by using high-strength steel plates for prestressing instead of cables, the structural detail of the anchor area becomes simple, and both dead load stress and live load stress can be reduced in steel bridge girders[1],[2].

In this study, we investigate the behaviour of a reinforcing steel plate and a reinforced steel beam during pre-stressing and live-loading.

2. Experimental Method

2.1 Specimen

Fig. 1 shows the steel beam specimen reinforced by pre-stressed steel plate and locations of strain gauges to measure stresses. An H-beam (JIS-SS400 ; $\sigma_y = 338$ MPa) is used as a reinforced girder with a cover plate (JIS-SM490 ; $\sigma_y = 425$ MPa) on the top flange, and with a pre-stressed reinforcing plate (JIS-SM570 ; $\sigma_y = 545$ MPa) under the bottom flange.

By using strain gauges, longitudinal stresses are measured on the top and bottom flanges and reinforcing plate in the sections A and B.

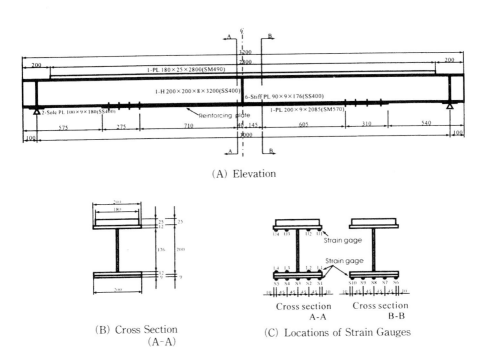

(A) Elevation

(B) Cross Section (A-A)

(C) Locations of Strain Gauges

Fig. 1 Steel Beam Specimen with Prestressed Reinforcing Steel Plate (unit of length : mm)

2.2 Method of loading

To investigate the behaviour of reinforced steel beam and reinforcing steel plate, we set up following three reinforcing conditions :

(1) no reinforcement
(2) reinforcement without pre-stressing
(3) reinforcement with pre-stressing

Pre-stresses are introduced by using two hydraulic jacks. During live-loading, the steel beam specimen is supported simply, and loaded at the center of the span. The magnitude of load for pre-stressing and live-loadings are designed so that the stress might be less than static allowable stresses for both the reinforced beam (JIS-SS400) and reinforcing plate (JIS-SM570).

3. Experimental Results

(1) No reinforcement

Figs. 2 and 3 show, in the case of no reinforcement, the change of stresses

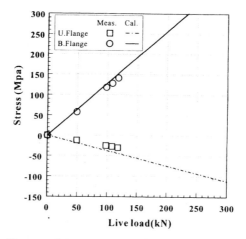

Fig. 2 Change of Stresses under Live-loading (No reinforcement)

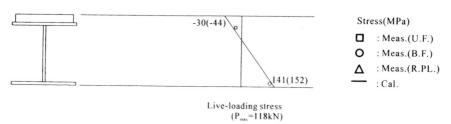

Fig. 3 Stress Distributions (No reinforcement)

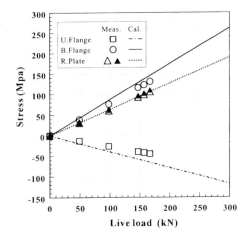

Fig. 4 Change of Stresses under Live-loading (Reinforcement without Pre-stressing)

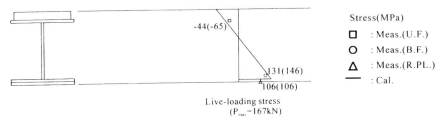

Fig. 5 Stress Distributions (Reinforcement without Pre-stressing)

in the top and bottom flange during live-loading, and stress distributions at the maximum load. Measurements and calculations almost correspond with each other. As the load increases, however, measurements tend to be less than calculations at both the top and bottom flange. It can be considered that deformation of the specimen is reduced because of friction resistance on the supporting points.

(2) Reinforcement without pre-stressing

Figs. 4 and 5 show, in the case of reinforcement without pre-stressing, the change of stresses in the top and bottom flange and reinforcing plate during live-loading, and stress distributions at the maximum live load. Measures and calculations correspond well in each figure. As the load increases, measured

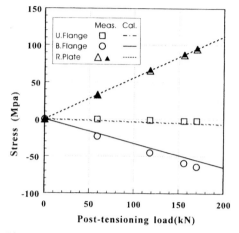

Fig. 6 Change of Stress during Pre-stressing

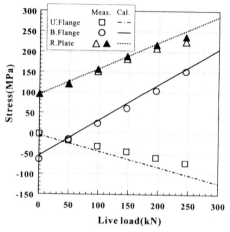

Fig. 7 Change of Stress under Live-loading (Reinforcement with Pre-stressing)

values are tend to be a little smaller than calculated values. We considered that this must also be caused by friction resistance on the supporting sections, similarly to the case of no reinforcement (Fig. 2).

(3) Reinforcement with pre-stressing

Figs. 6 and 7 show, in the case of reinforcement with pre-stressing, the change of stresses in the top and bottom flange and reinforcing plate during

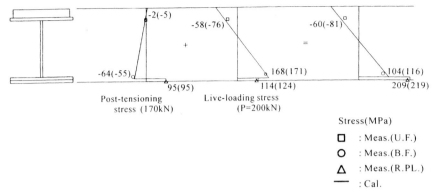

Fig. 8 Stress Distributions (Reinforcement with Pre-stressing)

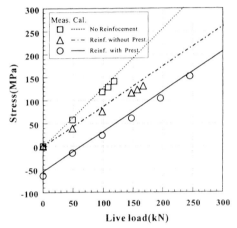

Fig. 9 Change of Stress under Live-loading in Three Cases of Reinforcement

pre-stressing and live- loading, respectively. Fig. 8 shows stress distributions. In each figure, calculated values well correspond to measured values. It can be confirmed, therefore, that this new reinforcing method using a high-strength steel plate for pre-stressing in stead of cables are very effective in reducing live load stress in the tensile flange.

Figure 9 shows the change of stresses under live loading in the bottom flange in three cases of (1) No reinforcement, (2) Reinforcement without pre-stressing, and (3) Reinforcement with pre-stressing. Compared with the case of no reinforcement, only composite effect by the reinforcing steel plate can be seen in the case (2), and moreover, in the case (3), both composite and pre-stressing effects can be observed remarkably. When the bottom flange (JIS-SS400) reaches its allowable stress (140 MPa), the live load capacity is 110 kN in the case (1), 160 kN in the case (2) (about 1.5 times), and 220 kN in the case (3) (about double).

4. Conclusions

Through this experimental study, a new reinforcing method for steel beams using high strength steel plates with pre-stressing was confirmed to be very effective in reducing both dead load stress and live load stress.

References

1) Official Report of Patent (A), *Japanese Patent Office*, 190009 (1999)
2) Sakano. M. et al. : Effect of Pre-stressing by using steel plate on Steel Beam, Proc. of the 7th symposium on Repair and Rehabilitation of Steel Structures, *JSSC*, pp. 57-64, 2000 (in Japanese)

Vibration Induced Fatigue of Overhead Sign Structures on Elevated Highway Bridges

YAMADA, Kentaro*, OJIO, Tatsuya*, LEE, Sanghun*, XIAO, Zhigang*, and YAMADA, Satoshi **
*Department of Civil Engineering, Nagoya University, Chikusaku, Nagoya 464-8603, Japan
**Topy Industries, Ltd., Toyohashi, Japan (Presently Doctoral Student of Nagoya University)

INTRODUCTION

On urban elevated highway bridges various overhead sign structures, such as sign poles for guidance and lighting poles for safe driving, exist. More is to be added in the near future in order to accommodate ITS systems in Japan. Such overhead sign structures were basically designed for wind loading. However, fatigue cracks were found recently in a few overhead sign structures installed on elevated highways. In one case a single pole with rather heavy traffic sign developed fatigue crack at rib ends, and dropped to underneath roadway in Tokyo in 1999. It suggested that design procedure and inspection manual for overhead sign structures are to be reviewed.

Fatigue cracks were found at a corner of an overhead sign frame installed on an elevated highway bridge in Nagoya Expressway. The cracks were found during special inspection carried out after the accident in Tokyo. The cracked sign frame was promptly removed and a newly designed overhead sign structure was installed. Stress and vibration measurements were carried out on the new overhead sign frame in order to see the vibration characteristics and to assess durability against fatigue.

DESCRIPTION OF OVERHEAD SIGN STRUCTURES AND FATIGUE CRACKS

The bridge, where the overhead sign frame was installed, is a part of inner circle of Nagoya Expressway near Nagoya Castle. It is a-three span continuous steel box girder bridge of equal span of 50 m long, as shown in Fig. 1. It carried originally two lane traffics supported by two steel box girders symmetrically placed over the steel piers. However, a new on-ramp lane was added in 1999, and the roadway was widened in order to accommodate three lane traffics by adding one more steel box girder. Steel piers were placed to support the additional steel box girder, but not at the last one where the sign frame was installed, due to the limitation of the roadway underneath. The overhead sign frame was widened accordingly in January, by adding a new part, as shown in Fig. 2. The sign frame carried traffic-counting sensors and a signboard. Since both were rather light in weight, the overhead sign frame was made of welded steel pipes in one plan. When heavier sign board or information board is attached, four columns support a more rigid beam.

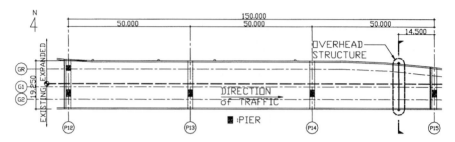

Fig. 1 Plan of bridge and location of overhead sign frame

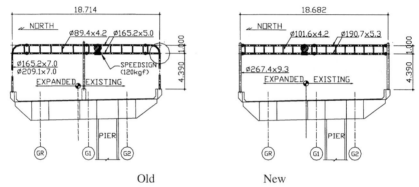

Old   New
Fig. 2 Cracked and replaced overhead sign frame

In July 1999, two cracks were found at the corner of the sign frame where a beam was fillet-welded to a column, as shown in Photo 1. The sign frame was removed promptly, and then fatigue-cracked part was cut and brought to lab investigation. Through the visual and SEM (Scanning Electron Microscope) observation, it was found that one crack initiated and propagated from the fillet weld, and the other from gas-cut hole of the column, where electric wires went through from the column to the beam. The gas-cut hole was poorly made so that some notch-like flaws were observed.

A newly designed sign frame was fabricated and installed in September 1999. It has the same span as the old one, but larger steel pipes were used to

Photo 1 A fatigue crack found at the corner of steel sign frame

increase the stiffness. The rib ends of the lower ends of the column was ground flush in order to increase fatigue strength over as-welded ones for precaution.

STRESS AND VIBRATION MEASUREMENS

The stress and vibration measurements were carried out on the newly installed sign frame. In order to measure vibration characteristics of the sign frame ten accelerometers of 1 G capacity were attached at various places, as shown in Fig. 3. Five accelerometers were to measure vibration characteristics of the main girder where the sign frame was attached. Another five were attached at the top of the sign frame. In Fig. 3, the symbols, -X, -Y and -Z show directions of movements, i.e. longitudinal, transverse and vertical direction with respect to the bridge. Five strain gages were attached near the ends of the stiffening ribs at base plate of one of the column, as shown in Fig. 4. These strain gages were attached between the ribs and at 15 mm above the rib ends. They represented about 90 percent of nominal stresses. Vibration measurement was carried out using digital dynamic recorder. Then, the strain gages and accelerometers were connected to so-called histogram recorder, where strain signals were automatically analyzed by Rain Flow counting method. This histogram records were carried out for 24 hours in a weekday. A tank truck of 196 kN was used as a test truck. It ran eight times in the night when traffic was rather scarce. Measurement was also carried out in service condition in daytime. For long-term measurement so-called trigger system was used, where data was recorded only when one of the accelerometers showed over 200 cm/s^2. In this way we could use effectively the limited memory of the system. Thus, 34 groups of data were recorded for about two hours of measurement. Fig. 5 shows an example of the vertical vibration of the beam center, where quite a few trucks ran over the period of 2.5 minutes.

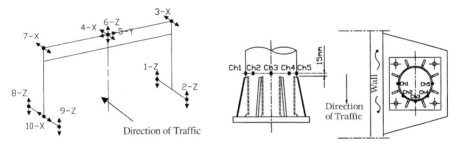

Fig. 3 Accelerometers attached at sign frame and bridge

Fig. 4 Strain gages attached near ends of stiffening ribs

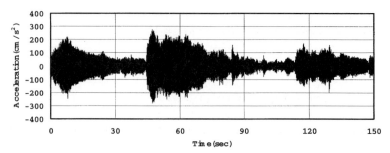

Fig. 5 Example of vertical vibration of beam center

VIBRATION CHARACTERISTICS OF SIGN FRAME DUE TO TRAFFICS

From the comparison between the measured vibration characteristics and eigenvalue analysis, vibration modes of the sign frame were estimated as shown in Fig. 6. The sign frame had rather small bending stiffness in the bridge axial direction as well as the vertical direction of the beam.

In Fig. 7, the typical vibration spectra of the base of the sign frame show that frequency between 2 and 4 Hz was dominant. The previous measurement indicated that trucks themselves vibrate with frequency of around 3 Hz. Fig. 8 shows that the vibration spectrum of the top of frame have two sharp peaks in 3.69 Hz and 4.32 Hz. Therefore, we concluded that the sign frame base was oscillated mainly by running trucks, and the twisting mode of 3.69 Hz and the vertical beam vibration mode of 4.32 Hz are dominantly occurred. Measurement was also carried out for service loading condition.

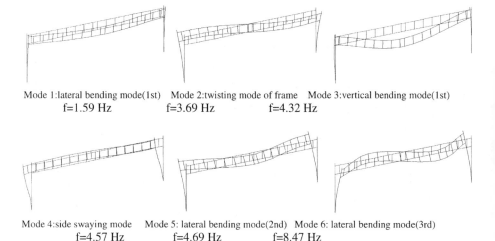

Mode 1:lateral bending mode(1st)   Mode 2:twisting mode of frame   Mode 3:vertical bending mode(1st)
f=1.59 Hz   f=3.69 Hz   f=4.32 Hz

Mode 4:side swaying mode   Mode 5: lateral bending mode(2nd)   Mode 6: lateral bending mode(3rd)
f=4.57 Hz   f=4.69 Hz   f=8.47 Hz

Fig. 6 Vibration modes and measured frequencies of sign frame

The maximum accelerations of the base of the sign frame and the frame itself were picked up from the data and are plotted in Figs. 9. The average of the measured maximum accelerations for trucks in service was 1.2 to 1.8 times larger than that for the test truck of 196 kN. It indicates that under service condition trucks heavier than the test truck are running on this bridge simultaneously, which caused larger vibration of the sign frame.

Histogram recorder can record the value of acceleration and its number of cycles for every hour. The maximum acceleration was compared with the hourly traffic volume, as shown in Fig. 10. The hourly traffic volume was measured near the sign frame by Traffic Control center of Nagoya Expressway. The hourly traffic volume was about 3,000 during the daytime, and was less than 1,000 during the night. The hourly maximum acceleration vaguely corresponds with the traffic volume in the morning hour, and less in the evening hour. On the contrary the hourly traffic volume seems to correspond better with the number of cycles of acceleration, as shown in Fig. 11. It implies that each vehicle, possibly truck, contributes a certain cycles of vibration to the sign frame. For example, hourly vertical vibration of the sign frame was about 35,000 during the daytime, while hourly traffic volume was around 3,000, roughly 10 to 1 ratios.

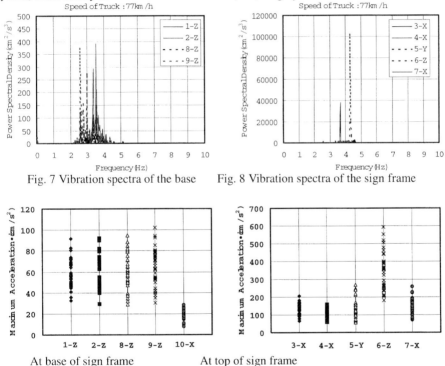

Fig. 7 Vibration spectra of the base   Fig. 8 Vibration spectra of the sign frame

At base of sign frame   At top of sign frame
Fig. 9 Measured maximum acceleration

304 MAINTENANCE PRACTICE

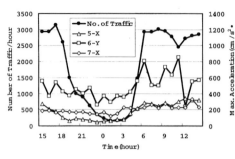

Fig. 10 Maximum acceleration and hourly volume of traffic

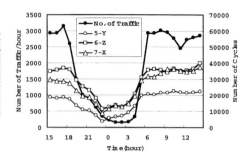

Fig. 11 Number of cycles of acceleration and hourly volume of traffic

STRESS RANGE HISTOTRAM AND FATIGUE LIFE EVALUATION

The hourly traffic volume is also compared with the stress range histogram recorded near the ends of the stiffening ribs, as shown in Fig. 12. It corresponds well with the number of stress cycles, and in this case stress cycles of five times larger than that of traffic volume was observed. The stress range histogram can be used to estimate the fatigue life of the base of the sign frame. Fatigue life was estimated by Miner's rule based on the JSSC Fatigue Design Recommendation of Steel Structures (1993). Since strain gages were attached near the stiffening rib ends, the structural detail can be classified as JSSC-G for as-welded rib ends. When the old sign frame was replaced due to fatigue cracking, the rib ends of the new sign frame were ground flash to increase fatigue strength for precaution. The structural detail of this type is not specified in the JSSC Fatigue Design Recommendation, and we assumed that this detail has higher fatigue strength than the as-welded rib ends by one rank, that is JSSC-F.

The fatigue life was computed and summarized in Table 1. It was found that the shortest fatigue life was 87 years; even we assume the rib end as as-welded one. Therefore, it was concluded that the rib ends of the new sign frame do have sufficient durability against fatigue, provided that the traffic condition is as it is in the future.

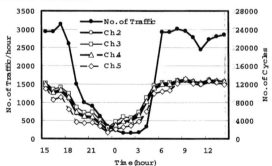

Fig. 12 Hourly traffic volume and number of stress cycles

Table 1 Summary of fatigue life at the rib ends

Channel No.	Ch2	Ch3	Ch4	Ch5
JSSC-F	150	87	126	497

SUMMARY OF FINDINGS

Fatigue cracks were found at the joint of the tubular sign frame during inspection and it was replaced promptly. Vibration and stress measurement was carried out for a newly installed sign frame. The followings summarize the results.
1) The dominant vibration of the base of the sign frame was between 2 and 4 Hz, which seems to be mainly due to the running trucks in service.
2) In service condition the maximum acceleration was 1.2 to 1.8 times more than that due to the test truck of 196 kN.
3) Hourly number of cycles of acceleration and stresses were 5 to 10 times of the hourly number of traffics in service. It indicates that the number of cycles needed for fatigue life estimation can be estimated by daily traffics or daily truck traffics.
4) The stiffening rib ends near the base of the sign frame, which seems most critical in terms of fatigue cracking, showed over 80 years of fatigue life, and we concluded that they have sufficiently long fatigue life, provided that the traffic condition would be the same in the future.

ACKNOWLEDGEMENT

The authors are grateful to Mr. Nariaki Mori and Nobuaki Morishita of Nagoya Expressway Public Co. who made it possible to publish this paper.

REFERENCES

1. DOT, NY; Bridge Inspection Manual, 1997.
2. DOT, NY; Overhead Sign Structure, Inventory and Inspection Manual, 1999.
3. Yamada, K. et al.; Fatigue Strength of Steel Lighting Pole's Tubular Flange Joints, Proceedings of Structural Engineering, Vol. 28A, JSCE, March 1992.
4. Research Group of Bridge Vibration; Measurement and Analysis of Bridge Vibration, Gihodo, October 1993. (in Japanese)
5. JSSC; Fatigue Design Recommendations for Steel Structures, Gihodo, 1993. (in Japanese)
6. OJIO, T. et al.; Fatigue Durability of Sign Frames Due to Traffic Induced Vibration, Proceedings of Structural Engineering, Vol. 47A, JSCE, March 2001.

Design and theory

Design for Durability – A Maintenance Engineer's Viewpoint

RICHARD J FEAST
TMML (seconded from AMEC International Construction)
Hong Kong, China

Background

From the early and mid 1980's, bridge maintenance engineers were concerned about the high level of remedial repairs bridge structures in general were requiring along with the rising maintenance budget required to resource this element of work. It was clear many concrete structural elements to existing bridges were either showing severe serviceability problems or even failing after just 15 to 20 years' service life, far less than the 120-year design. This poor performance was very concerning, from a structural material such as reinforced concrete, that up to then, was considered by designers to be relatively maintenance free.

In 1989 G. Maunsell & Partners produced a report on a survey of the performance of 200 concrete highway bridges they had been commissioned to undertake as a result of bridge owners' concern. The findings of the report statistically confirmed what many bridge maintenance engineers had suspected for a number of years: that concrete bridges were suffering significant durability problems and these problems were common throughout the geographical distribution of the bridges surveyed. Therefore the problems were not confined to particular geographical locations, particular design types or particular loading conditions. There appeared to be a fundamental problem that the current design and detailing of concrete bridges was not providing the durability required for them to attain their 120 year design life.

The summarised findings of The Maunsell Report were:

- Most bridges surveyed showed that the commonest cause of spalling or rust to reinforcing bars was inadequate cover.

- Spalling concrete was also significant where the specified cover was correct due to high levels of chloride ingress.

- The majority of cracking was caused by early thermal movements or shrinkage, to abutments, wing walls, piers, deck slabs and cantilever parapet edge beams.

- The majority of bridge joints leaked allowing water and contaminants onto bearing shelves, abutments and piers.

- There were also signs of leakage through a number of bridge decks.

- Structural elements adjacent or in close proximity to marine environments were suffering chloride contamination to abutments, piers, wing-walls and generally most exposed concrete surfaces.

- See Figure 1 for typical problem areas.

The Maunsell Report made the following recommendations with regard to improving the situation of the existing concrete bridge stock:

- Repair or replace all leaking bridge joints.

- Clean all bearing shelf drainage systems.

- Prevent contamination of concrete surfaces from spray by an impregnation of silane.

- Annually wash down coin surfaces.

- Instead of concrete patch repairs to structural elements subjected to chloride induced corrosion, consideration should be given to providing cathodic protection systems or to the complete replacement of contaminated structural elements.

- Apply waterproof membranes to all bridge decks, whether as new build or refurbishment works.

The Maunsell Report then made the following recommendation with regard to avoiding repeating mistakes in new construction *"The designer should be aware of maintenance problems and design accordingly"*.

Thus started the impetus for the philosophy of "Design for Durability".

Design for Durability

Definition of Bridge Durability:

Durability is the ability of materials or structures to resist, for a certain period of time and with only routine maintenance, all the effects to which they are subjected, so that no significant change occurs in their serviceability.

Durability is influenced by the following factors:

1. Design and detailing.

2. Specification of materials used in construction.

3. Quality of construction (workmanship).

The control of items 2 and 3 are achieved through the use of accepted standards and procedures. However, the design of structures is not so readily associated with the achievement of durability, beyond such considerations as cover to reinforcement, crack width limitation or minimum steel plate thickness. This lack of attention to the durability aspect has resulted in the premature loss of serviceability in many highway structures.

Improved Durability – Details to consider at conceptual design stage

From looking at types of structures that have historically performed well the following construction details should be considered to minimise future durability problems:

- **Bridge Deck Continuity** – Make bridge decks continuous over bridge intermediate supports. Continuous bridge decks have been far more durable structures than structures with simply supported decks because bridge joints have failed allowing water and contaminants to leak through to piers and abutments.

- **Integral Bridges** – Where possible design bridge decks integral with the abutments. An extension of deck continuity concept with decks being designed to be connected to the abutments and thus removing the need for movement joints or bearings. Load effects due to temperature changes, shrinkage and creep are considered in conjunction with soil / structure interaction. This concept is only effective for simply supported spans of up to 60m and with minimum skew.

- **Post-tensioned Structures** – Designers should ensure that the layout of the anchorage zone reinforcement is not congested or likely to cause difficulty in placing and compacting the concrete. Increased cover should be provided to ensure effective protection to the steel.

- **External Post-tensioning** – The use of external post-tensioning has a number of advantages for both the Contractor and the Maintenance Authority over internal bonded tendons:

    - Tendons are easy to install and inspect.
    - If suitably detailed the external post-tensioning tendons can be restressed or replaced if necessary.
    - The provision of special monitoring devices to detect loss of post-tensioning or corrosion could be considered.
    - By moving tendons outside the concrete section for most of their length, especially in the webs, concreting becomes easier.
    - Reinforcement detailing is more straightforward as tendons do not have to enter and leave the web shear cage.
    - The activities of the main contractor and the specialist prestressing sub-contractor are separated.

- **Box Sections** – Access to inner cells to box beams, abutments or piers must be such that they provide a safe working environment for inspection and routine maintenance activities. Large access openings should be provided where possible along with natural lighting and good ventilation. Artificial lighting should also be provided including power points for inspection and routine maintenance works.

- **Plain Concrete** – Remove the need for ferrous reinforcement where possible. Consideration should be given to the use of plain concrete construction by the choice of suitable types of structure i.e. abutments, wing walls, retaining walls, arch structures etc.

- **Non-ferrous Reinforcement** – Used at present to control early thermal cracking in plain concrete members. Research and development is now ongoing to use non-ferrous reinforcement in the design and construction of primary structural members, i.e. Kelvar fibres encased in epoxy-compound or carbon fire reinforced plastic reinforcement bars.

- **Inspection and Maintenance** – Designers should ensure all bridge elements and components can be effectively inspected and maintained. Designs should also cater for the replacement of items, especially those that are subject to wear, bearings, movement joints etc. – see also **Box Sections** above.

- **Bridge Abutment Galleries** – Where conventional abutments are used, inaccessible vulnerable areas, i.e. end anchorage blocks, are subject to water ingress from leaking bridge joints and are difficult to inspect and maintain, see Figure 2. Therefore the designer should consider providing room for inspection and maintenance operations by providing abutment galleries, see Figure 3. Abutment galleries allow for the effective and safe inspection and maintenance of the following:

    - Bridge joints – expansion and rotation.

    - Bearings – routine maintenance, cleaning and greasing.

    - Abutment curtain walls – inspection.

    - Bridge deck ends – post-tension anchorage inspection and maintenance.

    - Drainage – ample room to discharge road drainage into and for cleaning and rodding purposes.

    - Jacking – provide room for bridge deck jacking facilities for the provision for the replacement of bearings.

- **Waterproofing Systems** – A good deck waterproofing membrane is a vital defence against the ingress of water and contaminants. The membrane should cover the entire deck between parapets, with careful detailing to ensure continuity under kerbs, verges and central reserves. The following areas of concrete should also be waterproofed:

    - Vertical faces at deck ends and abutment curtain walls.

    - Top faces of piers and abutments bearing shelves.

    - Inaccessible areas which may be subject to leakage, i.e. beam ends.

- **Drainage Systems** – Should be designed to minimise the risk of blockage and be accessible for cleaning. The system should be robust enough to resist damage from cleaning operations and from chemical spillage on the road surface. Drainage downpipes cast into columns should be avoided if possible. Cast-in downpipes make for very difficult cleaning operations.

- **Impregnation of Concrete surfaces** – This provides effective protection against the ingress of air-borne chlorides from a marine environment.

- **Early Thermal Cracking** – Design to Clause 4.3 of the "Structures Design Manual for Highways and Railways".

- **Half Joints** – Both in steel and concrete usually present severe maintenance problems. They are very difficult to inspect and repair and should be avoided in new design.

- **Concrete Hinges** – Highly stressed areas where, because of the large amounts of reinforcement present, compaction of concrete is difficult. The steel reinforcement in the hinges is vulnerable to the ingress of water and hence corrosion. Concrete hinges should be avoided in new construction but if there is absolutely no alternative ensure that the hinges are visible for inspection and maintenance.

Design for Durability and the Highway Structures within the Tsing Ma Control Area

A number of the above durability details have been incorporated by the various designers to ensue the long term performance and durability of their structures that form the Tsing Ma Control Area. The author would like to review a sample of these "durability details" to illustrate how the philosophy has been adopted in a number of the contracts that form this section of the Airport Core Project.

Kwai Chung Viaduct *(illustration of deck continuity using precast beams)*

The Kwai Chung Viaduct is a dual 3 to 4 lane expressway of approximate length 3km, which links West Kowloon Expressway at Mei Foo with the new Rambler Channel Bridge approach viaducts at Lai King.

The structure is supported on columns with bored piles and barrette foundations. The deck of the expressway varies in width between 40m and 65m at ramp and slip road junctions. There are a total of 100 spans. The typical length of span is approximately 35m, but there are large variations from 12m to 40m. The height varies between a maximum of 28m above the existing ground level at Lai King and 18m above ground level at Mei Foo.

Sections of Kwai Chung Viaduct interface with various railway viaduct for the Airport Railway Express from Mei Foo to the Southern Approach of the southern approach of the Airport Railway Express at Lai King Tunnel and from Lai King Station to Kwai Chung Park.

As a result of the very onerous construction constraints that the contractor had to overcome, the viaduct was designed and constructed of pre-cast, pre-stressed 'U' beams with an in-situ reinforced concrete deck. The designer then took the decision to make individual spans continuous, usually over three spans. The spans were made integral over the intermediate supports by longitudinal post-tensioning through the 'U' beams and the intermediate transversely post-tensioned deck diaphragms. The deck diaphragms are then supported on column supports.

314 DESIGN AND THEORY

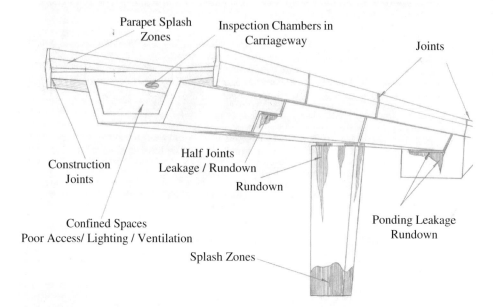

Figure 1 – Typical Problem Areas

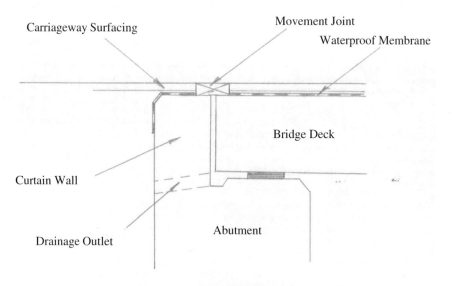

Figure 2 – Inaccessible Bearing Shelf

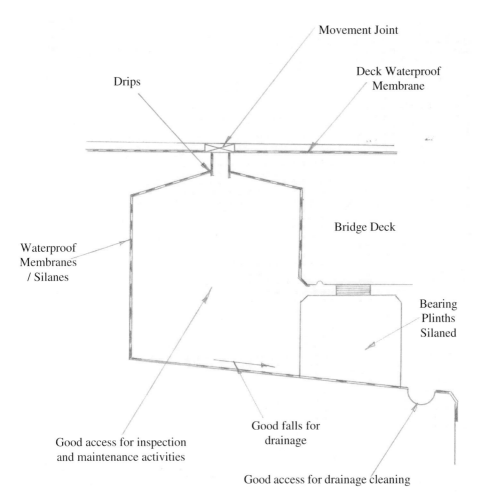

Figure 3 – Abutment Gallery

The continuity details have minimised the number of bridge joints required and hence the future maintenance liabilities along with providing an improved ride quality for the travelling public. The reduced maintenance liability is of particular importance with this structure due to the onerous inspection and maintenance restrictions imposed by the interface with the Airport Expressway Railway and the very limited "windows of opportunity" allowed of inspection and maintenance works adjacent and over the railway envelope.

Rambler Channel Bridge *(illustration of external post-tensioning, deck waterproofing and good inner cell access for inspection and maintenance)*

Rambler Channel Bridge consists of a dual 3-lane bridge of 3 no. spans of approximately 121m and 2 no. end spans of approximately 65m long. The bridge deck box pre-cast segments are post-tensioned by external tendons and were constructed using the counter-cast segmental balanced cantilever system. Being of the segmental construction the designer specified a waterproof membrane to be applied to the bridge deck to prevent the ingress of water leaking through the segment joints.

Access within the spine box beams is very good with specially constructed walkways and lighting to ensure inspection and routine maintenance operations are carried out in a comfortable and safe working environment. The post-tension cables, being external, are readily inspectable along with their anchorages. Space has also been allowed for additional post-tension cables to be added if required during the bridge design life and, being external, the existing post-tensioned cables can be replaced if necessary.

Good, safe access facilities have been provided to the pier tops and bearing shelves for inspection and routine maintenance operations and provision has been built in for future jacking operations for bearing replacements should the need arise.

Tsing Yi West Viaduct – Mainline *(illustration of external post-tensioning and bridge inspection galleries)*

The Cheung Tsing Viaduct – Mainline consists of a 3 to 4 lane viaduct of total length approximately 510m with spans varying from 51m to 66m. The viaduct construction is similar to that of Rambler Channel Bridge, pre-cast deck box segments, externally post-tensioned with a waterproof membrane spray applied to the bridge deck.

Access into the internal cells of the box beams is via abutment galleries that also provide very good access for:

- The cleaning and jetting of the bearing shelf drainage systems.
- Inspection and routine maintenance of the bearings.
- The inspection of the post-tensioned end anchorages.

General Access for the External Inspection of Structures *(illustration of planned access for inspection and maintenance of the external structural elements)*

Specifically designed and constructed access gantries and gondolas have been provided for the external inspections of the Long Span Supported Bridges. The design for the access gantry to Ting Kau Bridge has also include cradles that allow inspection from the access gantries of the lower concrete tower elements and the transverse tower cable stay anchorages.

Further, a specifically designed Bridge Inspection Vehicle has been supplied for inspection works to the structures within the Tsing Ma Control Area with a very flexible working envelope that incorporates a vertical reach of 20m as well as an underbridge deck soffit inspection capacity.

The Benefits of Design for Durability to date for the Highway Structures of the Tsing Ma Control Area

TMML inspect all the highway structures within the Tsing Ma Control Area in accordance with *"The Inspection Manual for Highway Structures"* produced by the Highways Department – Structures Division – of the SAR. TMML have been able to meet the required inspection requirements because of the access facilities built in, and provided, by the designers and our Client "The Highways Department".

Resulting from this thorough inspection regime TMML have been able to identify problems at a very early stage in a structures design life that have since been rectified before any resulting serious damage could occur, i.e. a filed internal drainage joint within the inner cell of the Rambler Channel Bridge. This particular problem was identified in one of the 6 monthly visual inspections and because of the good internal access to the inner bridge cells the fault will shortly be repaired with the maintenance team working within a good and safe environment and with no disruption to the travelling public.

Conclusion

As can be seen from the above examples, the philosophy of "Design for Durability" has been adopted on a number of the individual contracts that form the infrastructure network of the Tsing Ma Control Area and can be seen to be working. These design details will improve the "whole life costs" of the structures and will allow the bridge maintenance engineer's role to be far more focused on routine inspection and maintenance work rather than becoming preoccupied with un-programmed emergency remedial works. This is now unacceptable in today's climate of "no tolerance" for infrastructure downtime.

Review of Design Thermal Loading for Steel Bridges in Hong Kong

F.T.K. Au[1], L.G. Tham[2] and M. Tong[3]
[1] Associate Professor, [2] Senior Lecturer and [3] Former Research Student
Department of Civil Engineering, The University of Hong Kong, Pokfulam Road, Hong Kong

Abstract

The relationship between temperature distribution in steel bridges and climatic conditions in Hong Kong has been investigated in the present study. The temperature measurements at Tsing Ma Bridge have provided very useful data for understanding the thermal behaviour of steel bridges. Three steel segmental models with instrumentation have also been set up since 1993 on the roof of Runme Shaw Building on the campus of the University. On the basis of the measured data, finite element models for different structural components of the bridge deck have been developed to predict the temperature variation. Good agreement with field data was observed. Comparison of temperature gradients among different structures has shown that they can generally be classified into two categories depending on the number of web faces exposed to convection. Values of design thermal loading for a 50-year return period are determined from the statistics of extremes over 40 years of meteorological information in Hong Kong. The design temperature profiles for various types of steel bridge deck with different thickness of bituminous surfacing are proposed. The factors affecting the design effective temperature are also reviewed and suitable values for Hong Kong are proposed. Results are compared with recommendations of the local code.

Keywords

Extreme analysis, steel bridge, temperature, temperature gradient, thermal response

1. Introduction

Researchers such as Zuk (1965) began to study the thermal behaviour of highway bridges since 1960. From a number of bridges investigated, it was concluded that the air temperature, wind, humidity, intensity of solar radiation and material type would all affect the temperature distribution. On that basis, Reynolds and Emanuel (1974) further investigated the properties of the temperature differential and the effect of asphalt surfacing. Subsequent research work on the topic mainly concentrated on the prediction of temperature distribution for different forms of bridges. They included the work on steel decks by Capps (1968), concrete deck by Hunt and Cooke (1975) and composite deck by Emerson (1973) as well as Kennedy and Soliman (1987).

Engineers in Hong Kong have traditionally followed the recommendations of the British codes. However, due to difference in geographical location, the climatic conditions in Hong Kong are quite different from those of the U.K. In 1980, the Department of Civil Engineering of the University of Hong Kong initiated, upon the request of the then Highways Office of the Hong Kong Government, various programmes on the measurements of temperature distributions in highway bridges in Hong Kong. The Castle Peak/ Texaco Road Flyover and the Cornwall Road/ Waterloo Road/ Junction Road Interchange have been studied in detail (Liu, 1985; Ho and Liu, 1989). Their results actually formed the basis for the design code on thermal effect of concrete bridges later in Hong Kong. However, the work carried out so far has been limited only to concrete bridge decks. It is therefore useful to carry out similar investigations on steel bridges for updating the local design codes.

2. Review of design codes

Bridges in Hong Kong have traditionally been designed according to various standards and codes of practice published by the British Standard Institution (BSI) with certain parameters amended to suit local conditions. In *BS153* (BSI, 1972) for the design of steel girder bridges, the differential temperature specified was still linear. An investigation (Emerson, 1973) undertaken by the Transport and Road Research Laboratory (TRRL) reported that the temperature distribution down the depth was non-linear. It implies that thermal stresses will always be induced whether the structure is simply supported or continuous. In addition, the TRRL also studied the thermal movements as well as the estimation of temperature distribution from the shade temperature (Capps, 1968; Emerson, 1968; Emerson, 1976). These findings were later incorporated into the code *BS5400* (BSI, 1978) which encompassed the design of steel, concrete and composite bridges.

The use of the British codes on the design of bridges in Hong Kong since the 1970s and until early 1990s was regulated by the *Civil Engineering Manual* (Public Works Department, 1979). The 1979 version of the manual specified the effective bridge temperatures for use in various groups of superstructures in Hong Kong, while recommendations in the British code were followed in all other aspects. To provide data for drafting the local code in this respect, various programmes on the measurements of temperature distributions in highway bridges in Hong Kong were started. Two concrete bridges were selected for the field measurements, which lasted several years. The findings from this comprehensive research project (Liu, 1985; Ho and Liu, 1989) were extensively used in the later versions of the *Civil Engineering Manual* as well as the currently used *Structures Design Manual for Highways and Railways (SDMHR)* (Highways Department (HyD), 1997). However as the field measurements in the 1980s were carried out only on concrete bridges, the specified thermal loading on other types of bridge decks are still largely based on recommendations given in *BS5400:Part 2* (BSI, 1978).

Different approaches have been adopted for the specification of thermal loading for bridges in different countries. In the United States, detailed specification of the positive and reverse temperature gradients are given for concrete and composite bridges by the *American Association of State Highway and Transportation Officials (AASHTO) Code* (AASHTO, 1998). However, only the extreme effective bridge temperatures are given for steel bridges (AASHTO, 1996). This is probably due to the smaller temperature gradients caused by the higher thermal conductivity of steel. Investigations in New Zealand were mainly focused on the temperature-depth relationship within the bridge deck. Priestley (1978) proposed a non-linear distribution represented by a fifth-order polynomial at the top, a uniform segment in the middle and a linear distribution at the bottom. In Canada, the *Ontario Highway Bridge Design Code (OHBDC)* (Ministry of Transport, 1991) specifies the effective temperature for different types of bridges. Apart from that, an effective construction temperature of 15°C is also assumed in the calculation of thermal expansion and contraction.

The relevant section of the local *SDMHR Code* (HyD, 1997) on thermal loading has been written generally in line with the British code, except that certain parameters have been adjusted to account for the local climatic conditions. Depending on the type of construction, bridge decks are divided into four different groups. Effective bridge temperatures and temperature differences are specified for various groups of structures. The necessary adjustments for thickness of surfacing and height above mean sea level are also specified. However these recommendations have been drafted on the basis of field measurements in concrete bridges only. It is therefore useful to carry out similar investigations on steel bridges for updating the local design codes.

3. Theoretical Background

The temperature distribution at a section is normally described by its effective temperature and the temperature difference there. The effective temperature is the weighted mean value of temperature at the section defined as

$$T_e = \int E\alpha_e T \, dA \bigg/ \int E\alpha_e \, dA \qquad (1)$$

where E is the Young's modulus, α_e is the coefficient of thermal expansion, T is the temperature and dA is the differential cross-sectional area. The temperature difference refers to the difference in temperature from that at a chosen reference point in the section. The daily extreme air temperatures usually affect the effective temperature while the solar radiation usually contributes to the temperature difference in the section. The temperature difference would tend to induce a bending moment given by

$$M_0 = \int E\alpha_e Ty \, dA \qquad (2)$$

where y is the distance from the centroidal axis of the section.

The factors affecting the temperature distribution in a bridge deck include the thermal conductivity of the material k, the specific heat capacity of the material c, the density ρ, the absorptivity of the surface α, the emissivity of the surface ξ and the film coefficients h. The absorptivity α and the emissivity ξ depend on the colour and texture of the material surface and they govern the proportion of solar radiation absorbed during daytime and the outgoing radiation emitted at night. The film coefficient h at a surface represents the total amount of thermal energy lost per unit area to the surrounding caused by a unit temperature difference between the surface and the surrounding. Such losses are mainly due to convection and radiation, and hence the film coefficient h depends on the wind speed that would promote the heat loss from the surface.

The temperature distribution within a steel section is governed by heat conduction inside its body, and the convective and radiative heat exchange with the surrounding environment. Under normal circumstances, the longitudinal heat flow in a prismatic steel section can be neglected. Because of the high thermal conductivity and small thickness of steel plate, the temperature gradient across the thickness is neglected. Therefore, conductive heat flow is only considered along the width of each plate. Heat exchange with surrounding through convection and radiation takes place on the two surfaces only. The transient one-dimensional heat flow by conduction in a homogeneous solid follows Fourier's law

$$k\frac{\partial^2 T}{\partial x^2} = \rho c \frac{\partial T}{\partial t} \tag{3}$$

where T is the temperature at position x at time t. At the interface with surrounding, the heat flow relationship with convection and radiation during daytime is given by

$$-k\frac{\partial T}{\partial n} - h(T - T_a) + \alpha I = 0 \tag{4}$$

where T_a is the ambient temperature, I is the solar intensity, and n is the normal to the surface. A similar equation at night when there is outgoing radiation emitted from the bridge can be written, and it depends on the emissivity of the surface ξ. To apply the above theory to the bridge section, the section is discretized into finite elements. Following the standard procedure, one can easily assemble the characteristic matrices and solve for the temperature distribution of the section by carrying out time marching (Tong et al., 2001).

4. Investigations on Campus and Field Measurements

The most ideal method of investigation is to instrument prototype bridges for temperature monitoring. However, for various practical reasons, it may not be possible to put instrumentation in all the components of interest. For example, trough stiffeners and box girders cannot be drilled through; otherwise the warranty will become void. It is therefore necessary to complement such on site monitoring of actual structures with experiments on scale models. A small-scale experiment has been set up on the campus since 1993 (Au et al., 1999). Three steel segmental models with instrumentation have been set up on the roof of Runme Shaw Building on the campus of the University, as shown in Figure 1. The models include two box girder sections and a π-girder section, all of which have been fabricated from steel plates. One of the box girder sections has the ends open so that the cell is freely ventilated while the other has both ends blocked by plywood to simulate normal box girders in which the interior is not freely ventilated. Platinum resistance type detectors are used to measure temperatures of the steel plates and shade temperatures at various locations.

Statistical correlation was carried out between the field measurements and the meteorological data given by the Hong Kong Observatory. Finite element analysis was carried out for the models using certain initial parameters. Sensitivity analysis and calibration of these parameters were performed. Good agreement between the numerical results and the field measurements was observed (Au et al., 1999).

Before the methodology developed at the campus investigations can be generally applied to real life situations, it is necessary to test it with actual structures. Tong et al. (2000) reported some results of the temperature measurements taken at various superstructure components of Tsing Ma Bridge. A continuous 30-month temperature record on Tsing Ma Bridge up to December 1999 has been used as a database for the investigation, and hourly global solar radiation data are provided by the Hong Kong Observatory. The data have been analysed and various numerical models have been set up and calibrated against the field data. In particular, the meteorological conditions on one summer day (4 August 1998) and one winter day (17 January 1998) have been chosen for analysis and comparison with field measurements. The values for film coefficients, absorptivity and emissivity have been worked out after a series of parametric studies and they are listed in Table 1. The validity of the one-dimensional temperature profile is confirmed. The numerical results and field measurements also

demonstrate that temperature gradient along a steel web can be divided into two categories according to the thermal environment (Tong et al., 2000):
- Normal heat exchange conditions on both faces, e.g. an I-beam section; and
- Normal heat exchange conditions on one face only, e.g. a box section with an enclosed air void.

It is noted that webs of a closed box section normally have steeper temperature gradients than those where both faces can lose heat.

5. Parametric Studies

Figure 2 shows the notations and default dimensions of a plate girder and a box girder chosen to verify the above observations. The section depth d_t is taken as 1500mm so as to allow sufficient depth for the girder web to return to the ambient air temperature. The meteorological conditions of the chosen summer day are assumed, and the solar radiation is taken to act only on the top surface of the top flange. In addition, a hypothetical "partially exposed" I-section is considered. It has the same section as the plate girder but one face of the web, the bottom face of the top flange and the top face of the bottom flange on one half are assumed to have no heat exchange with the environment at all. The results shown in Figure 3 suggest that the temperature distribution depends primarily on the heat exchange conditions of the web. The positive temperature profiles developed in the plate-girder and box-girder sections are different. However the results from the "partially exposed" I-section are good enough to predict the behaviour of the box-girder section.

Strictly speaking the temperature distribution of a steel bridge deck depends on all the geometric and material parameters in addition to the environment. However most design codes classify sections into categories, each of which has one single design profile. Steel sections may be classified into closed and open sections for thermal loading specifications. The thermal properties of steel and bituminous surfacing are known with little uncertainty. A study on the influence of geometric parameters on the temperature distribution of steel sections has been carried out (Tong, 2000) by varying only a parameter each time. The effects caused by changes in girder spacing, web thickness and flange thicknesses are minimal. Therefore for practical design purposes, the only significant factor is the thickness of the bituminous surfacing. The effects of varying the thickness of bituminous surfacing as shown in Figure 4 are substantial.

6. Extreme Distributions of Climatic Factors and Thermal Loading

The thermal loads on a bridge may be considered as random functions of the random variables T_{max}, the daily maximum air temperature, T_{min}, the daily minimum air temperature and Q, the daily global solar radiation. The extreme thermal loads are therefore random variables. To predict the probability of occurrence of their maxima and minima, their frequency distributions are required. It is assumed that the extreme thermal loads are functions of the extreme values of Q and T_{max}.

To determine the extreme thermal loads at a geographical location, it is necessary to carry out the extreme analysis of the climatic factors first. Gumbel (1960) has observed that the tails of the frequency distributions of most climatic factors decay in an exponential manner. Among

n observations of a random variable y, the cumulative distribution function of the largest value Y_n follows a double-exponential Type I asymptotic distributions, namely

$$P_{Y_n}(y) = \text{Prob}(Y_n < y) = \exp\left[-e^{-a_n(y-u_n)}\right] \tag{5}$$

where a_n is the scale parameter and u_n is the location parameter. When the above extreme distribution is defined, the design value for any return period can be calculated from the statistics of extremes. In the study, the parameters a_n and u_n are obtained from the linearised plot of the climatic factor y_i from n observations and their corresponding probabilities p_i using the reduced variate x_i, defined as (Ho and Liu, 1989):

$$x = -\ln\left[-\ln(P_{Y_n}(y))\right] \tag{6}$$

The corresponding cumulative distribution function for the smallest value Y_1 among the n observed values is

$$P_{Y_1}(y) = \text{Prob}(Y_1 < y) = 1 - \exp\left[-e^{-a_1(y-u_1)}\right] \tag{7}$$

Statistical analyses are carried out based on the meteorological records of the Hong Kong Observatory since 1958. Table 2 summarises the statistics of climatic factors adopted (Tong, 2000) together with the predicted values y_{50} and y_{120} for return periods of 50 and 120 years respectively. To facilitate subsequent extreme analysis of thermal loading, the variations of the instantaneous solar intensity I_t, the diurnal variation of air temperature T_a, and the daily maximum and minimum air temperatures for summer and winter are described by equations obtained by curve fitting from local meteorological records (Tong, 2000).

Theoretically, the extreme distributions of bridge thermal loading can be evaluated in a similar manner as the climatic factors. However, this approach requires a database with long period of field measurements, which is not available. A convenient, though not mathematically rigorous, alternative is to evaluate the extreme thermal loading by using numerical integration (Ho and Liu, 1989). The statistical parameters, including the mean μ, standard deviation σ, skewness θ and kurtosis κ, which define the extreme distributions of air temperature and solar radiation are calculated and utilised in the prediction of extreme thermal loading (Tong, 2000).

7. Design Temperature Profiles

To generate design temperature profiles for general use, only the more important parameters are considered. The design profiles for steel sections are divided into two categories: closed and open sections. For each category, design profiles predicted for different thickness of bituminous surfacing are also given. A fifth-order polynomial (Priestley, 1978) is used to model the distribution of temperature. It is assumed that the web will approach the ambient air temperature if it is deep enough. The profile is given by

$$T = \begin{cases} T_D(1-h/h_0)^5 + T_b & 0 \leq h \leq h_0 \\ T_b & h > h_0 \end{cases} \tag{8}$$

where T is the temperature at a depth h from the top, h_0 is the depth below which the web temperature becomes a constant value T_b, T_D is the difference between the top flange temperature and the constant temperature T_b. Then the values of the parameters T_D, T_b and h_0 are determined so that the fifth-order profile generates the 50-year return values of maximum effective temperature T_e, positive temperature difference ΔT, top surface temperature T_s and thermal moment M_t which is given by

Table 1. Parameters adopted in full-scale bridge study.

		Steel	Asphalt
Density (kg/m^3)		7854	2100
Specific Heat Capacity (J/kg/K)		434	840
Thermal conductivity (W/m/K)		60.5	0.75
Film coefficient (W/m^2/K)	(Upward heat loss)	-	18.5
	(Downward heat loss)	6.8	-
Absorptivity		-	0.95
Emissivity		-	0.9

Table 2. Statistics and predicted values of climatic factors.

	a_n or a_1	u_n or u_1	Statistics				Tong (2000)		Liu (1985)	
			μ	σ	θ	κ	y_{50}	y_{120}	y_{50}	y_{120}
Daily maximum air temperature (°C)	1.6	34.0	34.33	0.78	1.14	5.40	36.4	36.9	36.3	36.9
Daily minimum air temperature (°C)	-0.7	7.5	6.74	1.76	-1.14	5.40	2.2	1.0	2.1	1.1
Max. global solar radiation (MJ/m^2/day)	0.5	26.2	27.43	2.81	1.14	5.40	34.7	36.7	37.1	38.6

Table 3. Positive temperature difference for closed section with values from SDMHR (HyD, 1997) shown in brackets.

	Thickness of surfacing (mm)							
	0		40		50	100	150
T_1 (°C)	28	(39)	24	(33)	23	18	15
T_2 (°C)	22	(21)	19	(19)	18	15	13
T_3 (°C)	14	(8)	12	(11)	11	10	8
T_4 (°C)	7	(4)	5	(6)	5	5	4

Table 4. Positive temperature difference for open section with values from SDMHR (HyD, 1997) shown in brackets.

	Thickness of surfacing (mm)							
	0		40		50	100	150
T_1 (°C)	22	(39)	17	(33)	16	14	12
T_2 (°C)	17	(21)	14	(19)	13	11	10
T_3 (°C)	11	(8)	9	(11)	8	7	7
T_4 (°C)	5	(4)	4	(6)	4	4	4

Table 5. Comparison of temperature differentials with various code requirements.

Temperature differential	Closed section	Open section	SDMHR	BS5400	OHBDC
Maximum positive	24 °C	17 °C	33 °C	24 °C	10 °C
Maximum reverse	-3 °C	-2 °C	-3 °C	-6 °C	-5 °C

Table 6. Comparison of effective temperatures with various code requirements.

Effective temperature	Closed section	Open section	SDMHR (50-year)	AASHTO Code	BS5400	OHBDC
Maximum	47 °C	44 °C	44 °C	49 °C	46 °C	59 °C
Minimum	1 °C	1 °C	0 °C	-18 °C	-	-6 °C

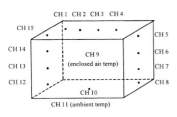
Figure 1(a). Box girder section.

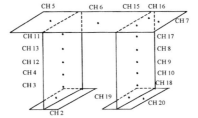
Figure 1(b). π-girder section.

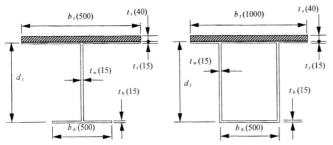
Figure 2. Sections adopted in extreme analysis, with default dimensions in mm for sensitivity studies shown in brackets (not to scale).

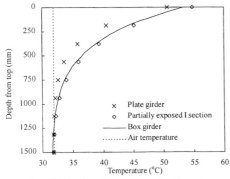
Figure 3. Comparison of temperature profiles along web.

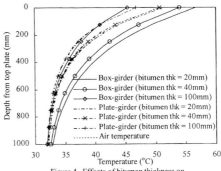
Figure 4. Effects of bitumen thickness on temperature profile.

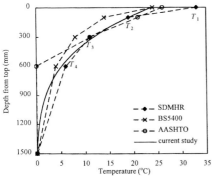
Figure 5. Design temperature profiles for box-girder.

$$M_T = \int Ty\,dA/I \qquad (9)$$

in terms of the second moment of area I. In the evaluation of T_e and M_t for the determination of design temperature profile in this section, only the web is taken into account. However it will not affect the accuracy as T_s is treated as a key parameter and the top flange temperature is effectively constant. When the parameters T_D, T_b and h_0 are calculated based on the extreme thermal loading, the corresponding extreme temperature profile can be defined. Figure 5 shows the temperature profile calculated based on the thermal loads of 50-year return period. The calculated profile is compared with those given in the *SDMHR Code* (HyD, 1997), BS5400 (BSI, 1978) and other design codes by aligning the profiles at soffit. It is observed that, in general, the calculated profile is close to other specifications except the local *SDMHR Code*, which tends to overestimate the temperature especially at the top. Table 3 shows the values of T_1 to T_4 for different thicknesses of surfacing and a 50-year return period. The corresponding results for the open section are shown in Table 4. The reverse temperature profiles can be obtained similarly. Table 5 compares the temperature differentials of closed and open sections with 40mm bituminous surfacing with recommendations from other codes.

In the search for a realistic design effective temperature, one may for the time being take the average value of top flange temperature T_s and bottom flange temperature T_l as a reasonable estimate of the effective temperature. In view of the shape of the temperature profile, the design effective temperatures actually depend on the section depth. The design effective temperatures assuming a section depth of 1000mm with 40mm bituminous surfacing for a return period of 50 years are compared with other recommendations in Table 6.

8. Conclusions

After a comprehensive study of the thermal behaviour of steel bridges, it is confirmed that the shade air temperature and solar radiation are the most influential factors affecting the temperature distribution of steel structures. Both the field measurements and the numerical results from validated mathematical models have indicated that the temperature gradients induced in a closed steel section are steeper than those in an open steel section under the same meteorological conditions.

By using the mathematical models, extreme analysis is carried out based on the statistics of extremes and numerical integration. Some design temperature profiles have been developed for both closed and open steel sections, with allowance for various thicknesses of bituminous surfacing. The design temperature differences and effective temperatures have been obtained and compared to the recommendations of the local SDMHR code (HyD, 1997). It confirms that the existing requirements are reasonable, although slightly conservative in respect of the temperature difference.

References

1. American Association of State Highway and Transportation Officials (1996) "Standard Specifications for Highway Bridges", Washington, D.C., U.S.A.
2. American Association of State Highway and Transportation Officials (1998) "AASHTO LRFD Bridge Design Specifications, 2nd Ed.", Washington, D.C., U.S.A.

3. Au, F.T.K., Tham, L.G., Tong, V.M. and Lee, P.K.K. (1999) "An Initial Review of Design Thermal Loading for Bridges in Hong Kong", Proceedings of Seminar of Structural Aspects of Airport Core Program projects, May, 1999, pp. 89-103.
4. British Standard Institution (1972) "BS153, Specification for Steel Girder Bridges", London, U.K.
5. British Standard Institution (1978) "BS5400, Steel, Concrete and Composite Bridges", London, U.K.
6. Capps, M.W.R. (1968) "The Thermal Behavior of the Beachley Viaduct/Wye Bridge", Ministry of Transport, Road Research Laboratory, RRL Report LR 234.
7. Emerson, M. (1968) "Bridge Temperatures and Movements in the British Isles", Ministry of Transport, TRRL Report LR 228.
8. Emerson, M. (1973) "The calculation of the distribution of temperature in bridges", Department of the Environment, TRRL Report LR 561.
9. Emerson, M. (1976) "Bridge Temperatures estimated from the shade temperature", Department of the Environment, TRRL Report LR 696.
10. Gumbel, E.J. (1960) "Statistics of Extremes", Columbia University Press, New York.
11. Highways Department (1997) "Structures Design Manual for Highways and Railways, 2nd Ed.", Hong Kong Government.
12. Ho, D. and Liu, C.H. (1989) "Extreme Thermal Loadings in Highway Bridges", American Society of Civil Engineers, Journal of Structural Division, Vol. 115, No. 7, pp. 1681-96.
13. Hunt, B. and Cooke, N. (1975) "Thermal Calculations for Bridge Design", American Society of Civil Engineers, Journal of Structural Division, Vol. 101, pp. 1763-81.
14. Kennedy, J.B. and Soliman, M.H. (1987) "Temperature Distribution in Composite Bridges," Journal of Structural Engineering, Vol. 113, No. 3, pp. 475-82.
15. Liu, C.H. (1985) "Investigation of Temperature Distribution in Highway Bridges", M.Phil. Thesis, The University of Hong Kong, Hong Kong.
16. Ministry of Transport (1991) "Ontario Highway Bridge Design Code and Commentary," Canada.
17. Priestley, M.J.N. (1978) "Design Thermal Gradient for Concrete Bridges", New Zealand Engineering, Vol. 31, Part 9, pp. 213-219.
18. Public Works Department (1979) "Design of Highway Structures", Civil Engineering Manual, Vol. V, Chapter 4.
19. Reynolds, J.C. and Emanuel, J.H. (1974) "Thermal Stresses and Movements in Bridges", American Society of Civil Engineers, Journal of Structural Division, Vol. 100, pp. 63-78.
20. Tong, M.V. (2000) "Temperature Distribution in Highway Bridges", M.Phil. Thesis, The University of Hong Kong, Hong Kong.
21. Tong, M., Au, F.T.K., Tham, L.G. and Wong, K.Y. (2000) "A study of temperature measurement in long span steel bridge", Proceedings of workshop on research and monitoring of long span bridges, The University of Hong Kong, 26-28 April 2000, pp. 188-195.
22. Tong, M., Tham, L.G., Au, F.T.K. and Lee, P.K.K. (2001) "Numerical modelling for temperature distribution in steel bridges", Computers and Structures, Vol. 79, No. 6, pp. 583-593.
23. Zuk, W. (1965) "Thermal behaviour of Composite Bridges – Insulated and Uninsulated", Highway Research Record, No. 76, pp. 231-253.

Decision Support System for Bridge Aesthetic Design Using Immune System

HITOSHI FURUTA, MICHIYUKI HIROKANE AND KAYO ISHIDA
Department of Informatics, Kansai University, Takatsuki, Osaka 569-1095, Japan

ABSTRACT

Aesthetic design of bridge is becoming more and more important to establish a beautiful scenery and landscape. However, it is not easy for bridge engineers to achieve a satisfactory aesthetic design of bridge. In this paper, an attempt is made to develop a decision support system that a bridge engineer without experience and education on the aesthetics can easily use to obtain several candidates for designing bridges. It is concluded that the proposed system can provide us with several design solutions with different characteristics that are useful for practical cases.

INTRODUCTION

Aesthetic design of bridge is becoming more and more important to establish a beautiful scenery and landscape. However, it is not easy for bridge engineers to achieve a satisfactory aesthetic design of bridge. In this paper, an attempt is made to develop a decision support system that a bridge engineer without experience and education on the aesthetics can easily use to obtain several candidates for designing bridges. The decision support system consists of three subsystems such as evaluation system of aesthetics, optimization system, and describing system using computer graphics. As an optimization tool, immune system is employed, because the immune system has such advantages that it can solve a combinatorial problem with discontinuous objective functions, and in addition, it can provide several solutions with different characteristics. Using these advantages, it is possible to develop a practical design system for realizing the aesthetics in the design of bridge structures. Some candidates given by the system are compared and checked with visual expressions through computer graphics. If all candidates are not satisfactory, the system can provide more suitable solutions by modifying the design concept. Several illustrative design examples are presented to demonstrate the efficiency and applicability of the decision support system developed here. It is concluded that the proposed system can provide us with several design solutions with different characteristics that are useful for practical cases.

IMMUNE SYSTEM

In immune systems of a living body, there are immune tissues which have a function that can distinguish itself or not. In immune responses, tissues of a living organism are distinguished from foreign substances (antigens) by immunocytes that protect the organism from antigens, eliminating such foreign substances as viruses and malign parts of the organism's own tissue. In order to respond to the antigens, the immune system reconstructs the genes of tissues and produces the antibody to eliminate the antigens. Furthermore, the immune system has also the immunity to self-organism. When a large amount of antibodies of the same type are produced, their production is suppressed by the so called "suppressor cell" to restore a regular state from unbalanced state.

The immune system can be applied to an optimization problem as follows:

Step 1 : Recognition of antigens

Antigens are recognized as input data by the system. In an optimization problem, antigens correspond to objective functions and constraints. On the other hand, antibodies correspond to the optimal solutions.

Step 2 : Determination of antibody genotype

Gene elements of antibodies are expressed with symbol strings. Input data of the problem are necessary to be converted to symbol strings. In usual, each antibody has a one-dimensional genotype and each element is expressed by binary figures, i.e., $\{0,1\}$.

Step 3 : Production of the initial group of antibody

A number of strings are generated based upon the model given at Step 2. An initial antibody group is formed by determining each antibody gene at random so as to generate antibodies of different genes.

Step 4 : Calculation of affinity between antibodies

Affinity or similarity is calculated among all the antibodies. The affinity between two antibodies, $a_{y,w}$, is obtained as

$$a_{y,w} = 1 / (1 + H_{v,w}) \tag{1}$$

where $H_{v,w}$ is the distance between antibodies v and w. The two antibodies perfectly match when $H_{v,w} = 0$. As the distance, Hamming distance or information entropy is employed.

Step 5 : Calculation of affinity between antigens and antibodies
The affinity of each antibody with an antigen is calculated as

$$ax_v = opt_v \qquad (0 \leq opt_v \leq 1) \qquad (2)$$

where opt_v means the intensity of connection between antigen and antibody v, which is defines as the optimal value is obtained when $opt_v = 1$. The larger the value of ax_v is, the stronger the connection between the antigen and antibody is. The elimination of antigens is succeeded when $ax_v = 1$.

Step 6 : Differentiation into memory and suppressor cells
Antibodies obtained during the search process are produced as memory cell or suppressor cell. The memory cells become the candidates for the optimal solution. First, concentration of all the antibodies are calculated. Then, the antibody v with the concentration larger than the threshold value of Tc is differentiated into memory cells. Since the total number of the memory cells is limited, the new memory cell is exchanged with the stored memory cell with the largest affinity when the total number reaches the prescribed number. Next, the suppressor cell s with the same genotype as that of the new memory cell is differentiated. The memory cell to be differentiated at a generation is at most one and selected as the cell with the maximum expected value given in Eq. 5. The concentration of antibody v, cv, is calculated as

$$ac_{v,w} = \Sigma \ ac_{v,w} / N \qquad (3)$$

$$c_v = \begin{cases} 1 & ay_{v,w} \leq Tac1 \\ 0 & \text{otherwise} \end{cases} \qquad (4)$$

where Tac1 is the threshold value of affinity, and N is the total number of antibodies. This implies that antibodies v and w can be equivalent when they have a large affinity.

Step 7 : Promotion and suppression of antibody production
    1) N/2 antibodies having the lowest affinity with the antigen are extinguished.
    2) Each of the remaining antibodies is extinguished if its affinity with a suppressor cell exceeds a threshold. This is for averting duplicative production of memory

cell.

3) The expectation of an antibody to survive into the next generation is given by

$$e_i = a_{xv} \prod (1 - a_{skv,s}) / c_v \sum a_{xv} \qquad (5)$$

$$a_{sv,s} = \begin{cases} a_{yv,s} & a_{yv,s} \geq Tac2 \\ 0 & \text{otherwise} \end{cases} \qquad (6)$$

where Tac2 is the threshold value of affinity, S is the total number of suppressor cells, k is the suppress power.

Step 8 : Production of antibodies
1) Instead of the antibodies extinguished at Step 7 (2), new antibodies are produced randomly.
2) Among the antibodies remaining after Step 7 and the newly produced antibodies, N/4 pairs are selected while allowing duplication. Antibodies with higher expectation values are more likely to be selected. Genes are crossed over for each antibody pair, and N/2 antibodies are newly generated. Mutation is induced in the generated antibodies so as to alter genes.

Step 9 : Termination of calculation
The calculation is terminated when the number of generation reaches the prescribed number or the solution searched is thoroughly satisfactory.

DECISION SUPORT SYSTEM FOR BRIDGE AESTHETIC DESIGN

In this study, an attempt is made to develop a decision support system for the aesthetic design of bridges by applying the immune system to the optimization procedure. Introducing the immune system, it is possible to obtain several design candidates with different characteristics.

Aesthetic Factors for Bridge Design

For the aesthetic factors of bridge design, the following items are considered: 1) overall configuration of bridge, 2) configuration of pier, 3) configuration of main girder, 4) configuration of handrail, and 5) colors of main girder and handrail. To realize the aesthetic design of bridges, several concepts are prepared, which are summarized in Table 1.

Table 1 Design Concepts

Harmony to environment	Necessary condition
Symbol	Selected condition
Uniqueness	Selected condition
Reliability (Peaceful)	Selected condition
Friendly	Selected condition
Dignity	Selected condition
Amenity	Selected condition
Locality	Selected condition
Internationality	Selected condition

Representative environmental conditions prepared here are summarized in Table 2.

Table 2 Environmental Conditions

Clear sky
Cloudy
White cloud
Mountain (Green Leaves)
Mountain (Dead Leaves)
Rock or Soil
Pavement
River or Sea
Urban (Buildings)
Urban (Residences)

Parameters for Immune Calculation

The aesthetic factors are encoded to a string consisting binary figures, i.e., {0,1}. For overall bridge configuration, configuration of piers, configurations of main girders and handrails, and colors of main girders and handrails, 4 classes, 16 configurations, 4 configurations, and 64 colors are prepared in this study, respectively. Then, the total length of the string becomes 22 bits, and the number of the all combination is 4,194,304.

Parameters necessary for the immune calculation are summarized in Table 3.

Table 3 Parameters for Immune Calculation

Population size	50
Limited number of memory cell	5
Crossover	0.6 (2-point crossover)
Mutation	0.125

To evaluate the color harmony, the theory of Moon and Spencer is used with considering the effects of area. The relations between configuration and color are evaluated by introducing the measure of beauty proposed by Birkhoff.

In order to realize a good aesthetic design of bridges, it is necessary to establish an evaluation method of aesthetics. Using the above methods, the aesthetics of bridge is evaluated by applying fuzzy logic, e.g., fuzzy reasoning. To choose appropriate colors and configurations, 14 pairs of adjectives and 6 pairs of adjectives are used as shown in Tables 4 and 5.

Table 4 Adjectives for Colors

1	Warm ⟵⟶ Cool
2	Bright ⟵⟶ Dark
3	Showy ⟵⟶ Plain
4	Heavy ⟵⟶ Light
5	Stable ⟵⟶ Unstable
6	Harmonic ⟵⟶ Not harmonic
7	Balanced ⟵⟶ Unbalanced
8	Stimulating ⟵⟶ Modest
9	Soft ⟵⟶ Hard
10	Dynamic ⟵⟶ Static
11	Modern ⟵⟶ Classic
12	Unique ⟵⟶ Normal
13	Fresh ⟵⟶ Common
14	Strong ⟵⟶ Weak

Table 5 Adjectives for Configurations

1	Strong ⟵⟶ Weak
2	Simple ⟵⟶ Complicated
3	Hard ⟵⟶ Soft
4	Open ⟵⟶ Close
5	Unified ⟵⟶ Scattered
6	Continuous ⟵⟶ Discrete

Fuzzy reasoning is executed by using the following If-Then rules:

$$\text{If } X \text{ is } A \text{ Then } Y \text{ is } B. \qquad (7)$$

where A and B are fuzzy sets defined in terms of membership functions. Through the inference process, the final evaluation is obtained using the If-Then rules that relate the design concepts and the evaluation of colors and configurations.

APPLICATION EXAMPLES

To demonstrate the efficiency of the system developed here, consider a design example of arched bridge. For the calculation, the following conditions are assumed:
1) The number of antibodies is 20.
2) Crossover rate is 0.6.
3) Mutation rate is 0.125 to avoid the trap of local optimum.
4) The calculation is terminated after 100 generations.
5) The threshold values are assumed that $Tac1 = 0.8$, $Tc = 0.35$, $Tac2 = 0.95$.

As the design concept preferred for this plan, the harmony with the environment, reliability, friendly, and locality are selected, and as the landscape factors, clear sky, mountain (green leaves) are chosen. For other design conditions, the followings are assumed: the clearance below the girders is low and the road width is narrow.

Implementing the proposed system, five design alternatives are obtained as shown in Table 6.

Table 6 Design Alternatives

	Bridge	Pier	Main Girder	Handrail	Main Girder	Handrail	Degree of Beuaty
Design 1	Cable-stayed	D	Varied(slow)	c	Purple	Purple	1.018519
Design 2	Arched	O	Varied(Abrupt)	b	Yellow	Purple	0.806250
Design 3	Plate Girder	C	Varied(Abrupt)	c	Purple	Cream Yellow	1.071875
Design 4	Trussed	O	Varied(Abrupt)	c	Light Green	Purple	0.878125
Design 5	Cable-stayed	C	Constant	a	Blue	Dark Blue	0.666129

It is said that the color harmony is satisfactory, if the degree of beauty is larger than 0.5. The five design plans obtained are considered to satisfy the minimum requirement. Design plan 1 has purple colors for both main girder and handrail. The combination of the same color provides the feeling of good harmony which leads to friendly and reliable images. Munsell value of purple is 8P2/4 and this color shows the dignity and plain. On the other hand, design plan 2 has yellow color with 8YR7/6 of Munsell value. This color gives the warm and reliable feeling and the configuration of arched bridge provides soft image. Through computer graphics, it seems that the design plan looks more fresh and modern than expected. Design plan 3 has purple for main girder and cream yellow (Munsell value = 10YR7.5/13) for handrail. Since the color of handrail (cream yellow) is much brighter than the color of main girder (purple), it looks stable and reliable. From the configuration selected, it is considered that the bridge has the unification. Computer graphics provides a feeling of unified, hard, continuity and strong, though the bridge is simple. Design plan 4 has light green (Munsell value = 2.5G8.5/2.5) for main girder, which shows the locality due to the plain and light feelings. Since the plan is trussed bridge, it is likely to have the hard feeling. However, the computer graphics does not show the strong and hard image, but open and continuous feelings. Finally, design plan 5 has blue color (Munsell value = 9B7.5/5.5) for main girder and dark blue (5PB3/4), which show the image of brightness and fresh with friendly feeling. The configuration chosen provides the unification. Visualization through computer graphics gives good and strong feelings. It is concluded that the proposed system can provide us with several satisfactory design alternatives with completely different characteristics.

At last execution results are shown in figure 1 below.

Fig. 1 An execution result

CONCLUSIONS

In this study, an attempt was made to develop a decision support system for the aesthetic design of short and middle spanned bridges. In order to obtain several satisfactory design alternatives among the great deal of design combination, the immune system was applied to the optimization procedure. For evaluating the color harmony, the theory of Moon and Spencer was used, and the measure of beauty proposed by Birkhoff was employed so as to account for the interrelation between colors and configurations. Furthermore, the aesthetics of bridge was evaluated through the fuzzy reasoning using If-Then rules.

Through several design applications, the following conclusions were derived:

1) Using the proposed decision support system, it is possible for a bridge engineer without experience and education on the aesthetics to obtain several design alternatives.
2) As an optimization tool for the aesthetic design of bridges, immune system is quite useful and powerful, because the immune system has such advantages that it can solve a combinatorial problem with discontinuous objective functions, and in addition, it can provide several solutions with different characteristics. Using these advantages, it is possible to develop a practical design system for realizing the aesthetics in the design of bridge structures.
3) Some design alternatives given by the proposed system are compared and checked with visual expressions through computer graphics. If all candidates are not satisfactory, the system can provide more suitable solutions by modifying the design concept. Illustrative design examples showed the efficiency and applicability of the decision support system developed here. It

is concluded that the proposed system can provide us with several design solutions with different characteristics that are useful for practical cases.

References

Ishida.Y, Hirayama.H, Fujita.H, Ishiguro.A, Mori.I (1998) : *Immune System and The applications,* Corona-sha

Nishida. T (1981) : *The color psychology learning*, Zokei-sya

The Japanese standard society (1996) : *Hand Book of JIS*, The Japanese standard society

Kondo. T (1986) : *The view and color learning*, Rikou-Book

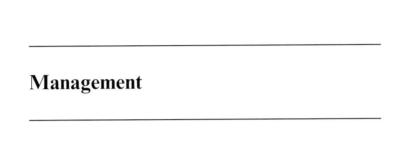

OPTIMUM MAINTENANCE STRATEGIES FOR TRUNK ROAD BRIDGES IN ENGLAND AND ITALY

PARAG C DAS OBE, PhD, CEng, FICE
Project Director Bridge Management, Highways Agency, London, UK

LIVIA PARDI Dott Ing
Senior Maintenance Engineer, Autostrade, Rome, Italy

INTRODUCTION

The Highways Agency is responsible for approximately 15000 structures on the trunk road network in England. Of these, 10,500 are bridges. Similarly, Autostrade manages some 3000 bridges on the trunk roads in Italy. The compositions of the two bridge stocks are shown in Figures 1 and 2 respectively.

A co-operative project was undertaken in 1999, which is still continuing, to study the maintenance problems faced by the bridge managers in the two organisations and to develop a common methodology for predicting maintenance costs.

The paper contains the preliminary findings from the co-operative project. It is hoped that, in the next stage, the work will be progressed to produce a general methodology which will not only be acceptable to the two agencies, but also to other bridge owners and operators.

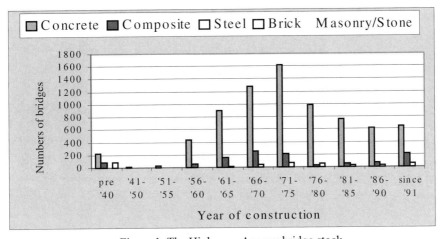

Figure 1. The Highways Agency bridge stock

342 MANAGEMENT

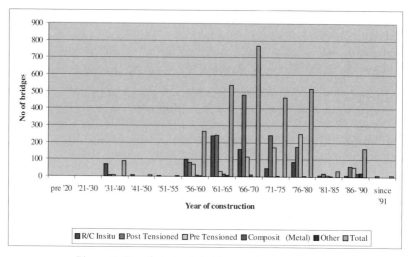

Figure 2. The Autostrade bridge stock

OPTIMUM MAINTENANCE STRATEGIES

The necessity and options for the essential work required when a structure is assessed to be sub-standard are by and large obvious. On the other hand, when a structure is assessed to be adequate at present, preventative work is more difficult to justify since future predictions invariably contain considerable uncertainty. It is therefore necessary to develop standard recommendations regarding appropriate preventative methods for different types of structures based on optimum strategies determined using whole life costing. The following procedure which was used in the case of the Highways agency's composite bridges, is recommended for this purpose.

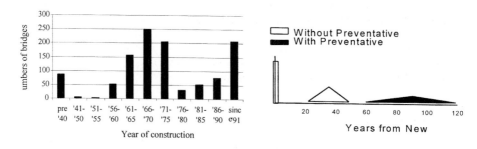

Figure 3. Composite bridges on the trunk road network in England

Figure 4. Bridge rehabilitation rates

First, the numbers of bridges in the group in the network need to be tabulated for each year of construction, as shown in Fig.3. Then the year by year rates of rehabilitation of the bridges in terms of their age, in the absence of periodic repainting, the preventative method under consideration, as well as when regular repainting is carried out, need to be determined as shown in Fig.4 (estimated by experts in this case).

The rate of application of painting in terms of bridges per year is also similarly estimated. Multiplying these rates of rehabilitation and painting with the numbers of bridges built in different years, the profiles of future numbers requiring rehabilitation with or without painting can be determined as shown in Fig.5. It has also been found that approximately 7 % of the bridges will need repainting each year.

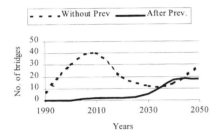

Figure 5. Predicted bridge rehabilitation numbers with and without preventative maintenance

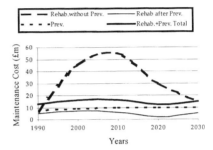

Figure 6. Predicted total maintenance costs for two options

Multiplying the above annual numbers by unit costs for rehabilitation and repainting an average bridge, the cost profiles of rehabilitation with and without repainting can be determined as shown in Fig.6. It can be seen from Fig.6. that the cost of rehabilitation without using repainting as a preventative measure is considerably greater than when it is used. The recommended maintenance procedure in the case of composite bridges in the Highways Agency's network will therefore be to repaint as necessary as a preventative measure.

In the case of other preventative measures, the cost differences between the 'with or without' options may not be so obvious. In such cases, the discounted present value of each option should be determined for the choice to be clearer.

REHABILITATION RATES

It can be seen from the above methodology that rehabilitation rates with and without different preventative measures are necessary for determining bridge type specific optimum maintenance strategies. In the above example, the rehabilitation rates shown in Fig.4. are from a range covering different bridge types which were based on expert estimates.

A further consultation with experts has resulted in the probabilistic distributions of rehabilitation/replacement rates shown in Table 1.

	Without Prev.		With Prev.		
R/C Insitu	Weibul	4.3	51.1 Lognorm	89.2	17.35
Post Ten	Lognorm	50	10 Normal	105	45.4
Pre Ten	Weibul	5.1	61.2 Lognorm	97.5	17.1
Comp	Lognorm	35.9	6.2 Lognorm	89.2	17.35
Other	Lognorm	36	6.1 Normal	80	31

Table 1. Estimated rehabilitation rates of different bridge types

STRATEGIC PLAN

An important step in the management process is to produce a strategic plan which would provide the future expenditure profiles for different types of maintenance work covering a number of years representing an ideal mix of work in terms of logistics and funding.

A strategic plan was prepared by the Highways Agency in 1997 which is shown in Fig.7. It was based on the whole structures stock being divided into four bridge types. Optimum maintenance strategies were determined for all four types of Highways Agency's bridges (see Fig.1), using the method described earlier for the composite bridges. The overall plan was the summation of the four optimum strategies.

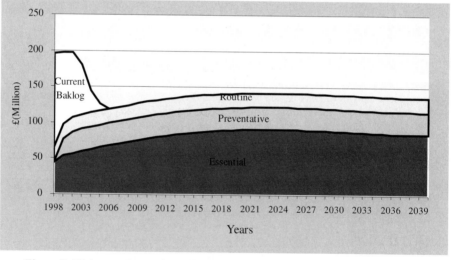

Figure 7. Highways Agency's predicted future expenditure on bridge maintenance

STRATEGIC PLAN FOR AUTOSTRADE BRIDGES

Using the above procedure, a strategic plan has been produced for the Autostrade bridges. The bridges were divided into 5 bridge types as in Fig.2.

The assumed rehabilitation rates are as shown in Table 1. The maintenance unit costs for preventative and rehabilitation work for all categories of the bridges were taken from the data-bank of Autostrade. The preventative maintenance cost per bridge of approximately £75,000 is assumed to apply to all bridge types. It is also assumed that only 6 % of the bridges will have preventative treatment per year, i.e. a work cycle of 15 years is implied.

Multiplying each bridge type numbers in Fig.2 with the rehabilitation rates and the unit costs, for each type a cost comparison was made. The comparison for reinforced concrete bridges is shown in Fig.8. In the figure,

R/C With = Rehabilitation when preventative work is done
R/C No = Rehabilitation without preventative work
R/C R+Pr = Rehabilitation with preventative cost total.

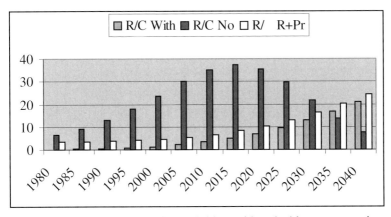

Figure 8. Comparison of cost profiles for RC bridges with and without preventative work

It can be seen from Fig.8 that rehabilitation with preventative maintenance should be the chosen option. Similar conclusions are also drawn form the comparisons involving the other four types of bridges.

Finally the Strategic Plan has been produced, as shown in Fig.9, by adding all the chosen optimum strategies for the five types of bridges.

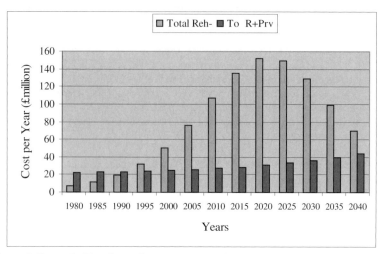

Figure 9. Strategic Plan for maintenance expenditure for the Autostrade bridges

The Strategic Plan shows that preventative maintenance should be carried out as assumed, and the recommended total expenditure level for the Autostrade bridges is about £ 20m - £30m per year for the foreseeable future. This is excluding routine maintenance costs and any backlog of rehabilitation which may be outstanding.

COMPARISON WITH AUTOSTRADE DATA

A study was carried out in Autostrade at the time of the renewal of the concession in 1997. As the period of the concession was extended from the original 2018 to 2038, the company was obliged to revise all its financial strategies according to the new deadline and taking into account two major constraints.

The first one is that the bridges must be returned to the state "in good conditions" at the end of the concession. The second one is that the fare policy that the company is allowed to pursue depends on "the level of quality" offered to the users, its values being updated every 5 years.

The results of this study are represented in figs.10-11. The level of quality remains almost constant if not only rehabilitation, but also preventative measures, are taken into account. This implies repairing all the bridges of the network.

The annual expenditure for maintenance is approximately £20m, matching substantially the value indicated by The Highway Agency method.

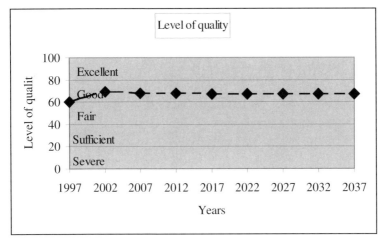

Figure 10. Strategic Plan for maintenance expenditure for the Autostrade bridges

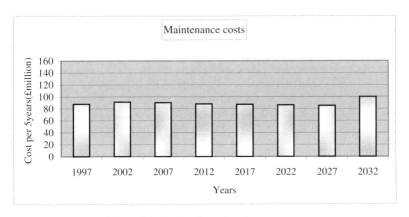

Figure 11. Evaluation of maintenance costs

CONCLUSIONS

The paper describes in broad outline a probability based methodology for determining optimum maintenance strategies for different bridge types. The individual recommended cost profiles are then added together to produce Strategic Plans for bridge maintenance expenditure for the whole stocks.

The methodology has been applied to produce Strategic Plan recommendations for the bridge stocks of the Highways Agency in England and Autostrade in Italy. It is demonstrated that the methodology is practical and rational. However, a number of simplistic assumptions have been made regarding rehabilitation rates and unit costs, which can and should be improved in the future for more accurate predictions.

ACKNOWLEDGEMENTS

This paper is being presented with the kind permission of the Chief Executives of the Highways Agency and Autostrade. The authors are is also grateful to Prof Mike Forde of Edinburgh University, Dr Toula Onoufriou of the University of Surrey and Dipl.Eng. Katja D Flaig of Messrs W S Atkins, for their very helpful comments and suggestions during the drafting of the paper. The authors are particularly grateful to Mr Camomilla, Head of Research at Autostrade for his encouragement for undertaking the co-operative project, which forms the basis of this paper.

REFERENCES
1. Das, P.C. Development of a comprehensive structures management methodology. Management of highway structures. Thomas Telford Publications, London, 1999.
2. Narasimhan, S. And Wallbank, W. Inspection manuals for bridges and associated structures. Management of highway structures. Thomas Telford Publications, London, 1999.
3. Shetty, N. et al. Advanced methods of assessment of bridges. Management of highway structures. Thomas Telford Publications, London, 1999.
4. Haneef, N and Chaplin, K. Bid assessment and prioritisation system. Management of highway structures. Thomas Telford Publications, London, 1999.
5. Das, P.C. New developments in bridge management methodology. Structural Engineering International. Vol 8 No 4., November 1998.
6. Das, P.C.. Maintenance planning for the trunk road structures in England. International bridge management conference. TRB Conference. Denver, Colorado, USA, April 1999.
7. Das, P.C. and Micic, T.V. Maintaining highway structures for safety, economy and sustainability. Eighth International conference on structural faults and repair. London, 1999.
8. BD 21. The assessment and strengthening of highway bridges and structures. Design manual for roads and bridges. The Stationery Office, London,1997.
9. QUADRO. Vol. 14 Section 1 Design manual for roads and bridges. The Stationery Office, London.
10. Frangopol, D.M. and Das, P.C. Management of bridge stocks based on future reliability and maintenance costs. International conference on current and future trends in bridge design construction and maintenance. Singapore, October 1999.
11. Thoft-Christensen, P. Estimation of bridge reliability distributions. International conference on current and future trends in bridge design construction and maintenance. Singapore, October 1999.
12. Frangopol, D.M. et al. Optimum maintenance strategies for highway bridges. International conference on current and future trends in bridge design construction and maintenance. Singapore, October 1999.

Inspection and Maintenance of Hong Kong's Long Span Bridges

JAMES D GIBSON
TMML (seconded from AMEC International Construction)
Hong Kong, China

Introduction

Prior to the completion of the construction contracts for the Lantau Fixed Crossing between Tsing Yi and Lantau Islands in May 1997 and several other projects within the Airport Core Programme, Tsing Ma Management Limited (TMML) was awarded a contract by the Hong Kong SAR Government to manage, operate and maintain an area of road network called the Tsing Ma Control Area (TMCA).

TMML were given approximately a four and half months gearing up period in which to make the necessary preparations in order to ensure that it was in a position to manage the TMCA from its opening to the public on 19 May 1997.

These preparations included: the fitting out of the various buildings with furniture and equipment, the recruitment and training of sufficient staff to fulfil the Government's minimum manning levels, the purchase of various vehicles (recovery, patrol, maintenance) the participation in all final commissioning and handover inspections between the various contractors and the Government's agents, the submission of various procedures such as those for Operating and Maintenance and procurement of the various permit and license applications.

Tsing Ma Control Area (TMCA)

The TMCA forms a critical essential link in both highway and railway access to the new international airport at Chek Lap Kok from the existing developed urban areas of Hong Kong, some 160km of lane length of carriageway is available to traffic.

This area is composed of some rather impressive facilities and major components as follows:

- Kwai Chung Viaduct
- Cheung Tsing Bridge
- Cheung Tsing Tunnel
- North West Tsing Yi Interchange
- Lantau Toll Plaza
- North Lantau Expressway (Part)
- Ting Kau Bridge and Approach Viaducts

- Kap Shui Mun Bridge and Ma Wan Viaduct
- Tsing Ma Bridge

In addition to the above there are additional facilities such as two administration buildings, and numerous other smaller buildings. In total there are over 130 GEO registered slopes some over 100m high.

The E&M equipment provisions are also extensive and include Traffic Control and Surveillance Systems, Toll Collection System, Supervisory Control and Data Acquisition System (SCADA System), Lighting, Power Supply, Security and Wind and Structural Health Monitoring Systems, bridge inspection equipment.

Tsing Ma Management Limited (TMML)

TMML is owned by three companies Wilson Group (part of Sun Hung Kai), the leading shareholder, China Resources and AMEC International Construction and was formed specifically to undertake the Operation, Management, Inspection and Maintenance of the TMCA.

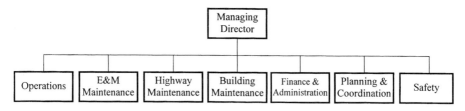

Figure 1 TMML Organisation Chart.

The company at present employs approximately 640 staff who are assigned to the various operating sections.

The Highway Maintenance Department of TMML is tasked with the job of inspecting, reporting and maintaining the bridges, structures, slopes and roads within the TMCA. One of the sections within this department is responsible for the inspection and maintenance of the three long span bridges.

The Building Department is responsible for monitoring the upkeep of all buildings, cleansing and the gardening of the planted beds.

E&M maintains some 102 vehicles, all the electronics, software and services within the TMCA.

Operations have within their scope of work traffic control and surveillance, vehicle recovery, security control, toll collection, lane enforcement and operation of various E&M systems.

Both E&M and Operations operate around the clock on a 3-shift roster basis.

Lastly, the F&A Department has responsibilities for personnel, general administration, canteen services, purchase and supplies, accounting and the toll money.

Government Monitoring Team

The Government has appointed several teams tasked with monitoring and auditing TMML's performance in respect of the safe and efficient management operation and maintenance of the TMCA.

To date the following Government Departments have been involved: Architectural Services, Electrical and Mechanical Services, Transport Department and the Highways Department.

All of the above have direct liaison with each of the mirror departments within TMML and thus prolonged lines of communication have been greatly reduced.

There are times however when more than one department is involved with a particular matter, for example, alteration of signage and road markings.

Highway Maintenance

It is TMML's policy to undertake inspection and maintenance in-house as far as possible in order to achieve better control of work quality as well as response and completion times. For those works that fall beyond TMML's own expertise or those of an infrequent nature or perhaps not economical, then TMML engages specialists to carry out the work under TMML supervision.

Such examples would be any carriageway resurfacing, surveying or Engineering Inspections of slopes.

The HM department is divided into five sections: roadworks, geotechnical works, structures, long span cable supported bridges and works.

Each of first four sections is led by an experienced professional engineer who reports to the Highway Maintenance Manager. Depending on the workload, each technical section has a number of deputy engineers, supervisors and inspectors. There has been a very limited amount of staff movements between sections and it can not be considered as significant, the main reason being the vast variation in skills that are required in each discipline. Labour as required by each section is drawn from a pool controlled by the works section. Currently there are:

Deputy Engineers	5	Technical Supervisors	5
Inspectors	20	Works Supervisors / Inspectors	3
Labour	30	Senior Drivers	10

The last item above, that of Senior Drivers requires further explanation.

Due to the low levels of car ownership in Hong Kong, not every person has either the confidence or a driving license and to guarantee driving resources additional staff have been employed. In addition within the Highways section there are several vehicles that require special licenses to operate them such as the Bridge Inspection Vehicle, Tunnel Washing Vehicle, Road Sweepers, Crane Lorry, Gully Emptier, Hoists, in total there are some 22 vehicles.

352 MANAGEMENT

Inspection: Long Span Cable Supported Bridges

The following structures are included in the scope of works
- The Tsing Ma Bridge
- The Kap Shui Mun Bridge (excluding Ma Wan Viaduct)
- The Ting Kau Bridge (excluding the approach viaduct and ramps)

The following documents specify the inspection requirements:
- Inspection Manual for Highway Structures
- Port Works Manual
- Master Maintenance Manual (MOM Contract Document)
- Maintenance Manuals for each of the three bridges

The MOM Contract requires six basic types of inspection to be carried out with the following frequencies:
- Routine Superficial Inspections (RI) – Weekly – Visual
- Routine Statutory Inspections (Stat) – Annual or as required
- Routine Functional Inspections (FI) – Annual or as required
- Routine General Inspections (GI) – Annual or Two Yearly – Close Visual
- Special Inspections (SI) – As required – Visual / Close Visual
- Routine Principal Inspections (PI) – as GI

The scope of each of the above is as follows:

Routine Superficial – to identify any defects, damage, debris or items that may present hazards to the public or requires further investigation.

Routine Statutory – to ensure that various facilities are fit to be used and may include underdeck maintenance gantries, suspended working platforms and lifts.

Routine Functional – to ensure that various equipment is safe to use.

Special Inspections – non routine inspections following severe climate or environmental events, incidents that have threatened the integrity of a particular structural element and lastly the need to acquire additional information necessary for further investigation. Only two typhoons have been severe enough to warrant Special Inspections of the bridges during the last four years.

Routine General – the scope of each inspection for a particular bridge element is varied between a visual or close visual inspection (within touching distance), the latter being the more common. It is dependent on the need to maintain structural integrity (criticality) and the risk to damage or deterioration (vulnerability). The majority of elements that go to make up each bridge have been given appropriate ratings.

Criticality	Vulnerability	Interval Between Inspections
A	1	6 months
B	2	1 year
C	3	2 years
D	4	6 years

Table 1 Inspection Intervals

Some Typical Elements	Criticality
Bearings	B
Suspender Wire at Bottom	B
Main Cable Wires	B
Ortho Deck PLT Trans Joint	C
Ortho Deck to X Frame	B
Concrete to Piers	C

Table 2 Typical Inspection Intervals

Vulnerability ratings indicate the propensity to deterioration due to corrosion, accidental damages or wear. Obviously, with new structures and the very thorough and regular inspection regime, one would not expect to be finding many examples of severe damage.

During the current contract some 40 number detailed specifications for Inspection and Maintenance (IMS) relevant to each major structural element have been developed eg. Deck Movement Joints, Bearings, Main Towers, Suspension Systems and Deck sections. In conjunction with the IMS documents some 25 Inspection Procedures have also been prepared to control the inspection process and provide guidance to the inspection teams on the frequency, scope inspection equipment, safety procedures and traffic management requirements.

Some examples are:-
- Reinforced Concrete – Close Visual Inspection
- Stay Cables – Close Visual Inspection
- Structural Steelwork – Close Visual Inspection

Access for Inspection

All of the bridges have been fitted with access walkways ladders and platforms. However, it would be impossible to cater for every eventuality and purpose. One of the obvious methods used to overcome this is the installation of scaffold platforms. Steel scaffolds are the preferred method. As an inspector undertakes his inspection he has to have his hands free to write and collect any findings, a bamboo scaffold has the tendency to be less rigid and require the user to hang on. Additionally if any load has to be carried such as during some maintenance activities then again, steel tube scaffolds would be the norm.

Other means of access available are lifts in the tower legs of each bridge, underdeck maintenance gantries, hydraulic aerial platforms, scissor lifts, a double decker bus, MTRC works train, a modular suspended access platform system, stay and main cable gondolas. Unfortunately industrial roped access methods have recently been disallowed after a change in the legislation for normal use and can only be permitted under very exceptional circumstances, and not for general inspection.

Probably one of the most difficult inspections is that of the underside of the Tsing Ma Bridge upper deck. The height above the lower deck floor or road level varies from 5.3m to 8.5m at the Tsing Yi abutment. Under the MOM Contract, there is a requirement that certain welds receive an annual inspection. It should be remembered that the lower deck of the bridge is divided up into a series of zones delineated by fences i.e. the fairing services envelope, the two lower deck carriageways and the railway reserve. The inspection route has therefore been delineated by the fences and is along the length of the bridge in each zone.

354 MANAGEMENT

Service Envelope– Permanent walkways and ladders supplemented by Youngmans boards and support frames and a temporary scaffold platform within the high area at Tsing Yi.

Lower Deck Carriageway– By a jet fan scissor lift, it provides a large area zone 3m x 7m is mobile and can easily be set at the correct height. Sometimes an aerial platform is used for localised access, however it is rather slow as the outriggers have to be set down before the arm can be used. TMML also has a double decker bus in which the upper deck has been modified to provide an access platform.

Rail Envelope – By use of the MTRC's overhead power line works train and use of their MOOG bridge inspection vehicle again for the high area at Tsing Yi. Inspection in this area is further complicated by the overhead catenary cable support system and power lines. These cables are generally 1.5m below the area of inspection and have to be physically squeezed past by the inspectors.

Road & Rail Access Equipment	Owner
5m aerial platforms (2 No)	TMML
20m aerial platform	TMML
34m aerial platform	TMML
Palfinger – Bridge Inspection Vehicle	Gov. HK for use in TMCA
Works Train	MTRC
MOOG Bridge Inspection Vehicle (Rail Mounted)	MTRC
MOOG Bridge Inspection Vehicle	HyD

Bridge Access Equipment		Number
Underdeck Gantries	TMB	9 x 3 sets
	KSMB	3 sets
	TKB	4 x 2 sets
Gondolas	TKB	4-Transverse & 1-Longitudinal
	KSMB	4
	TMB	1-Suspender
		1-Main Cable
Cradles	TKB	1-Solsit, 1-Semi-circular,1-6m (1-3m)
		2-U/deck gantry
	KSMB	4-Solo, 1-10m (6 x 3m)
	TMB	2-Semi-circular, 8m, 6m, 3m

Table 3   Access Equipment

Inspection Findings

As can be expected with new structures, the defects found have been very minor indeed and generally fall into one of the following categories.

- Cracking of the paint system where it was applied too thick.
- Localised paint damage due to wear and tear.
- A few weld cracks at the stop/start positions on ends of stiffeners and on edges of plates at cope hole positions.
- Exposed reinforcement, usually those bars that were only covered with a film of grout that has subsequently been eroded away by rain.

- Cast in sockets used for temporary works during construction that proved to be too difficult to remove or were overlooked i.e. threaded rods that were cut and not removed and are now corroding.
- Loose bolts in secondary steelwork.
- Water leakage through cable penetrations.

In terms of damage due to accidents or typhoons, again only minor damage has generally been experienced such as sign plates becoming detached from their support frame, a tower hatch damaging its lifting mechanism when blown open. A typical vehicle impact on one of the tensioned wire parapets usually results in displacement of the spacer battens, a couple of bent bolts and broken support shoes and of course damaged paint.

There has of course been some defects that were related to the unexpected performance or anomalies of installation and these have all been repaired by the original contractors as part of their defects liability period, to the satisfaction of all concerned parties.

Wind and Structural Health Monitoring System (WASHMS)
The Highways Department of the Hong Kong SAR has had a series of sensors installed on all three bridges to monitor the health and performance of each structure. Generally there are temperature sensors in the main cables and the decks, strain gauges in longitudinal girders and cross frames, and a level sensing system, displacement sensors, accelerometers, anemometers, a weigh-in motion system and lastly a GPS system for TKB, KSMB and TMB. Data is acquired initially at one of a series of outstation computers and is then forwarded via a fibre optic network to a central computer for processing.

The analysis and publication of results is all managed and controlled by the Highways Department of the Hong Kong SAR.

Maintenance

The above topic is deemed to cover monitoring, inspection, remedying, maintenance and repair together with the preparation of reports and records. Under the MOM Contract, there are two types of maintenance.

The first being Scheduled Maintenance (i.e. part of TMML's obligations under the MOM Contract) and includes:
- Maintaining the cleanliness of the sliding surfaces of the bearings.
- Repair of accident damage to safety barriers.
- Checking the bolt tension and undertaking bolt tightening of the cable band bolts of Tsing Ma Bridge.
- Removal of minor objects trapped in the road movement joints.
- Removal of water that has accumulated or is trapped in parts of the bridge structure.
- General cleansing of the carriageways and removal of litter and debris.

TMML is also obliged under the MOM Contract to provide a painting gang and welding gang. The scope of these services is limited to repairs of defects that can be reasonably handled by the said gangs. The service is also deemed to include the cost of paint, labour, plant, access, non-destructive testing etc.

The second being Non Scheduled Maintenance (i.e. work outside the MOM Contract) and might cover:
- Major repainting of section of steel deck or the main cable.
- Repair of concrete cracks and the treatment of corroded reinforcement.
- Replacement of traffic signs.
- Replacement of bearings or parts.
- Replacement of movement joint control springs.

Additional Works

Sometimes additional work, usually improvement works, is instructed by the Highways Department of the Hong Kong SAR to facilitate an enhancement or change to the structures.

Some examples of these are:-
- Changing of the sign faces on sign gantries due to the opening of a new section of highway adjacent or connecting with the TMCA.
- Installation of crash cushions at bifications between the main carriageways and ramps.
- Installation of additional access ladders/platforms to improve access to a particular area.
- Application of a silane paste to the towers of Tsing Ma and Kap Shui Mun bridges.

> This last item requires further explanation. Due to the rather humid nature of Hong Kong's climate at certain times of the year and the levels of pollution generally in the Pearl River Delta, this mix of pollution and moisture provides an ideal breeding environment for mould to grow on concrete surfaces. This growth appears as very unsightly dark staining. It was decided that such prestigious bridges should look good. The concrete was first cleaned by water jetting and once dry a silane paste was applied to severely limit any water drops from remaining on the structure, thus preventing a re-occurrence of the mould. This has proved to be the case. A decision on whether to tackle some of the other elements such as anchorages will be made at a later date.

Maintenance Database

TMML has the responsibility of implementing and developing a maintenance database under the MOM Contract and this was done in collaboration with EXOR Corporation Limited.

The software package is based on the Oracle Relational Database Management System and was primarily developed as a road management system. This software has since been modified to accept the complex data models needed to handle the long span bridges, slopes and structures of the TMCA. The programme is modular in design and thus far, TMML are currently using the Network Manager, Maintenance Manager and Document Manager modules. The Public Enquiry Manager and Spatial Data Manager are being considered for future incorporation into the system.

In brief, each module will now be described.

Document Manager allows for the definition, storage, management and retrieval of documents of any media type eg. scanned documents, digital photographs, documents videos etc. These items can then be attached in a "many to many" relationship with either a section of network or an item of inventory or even a defect.

Network Manager provides the key referencing system by which all other items are linked. This permits all operational data to be shared and provides management and staff with the most up to date information. The network has been considered as four linear networks one for each engineering discipline as illustrated in Figure 2.

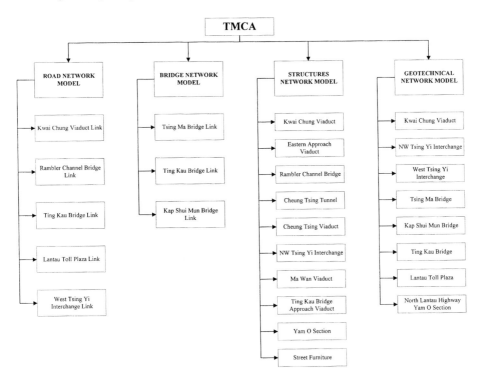

Figure 2 TMCA Maintenance Database Referencing System

Each network is further subdivided into links and sections i.e. major and minor parts and it is on these minor parts that the assets or inventory are attached.

The Maintenance Manager is the application module and allows TMML to perform and monitor the maintenance activities from viewing or loading of assets, through their inspection, the loading of defects and the management of their repair through works orders. The bridge assets have been defined as hierarchical assets principally because of the need to use part of the hierarchy for storing a reference for the asset and the others to store the actual asset. A typical asset structure is presented in Figure 3.

One point to note with long structures such as suspension bridges, is that unless the referencing system is robust, simple, easy to follow it can be very nearly impossible to return to the exact same spot where a defect is located.

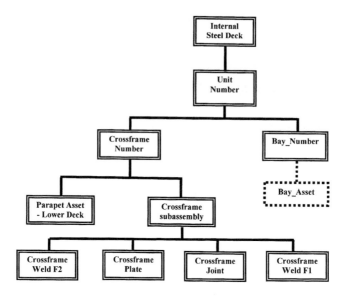

Figure 3 Hierarchical Asset Structure

The Future

The first four years have proved to be very intensive for all those concerned. There was the initial period of familiarisation prior to the opening of all the structures and facilities. This being followed by the post opening era, where the structures and facilities were in use and the contractors were finishing off their contractual works. At the same time, TMML were involved with handover inspections, receipt of equipment and spare parts. It actually proved to be a useful period in which information and knowledge could be gleaned from the departing contractors and consultants staff. During the latter part of the contract, the nature of the work process has now settled down into more of a routine.

TMML was recently successful in securing the next MOM Contract for a further term of six years. The intensive inspection regime has not found any significant defects and the future looks good for these major structures.

Acknowledgements

The author wishes to thank the Highways Department of the Hong Kong SAR for permission to publish this paper. Any views expressed are entirely those of the author.

How effective is bridge posting in enhancing reliability?

SOLOMON B. A. ASANTEY and F. MICHAEL BARTLETT
Department of Civil and Environmental Engineering, University of Western Ontario, London, Canada.

Abstract

In practice, compliance to bridge load restrictions is imperfect. This paper considers the effect of imperfect load compliance on the member reliability of a 30-m span bridge, using Monte Carlo simulation and sensitivity analysis. Three annual volumes of a governing truck are considered. Results show that under a stringent load restriction, even a small violation rate, in the order of 2.5%, leads to a significant loss of reliability. In other words, if load restrictions are not strictly enforced, the increase in reliability associated with posting a bridge can almost vanish. The loss in reliability is more pronounced for larger annual truck volumes. This study, therefore, questions the effectiveness of load restriction enforcement as a bridge management tool.

Introduction

Where growth in payloads and volumes of heavy commercial vehicles impart excessive loading on an aging bridge stock, bridges may be rendered structurally deficient (Stewart and Val 1999). The usual response of the bridge owner, who often has to manage the bridge inventory with limited resources, is to post a bridge to restrict its use to lighter trucks or, in extreme cases, to close it to all traffic. Bridge evaluators are then called upon to make recommendations for three options: (a) to replace the bridge, (b) to rehabilitate the bridge, or (c) to post the bridge indefinitely with a weight-limit. For bridges that carry relatively low volumes of heavy truck traffic and have convenient bypass routes, indefinite bridge posting is a valid option. However, in practice, circumstances that make long-term posting attractive are not conducive to cost-effective weight restriction enforcement. Of the bridge management alternatives currently in use, the effectiveness of load restriction enforcement as a bridge management tool has not received detailed investigation.

From the bridge evaluation viewpoint, the impact of changes in both the regulatory limit and accompanying violation rates on bridge reliability requires investigation. An efficient enforcement strategy that ensures effective compliance to load restrictions is a critical element of any plan for controlling vehicle weights on a bridge. Without effective enforcement, including the certainty of penalties and sanctions sufficient to deter violation, load postings can become ineffective.

The objective of this study is, therefore, to investigate the impact of incomplete or imperfect compliance to posted loads and annual traffic volumes of a governing heavy truck on bridge member reliability. The scope of the overall investigation is limited to maximum one-lane bending moments for simple spans 5-40 m, although the findings presented in this paper pertain only to a 30-m span. A reference period of one year is used for the reliability analyses.

Research significance

Governing load effects for short span bridges (5-40m) are caused by the passage of a single truck, a heavy axle group, or a group of axle configurations with dynamic effects (Harman and Davenport 1979). This implies that the passage of one maverick truck, out of a good number of legal heavy trucks that ply a short span bridge, could significantly reduce the reliability index of a bridge component, perhaps rendering it substandard, depending on the consequence of failure.

Results from the present study show that for a stringent posted load, even a relatively small violation rate, in the order of 2.5%, leads to excessive loading and a significant reduction in reliability. It is also demonstrated that the combined adverse effects of a less stringent restriction load and a high violation rate could significantly reduce the reliability index, especially for larger annual truck traffic volumes.

Imperfect load compliance

Figure 1 is a schematic representation of the gross vehicle weight (GVW) cumulative distribution function, for various compliance conditions. The smooth s-shaped curve, shown as a thin full line, corresponds to the original GVW cumulative distribution without any load restriction. Under conditions of perfect compliance to a load limit, GVW_L, the original GVW cumulative distribution is transformed into a truncated distribution shown by the thick full line. For the imperfect compliance conditions, the cumulative distribution falls between the limits for no compliance and perfect compliance, as shown by the dot-dashed line in Fig. 1.

In Fig. 1, the fraction of a given truck population with gross vehicle weights (GVW) exceeding the GVW load limit (GVW_L) that pass over a bridge is shown by the interval L_1, and the fraction of the total population of the truck type on the highway section with GVW exceeding GVW_L is shown as L_2. In this study, violation rate (VR) is defined as the ratio of the number of trucks with $GVW > GVW_L$ that pass over the bridge to the total number of trucks on the highway section whose weights exceed GVW_L (i.e. $VR = L_1/L_2$). It is assumed that violation rate is constant for all GVW exceeding a given GVW_L.

Live Load Data and Model

The static live load data used in this study is based on parameters of a governing 6-axle "extra-large" (XL) truck configuration obtained from a traffic load survey of unrestricted normal traffic conducted in Ghana (GHA 1995). Along the 1500-km primary highway route, an annual volume of 116 XL trucks was estimated. The population of XL trucks observed had a mean GVW of 526 kN with a standard deviation of 172 kN. The legal limit at the time of the survey was a GVW of 550 kN.

Dynamic amplification parameters are obtained from Cooper (1997), who reported a statistical model for evaluating dynamic amplification factors (*DAF*) imposed on short span bridges based on heavy traffic data from various bridge sites in the UK. The use of UK-based *DAF* parameters to amplify static loads on Ghana bridges is appropriate because most commercial vehicles operating in Ghana originate from the UK and other European countries.

The annual maximum total live load (L_A), which represents the maximum live load, including dynamic amplification that would occur during a 1-year period, is modeled as

$$[1] \qquad L_A = GVW_A \cdot DAF_{CA}$$

where GVW_A is a value of the annual maximum *GVW* distribution. Figure 2 illustrates the transformation of the probability density function of the event *GVW* to the annual *GVW* distribution. If the number of events per year (N_a) is known, a GVW_A value can be derived from the event distribution (Kennedy et al. 1992). The resulting annual maximum *GVW* distribution has a relatively higher mean value and a smaller variability than the corresponding event *GVW* distribution, as shown in Fig. 2. In Equation 1, the DAF_{CA} value is a "companion action" dynamic amplification factor. Following the work of others, it is assumed that normal distributions can represent the event *GVW* (Harman and Davenport 1979, Heywood 1995) and event *DAF* (Cooper 1997) of unrestricted traffic. The probability of the occurrence of more than one truck in the critical region of a short span bridge is assumed negligible.

The annual maximum total live load effect (LE_A) on a bridge member, required for reliability analysis, requires modifying the annual maximum total live load using a load-to-load effect conversion factor (K_{LE}) and a load distribution factor (K_{DF}). To evaluate K_{LE}, a truck is positioned at a critical location on the bridge to induce the maximum load effect. If each axle load is assumed to be a constant fraction (F_i) of the *GVW*, the maximum load effect (*LE*) is the sum of the products of the *ith* influence line coefficient (IFL_i) and the respective axle load ($F_i \cdot GVW$). If the axle spacing for a given heavy vehicle configuration type is assumed constant, the set of IFL_i values remain constant for a given truck type positioned at the critical location. The maximum load effect (LE_j) on a short span bridge due to the *jth* truck with *n* axle loads could, therefore, be formulated as

$$[2] \qquad \begin{aligned} LE_j &= \sum_{i=1}^{n} \left(IFL_i \cdot F_i \cdot GVW_j \right) \\ &= GVW_j \cdot \sum_{i=1}^{n} \left(IFL_i \cdot F_i \right) \\ &= K_{LE} \cdot GVW_j \end{aligned}$$

Here the load-to-load effect conversion factor, $K_{LE} (= \sum_{i=1}^{n} IFL_i \cdot F_i)$, remains constant for all vehicles with a given configuration applied to a given influence line.

The parameters of the load distribution factor (K_{DF}) account for the transverse distribution of the load effect and are evaluated using the simplified method in the Canadian Highway Bridge Design Code (CSA 2000).

Thus, the equation for the annual maximum total live load effect (LE_A) is

[3] $$LE_A = K_{LE} \cdot K_{DF} \cdot L_A$$

where L_A is the annual maximum total live load from [1].

Reliability analysis

Equation 3 is incorporated into the limit state function (G), which is formulated in terms of the resistance (R), the dead load effect (D) and the annual total maximum live load effect (LE_A) variables, as

[4] $$G = R - D - LE_A$$

When the combined effects of dead and live load (S) exceed the resistance (i.e. G < 0), failure occurs. The probability of failure (P_f) can be expressed in an integral form as

[5] $$P_f = \int_{-\infty}^{\infty} F_R(x) f_S(x) \, dx$$

where $F_R(x)$ is the cumulative distribution function of the resistance at a value x, and $f_s(x)$ is the probability density function of the combined effects of dead and live load at x, as illustrated in Fig. 3. Load compliance, whether perfect or imperfect, renders the upper tail of the total load curve somewhat irregular as shown in Fig. 3.

The First Order Second Moment reliability method (FOSM) has been extensively used for reliability analyses in the literature (Nowak and Lind 1979, Verma and Moses 1989). This method is not appropriate for use in the present study, due to the irregular nature of the upper tail of the event static live load distribution. Because of the difficulty of assigning an appropriate analytical function to the annual live load distribution under imperfect compliance conditions, the use of the efficient first order reliability method (FORM) (Park et al 1998), which is sensitive to tail approximations, is also ruled out.

For these reasons, Monte Carlo simulation has been used to generate discrete values of the limit state function. The annual probability of failure, P_f, is determined directly from the simulation results for a one-year period as

[6] $$P_f = \frac{\text{Number of discrete values of } G \text{ less than zero}}{\text{Total number of discrete values of } G}$$

An equivalent annual reliability index (β), used in this study, is

[7] $$\beta = -\Phi^{-1}(P_f)$$

where P_f is obtained from [6] and Φ^{-1} is the inverse normal distribution function.

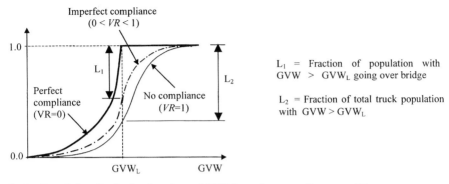

Fig. 1. Cumulative distribution functions of *GVW* for various compliance conditions

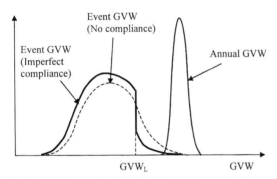

Fig. 2. Probability density functions for event GVW and annual GVW for imperfect load compliance

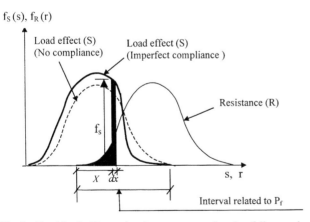

Fig. 3. Total load effect and resistance curves showing failure region for imperfect load compliance

The reliability index is computed using Monte Carlo simulation as follows:

1. Generate random values of resistance and dead load effect based on given parameters.

2. Retrieve a random value of stored annual maximum total live load (L_A) from a live load simulation analysis process based on Equation [1], accounting for the level of violation to a given load limit. Apply a load-to-load effect factor (K_{LE}) and a randomly-generated component load distribution factor (K_{DF}) to obtain the member annual maximum total live load effect.

3. Evaluate the limit state function, [4], using resistance and loads obtained from Steps 1-2.

4. Repeat Steps 1-3 for N simulation runs, for a given load limit, violation rate and annual traffic volume. $N = 10^5$ ensures reasonable convergence for the failure probabilities calculated.

5. Calculate P_f from [6] and β from [7].

Scope of parametric study: Resistance, dead load and live load effect parameters

A simply-supported multi-girder 30-m span concrete bridge was investigated in this study. The bridge has five parallel girders, equally spaced at 2.6-m on centre. The parameters for resistance, dead load moments, event static and dynamic live loads, load-to-load effect conversion factor (K_{LE}) and load distribution factor (K_{DF}) assumed are as given in Table 1. The coefficients of variation for resistance and dead load moment were based on statistical parameters reported in Kennedy et al (1992) and the Canadian Highway Bridge Design Code (CSA 2000), respectively. The mean dead load moment was estimated using approximate equations recommended by ACI Committee 343 (ACI 343, 1995). The mean resistance was assigned to achieve a reliability index of 3.0 at the current legal load of 550 kN for an annual traffic volume of 200 *XL* trucks. The impact of changes in load limits, violation rates and annual traffic volumes on the reliability index were then investigated.

In presenting the results, the posting loads have been normalized by the current *XL* truck legal limit (GVW_{L_o}) of 550 kN. A posting load range of $GVW_L / GVW_{L_o} = 0.8 - 1.2$ is investigated. A posting load of $GVW_L / GVW_{L_o} < 1$ reflects a decision by bridge managers to imposed a more stringent load limit, if the current load limit does not ensure the economic viability of investments in bridges and road pavement infrastructure on a given route. A posting load range of $GVW_L / GVW_{L_o} > 1$, however, simulates possible changes in the posting load that may result from the effect of mounting pressure from the trucking industry to increase load limits in order to ensure productivity gains.

Violation rates of 0% (perfect compliance), 2.5%, 5%, 10%, 30%, 50% and 100% were considered.

Effect of imperfect compliance on annual reliability

Figures 4, 5 and 6 illustrate the sensitivity of the reliability index (β) with posting loads and violation rates (VR) for annual volumes of XL trucks of 20, 200 and 2000, respectively, for 30-m single span bridge. In all cases, there is no change in the reliability index when the violation rate is 100%, irrespective of the posting load, as expected.

Figure 4 illustrates that for a very low volume of trucks, the violation rate must be 5% or less, if a minimum reliability index of 3.0 is desired. For this violation rate, virtually no gain is achieved by posting a bridge at a stringent load limit of $GVW_L/GVW_{L_o} = 0.8$, rather than at a less restrictive limit of $GVW_L/GVW_{L_o} = 1.2$. An improvement in load enforcement corresponding to a violation rate of only 2.5% at $GVW_L/GVW_{L_o} = 0.8$ or 1.0, however, increases the reliability index to 3.6 and 3.5, respectively.

For a bridge girder with a reliability index of 2.3, Fig. 5 indicates that, for a moderate traffic volume, there is no significant gain in reliability index due to posting at a stringent load if the violation rate exceeds 2.5%. If perfect compliance is ensured, however, posting at $GVW_L/GVW_{L_o} = 1.0$ increases the reliability index to 3.0 and posting at $GVW_L/GVW_{L_o} = 0.8$ achieves a reliability index of 3.8. But, even for a stringent load restriction of $GVW_L/GVW_{L_o} = 0.8$, a relatively small violation rate of 2.5% or 1- in - 40, reduces reliability from 3.8 to 2.3.

Figure 5 also shows that if perfect load compliance is possible, the reliability index at the current legal limit of $GVW_L/GVW_{L_o} = 1.0$ is 3.0. At a less stringent posting load of $GVW_L/GVW_{L_o} > 1.2$ and a violation rate of 50%, however, the reliability index drops to 1.2. This demonstrates the combined adverse effects on the reliability index of a less stringent restriction load and a high violation rate.

For a given load restriction and violation rate, Figs. 4-6 illustrate that the reliability index decreases for larger annual truck traffic volumes. Larger annual volumes (N_a) result in more severe annual maximum live load effects and, consequently, lower reliability indices. For example, a load restriction of $GVW_L/GVW_{L_o} = 1.2$ ensures β of 3.0 for N_a of 20 trucks per year (Fig. 4). An increase of N_a to 2000 (Fig. 6), however, reduces β to a value of 2.0 for this load restriction.

It is inferred from these observations that, if posting loads are not strictly enforced, the increase in reliability associated with posting a bridge can almost vanish. The loss in reliability is more pronounced for bridges that carry larger annual truck volumes.

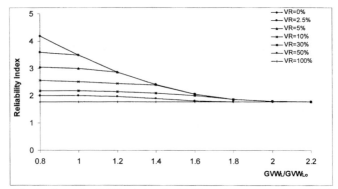

Fig. 4. Annual reliability index for 20 *XL* trucks per annum, considering various violation rates (VR).

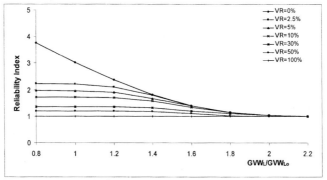

Fig. 5. Annual reliability index for 200 *XL* trucks per annum, considering various violation rates (VR).

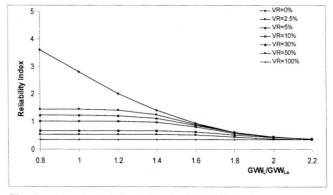

Fig. 6. Annual reliability index for 2000 *XL* trucks per annum, considering various violation rates (VR).

Summary and conclusions

The sensitivity of member reliability of a short span bridge to load restrictions, violation rates and annual truck traffic volume has been investigated. Conclusions of the study are as follows:

1. For a stringent load restriction, even a relatively small violation rate reduces the member reliability of a posted bridge significantly. For example, for a stringent load restriction of $GVW_L/GVW_{L_o} = 0.8$, a violation rate (VR) of 2.5% reduces the reliability index at $VR = 0\%$ from 4.2 to 3.8, for an annual volume of 200 XL trucks on a 30-m span.

2. If perfect load compliance is possible, the reliability index at a more stringent posting load of $GVW_L/GVW_{L_o} = 1.0$ is 3.0. The combined adverse effects of a less stringent posted load range of $GVW_L/GVW_{L_o} = 1.2$-1.4 and a high violation rate of 50% significantly reduce the reliability index to a value as low as 1.2.

3. Under imperfect compliance conditions, the reliability index is relatively insensitive to restrictive posting loads. Conversely, ensuring strict compliance to stringent load restrictions could enhance reliability greatly, especially for larger truck volumes.

Acknowledgement

The financial assistance provided for this research and for the first author to attend this conference by the Canadian Commonwealth Scholarship and Fellowship Plan and the Natural Science and Engineering Research Council of Canada is gratefully acknowledged.

References

ACI Committee 343, 1995. Analysis and Design of Reinforced Concrete Bridge Structures (ACI 343R-95). In ACI 1997. Manual of Concrete Practice, Part 4, page 343-R51. American Concrete Institute, Farmington Hills, Michigan.

Canadian Standards Association (CSA), 2000. Canadian Highway Bridge Design Code, Canadian Standards Association, Rexdale, Ontario, 742 pp.

Cooper, D. I. 1997. Development of short span bridge-specific assessment live loading. Safety of Bridges. Thomas Telford, London, 64-89.

Ghana Highway Authority (GHA), 1995. Report on Axle Loads on Ghana Roads. Bridge Report No. 23. Ghana Highway Authority, Accra.

Harman, D. J. and Davenport, A. G. 1979. A statistical approach to traffic loading on highway bridges. Can. J. Civ. Eng., **6**, 494-513.

Heywood, R. J. 1995. Live loads on Australian bridges-statistical models from weigh-in-motion data. Australian Civil Engineering Transactions, CE37 (2): 107-116.

Kennedy, D. J. L., Gagnon, D. P., Allen, D. E., and MacGregor, J. G. 1992. Canadian highway bridge evaluation: load and resistance factors. Canadian Journal of Civil Engineering, **19** (6): 992-1006.

Nowak, A. S. and Lind, N. C. 1979. Practical bridge code calibration. ASCE Journal of Structural Division, **105** (ST12): 2497-2510.

Park, C. -H, Novak, A. S., Das, P. C. and Flint, A. R. 1998. Time-varying reliability model of steel girder bridges. Proceedings of the Institution of Civil Engineers. Structures and Buildings, **128**: 359-367.

Stewart, M. G. and Val, D. V. 1999. Role of load history in reliability-based decision analysis of aging bridges. ASCE Journal of Structural Engineering, **125** (7): 776-783.

Verma, D. and Moses, F. 1989. Calibration of bridge-strength evaluation code. ASCE Journal of Structural Engineering, **115** (6): 1539-1554.

Authors Contact Information

Solomon B. A. Asantey, PhD Candidate
Dept. of Civil and Environmental Engineering
University of Western Ontario
London, Ontario, Canada N6A 5B9
Email: sbasante@julian.uwo.ca
Tel: 519-645-2643 Fax: 519-661-3779

F. Michael Bartlett PhD, P. Eng.
Associate Professor
Dept. of Civil and Environmental Engineering
University of Western Ontario,
London, Ontario, Canada N6A 5B9
Email: f.m.bartlett@uwo.ca

Table 1. Resistance, dead load moment and live load effect parameters for 30-m span

Variable (1)	Description (2)	Units (3)	Mean (4)	Cov (5)
R	Resistance	kN·m	8495	0.12
D	Dead load moment	kN·m	2450	0.10
GVW_E	Event GVW	kN	526	172
DAF_E	Event DAF	-	1.21	0.15
K_{LE}	Load-to-load effect Factor	m	5.6	0.0
K_{DF}	Load distribution factor	-	0.63	0.08

LIFE CYCLE COST OF POST-TENSIONED T-SECTION GIRDER BRIDGES

TAMIO YOSHIOKA, SHOICHI OGAWA, CHENGNING WU
Research Laboratory, Oriental Construction Co., Ltd.
5, Kinugaoka, Moka-shi, Tochigi-ken, 321-4367 JAPAN

TAKAFUMI SUGIYAMA
Department of Civil Engineering, Gunma University
1-5-1, Tenjin-cho, Kiryu-shi, Gunma-ken, 376-8515 JAPAN

1. INTRODUCTION

Prestressed concrete bridges, which are located in very severe hot marine environments and not suitably protected against chloride attacks, can be damaged in an unexpected short time. Such damage may result in corrosion of re-bars and the delamination of cover concrete. Consequently, decrease in the sectional areas of re-bars and degradation of bonding between corroded re-bars and concrete increase the risk of structural failure. As a result very costly repair and strengthening are demanded, skyrocketing life cycle cost(LCC).

The common protection measures against chloride attacks consist of: 1) thickening the cover concrete, 2) decreasing the diffusion through the cover concrete, 3) applying a paint layer on the surface of concrete to limit the penetration of chloride into concrete, 4) using non-corrosive materials, such as coated re-bars, plastic sheaths, FRP materials etc. and 5) using cathodic protection.

The durability design and performance design concept allow for the consideration of LCC, which is evaluated by adding maintenance, repair, strengthening and demolition costs to initial construction cost(ICC).

Mathematical models governing the chloride penetration process are used to simulate the effect of chloride attacks on the durability of post-tensioned T-section bridges. Such bridges are assumed to be protected against chloride attacks using different protection measures. The protection measures considered in the present study consist in thickening cover concrete, applying a paint layer at the initial construction and reduction of the diffusion coefficient of patching concrete by adding polymer admixture or combining pozzolanic admixtures.

In the present study, the life span of painting is assumed to be 20 years, considering degradation of paint performance with age. The repair work is planned whenever the chloride concentration reaches the threshold level, which causes structural damage due to severe corrosion of re-bars. The repair work is here defined as the replacement of contaminated cover concrete. By combining painting and repair(concrete replacement), the life-span of bridges under consideration is extended over 100 years.

It is studied how thickening cover concrete and initial painting contribute to LCC and furthermore how characteristics of patching materials affect to reduce LCC. These results are compared with those obtained from similar bridges without any protection measures.

2. ANALYTICAL METHODS

2.1 Analysis of chloride penetration

This study aims to develop a numerical algorithm that can be used to predict chloride ingress into concrete. Two basic solutions are used to analyze the chloride ingress into concrete, fixing boundary conditions at surface and calculating diffusion into concrete.

2.2 Boundary conditions

Boundary conditions are fixed by assuming that only a part of air-born chloride reaching the surface of concrete can penetrate into concrete. Therefore, Langmuir's adsorption theory is adopted to differentiate the amount of chloride ingress into concrete from that reaching the surface.

The air-born chloride and chloride penetrating into concrete is governed by the following relationships as proposed in [1]:

$$q(t,s) = \frac{3.51 \cdot p \cdot V_0 \cdot Q(s)}{V_0 + 3.51 \cdot Q(s)} - \delta \cdot p \cdot C_0(s) \qquad (1)$$

$$p = \frac{0.526(W/C) - 7.61}{57.4} \times 100 \qquad (2)$$

where, V_0 is the desorption of the chloride per unit of time and is assumed to be 42.1 mg/cm²/year, δ is the wash out of chloride due to rainfall and is assumed to be 0.58 mg/cm²/year, $C_0(s)$ is the chloride content at the surface, $Q(s)$ is the chloride ions reaching the surface of concrete and is assumed to be 20 mg/cm²/year and p is the ratio of the saturated area at the surface of concrete. The girder is divided into two halves, one on the sea side, the other on the land side. It is assumed that lesser chloride ions reach the concrete surface on the land side than those on the sea side.

2.3 Paint performance

It is assumed that chloride penetration is completely prevented just after painting and the performance of painting degrades with age. The degradation rate is calculated using a proposed equation as follows:

$$F = [2 - \exp(\lambda \cdot t^N)] \qquad (3)$$

where, F is the rate of paint performance, λ and N are the characteristic values of paint material, and assumed to be 7.35e-5 and 3, respectively. The rate of paint performance in the present study is shown in Fig.1.

2.4 Diffusion equation

The following equations developed by Saetta et al. [2] are used to evaluate the diffusion of chloride into concrete, considering the influence of relative humidity, temperature and the hydration rate.

$$\frac{\partial Ct}{\partial t} = div[D_a grad(Ct)] - \bar{u}_t \frac{\partial C_f}{\partial x_t} \quad (4)$$

$$C_t = \alpha \cdot C_f \quad (5)$$

$$\alpha = w + (1 - w_{sat}) \cdot \gamma \quad (6)$$

$$\gamma = \frac{C_s}{C_f} = 0.7 \quad (7)$$

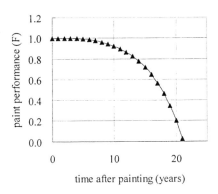

Fig.1 Degradation of paint performance

where, D_a is the apparent diffusion coefficient, and is evaluated as the ratio D_t/α (D_t is the intrinsic diffusion coefficient and α is the capacity factor), C_t is the total chloride amount, $\bar{u}_t$ is the flow velocity and is assumed to be zero in this study, C_f is chloride concentration in water, w is the water content and is assumed to be 5%, w_{sat} is the void ratio and is assumed to be 7%, C_s is the chloride concentration in the solid volume. The diffusion coefficient D_t depends on temperature, humidity, cement hydration rate and age. Calculation process of chloride diffusion through concrete is carried out using the following equations:

$$D_t = D_{t,rf} \cdot f_1(T) \cdot f_2(t_e) \cdot f_3(h) \quad (8)$$

$$f_1(T) = \exp\left[\frac{U}{R}\left(\frac{1}{T_0} - \frac{1}{T(t)}\right)\right] \quad (9)$$

$$f_2(t_e) = \zeta + (1-\zeta) \cdot \left(\frac{28}{t_e}\right)^{\frac{1}{2}} \quad (10)$$

$$f_3(h) = \left[1 + \left\{\frac{1-h(t)}{1-h_c}\right\}^4\right]^{-1} \quad (11)$$

where, $D_{t,rf}$ represents the reference value of the intrinsic diffusion coefficient, U/R is fixed at 3000degK, T_0=296degK, $T(t)$ is the temperature, ζ is a constant depending on type of concrete, t_e is equivalent maturation time in days, $h(t)$ is the relative humidity, h_c is the critical relative humidity and is fixed at 75%.

The W/C ratio of concrete is assumed to be 40%. $D_{t,rf}$ value is decided by the relationship between the W/C and diffusion coefficient [3].

The mean monthly value for temperature $T(t)$ and relative humidity $h(t)$ obtained in Okinawa - south islands surrounded by sea, Japan, is used in this study.

The relative humidity in concrete is determined using the humidity diffusion in concrete. To know the relative humidity in concrete following equation is proposed(refer to [2]):

$$h(t,d) = h_0(t) + [h_{ini} - h_0(t)] \cdot \left[1 - \exp\left(-\frac{d}{d_0}\right)\right] \quad (12)$$

where, $h(t,d)$ is the relative humidity at a distance d from the surface at time t, $h_0(t)$ represents environmental relative humidity, d_0 is fixed at 5cm, h_{ini} is the initial humidity and is assumed to be 90%.

2.5 Comparison between analytical diffusion models and existing bridges

In order to know the accuracy of the analytical models they are compared with existing structures under real environments or specimens exposed to marine environments for measuring chloride ingress. Data for existing structures and exposed specimens are tabulated in Table 1. The chloride ions reaching the concrete surface are assumed as to be 20mg/cm^2/year at the sea side and one third at the land side. In this comparison V_0 is assumed as to be 11.0mg/cm^2/year.

Fig.2 shows comparison between analysis and measured values. The broad lines are analytical results. It can be said that proposed diffusion models can simulate the chloride concentration measured in existing structures and exposed specimens.

3. MODELS AND PROTECTION MEASURES

The concrete bridge considered in this study consists of two span post-tensioned T-section girders, which are located in the Okinawa islands. Each span is 33 m in length and 15.8 m in width.

All girders are made with W/C=40% concrete. As initial protections against chloride attacks, thickening the cover concrete and initial painting are employed. B35 girder has the cover concrete of 35 mm. This girder is not protected against chloride attacks. In B50

Table 1 Data for existing structures and exposed specimens

Example	Literature	Structure	Age(year)	W/C(%)	Location
a	[4]	Pre-tensioned T-section bridge	10,25	40	Kanazawa
b	[5]	Post-tensioned T-section bridge	12	43	Sakata
c	[6]	Post-tensioned I-section bridge	17,27	43	Sakata
d	[7]	Post-tensioned T-section bridge	27	40	Sakata
e	[8]	Exposed specimen	10	40,50	Hamamatu
f	[9]	Exposed specimen	7	45-65	Kanazawa

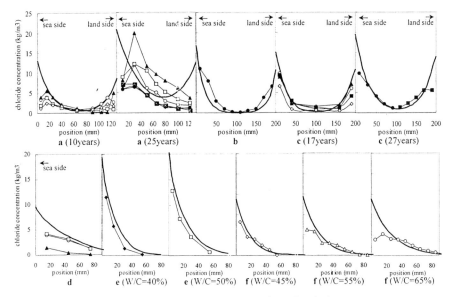

Fig.2 Comparison between analytical models and existing structures

girder, the concrete cover is increased by 15mm to 50 mm. The girder B35P and B50P are initially painted.

In Case 1, how thickening cover concrete and initial painting contribute to LCC is studied. In this case, diffusion coefficient of the new patching concrete is assumed to be the same as that of the existing concrete.

In Case 2 the contribution of diffusion coefficients of patching materials to LCC is investigated. Diffusion coefficient can be decreased by adding polymer admixture or pozzolanic admixture such as silica fume to concrete. Three types of patching materials are employed. Patching material A is concrete with W/C=40% and the same diffusion coefficient as existing concrete, B concrete with polymer mortar, replacing cement with polymer by 20% in weight of cement and C concrete with W/C=30%, replacing cement

Table 2 Assumed properties of patching materials for repair

Patching material	Admixture	Water Binder Ratio (%)	Cement content (kg/m^3)	Diffusion coefficient (m^2/sec)	Coefficient ζ
A	None	40	400	2.85e-12	0.65
B	Polymers (20W/W%)	40	400	1.0e-12	0.65
C	Silica Fume (5W/W%)	30	512	0.5e-12	0.80

with silica fume by 5% in weight of cement. Diffusion coefficients of patching materials used in the analysis are tabulated in table 2 [3],[10].

4. REPAIR/PAINTING SCHEDULE AND LIFE SPAN

In the present study, painting is employed at construction, at repair and every 20 years in general because of degradation of paint performance. However if repair work is expected 10 years after painting, painting is suspended till the next repair. All surfaces of girders are painted except for the top and side surfaces of the upper flange.

Repair work consists of the removal of the cover concrete and section patching by new concrete after well cleaning re-bars. Section patching is schematically shown in Fig.3. It is reported that the concrete behind the outermost re-bars should be sufficiently removed to take out highly contaminated concrete by chloride [11], [12]. In this study the concrete is removed up to 5mm behind the outermost re-bars.

Re-bar corrosion is normally observed at the corner of the lower part of the girder because of chloride ingress from both the side and bottom surfaces. Therefore, chloride concentration is investigated only at the view point shown in Fig.3. It is assumed that re-bar corrosion and structural damage develop when chloride concentration reaches the threshold value of 1.0% to the weight of cement [13]. Repair/painting works are repeated whenever chloride concentration reaches the threshold value or every 20 years. The required repair/painting works are repeated until year 100. The bridge reaches the life span when the chloride concentration first reaches the threshold value after 100 years.

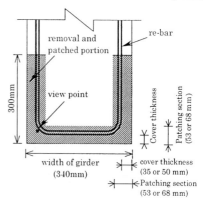

Fig.3 Section patching

5. COST ESTIMATION

ICC of the non-protected girder is estimated, based on existing bridges. The others are modified considering different mix proportions of concrete and initial painting. ICC, repair/painting costs are summarized in Table 3 as the ratio to ICC of B35. Repair and painting costs include the cost to remove old paint layers when the girder is previously painted. LCC is estimated as the sum of ICC, repair and painting costs.

Table 3 Costs as the ratio to ICC of B35

	Type of girder	Cost
ICC	B35	1.000
	B35P	1.229
	B50	1.018
	B50P	1.243
Painting	B35,B35P	0.579
	B50,B50P	0.568
Repair (non-painting before)	B35	1.191
	B50	1.336
Repair (existence old paint layer)	B35,B35P	1.382
	B50,B50P	1.525

6. RESULTS AND DISCUSSIONS

6.1 Chloride penetration and repair/painting works(Case 1)

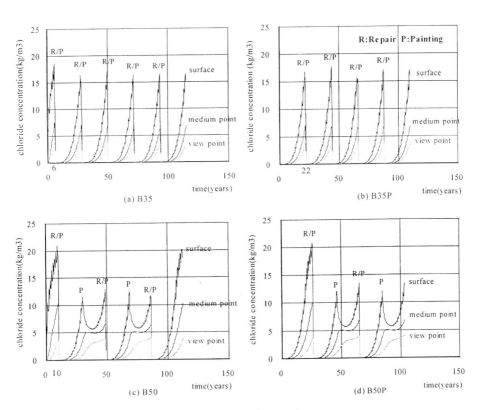

Fig.4 Results of analysis

Table 4 Repair/painting schedule for B35 with different patching

Patching		times	Rehabilitation schedule (Year)					Life-span (years)
A (B35)	Paint	0						115
	Repair	5	6	28	50	72	94	
B	Paint	0						115
	Repair	4	6	33	61	88		
C	Paint	2		26		74		102
	Repair	2	6		54			

Chloride concentration of each girder is shown over time in Fig.4. In the figure, R and P mean "Repair" and "Painting" respectively. In B35, which is not protected against chloride attacks, chloride concentration reaches the threshold value of structural damage just 6 years after construction and repairs are required five times to reach a life span of more than 100 years. In B35P, which is painted at construction, the first repair is delayed to 22 years after construction comparing with the 6 years of B35. Repairs are reduced to four times.

In B50 with increased cover concrete the first repair is slightly delayed to 10 years. Further painting is required twice and repairs are decreased to three times from five times in B35. In B50P, which is painted at construction and increased in cover concrete, repairs are reduced to twice from three times in B50.

From Table 3 it can be known that while the initial painting at construction increases ICC more than 20%, thickening cover concrete by 15mm increases ICC less than 2%.

Repair/painting cost for B35P is decreased by around 18% from that for B35 thanks to initial painting and LCC by 12%. Repair/painting cost for B50 is decreased by 18% from that for B35 and LCC by 15%.

Initial painting and thickening cover concrete by 15mm in B50P can decrease repair/painting cost by 38%, LCC by 30%, compared with those of B35.

Both initial painting and thickening the cover concrete are effective to decrease LCC.

6.2 Diffusion coefficient of patching concrete(Case 2)

Repair costs for three types of patching materials are estimated. Adding polymer or silica fume to the patching concrete increases repair cost by 0.9% or 0.8% respectively.

Table 4 shows a repair/painting schedule for B35 with three different patching materials. In the Table when the bridge

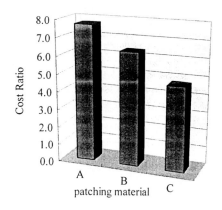

Fig.5 LCC for different patching materials

should be repaired and painted within its life span is shown. When patching material A is employed, repair work is required five times. In the case of patching material B with polymer mortar repair work is decreased to four times. Patching material C with silica fume requires repair work twice and painting alone twice.

Fig.5 shows LCC of the three cases, considering cost increase of this innovative patching materials. When patching material B and C are employed, LCC decreases by 17% and by 38% respectively, compared with patching material A.

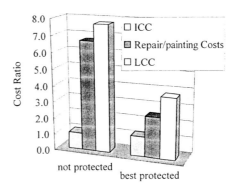

Fig.6 Cost comparison

These results indicate that LCC is dependent on the characteristics of the patching materials. Innovative patching materials with a smaller diffusion coefficient contribute to LCC drastically.

6.3 Best protection against chloride penetration

From discussions mentioned above, thickening cover concrete, initial painting and use of innovative patching materials with a smaller diffusion coefficient are very useful against chloride attacks. ICC, repair/painting cost and LCC are shown in Fig.6 for the best protected girder, compared with non-protected one. While ICC increases by 24% due to initial painting, repair/painting cost decreases by around 63%. LCC can be decreased up to half that of non-protected girder.

7. CONCLUSIONS

Based on the results obtained in this investigation regarding the life cycle cost (LCC), we can reach the following conclusions:

(1) Proposed analysis can simulate chloride ingress measured in existing structures in real environments.

(2) Increase of cover concrete forms a very small part of ICC, decreasing LCC by 15%. Initial painting at construction increases ICC by more than 20%, decreasing LCC by 12%.

(3) Performance of patching material used in sectional repair influences LCC drastically. Increase of repair cost for innovative patching materials is less than 1%, decreasing LCC at maximum by 38%.

(4) Increase of cover concrete, painting at construction and use of innovative patching material with a smaller diffusion coefficient can decrease LCC by up to half, compared with one without protection measures.

REFERENCES

[1] Yamada, Y., Oshiro, T. and Masuda, Y. : Analytical study on chloride penetration into concrete exposed to salt-laden environment. *J. Struct.Constr. Eng., AIJ*, No.501, pp.13-18, Jul., 1997 (Japanese)

[2] Saetta, A.V., Scotta, R.V. and Vitaliani, R.V. : Analysis of chloride diffusion into partially saturated concrete. *ACI Materials Journal*, Vol.90, No.5, Sep-Oct, pp.441-451, 1993

[3] Sugiyama, T., Tsuji, Y., Bremner, T. W. and Hashimoto, C. : Determination of chloride diffusion coefficient of high-performance concrete by electrical potential technique, *Performance of Concrete in Marine Environment*, Spec. Publ. American Concrete Institute SP-163, pp.339-354, 1996

[4] Kanaumi, Fujiwara, Tabiguchi, Ishimura : Damage investigation of PCT girder bridge exposed to chloride attacks for 25 years, Proceedings on rehabilitation of concrete structures, pp.55-60, 1998.10 (Japanese)

[5] The ministry of construction : Cathodic protection for Shin-isuzu bridge, The technical report on measures for chloride damaged bridges in Atsumi region, pp.57-89, 1998.10 (Japanese)

[6] Matsuda, Y., Ishibashi, T., Toyooka, A. and Amaki, G. : Repair performance of prestressed concrete bridge damaged due to salt attack, *Proceeding of the Japan Concrete Institute*, Vol.21, No.2, pp.1015-1020, 1999 (Japanese)

[7] Ishimura, K., Yamaguchi, M., Watanabe, H. and Hosaki, T. : Repairing and retrofitting of prestressed T-girders damaged by salt-induced corrosion, *Concrete Journal*, Vol.38, No.7, pp.34-40, Jul., 2000 (Japanese)

[8] Public Works Research Institute of the ministry of construction, Prestressed Concrete Contractors Association : Report on technology to improve the durability of marine structures(Corrosion protection for marine structures in the splash zones), 1995.12 (Japanese)

[9] Sasatani, T., Torii, K., Mitsunori, K. and Kajikawa, Y. : Chloride ion penetration into the concretes exposed to a marine environment for a long period, *Proceeding of the Japan Concrete Institute*, Vol.18, No.1, pp.957-962, 1996 (Japanese)

[10] Ohama, Y. et.al. : Properties of chloride ion diffusion in polymer cement mortar and concrete, *CAJ Review of the 40th General Meeting/Technical session*, Vol.40, pp.87-90, 1986 (Japanese)

[11] Shoto, K., Kamihigashi, Y., Nojima, S. and Yoshida, A. : Rehabilitation methods for concrete structure, *Highway Technology*, No.17, pp.33-43, Oct., 2000 (Japanese)

[12] Unisuga, Y., Masuda, Y., Jitousono, H. and Fujii, K. : Repair system for reinforced concrete structure damaged by chloride corrosion, *Proceeding of the Japan Concrete Institute*, Vol.19, No.1, pp.1165-1170, 1997 (Japanese)

[13] Bentur, A., Diamond, S. and Berke, N.S. : Steel Corrosion in Concrete, E & FN Spon, London, 1997

West Rail Viaducts – An Overview

CHRIS CALTON and SAMUEL LO
Kowloon-Canton Railway Corporation, Hong Kong

Introduction

The Kowloon-Canton Railway Corporation's (KCRC) Phase I of the West Rail project is a 30.5km long, 9 stations, high-capacity passenger line which will form a vital link between the North West New Territories and urban Kowloon.

The design and construction of the West Rail Phase I Works, with an estimated value of over HK$46.4 billion, commenced with the award of five detailed design packages in early 1998. Construction of this fast-track project quickly followed with the award of two 'design-build' contracts for the main tunnel works later in 1998. The majority of the fifteen 'construct-only' contracts for the stations and remaining line works, which comprise a considerable length of viaducts, commenced in mid-1999.

Figure 1 shows the significance of the West Rail Phase 1 project as a key element of Hong Kong's railway network.

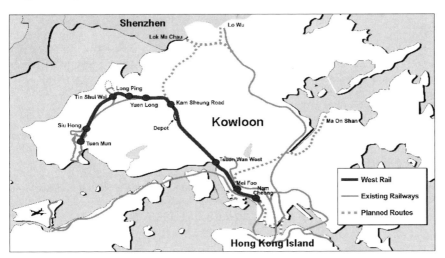

Figure 1: West Rail – A Key Element in the Hong Kong Railway Network

This paper provides a brief description of this ambitious transport infrastructure development undertaking in Hong Kong; an introductory overview to three more detailed companion papers on the subject of West Rail's Viaducts provisions and the project management approach adopted by KCRC.

West Rail Phase I Project

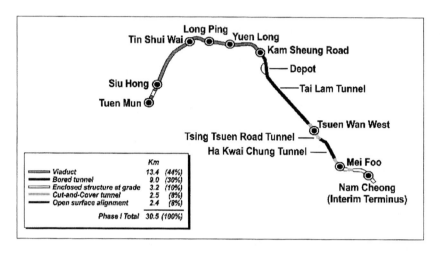

Figure 2: Types of Construction

West Rail is an ambitious mass transit railway development project undertaken by KCRC in line with the Government's policy to make railways the backbone of the territory's transport system into the new millennium. The plan for a mass transit railway linking Urban Kowloon and North West New Territories was conceived in 1994. After strategic development, feasibility studies, and detailed design, the Executive Council of the HKSAR Government gave the final authorisation in September 1998 for the Kowloon-Canton Railway Corporation to proceed with the construction of West Rail.

The new mass transit line will pass through five districts: Sham Shui Po, Kwai Tsing, Tsuen Wan, Yuen Long and Tuen Mun. It has nine stations, a depot, and a headquarters building housing a central operations control centre.

The route commences in the south at the Phase I terminus at Nam Cheong Station in Sham Shui Po district. The alignment parallels the West Kowloon Expressway and runs to the northwest before curving northwards to Mei Foo Station, an enclosed, landscaped station within Lai Chi Kok Park. From Mei Foo the alignment continues through the Ha Kwai Chung Tunnel, under a portion of Kwai Fuk Road, and then through the Tsing Tsuen Tunnel before entering the Tsuen

Wan West Station, which is situated on new reclamation in Tsuen Wan Bay. The tracks then continue northwestwards and enter the Tai Lam Tunnel, gradually rising to the north portal.

From the north portal, the alignment continues into the Kam Tin Valley where a 32.5ha site houses a depot, which will provide maintenance and stabling facilities for the entire West Rail fleet. Immediately beyond the depot, the track will enter the elevated Kam Sheung Road Station.

The alignment from Kam Sheung Road Station to Tuen Mun is for the most part on viaduct. After turning to the west at Au Tau, the railway crosses over Route 3 and Castle Peak Road to enter Yuen Long. It continues further westward with elevated stations in Yuen Long and Long Ping. The alignment remains elevated before reaching Tin Shui Wai Station, which will be a key interchange facility between the Corporation's West Rail and Light Rail operations and other modes of public transport.

From Tin Shui Wai, the alignment then turns south, heading for Siu Hong Station, and onward to the Tuen Mun Station terminus, with both stations built over the Tuen Mun nullah.

The following sections describe the considerations given to the adoption of viaducts for this 13.4km section of railway, interspersed with six elevated stations. The section of railway on viaducts represents approximately 44% of the overall route length.

West Rail Viaducts

Emerging from Tai Lam Tunnel, the alignment from Kam Sheung Road Station through to Tuen Mun Station is almost entirely on viaduct. With a total length of 13.4 km and passing through six elevated stations, the West Rail Viaduct is the longest in Hong Kong.

The viaduct design has been carefully considered to ensure visual integration into the environment, both in the rural and in the developed residential areas. The design has been vetted by HK SAR Government's "Advisory Committee on the Appearance of Bridges and Associated Structures" (ACABAS). The viaduct will also incorporate an innovative combination of engineering measures to mitigate the operational noise of the trains and make West Rail one of the quietest operating railways in the world.

Running the railway on viaduct through the flood plains of the North West New Territories will keep trains and associated facilities clear of any flooding risk and will not obstruct existing drainage control facilities in flood-prone areas. Construction of an elevated railway rather than simply building at-grade, through this particular area of HK, imposes less disruption to existing traffic, less severance of the community, and promotes flexibility for future urban development given that elevated structures pose minimal planning constraints.

Using pre-cast segment construction methods, deck sections for the viaducts will be manufactured at a centralised casting yard in the Mainland and then transported by road to the West Rail works sites. The land required in the corridor for the viaduct construction will therefore be minimal, and the impact to the community reduced.

Environmental Constraints and Considerations

Environmental consideration has greatly influenced the design of the viaducts. KCRC is committed to protecting and improving Hong Kong's environment and has adopted a proactive approach to environmental management. A comprehensive Environmental Impact Assessment (EIA) has been completed to identify potential environmental issues arising from construction and operation of West Rail and to recommend appropriate mitigation measures. The EIA covers the full range of environmental concerns including noise and vibration, air and water quality, land use and landscape, visual impact, archaeology and culture, resources, ecology, waste management, and man-made hazards.

The HK Noise Control Ordinance places a legal obligation on the KCRC to control noise levels emanating from the railway during night-time operations to less than an average continuous energy level (Leq(30mins)) of 55 dB(A) at the nearest sensitive receiver. West Rail is among the first railways in the world to be constrained by such strict noise limits. By comparison, a modern railway operating at high speeds would normally generate wayside noise of around 88dB(A).

To address this challenging issue, the West Rail Project Team developed an integrated system-based approach, combining the collective expertise from civil, structural, environmental, permanent way and train designers, in conjunction with noise specialists. The combination of engineering measures to minimise train induced noise included:

- A multi-plenum noise attenuation system comprising under-car sound absorbing body panels on both sides of the train to trap the noise, and noise barrier along the viaducts.
- Noise enclosures at points and crossings
- "Floating" slab track supported by rubber bearings
- Vibration absorbing rail fasteners
- Stiffening up the main supporting structures

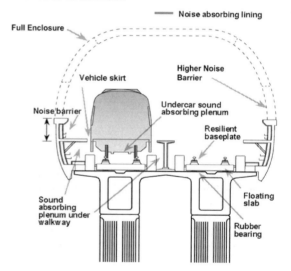

Figure 3: Noise Attenuation

West Rail will help reduce air pollution from vehicle exhaust emissions, one of Hong Kong's most serious environmental problems, by eliminating road traffic equivalent to 2,500 bus trips per day. A regenerative braking system will enable a significant amount of the power required to drive the trains to be recovered and fed back into the power transmission system, thus keeping energy demands to a minimum.

A more detailed description of the factors affecting the design development of the viaducts is described in the companion papers entitled "KCRC West Rail Viaducts-Design Development".

Public Consultation and Community Involvement

Partnership with the community is the basic philosophy underlying the Corporation's efforts throughout the design and the construction phases of West Rail. Every effort is being made to ensure that the views of the community are taken into full consideration and their concerns are addressed to the fullest extent practicable.

Throughout the project planning process, KCRC has maintained a close dialogue with those members of the community that West Rail will serve. Public consultations have been held with the municipal councils, relevant District Boards and their subcommittees, rural committees and other statutory bodies to ensure that planners and engineers fully understand and address community issues. Public views on the alignment, interface issues and results of the environmental, traffic and drainage impact assessments have been carefully considered in the implementation plan for the railway.

As part of the KCRC's commitment to ensuring a community and environmentally friendly project, the Corporation will continue to adopt a proactive approach to understanding and resolving public concerns throughout all its new project developments and ensuing operations.

Contractor's Alternative Viaduct Designs

KCRC has established a contract strategy for West Rail following fair, open and competitive tendering practices in accordance with the World Trade Organisation procurement rules.

Particular emphasis has been placed on the importance of value engineering and partnering arrangements throughout the design and construction stages of the project. In addition to submitting conforming tender bids for the construction contracts, the tenderers were able to submit alternative tenders, which conform to the requirements of the contract documents. Sufficient supporting evidence was required to be tendered to enable the Corporation to fully assess the technical acceptability of the alternative tender and its effect on time for completion and costs of the Works, including the identity of any provisions of the tender documents in respect of which the alternative tender would not comply, and the advantages and disadvantages of such alternative provisions.

For the two viaduct construction contracts, CC201 and CC211, several conforming tenders and alternative tenders were received. Both contracts were awarded on the basis of an alternative superstructure design proposed and developed by the Maeda-Chun Wo Joint Venture.

The alternative revised the engineer's conforming design solution of the twin concrete box girders for the superstructure but keeping the parapet, noise barriers, walkway and other accoutrements to the main structure unchanged. The Engineer's conforming design spaced the girders wide enough apart to provide stability under the overturning high typhoon wind forces acting on the deck and noise barriers. The alternative revised the superstructure by placing the girders closer together beneath the floating slab track bearings, which effectively increased the mechanical impedance of the structure and facilitated a reduction in the noise radiation from the deck. The flange and web panel thickness could therefore be sized to give an overall benefit in material cost savings without compromising the environmental noise specifications.

Placing the webs directly under the floating slab track bearings would lead to a box girder, which would be too narrow to be stable against overturning under wind loading if supported conventionally on bridge bearings. The alternative solution was to make the narrow box girder monolithic with its supporting columns at each end to transfer the overturning load into the foundations. The significant saving of material in the superstructure enabled a corresponding reduction in the foundation costs.

A more detailed description of the alternative design of the viaducts is contained in the companion paper entitled, "The Alternative Design of the West Rail Viaducts".

Project Management

West Rail Phase I is currently the largest stand-alone infrastructure project undertaken in Hong Kong and is the first major capital expansion project for KCRC.

To control costs and to ensure the aggressive programme is achieved, the Corporation has assembled an in-house team of experienced professionals to undertake the project management of West Rail, recognizing the need to maximize the use of its resources in producing an excellent end product. A baseline programme and cost estimate have been established against which all progress and potential changes are monitored throughout design and construction. As soon as potential deviations are identified, corrective action is taken to keep the programme and costs on track.

As a demonstration of effective cost management, the estimated cost of commissioning West Rail was reduced from HK$64.4 billion (money of the day) pre-tender to HK$51.7 billion following a major value engineering effort just prior to award of the main civil contracts. The Project estimate now stands at HK$46.4 billion resulting from further value engineering and partnering during construction. These value engineering initiatives have also enhanced the level of service expected for the new railway operations.

In addition to managing the construction works on West Rail, a host of other areas need to be managed. These include: the design process, land matters, contract formulation and procurement, risk management, safety management, quality assurance, environmental issues, community relations, testing and commissioning.

To manage these elements, the West Rail Project Team is divided into three departments. The Construction Department is responsible for civil design management, land matters, environment, safety, construction planning, and construction management.

The Railway Systems Department is responsible for all railway systems contracts, design management, manufacturing, installation, systems planning and control, systems integration, testing and commissioning.

The Project Support Department is responsible for all support functions including financial, human resources, legal, cost control, procurement, information technology, and public affairs.

The West Rail civil construction management approach is based upon Detailed Design Consultants providing resident site staff, who integrate with KCRC staff as shown below, to cover both the design and construction processes.

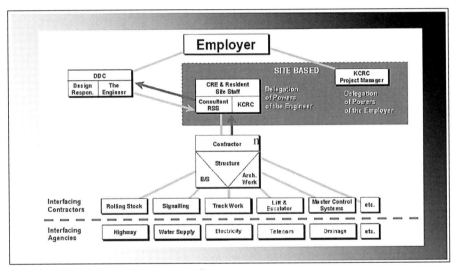

Figure 4: Construction Management Structure

The ultimate aim of KCR's West Rail project management is to complete the project safely, on time, within budget, and in an environmentally responsible manner.

Safety

The existing KCRC East Rail service has the enviable reputation of being one of the safest railway operations in the world, with an accident rate as low as 0.05 accidents per million passengers carried. With the aim of matching or bettering this safety standard, West Rail has adopted a risk-based approach to safety management. This approach allows for potential operational safety issues to be identified prior to or during design and for those problems to be effectively designed out in a cost effective manner. Recognising that the risk can never be zero in an operating railway, the risk is minimised to a level 'as low as reasonably practicable' (ALARP). All designers, contractors and suppliers must adhere strictly to the safety management process throughout all stages of the scheme implementation from design to commissioning.

The safety policy set out by KCRC demonstrates the Corporation's commitment towards safety throughout the construction of West Rail and meets the construction safety standards of Hong Kong SAR Government. The senior management of KCRC is committed to a target of zero fatalities and an accident rate comparable with the highest international standards.

A team of highly experienced safety practitioners ensures construction safety standards are strictly enforced on the West Rail project. It will continuously review and improve the Corporation's safety requirements throughout the project term, and continue to audit the contractors' performance at regular intervals for compliance.

Conclusion

Hopefully, this paper has provided an introduction and a background to the three companion papers on the subject of KCRC's 13.4 km of viaducts for the West Rail project. Having combined the collective experience from civil, structural, permanent way and train designers, in conjunction with significant input from noise specialists, the West Rail team developed a systems based approach to tackle an extreme problem in setting new standards in noise mitigation measures. The approach was further finessed during construction with the adoption of a contractor's alternative design.

Acknowledgement

The authors wish to thank Kowloon-Canton Railway Corporation for giving permission to publish this paper, and acknowledge that the views expressed in this paper are those of the authors and not necessarily those of KCRC.

Life cycle cost analysis of bridges where the real options are considered

YASUHIRO KOIDE[a], KIYOYUKI KAITO[b], MAKOTO ABE[a]
a Bridge Maintenance Consultant Co Ltd, Chiba Japan
b Dept. of Mech.& Civil Eng., Columbia Univ., New York USA

1. Introduction

A tremendous number of bridges have been energetically built in Japan since the advent of the high growth period of Japanese economy after the World War II until this point in time. In view of the mature expansion of road and railway networks, as well as the current social conditions, however, it is considered that future capital investment in building new bridges is to gradually decrease.

In contrast, there are existing a tremendous number of groups of bridges and on these bridges, due to the progress of aging thereof coupled by worsening traveling load conditions thereon, damage and deterioration are getting serious, and it looks like a future increase in maintenance and management costs to be incurred for those damaged and aging bridges will be difficult to be avoided [1) 2)]. It is statistically pointed out that in 2020, the number of aging bridges which will have been in service then for not less than fifty years will exceed the number of new bridges, this developing further serious concern over the damage and deterioration of those aging bridges [3)]. As described above, in view of the current social situations where the capital investment in building new bridges gradually decreases and the role that bridges play in the infrastructures, it looks difficult to adopt a so-called scrap and build scheme for aging bridges, and therefore, the rational prolongation of lives of the existing bridges needs to be sought within limited expenditure.

Life-cycle cost analysis (LCCA) is intended not only to think over the size of initial costs that would be involved in building a bridge but also to determine a capital investment so that the total costs that are to be incurred throughout a certain length of time to take care of the bridge is minimized [4)]. Given attention thereto, LCCA is regarded as one of methods for determining bridge management strategies. Conventionally, with the exception of road paving, it is believed difficult to make a model of quantitatively identifying the degree of deterioration of a structure, and this decreases the accuracy at which repair costs are estimated, thereby there having been less opportunity for the concept of LCCA to be used [5)].

In this study, the idea of LCCA is applied to the optimum maintenance and management of an existing bridge. In general, it is true that the managerial person can make only a 5-year or 10-year bridge maintenance and management plan, at the longest, for an existing bridge, and a budget that can be allocated for the maintenance and management of the bridge is very variable. Consequently, it is considered to be difficult to apply to the bridge maintenance and management the conventional cost-and-effect analysis in which maintenance and repair are treated as fixed events.

To cope with this, in this study, the total costs for the maintenance and management of an existing bridge for thirty years is analyzed using the concept of Real Options. A standard steel plate girder railway bridge which is assumed to have been in use for fifty years is taken for example, and in order to minimize the total costs for maintenance and management of the bridge for thirty years, effects resulting from main factors such as repair cost, regular maintenance and management costs for inspection and painting and replacement cost, will be studied.

Originally, in studying the life-cycle costs, the owner's costs including design cost, building cost, maintenance and management cost, repair cost and rehabilitation cost, and the user's costs including running cost and delay cost need to be considered. However, since the use's costs are difficult to be calculated, in this study, while paying attention to among the aforesaid items factors directly affecting the maintenance and management of the bridge, how the life cycle costs affect the investment effect and risk will be considered.

2. Proposed Effective Management of Bridges

A bridge used in a calculation example is assumed to be a steel plate girder railway bridge which has been in service for fifty years. Assuming that the bridge is to be managed for thirty years from now, how the postponement of investment affects the total costs will be studied. Here, there are raised two strategies, and the total costs that would be incurred throughout thirty years from now will be calculated for each strategy.

Under the two strategies, the existing bridge is to be maintained for service for thirty years. Under Strategy A, the bridge continues to be maintained and managed to a minimum level until the bridge becomes difficult to be maintained. On the other hand, under Strategy B, the bridge is studied and diagnosed for grasping the current residual life thereof in a quantitative fashion and required repair and reinforcement are assumed to be carried out in the initial year in order to prolong the life of the bridge for thirty years from now.

Under both the strategies, in the event that damage is found in the bridge while in service, replacement or repair may be selected. Details of the maintenance and management costs under the respective strategies are shown in Table-1 and Table-2. Details of repair and reinforcement costs in the initial year under Strategy B are shown in Table -3, and details of design, production and replacement costs for the respective strategies are shown in Table-4. In order to improve the reliability of this study, experienced engineers were asked for calculation.

Table -1 Maintenance costs on Strategy A

Works		Unit cost	Amount	Total
Painting (every 15 years)	Painting cost	$60 / m^2$	600 m^2	$36,000
	Scaffolding cost	$30 / m^2$	160 m^2	$4,800
				Subtotal $40,800
Inspection and Diagnosis (every 2 year)	Visual inspection	Engineer $465/day	2	$930
		Engineer $290/day	2	$580
		Guard man $232/day	2	$464
				Subtotal $1,974
Repair (for 10 years keep)	Bearings	$30,000 / 10 years	1	$30,000
				Subtotal $30,000

Table -2 Maintenance costs on Strategy B

Works		Unit cost	Amount	Total
Painting (every 15 years)	Painting cost	$60 / m^2$	450 m^2	$27,000
	Scaffolding cost	$30 / m^2$	160 m^2	$4,800

					Subtotal $31,800
		Engineer $465/day	1		$465
Inspection and diagnosis (every 2 year)	Visual inspection	Engineer $290/day	1		$290
		Guard man $232/day	1		$232
					Subtotal $987

Table -3 Initial Repair/rehabilitation costs on Strategy B

	Works	Unit cost	Amount	Total cost
Superstructure	Repair and improvement of bearings Repair of support bed (including scaffolding cost)	$20,000	4	$80,000
	Replacement of cover plates and Secondary members	$140,000	1	$140,000
	Replacement damaged rivets to high tension bolts	$10,000	1	$10,000
	Corrosion protection painting	$100 / m^2	300m^2	$30,000
			Subtotal	$260,000

Table -4 Reconstruction total cost

	Design	Construction	Replacement
Strategy A,B commonly	$50,000	$620,000	$140,000

3. Life-Cycle Cost Considering Real Options Theory

The option theory is a theory, which was developed in the financial engineering, and the option is understood as a "right to take a certain action in the future" [6]. In other words, the flexibility in investment can be taken into consideration which flexibility could not be taken into consideration with the maintenance and management scheduled in advance under the conventional cost-and-effect analysis. In this study, in the event that the damage propagates while the bridge is in service, replacement needing to be studied, it is made possible to select either replacement without any delay or partial repair to put off replacement.

First of all, a simple decision-making event tree is considered. ● indicates an investigation node where a legal inspection is carried out. In addition, the probability in which a damage is found during this investigation is represented by a probability density function $f(x)$, where x is the number of years needed since the bridge was built until the damage is found. When the bridge is found sound in this investigation, the flow advances to the following investigation node.

On the contrary, when a damage is found, then advance to a rehabilitation node indicated by ●, where bridges are replaced or the damaged bridge is repaired. A legal inspection is resumed from the following investigation node.

In this study, the probability density

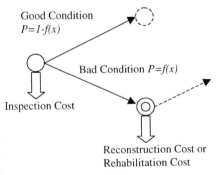

Fig.-1 Event tree of decision making

function of damage occurrence was estimated by obtaining damage rates by longevity of service based on the results of investigations carried out on some 10,000 railway bridges in Japan by the author and el and predicting the distribution of damages with an exponential function.

It is generally said that the deterioration of a bridge occurs in an exponential function fashion, and it is about eighty years when the accumulation density function of this distribution function becomes one. This distribution is found reasonable from the fact that the actual maximum service of a deck bridge plate girder is eighty-three years. [7]

4. Calculation of Total Costs Considering the Real Options Theory

Based on the decision-making event tree defined in the previous chapter and understanding that legal inspections are carried out every two years, an event tree for thirty years was prepared. In this study, the parameters of the probability density function are returned to zero every time bridges are replaced. In addition, in a case where the damaged bridge is repaired without replacement, based on an understanding that the repair effect lasts 10 years, a decision is to be made in ten years time on whether the bridge is repaired or replaced with another bridge. Furthermore, for the sake of simplicity, the probability to select either replacement or repair is 50%.

The event tree is supposed to be divided into branches of 2 to the 15th power, but in order to make calculations simple, calculations were made based on an understanding that once bridges are replaced as shown in Fig. 3, no more replacement occurs thereafter. Furthermore, while a social discount rate is taken into consideration in calculating life-cycle costs in general, in this study no such discount rate was taken into consideration.

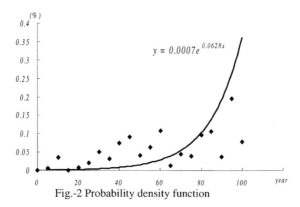

Fig.-2 Probability density function

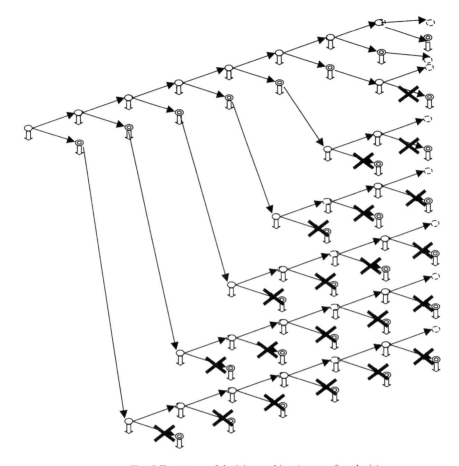

Fig.-3 Event tree of decision making (a part of analysis)

Based on the aforesaid premises, the total costs were calculated. Fig. 4-6 shows the results of the calculations. Fig. 4 shows the results of the calculation for Strategy A in which the real options were considered. A graph of broken lines shows probability in which deformations are found in respective years, while a graph of bars indicates powers of probabilities and costs or an expected value. The total costs for this repair strategy can be represented by a sum of expected values of respective years.

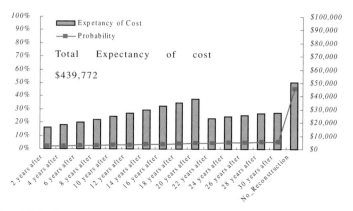

Fig.-4 Calculation Result with considering Real Options (Strategy A)

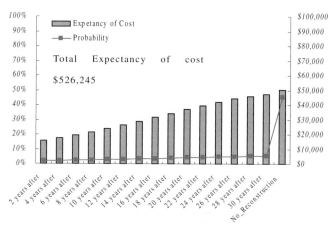

Fig.-5 Calculation Result without considering Real Options (Strategy A)

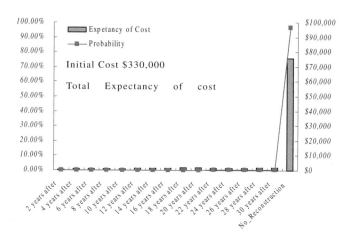

Fig.-6 Calculation Result with considering Real Options (Strategy B)

In addition, Fig. 5 is a graph showing the results of a case where no Real Options is considered. When comparing these two cases, a big difference was found, in particular, as to the expected cost for a case where deformations are found twenty
years after the start of the study. In addition, Fig. 5 is a graph showing the results of a case where no Real Options is considered. When comparing these two cases, a big difference was found, in particular, as to the expected cost for a case where deformations are found twenty years after the start of the study.

The reason why there was found little difference before then is because there was much opportunity for replacement to be selected when a decision was to be made between replacement and repair ten years after the initial option had been executed that the replacement of bridges was suspended for another ten years. However, when comparing the two cases as to the total costs, the case where the Real Options was considered was lower by about 85,000 dollars than the other case.

Furthermore, how different repair strategies produce a difference in total costs is studied. Fig. 6 is a graph showing a case where Strategy B was used for maintenance and management of the bridge. The deformation occurrence probability decreased due to a large-scale repair carried out in the initial year, and the maintenance cost also drastically decreased by that extent. This strategy was found less expensive by some 12,500 dollars the other even if the initial repair and reinforcement costs are included, thus indicating that proper repair leads to cost reduction.

5. Comparison with the Cost-applied Cost/Benefit Analysis

Here, the cost calculated using the event tree is compared with the cost calculated using a conventional cost/benefit analysis. The total costs were calculated as to Strategy B based on an assumption that the discount rate is 0%, no deformation has been found for thirty years, and hence only inspection and painting have been carried out to amount to 408,410 dollars, and the comparison disclosed that the calculation method using the decision-making event tree was higher by 18,780 dollars than the conventional calculation method. This is because since the damage discovery probability is used in decision making, a risk in which countermeasures are taken after a damage occurs was represented by money value. In other words, when compared with the conventional cost/benefit analyzing method, the method proposed in this study can be used to calculate the total costs which includes the risk of damage being made to the structure.

6. Conclusion

In this study, the life-cycle cost (total cost) analysis using the Real Options concept was carried out on the railway bridge, and the following are raised as the results of the analysis.
1.) The total cost calculation method is proposed which takes into consideration the options from the decision-making event tree using the deformation occurrence probability density function.
2.) Two maintenance and repair strategies for railways are proposed, and the strategy in which the options are taken into consideration was used for comparison to indicate the superiority of the initial repair.
3.) In the calculation of the total costs, a comparison is made between the method using the decision-making event tree and the conventional cost/benefit analysis to indicate that the total costs including the risk for the structure can be taken into consideration by using the decision-making event tree.

Additionally, an improvement in accuracy of a damage occurrence probability density function is raised as the future study direction.

References
1) K. Nishikawa, "A Concept of Minimized Maintenance Bridge", J. of Bridge and Foundation, 31, No.8, pp.64-72, 1997 (in Japanese)
2) Federal Highway Administration, "Life-Cycle Cost Analysis in Pavement Design", 1998
3) Japan Society of Steel Construction, "Steel bridge project manual (revised edition)", 1991 (in Japanese)
4) B.S. Yanev, "Infrastructure Management System Applied to Bridges", Operation and Maintenance of Large Infrastructure Projects, pp.11-22, Rotterdam, 1998
5) M. Abe, "Present and Future of Maintenance of Steel Railway Bridge", Proceedings of Railway Dynamics Symposium, JSCE, 1996 (in Japanese)
6) Avinash K.Dixit ,Robert S. Pindyck : The Options Approach to Capital Investment , Harvard Business Review ,1995.5-6
7) Japan Society of Steel Construction, "Report on Durability of Steel Structures ", J. of JSSC, 5, No.39, pp.1-30, 1969 (in Japanese)

SAFETY MANAGEMENT OF HIGHWAY STRUCTURES

A.J.WINGROVE
Senior Structures Advisor
Highways Agency, Birmingham, United Kingdom.

INTRODUCTION

Bridges and other highway structures are primarily designed to withstand the effects of the applied loading, i.e. the loading from self-weight, traffic, thermal effects, wind etc. In practice, however, due to the conservative design requirements and the sophisticated modern analysis techniques, the design strengths are such that structural failure from such loading, even in the extreme situations, is most unlikely. The probability of such failure is so small it exists only in the conceptual domain. Yet, there are always a number of other risk factors, which can endanger structural safety in a real sense. Some of these factors are as follows: -

(i) Faulty or inadequate materials and workmanship.
(ii) Inadequate design expertise or uneven application of the rules.
(iii) Gross errors in design and construction.
(iv) Undue relaxation of the rules in response to financial or other pressures.
(v) Unforeseen events such as terrorist bombs and unidentified defects.
(vi) Extreme events beyond what is possible or reasonable to design for, for example extreme flood situation, unusually extreme vehicle impact etc.
(vii) Risk factors identified after construction such as defective grouted duct post tensioned bridges, ASR and Thaumasite affected concrete etc.
(viii) Continuing deterioration of structures and inadequate or inappropriate maintenance or modification.
(ix) Structures assessed at or near a critical state reliant on monitoring other interim measure.

The highway authorities generally use four main weapons in guarding against these adverse possibilities:

- National and international design standards and codes,
- Material and workmanship specifications including supervision arrangements and quality assurance schemes,
- Inspection and other management procedures
- Technical approval and certification requirements.

The Highways Agency, responsible for the trunk road network in England, has long experience of using these tools. The paper explains, with illustration, the various structural risk types encountered by the Agency in recent times and how the procedures have been used to safeguard against these risks. In particular, the paper focuses on the experiences of the technical approval engineers.

The specific tasks of technical approval includes granting 'departures' from the standards in a rational manner, and also, giving jurisdiction on methods for dealing with 'aspects not covered' by the standards. The deliberations of the technical approval engineers take place within strict time limits dictated by scheme preparation or contract timetables, and with the background of the pressing needs of functional and other considerations. In all this, public safety is taken as being of paramount importance and the Agency's bridge engineers have to choose the best of the safe options that is consistent with other requirements. The paper discusses the issues involved using a number of examples.

The paper explains how research development has assisted the decision process in the past and continues to maintain and improve the safety management of highway structures.

Background To Technical Approval

The method of procurement of new highway bridges in the UK has evolved over the last century as the pressure to provide more roads has increased. In the first half of the Twentieth Century the majority of structures were designed and constructed by the then Ministry Of Transport with limited new road construction.

Between 1955 and 1972 the increasing demand for new infrastructure and road space lead to an expansive roads programme and construction boom, the spend on highway maintenance and improvement increased 2000% in real terms to £556 million in 1972, of which £100m accounted for structures. The increase in construction activity and advances in technology, together with increasing complexity of design, outpaced the ability of the Ministry's Bridges Engineering Division to continue its full checking responsibilities and they resorted to checking a limited sample of the submissions.

In the early seventies the collapse during construction of major highway structures at Yarra (Australia), Milford Haven (Pembrokeshire, Wales) Koblenz (Germany) and the Danube (Austria) highlighted a need to separate the design and checking process from the Departments responsibilities and liabilities. Proposals for new procedures were published in 1970/71 for consultation amongst consultants and highway authorities, and the resulting procedures defined the role of the TAA and clarified the responsibilities of the designer and checker. The main recommendations were:

The Department would examine design criteria but not computations.

A certificate of independent check of the design computations would be submitted to the Department.

The application of an Approval In Principal stage to all but minor structures would cover the selection of bridge type, suitability of function, materials of construction, methods of analysis and design.

These recommendations adopted by the then Secretary Of State for Transport, published in Technical Memorandum BE4/73 were consistent with, the underlying philosophy and objectives of the Technical Approval Procedures. To ensure highway structures will perform their intended function, are safe and serviceable, as well as economic to build and maintain.

Current Role Of Technical Approval Authority

The present day Technical Approval Authority, TAA is part of the Civil Engineering Division of the Highways Agency, an Executive Agency within the Department of Environment Transport and the Regions. The Agency discharges most of the Secretary Of State's (SoS) responsibilities for operation, maintenance and development of the trunk road network against an annual budget of £1.5 billion.

These responsibilities encompass the maintenance of existing highway structures and the provision of new structures in improvement schemes. In this respect the TAA, in the performance of its semi regulatory role, and by the advice it gives, is protecting the interests and acting on behalf of the SoS.

The requirements for Technical Approval are presently defined in the Departments design standards specifically BD2/89 that introduces the concept of Approval In Principle. This standard applies to all works or contracts on highway structures where structural integrity, long-term durability or safety is affected. One exception is Design, Build, Finance and Operate contracts where the operator is responsible for maintenance over the 30-year concession period. For these contracts the TAA function is appraisal and acceptance rather than approval, but the principles are otherwise unchanged.

For new works the Approval In Principle document is live throughout the design and construction process. As well as a record of the initial design principles agreed between the designer and the TAA, including design standards to be used and any requested departures from the standards, changes proposed and agreed throughout the contract are also recorded. In parallel with the AIP, procedures are in place for the approval of departures from the specification, to evaluate suitability of proposals including materials that are outside the current approvals process. Advances in technology, innovative ideas or omissions in the specification for particular circumstances may drive these departures.

Departures From Standards

The Highway Agency is responsible for promulgating and maintaining its own standards in consultation with the three other UK overseeing authorities for Scotland Wales and Northern Ireland. Whilst many of these standards are stand alone for maintenance and management of bridge structures other implement national standards (British Standards).
In this position of intelligent client the Agency is in a 'relatively' unique position of accepting and approving departures from its own standards (and those it implements). Without agreement designers are not only unprepared to design structures which depart from the Agency's standards they would also be in breach of contract. This is an untenable position for their Professional Indemnity insurers who would be exposed to unacceptable liabilities.

It follows that should the designer wish to submit a departure from standards for design or maintenance there should be a strong justification balanced against a risk assessment such that the Agency can balance risk against benefit before agreeing to any departure. The TAA is central in assessing probability and consequence in determining the contractual, political and engineering risks, and in recommending acceptance of all structural departures.

The following process of approving departures is simple in concept and a period of 42 days is normally allowed for completion;

- Designer proposes with justification.
- TAA recommends acceptance or rejection.
- Project Manager agrees and endorses.
- Standard Author accepts or rejects and returns to TAA.
- TAA submits to Head Of Division for Approval if neither rejected nor delegated.
- Designer is notified whether or not proposed departure is acceptable prior to commencement of design.

Where the pressures of Programme and Budget lead to the design commencing ahead of the final stage, the designer is proceeding at his own risk without formal approval of the TAA. During the process, progress is recorded and measured through a database tracking system. This is a record of each decision together with justification and history of the approval process is recorded on a national Departures From Standards database

Departures Database

An analysis of the current database indicates that over the last four years over 2080 structural departures have been processed through the DFS system, a breakdown of these is shown in figure 1 below.

Departure Type By Percentage

- Rejected Withdrawn 8%
- Safety Fence 12%
- Parapet 9%
- Assessment Bridges/gantries loading 12%
- Temporary Works 13%
- Temporary Barriers 5%
- Bearings Hinges 2%
- Waterproofing/Durability 3%
- Anchorage 4%
- Shear 7%
- Box Girder Buckling 2%
- Paint 3%
- Beam Steel BS5400 9%
- Headroom 1%
- Others 10%

Typical examples of the most commonly occurring departures shown in Figure 1 are: -
- Acceptance of temporary structures design based on load capacity testing rather than standards.
- Reduction in load factors due to certainty of loading but reliant on future management or monitoring of the structure.
- Reduction in lengths and alignment of safety barriers due to site constraints.

- Acceptance of structural components and systems proved by insitu or example test results.
- Design proposals using techniques/standards from other countries.
- Methods of design and assessment utilising Technical Papers and research.
- Acceptance of existing or new bridge specific parameters e.g. percentage overstress of elements allowing retention of structure.

Quantitative savings have been recorded from 1997 to date, the 470 departures on the database, recognised as having actual financial savings have a sum total of over £75 million over this 4-year period.

In addition to recording financial savings on departures the quantitative and qualitative benefits of general advice is reported quarterly, presently recorded at 27 per quarter. Of these 63% have no attributable value, although in theoretical terms they have saved large sums in reducing road user delays and contributed to improved safety. Of the remaining 37% these average £72k cost benefit per record.

Stages Of Risk Management in Steady State Assessment And Maintenance Of Existing Structures

To safeguard against risks the TAA applies its procedures in assessing the probability and consequence of each risk as it arises and applies to structures, whether it is part of a major new construction contract, assessment and strengthening of existing or steady state maintenance on the network. Procedurally and chronologically each is different and is dictated either by Parliamentary Acts or EU directives.

Steady State Maintenance requires the Agency' maintaining agents to carry out regular inspection of all structures on the network, put in place a programme of works based on severity of defect and priority of work. Minor work will be completed as routine maintenance albeit with informal risk assessment. Major schemes, generally over £200,000 are formally value engineered and value managed on a risk based scoring system defined as follows:

The Process: -

Stage 1	Identify need.
Stage 2	Identify risk on the basis of what would happen if, having identified this need, did nothing about it now.
Stage 3	Assess level of likelihood of risk occurring
Stage 4	Assess level of consequence if risk occurred
Stage 5	Assess level of overall risk and hence appropriate priority category for need.

Through stages 1 to 5 the following are considered.

The level of risk determined on the basis of "what would happen if did nothing now? Risks will include failure of a structure or a component, failure to meet certain obligations for freedom of movement on the network, the build-up of a future backlog of essential work, etc.

How likely is the risk associated with this need to occur? What would be the conditions needed for the risk to occur? How likely are they, given the history of the structure?

The likelihood of an event occurring will be influenced by: -
- The nature of the cause of the risk i.e. persistent, transient or accidental
- The level and type of traffic using a structure or route
- Features surrounding the structure which might affect visibility, vehicle performance/handling etc.
- The level of reserve safety within the structure (and the level of assessment used, redundancy of the structure and mode of failure/warning of failure) etc.

The consequences of an event will depend upon: -
- The level and type of traffic using a structure or route
- Features surrounding the structure i.e. what it crosses or supports
- The availability of alternative routes

Consequences assessed against 4 criteria: -
- Safety – of the public
- Functionality – level of service/availability of the network
- Sustainability – where the aim is for steady expenditure and workload, avoiding building a backlog of essential work by doing effective, targeted preventative maintenance, i.e. painting steelwork, silane of new concrete and management planning on all bridges in new/as new condition.
- Environment – including appearance of the structure

Identify the overall level of risk and appropriate priority category from Unavoidable to Highly desirable.

For each need, the level of likelihood and consequence (in terms of the 4 criteria) is assessed and a simple matrix is used to determine the overall (criteria-specific) level of risk.

This matrix is shown below: -

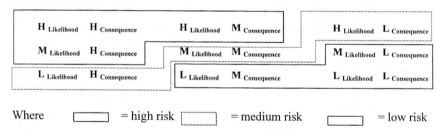

The criteria-specific level of risk and overall priority category for the need is then determined using the matrix shown below: -

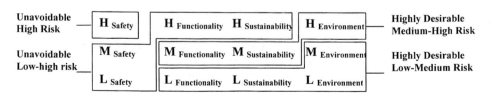

New scheme construction will pass through public enquiry and compulsory purchase order stages prior to construction and historically have taken 10 years to completion. Procedures to accelerate this process are currently being developed.

Schemes are justified through a Cost Benefit model (COBA) that considers a variety of risks including accidents and user delay but when a scheme is approved it is risk managed. The process is described in the Agency's Value For Money manual and defines the value engineering and risk assessment stage of each scheme or project based on cost and risk ranked options, assessed at a formal workshop using the Highway Agency's Risk Model (HARM).

Examples of a number of realised risks and their management

On all types of construction and maintenance the TAA has been instrumental in advising and directing the assessment of risks and safe management of structures. The following are examples of both ongoing historical and recent events that have occurred on structures on the network and how they were/are being managed:

Impact On Structures: Substructure Impact

Whilst this is relatively rare with only one complete bridge demolition on the M50 in the 1970s, a recent accident on the M42 at Junction 11 occurred where a Heavy Goods Vehicle driver fell asleep at the wheel and left the carriageway demolishing one of four column supporting the end span of overbridge (Figure 2). Sufficient load sharing capacity of the insitu deck prevented Global collapse. The agency operates a risk-based strategy taking into account location and strength factors and operates a policy of strengthening/protecting on a priority basis.

The TAA in conjunction with the Agent decide on the quickest safest means of ensuring the carriageway can be opened to the public and means of certifying the temporary support and protective barriers before the road is reopened.

Fig 2   Fig 3

Impact On Structures: Superstructure Impact

The list of over height vehicles damaging bridge structures is extensive and occurs for various reasons. However the majority of superstructure damage on structures managed by the Agency are caused by Vehicles being over the regulatory minimum maintained headroom height of 5.03m (16' 6") Figure 3. High load routes are maintained for vehicles over this height.

The 1987 Report 'A Strategy For The Reduction Of Bridge Bashing' concluded that it would cost £21million at 1987 prices to raise only 70 structures at the highest risk to the minimum height of 5.03m and that large-scale public expenditure was not justified. On average over 92% of bridge strikes occur on local roads that are not managed by the Agency

On a site specific basis crash beams, hazard warning lights actuated by overheight vehicles and signing at structures with less than 5.03m headroom are measures taken to prevent such accidents, although such measures are not mandatory on structures over 5.03m they are provided at vulnerable locations.

Gross Design Errors

The process of Approval In Principle is devised to ensure gross design errors are a very rare occurrence and Quality Assurance of Design Organisations provides additional assurances that a structure designs will be fit for purpose. The 15-year assessment programme, 1984-1999, has shown that many structures were designed below the loading level requirements of the time. However the TAA through its input to the assessment programme has helped attain an acceptable (40 Tonne Gross Vehicle Weight) load capacity for numerous bridges that initially failed their assessments. The in-built design and material factors of safety have been successfully mobilised to demonstrate very few are either unsafe or unserviceable; structures that do fall below acceptable levels are strengthened or replaced.

Post-Tensioned Structures.

After failure of the Ynys-y-Gwas bridge in Wales (1985) a moratorium was placed on the construction of new post-tensioned structures and a programme initiated to investigate the condition of all existing structures in the UK. Structures are only given a clean bill of health after three phases of investigation and inspection are complete and the final phase signed of by the TAA.
Through a programme of intrusive investigation many structures have been found with voided ducts, and corroded tendons, some have been re-grouted whilst others are being monitored, using a variety of methods such as Acoustic Monitoring, Vibrating Wire Strain Gauges, Radar Penetration, etc.

Extreme Flooding

Recent 1:100 year flooding throughout the UK has prompted a whole scale Inspection Programme examining scour, debris build up and damage to all the Agencies structures. Fortunately little damage has been found to date however Maintaining Agents have still to complete the programme. Regular diving inspection is undertaken as part of the Agency's Principal Inspection programme. In general bridges are assessed on a risk basis for: scour, hydraulic, bridge failure modes, the evaluation of demand models, through to interpretation of results on a spreadsheet.

Impact On Parapets

A programme of upgrading sub-standard parapets for Motorway and Trunk Roads with advice on implementation given at local level but with risk ranking to national standards.

Parapets are ranked on risk in accordance with the following criteria: Remnant Strength, type of highway carried, features below, road/structure layout, and containment feature. This currently takes no account of bridge specific traffic percentile traffic speeds and research is almost complete that will reassess impact requirements and consequently risk of the parapet being breached by impact, Figure 4.

Fig 4   Fig 5

Risk Factors Identified after construction

Two factors newly identified on UK structures over the last decade are Thaumasite sulfate attack and Alkali Silica Reaction, both phenomenon initiated research programmes resulting in new standards and programmes of re-assessment and investigation.

The management of programme for Thaumasite, in which sulfate attacks the calcium silicate paste and produces the Thaumasite mineral, resulting in the deterioration of the concrete, Figure 5, has reached Phase 2 (Site Investigation).
The management of structures affected by Alkali Silica Reaction is continuing, and advice on assessment of existing structures and control through constituent materials of concrete in new construction has been in existence for over 6 years.

Faulty Detailing And Structural Deterioration

Many structures have built in chlorides resulting from their use as accelerators for curing concrete or from contaminated water sources. On the M5 and M6 Motorways and on other structures around the network faulty or poorly maintained expansion joints, have failed in the early years of service. This resulted in salt laden water from de-icing salts soaking and contaminating the supporting structure with chlorides Figure 6.
Corrosion of reinforcement initiated by chloride attack deposited by salt spray from vehicles is the most commonly occurring problem with abutments and piers.

Preventative maintenance since 1990 has been through silane impregnation designed to prevent further chloride ingress and this has proved to be relatively successful to date. Existing structures with chloride contamination above the threshold of 0.3% and conditions favourable for corrosion of reinforcement will either have contaminated concrete replaced or a specific form of electro chemical treatment and long term monitoring. The decision on either is based on whole life costing.

Fig 6

Fig 7

Faulty Workmanship

Box Girders Constructed 30-40 years ago showing little signs of distress were assessed to need limited strengthening through additional bracing. However strengthening contracts uncovered, hydrogen cracking, inclusions and under-size/strength welds caused by poor workmanship during original construction, Figure 7. Similar occurrences of weld defects have been found in some sign gantries.

Future And Continuing Role Of Technical Approval In Safety Management Of Highway Structures

From 2001 the Agency is charged with delivering a £21 billion Ten Year Plan of improvements to the network, whilst continuing to manage its existing 16,000 + structures, and make year on year efficiency gains in its own working practices.
The challenge to the TAA is to ensure that safety is not compromised during this increased programme of new build and that an ageing and deteriorating bridge stock is managed properly.

The tools at the disposal of engineers have been developed through the Agency's Programme of research, and over the next ten years engineers will use the following methods to ensure that the United Kingdom's bridge stock is safe and serviceable;

- New Inspection Procedures that target critical vulnerable areas of structures.

- The steady state assessment of existing structures at a rate of 6% per year on a prioritised basis.

- Whole life assessment of existing structures: Through deterioration models, predicting future dates at which capacity of structure become critical and determining intervention dates.

- Innovative ways of delivering outcomes that matter to our customers. Through streamlined procedures, new materials and construction methods.

- The use of IT through Extranet/Intranets and a functional structures asset management information system, (SMIS).

- Through continuing self-assessment/benchmarking identifying areas for continual improvement and their implementation.

- From research and development and the identification of new research ideas.

Through the application of these techniques and the experience and effective application of the Civil Engineering Division and the Technical Approval engineer's, the Agency aims to continue the cost effective assurance of fitness for purpose of all new and existing structures.

References

1. Parag C Das et al (1997) Safety Of Bridges. Thomas Telford London
2. Working Party Report (1987) A Strategy For The Reduction Of Bridge Basing HMSO
3. Highways Agency Business Plan 2001/02, "Delivering the 10 Year Plan"

STAGED INVESTIGATIONS OF BRIDGES

DR DONALD PEARSON-KIRK
Parsons Brinckerhoff Infrastructure, Taunton, UK

ABSTRACT

Interactions between singular causes of deterioration of bridges are numerous and can greatly accelerate the rate of deterioration. Causes of deterioration may relate to inadequacies in design, in construction or to environmental processes which may lead to deterioration of concrete and/or steel components of the bridges. The unexpectedly early deterioration of bridges has been well documented and is of major concern in many countries, in all climatic zones. The dilemmas facing owners and managers of structures trying to plan the maintenance, repair and replacement of bridges, or elements of bridges, are outlined. A strategy is presented whereby the most cost-effective rehabilitation options for bridges may be selected, substantially based on staged investigations of those structures. In turn the presence of any problem is identified, the cause or causes of the problem determined, and the appropriate solution to the problem selected. Case studies are presented for bridges that demonstrate the benefits of staged approaches to investigations.

1.0 INTRODUCTION

Infrastructure is expensive, and so should be appropriate, economic, and lead to improved quality of life and safety, whilst giving the minimum of adverse environmental impacts. Problems resulting from inappropriate design and poor construction practices have become apparent in many countries. Even in countries where design procedures, construction practices, and maintenance of structures have been apparently well carried out, unpredicted deterioration of structures has occurred.

Many factors need to be considered in order to predict the durability and useful life of a structure. These include experience, results of assessments and tests, and the effects of the actions of agents on different components of the structure (British Standards Institution (BSI) 1992)(1).

The maintenance, repair and replacement of structures, or parts of structures, provide substantial information on durability, but little data is collected systematically, in a form that can be used, to predict durability with any certainty. Life cycle costing techniques are being developed in order to aid decision making when it may no longer be economical to maintain and repair a structure.

The predicted service life of a structure may be assessed (1) in one or more of the following ways: -

- By reference to experience with a similar structure in similar circumstances.
- By measuring the rate of deterioration over a short period and estimating when the limit of durability will be reached.
- By interpolation from accelerated tests that will shorten the response time to the action of the agent(s) causing deterioration.

Service life prediction and management strategies for structures can benefit significantly from: -

- Well targeted testing regimes.
- Proper execution of site investigations by appropriately trained technical staff.
- Appropriate interpretation of site and laboratory testing results by engineers experienced in testing and monitoring concrete structures.

2.0 CAUSES OF DETERIORATION

Interactions between singular causes of deterioration of concrete structures are numerous and can greatly accelerate the deterioration. Causes may relate to inadequacies in design and/or construction or may result from agents acting on structures. Various organisations have reported on causes of deterioration to concrete structures, for example the Organisation for Economic Co-operation and Development (OECD)(2) which reported in 1988 that singular causes of deterioration to 800,000 concrete highway structures in decreasing order of importance were as follows: -

Chloride contamination;
Sulfate attack;
Non-conformance to the project design;
Thermal effects;
Inadequate design;
Insufficient reinforcement/or insufficient size of reinforcement bars;
Insufficient quality of concrete;
Insufficient concrete cover;
Insufficient protection against rain water;
Lack of maintenance;
Alkali-silica reactions;
Flood, scour;
Change of water table;
Fatigue;
Earthquake;
Insufficient drainage.

3.0 THE NEED FOR TESTING

In-situ testing and sampling of structural materials can be time-consuming and expensive. It is of benefit if it is closely specified, controlled, properly organised and supervised at the appropriate level. Valuable information can be obtained for use in the assessment of the load carrying capacity of a structure, in the assessment of the state of deterioration of the structure and in formulating possible repair, strengthening and/or replacement strategies. (Frostick)(3)

The use of material strength testing is often undertaken to determine whether or not worst credible strength will result in capacity assessments higher than those achieved by using characteristic (default) strengths. The use of worst credible strengths permit reduced material factors and this may give a two fold benefit in possibly improving the load assessment capacity of a structure.

The present state of a concrete structure in terms of durability needs to be determined by condition testing of the concrete and the reinforcement.

Information from site testing of highway structures and laboratory testing of samples from those structures can prove valuable in the assessment of load carrying capacities, condition assessments and selection of any rehabilitation works. Decisions based on the consideration of visual assessments together with testing results, will be more reliable than decisions based on visual assessments alone.

Condition assessments can change when testing results are considered rather than relying solely on visual assessments. Table 1 shows the changes in condition classifications of 200 concrete bridges (Wallbank, 1989)(4) and 25 precast concrete beams in the deck of a bridge over a reservoir spillway (Craddy and Pearson-Kirk, 1996)(5). For the former the overall condition assessment improved when testing data was considered, whilst for the latter the condition assessment worsened.

CLASSIFICATION	200 CONCRETE BRIDGES		25 CONCRETE BEAMS	
	Without Testing Data	With Testing Data	Without Testing Data	With Testing Data
GOOD	25	59	3	3
FAIR	114	100	9	5
POOR	61	41	13	17

TABLE 1: CONDITION CLASSIFICATION OF CONCRETE HIGHWAY STRUCTURES

The classification of both observational and testing data has to lead to a more accurate estimate of the condition of the structures under review.

Better quality advice can then be given to owners or monitoring agents of structures, who can then manage their structures more cost-effectively.

4.0 ASSESSING THE DURABILITY OF CONCRETE STRUCTURES

Assessing the durability of concrete structures is essential for the cost-effective management of those structures, with assessments being improved by consideration of the results of testing and/or monitoring of the structures. In the last 15-20 years there has been a rapid growth of tests and testing and in 1997 the Concrete Bridge Development Group (CBDG) formed a task group to prepare the Technical Guide on "Testing and Monitoring the Durability of Concrete Structures" under the chairmanship of Mr J J Darby. The task group addressed the need for improvements in the specification of testing, carrying out of testing and in the interpretation of results. The guide (6) is now prepared for publication.

The structure maintenance process is improved by undertaking testing and/or monitoring in three stages (6), these being: -

- Condition monitoring phase in which irregularities are detected.
- Diagnosis phase in which causes of those irregularities are determined.
- Solution development phase in which the best course of action is selected.

The Technical Guide stresses the need for a desk study of available documents and for a preliminary site inspection prior to planning the testing and/or monitoring programmes. Detailed information is provided on site and laboratory tests both for determining physical structure and response and for determining corrosion activity and probability of corrosion (6).

The benefits of staged investigations are demonstrated in the case studies outlined below.

5.0 REINFORCED CONCRETE BRIDGE

A reservoir impounded behind a large mass concrete gravity dam which carries a five span reinforced concrete road bridge across the spillway crest was constructed in 1966. The dam is 182m long at crest level and the spillway bridge is 36m above the valley floor.

During a regular inspection, it was noticed that pieces of concrete had spalled off the underside of the deck and had rolled down the spillway. PB Infrastructure was appointed to investigate the problem and to assess the capacity of the bridge.

An inspection of the underside of the bridge deck with limited corrosion sampling and testing (BA35/90)(7) was undertaken (Stage 1 Investigation). This revealed that the main longitudinal beams showed evidence of extensive deterioration (5) including corrosion of reinforcement.

The deterioration appeared to be due to two main factors:

- The exposed position of the bridge.
- The nature of the deck construction.

The deck was designed and built as five simply supported spans with many elements being precast and assembled on site. In addition there was no waterproofing layer. As a result of these factors many leakage paths had developed in the deck through which road salts could penetrate and run down the faces of the main beams.

It was also found that there was deterioration to the pier supports and abutment faces. It was considered that some further investigations would be required to determine the extent and severity of the deterioration (Stage 2 Investigations).

Results of the Stage 2 corrosion testing and laboratory testing of concrete samples indicated extensive corrosion of reinforcement in 17 of the 25 beams, with chloride ion concentrations being at very high levels throughout the greater part of the beam cross sections. Exposures of reinforcement confirmed extensive pitting corrosion, with considerable loss of cross-sectional area of some bars. Diaphragms and deck slabs were similarly affected. Continuing deterioration would lead to a progressive reduction in structural capacity of the beams.

Testing of the substructure showed there to be significant concrete and reinforcement deterioration due to the effect of water all the year round and de-icing salts. Sulfate induced deterioration of bearing shelves to abutments and piers was confirmed

Petrographic examination and chemical testing of cores taken from the substructure (Stage 3 Investigations) confirmed considerable contamination of concrete to great depths. Alkali silica reaction (ASR) was found to be already occurring within the concrete and alkali contents in the concrete were high. It was considered that electrochemical desalination or cathodic protection, both of which were under consideration as remedial measures, could increase the risk of further ASR in the concrete.

The main conclusions and recommendations for the bridge based on the investigations carried out included:

- The extent and severity of deterioration to concrete and the reinforcement in the deck were of such magnitude that it would not be feasible to repair and/or strengthen the deck elements. However, the precast parapets could be retained for future use.
- Further deterioration of the sub-structure could be expected with the continuing ingress of water and salts.
- Chloride/sulfate contaminated concrete in the substructure should be removed, corroded reinforcement cleaned or replaced; and bearing shelves should be waterproofed.

For the remainder of the dam structure it was recommended that: -

- An impervious layer should be provided between the roadway surfacing and the top surface of the dam to prevent the further ingress of water and salts.
- Areas of the north face of the dam damaged by freeze-thaw action/or contaminated by chlorides/sulfates should be repaired.

6.0 POST-TENSIONED BRIDGE

The bridge is a four span structure which carries a trunk road at a skew of 46.5° over the electrified West Coast Mainline railway. The bridge was designed and constructed by British Rail circa 1960 and comprises simply supported spans of 6.25m, 18.6m, 18.6m and 6.25m with the central spans comprising precast, post-tensioned beams forming a shear key deck with the side spans being reinforced concrete slabs. The prestressed concrete deck contains 15 rectangular beams that are connected transversely by overlapping links projecting from the sides of the beams, through which longitudinal reinforcing bars were to have been threaded. The longitudinal spaces containing this reinforcement between adjacent beams are filled with insitu concrete to form the shear key deck. The beams are overlaid with a lightly reinforced insitu concrete topping of varying thickness to form the road profile. No waterproofing membrane was indicated on the drawings. The longitudinal and transverse joints were therefore poorly protected against surface water and salts. Drainage of the deck relies on the longitudinal and transverse falls to the surfacing.

PB Infrastructure were commissioned to undertake an inspection of the bridge. Evidence of significant flows of water through the longitudinal joints between beams and at beam ends was noted. Extensive severe deterioration of the piers was observed. Concrete testing and investigations were recommended. The results of Stage 1 investigations indicated that deck beam concrete and infill concrete had estimated insitu cube strengths of 50 N/mm^2 and 16 N/mm^2, respectively.

The Stage 2 Investigations involved the exposure and testing of the top surfaces of 8 No. post-tensioned beams, together with investigations of the infill concrete and its reinforcement. Excavation of the trial pits confirmed that the deck had not been waterproofed.

Testing to the top horizontal surfaces of the pre-cast beams showed half-cell potentials to be highly negative and concrete resistivities to be low/medium, indicating a potential for pitting corrosion of reinforcement. Chloride ion concentrations of concrete samples were below the accepted critical level and breakouts of concrete showed that reinforcement was in good condition at that time.

Infill concrete was in places poorly compacted with variable strength (range 5 to 47 N/mm^2). Half the samples of infill concrete tested for chloride ion concentrations had values exceeding the accepted critical level for the initiation of pitting corrosion, and severe pitting corrosion of infill reinforcement forming the hinges was observed. The longitudinal lacer bars forming the joint between adjacent beams was missing in four of the eight trial pits.

It was concluded that a condition factor of unity could be applied to the post-tensioned beams, although continuity between the beams should be disregarded. It was also concluded that a Special Inspection of the post-tensioned system should be undertaken. The lack of deck waterproofing had enabled water and salts to pass into the infill concrete and this resulted in the formation of stalactites to the main span soffits at the longitudinal joints between beams.

The Stage 3 investigations comprised a Post-Tensioning Special Inspection (PTSI) to BA 50/93(8)and also testing of the sub-structure. The PTSI comprised three phases: -

- Phase 1 Deck Study
- Phase 2 Preliminary Site Inspection and Preparation of Phase 3 Technical Plan
- Phase 3 Site Investigations

The investigations indicated that the post-tensioning ducts and tendons were in good condition. Of 19 separate exposures of ducted tendons in the beams, slight surface corrosion of ducts was noted in 1 of the exposures, with no significant corrosion to any ducts. At 11 locations surface corrosion of the pre-stressing strand was noted, but with no significant loss of section. The corrosion noted to certain strands may have been present at the time of construction. Voids were found in four ducts. Grout was hard, with no visible corrosion products. Chemical analysis of grout samples showed that chlorides were only present at concentrations far below the accepted level at which pitting corrosion of reinforcement would be expected to be initiated.

Corrosion testing and sampling showed that corrosion of reinforcement in the precast beams was highly likely in sections of beams adjacent to the longitudinal joints between beams in areas of water staining/stalactites. Chloride ion concentrations greater than the accepted critical level for the initiation of pitting corrosion of reinforcement were found in beam concrete. Infill concrete between beams had chloride concentrations of up to four times the critical level, and voids in the infill concrete between beams were observed to contain salt deposits.

Cement content of the beam concrete was high, alkali contents were below the upper limit recommended for the avoidance of alkali silica reaction and sulfate content was below the level at which it is generally considered that sulfate attack of concrete may be initiated.

It was considered that the effective control of water and salts gaining access to the deck was essential if pitting corrosion was not to progress in the precast beams. It was recommended (a) that drain holes approximately 300mm deep be provided between beams to ensure that water was drained from the insitu concrete and the voids between the beams, and (b) that the top surface of the deck should be waterproofed.

Based on the site testing undertaken and on the results of laboratory testing of samples taken from the piers, the capping beams were considered to be in poor condition with the potential for further deterioration of reinforcement and of concrete. It was considered essential to greatly reduce the further ingress of water and salts to the beams and columns from the roadway above in order that further deterioration was prevented. If this were to be achieved the capping beams needed to be monitored for corrosion activity and localised rehabilitation work be undertaken, if the beams were to be retained in the longer term.

It was recommended that consideration be given to: -

(1) Reducing/preventing the ingress of water by various measures including: -

- Providing drain holes through longitudinal joints between beams
- Waterproofing the deck
- Checking and repairing/replacing the transverse joints over the piers

(2) Carrying out concrete repair to damaged areas of the piers.

(3) Impregnating the exposed surfaces of the piers with a silane material to prevent the ingress of water and salts.

(4) Monitoring selected areas of the piers to demonstrate any changes in corrosion activity.

7.0 CONCLUSIONS

Particular attention has to be made to testing and monitoring the durability of concrete structures in order that those structures may be managed in the most cost-effective manner. Staged investigations enable irregularities to be detected, causes of deterioration to be diagnosed and the solution of problems to be determined.

It is essential that investigations be well targeted, be carried out by well trained engineering staff and that results of investigations be interpreted by experienced engineers.

8.0 REFERENCES

(1) British Standards Institution (1992) Guide to Durability of Buildings and Building Elements, Products and Components. BSI, London. pp43
(2) Organisation for Economic Co-operation and Development (1988). Durability of Concrete Road Bridges. Road Transport Research Programme. Paris. 136 pp.
(3) Frostick, I. (1996) The Use of Testing in the Determination of Material Strengths for Assessment and the Prediction of Durability. Proceedings of the Concrete Bridge Development Group Annual Conference. Dunchurch, UK March.
(4) Wallbank, E.J.(1989) The Performance of Concrete Bridges. A survey of 200 Highway Bridges. Report prepared by G Maunsell and Partners for the Department of Transport, HMSO, London. 96pp.
(5) Craddy, M.F. and Pearson-Kirk, D. (1996) The Restoration of Thrusscross Reservoir Spillway Bridge. Journal of Concrete Repair, London. April.
(6) Darby, J.J., Capeling, G, George C.R., Pearson-Kirk, D., Dill, M.J. and Hammersley, G.P. (2001) Testing and Monitoring the Durability of Concrete Structures. Concrete Bridge Development Group. Technical Guide No.2 pp 164 (in publication).
(7) BA35/90 (1990) Inspection and Repair of Concrete Bridges, Department of Transport Departmental Advice Note. London.
(8) BA 50/93 (1993) Post-tensioned Concrete Bridges. Planning, Organisation and Methods of Carrying out Special Inspections. Department of Transport Advice Note. London, July.

Verification of Girder Distribution Factors and Dynamic Load Factors by Field Testing

ANDRZEJ S. NOWAK and JUNSIK EOM
Department of Civil and Environmental Engineering,
University of Michigan, Ann Arbor, MI 48109-2125, USA

Introduction

The analytical studies performed in conjunction with the development of AASHTO LRFD Code (1998) indicated that the girder distribution factors (GDF) specified in AASHTO (1996) are not accurate for some groups of bridges. In particular, the analysis showed that GDF's are overly conservative for long spans and larger girder spacing, while they are too permissive for short spans and small girder spacing. Field testing is an increasingly important topic in the effort to deal with the deteriorating infrastructure. There is a need for accurate and inexpensive methods for diagnostics, verification of load distribution, and determination of the actual load carrying capacity. A considerable number of Michigan bridges were constructed in 1950's and 1960's. Many of them show signs of deterioration. In particular, there is a severe corrosion on many steel and concrete structures. By analytical methods, some of these bridges are not adequate to carry the normal highway traffic. However, the actual load carrying capacity is often much higher than what can be determined by analysis (Bahkt and Jaeger 1990), due to more favorable load sharing, effect of non-structural components (parapets, railing, sidewalks), and other difficult to quantify factors. Field testing can reveal the hidden strength reserve and thus verify the adequacy of the bridge.

Existing bridges are evaluated to confirm their adequacy for carrying traffic loads. The major parameters needed for evaluation include the actual loads and load carrying capacity. Based on the current evaluation procedures, many short span bridges are considered as deficient and in need of repair or replacement. Knowledge of the accurate girder distribution factors (GDF) and dynamic load factors (DLF) is needed to determine the actual value of live load (truck load) for bridge girders. Overestimation of GDF's and DLF's can have serious economic consequences, as deficient bridges must be repaired or rehabilitated. Therefore, the objective of this study is validation of the code-specified girder distribution factors (GDF) and dynamic load factors (DLF) for steel girder bridges with spans 10-45 m. The validation is carried out by field tests and finite element analysis.

About 20 structures were selected as representative for the bridge inventory in the State of Michigan. For each structure, field tests and analysis were performed. The girders were instrumented, and strains and stresses were measured due to heavy trucks (for some bridges deflections were also measured). GDF's were then calculated for one truck (one lane loaded) and two trucks side-by-side (two lanes loaded). The GDF's were also determined by the advanced structural analysis, based on the finite element method (FEM).

Code specified Girder Distribution Factors and Dynamic Load Factors

Measured girder distribution factors (GDF) are compared with the values calculated according to the current design codes. In this paper, the girder distribution factors are applied to the full truck rather than a line of wheel loads (half truck). For the bending moment in interior girders, the AASHTO Standard (1996) specifies GDF's as follows. For steel girder bridges and prestressed concrete girder bridges, with one lane, GDF is:

$$GDF = \frac{S}{4.27} \qquad (1)$$

and for steel girder bridges and prestressed concrete girder bridges, with multiple lanes,

$$GDF = \frac{S}{3.36} \qquad (2)$$

where S = girder spacing (m).

The AASHTO LRFD Code (1998) specifies GDF as a function of girder spacing, span length, stiffness parameters, and bridge skewness. For the bending moment in interior girders with one lane loaded, GDF is:

$$GDF = \left(0.06 + \left(\frac{S}{4300}\right)^{0.4} \left(\frac{S}{L}\right)^{0.3} \left(\frac{K_g}{Lt_s^3}\right)^{0.1}\right)\left(1 - c_1(\tan\theta)^{1.5}\right) \qquad (3)$$

and for multiple lane loading:

$$GDF = \left(0.075 + \left(\frac{S}{2900}\right)^{0.6} \left(\frac{S}{L}\right)^{0.2} \left(\frac{K_g}{Lt_s^3}\right)^{0.1}\right)\left(1 - c_1(\tan\theta)^{1.5}\right) \qquad (4)$$

$$c_1 = 0.25\left(\frac{K_g}{Lt_s^3}\right)^{0.25}\left(\frac{S}{L}\right)^{0.5} \qquad \text{for } 30° < \theta < 60° \qquad (5)$$

$$c_1 = 0 \qquad \text{for } \theta < 30° \qquad (6)$$

where S = girder spacing (mm); L = span length (mm); $K_g = n(I + Ae_g^2)$; t_s = thickness of concrete slab (mm); n = modular ratio for the girder and slab materials; I = moment of inertia of the girder (mm^4); A = cross section area of the girder (mm^2); e_g = distance between the centers of gravity of the girder and the slab (mm); and θ = skew angle in degrees. The AASHTO LRFD (1998) formulas were developed based on a NCHRP Project 12-26 (Zokaie et al. 1991). The formulas include the longitudinal stiffness parameter, K_g, and the span length, L, in addition to the girder spacing, S. AASHTO Guide for Load Distribution (1994) specifies similar load factors to those of AASHTO LRFD (1998).

Most bridge design codes specify the dynamic load as an additional static live load. In the AASHTO Standard (1996), dynamic load factors are specified as a function of span length only:

$$\text{DLF} = \frac{50}{3.28L + 125} \leq 0.30 \tag{7}$$

where DLF = dynamic load factor (maximum 30 percent); and L = span length (m). This empirical equation has been used since 1944. In the AASHTO LRFD (1998), the dynamic load factor is equal to 0.33 of the truck effect, with no dynamic load applied to the uniform loading.

Selected Bridges

This study is focused on steel girder bridges with simply supported spans from 10 to 45 m. The parameters considered include accessibility for testing equipment, traffic volume (ADT less than 15,000), skewness (no more than 30°), and existence of non-typical features (only typical and representative bridges were selected). All the selected bridges carry two lanes of traffic.

Instrumentation and Data Acquisition

The strain transducers were attached to the lower and/or upper surface of the bottom flange of steel girders at midspan. In addition, for some bridges, they were installed on selected girders at the ends to measure the moment restraint provided by the supports, and at intermediate span locations to measure variation in moment along the span.

Strain data for calculation of the girder distribution factors were taken from bottom-flanges of girders in the middle of a span. The measurements were taken under passages of one and two vehicles, each being a Michigan three-unit 11-axle truck with known weight and axle configuration. In Michigan, the maximum mid-span moment for medium span bridges is caused by 11-axle trucks, with gross vehicle weight (GVW) up to 730 kN depending on axle configuration. This is almost twice the allowable legal load in other states. Most states allow a maximum GVW of 350 kN with up to 5 or 6 axles per vehicle. The actual axle weights of the test trucks were measured at the weigh stations prior to the test for all bridges. Strain data was used to calculate load distribution factors.

Finite Element Analysis

The field test results were compared with analytical computations. The analysis was performed using ABAQUS finite element program available at the University of Michigan. Material and other structural parameters are based on the collected information about the bridge supplemented with engineering judgement.

For the purpose of finite element analysis, the geometry of the bridge superstructure can be idealized in many different ways. For this study, a three-dimensional finite element method (FEM) was applied to investigate the structural behavior of the considered bridges. The concrete slab is modeled using isotropic, eight node solid elements, with three degrees of freedom at each node. The girder flanges and web are modeled using three-dimensional, quadrilateral, four node shell elements with six degrees of freedom at each node (Tarhini and Frederic 1992). The structural effects of the secondary members, such as sidewalk and parapet, are also taken into account in the FEM models.

All investigated bridges were designed as simply supported. However, in older structures, corrosion of the bearings often causes additional constraints for both rotations and longitudinal displacements. It was observed, as also reported by other authors (Bakht and Jaeger 1988, Schultz et al. 1995), that even slight changes in boundary conditions have considerable effect on the results. Therefore, three cases of boundary conditions were considered in the FEM models: (a) the supports are represented by a hinge at one end, and a roller at the other end, (b) both supports are hinged, with no movement in longitudinal (horizontal) direction, (c) supports are assumed to be partially frozen, by applying elastic spring elements to the top and bottom flanges, with stiffness represented by K values. The magnitude of stiffness K was calibrated using field measurements.

Test and Analysis Results

For each bridge, the collected strain data served as a basis for the development of girder distribution factors.

Figure 1 shows GDF values obtained from the tests and code specified values, for the AASHTO Standard (1996) and AASHTO LRFD (1998). In calculation GDF according to AASHTO LRFD (1998), the actual value of the term $K_g / (Lt_s^3)$ (see Eq. 3-6) is used. The test GDF's are the maximum values from different truck loading positions. It is clear that code-specified GDF's for two lanes loaded are conservative in all cases. The discrepancy between the code-specified and test values indicates that the actual bridge condition can be different from what is assumed in the code. This can be due to deterioration. For comparison, the GDF's obtained in field tests as a part of this study are plotted versus analytical values calculated using AASHTO Specifications (1996), and AASHTO LRFD Code (1998), as shown in Figure 2. The results are presented for a single truck (one lane loaded), and for two trucks (two lanes loaded).

To verify the linearity of the bridge response to truck loads, the strains from single truck runs in two adjacent lanes were superimposed, and compared with strains obtained for two trucks side-by-side. The comparison of the superimposed GDF's and those for two trucks are also plotted in Figure 1. The differences are all within +/- 5%, and this is a good indication of linearity of the bridge behavior.

The relationship between DLF and static and dynamic strains is shown in Figure 3. The open circles correspond to static strain, and black solid squares correspond to dynamic strain. Dynamic strains remain nearly constant, while static strains increase as truck loading increases. This results in large dynamic load factors for low static strains. In all cases, DLF's corresponding to the maximum strain caused by two trucks side-by-side, are less than 0.10 at the most heavily loaded girders.

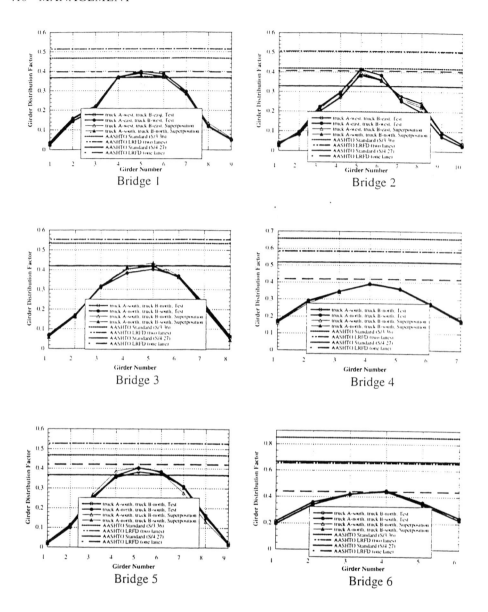

Figure 1. Girder Distribution Factor under Two Truck Side-by-Side Loading at Regular Speed.

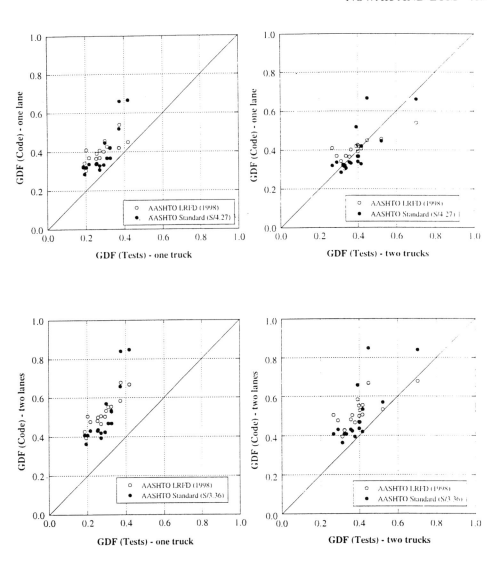

Figure 2. Test GDF's Versus Code GDF's.

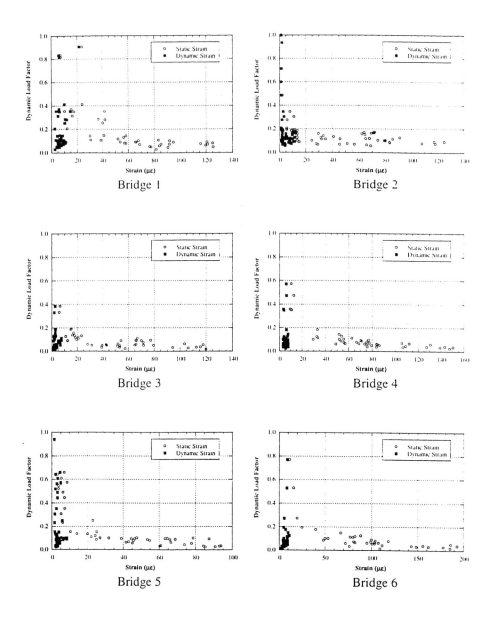

Figure 3. Strain Vs. Dynamic Load Factor.

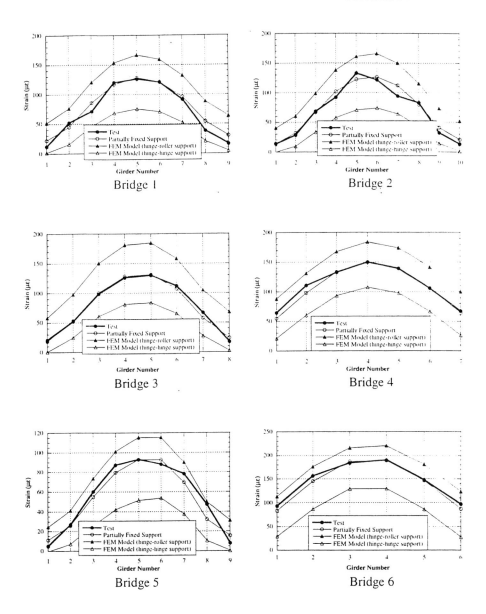

Figure 4. Results of the Finite Element Analysis, Strains.

The three-dimensional finite element method (FEM) was applied to calculate strains for the considered bridges. The FEM models were calibrated to accurately predict the test results. The parameters, including the spring coefficients, K, at the supports, were adjusted to match the test data. Some of the parameters had to be assumed using engineering judgement, such as the stiffness of the sidewalk and railings. After the models were calibrated, the FEM analysis was performed for three cases: (a) with the elastic springs removed from the model to simulate simple supports, (b) with longitudinal restrains to simulate the hinge-hinge condition, and (c) with elastic springs.

The FEM analysis results with three different boundary conditions for six tested bridges are shown in Figure 4. Strains are the largest for the hinge-roller supports simple support), and the lower bound is the hinge-hinge support (horizontal movement is restrained). The measured strains are between these two cases, and this means that the actual boundary conditions are somewhere between the simple support (with free longitudinal displacement) and the case where both girder ends are longitudinally fixed. The partially fixed support (with elastic springs) is adjusted to fit the test results. The spring coefficient K can be determined by trial and error.

Summary and Conclusion

Strains were measured for steel I-girder bridges in Michigan, with spans from 10 to 45 m. The test results were used to calculate the girder distribution factors. FEM models were developed and calibrated using the test data. Three different boundary conditions were considered in the FEM models; (a) roller-hinge supports, (b) hinge-hinge supports, (c) partially fixed supports. The girder distribution factors obtained from the tests and the FEM analysis are compared with the code specified values. The following conclusion can be drawn from the results of this study.

(1) The absolute value of measured strain is lower than predicted by analysis. One of the most important reasons is the partial fixity of support.
(2) The observed response is linear, which is confirmed by superposition of two truck effects. The comparison of strain values for a single truck indicates that for two trucks side-by-side, the results are equal to the superposition of the results for a single truck.
(3) Measured girder distribution factors are consistently lower than those of the AASHTO code specified values.
(4) Dynamic load is lower than specified values by AASHTO Standard (1996) and AASHTO LRFD (1998). For two trucks side-by-side it is less than 0.10. Dynamic load decreases with increasing static load effect.
(5) In evaluation of existing bridges, use of the code specified GDF values in the rating equation, without considering the effect of possible partial support fixity, can be too conservative.
(6) If the reduction of stress due to the partial fixity of supports is taken into account, then the code specified girder distribution values are suitable for use in the rating equations. However, it must be done cautiously because the support restraint can be broken when extremely high loads are present.

Acknowledgements

The presented research has been partially supported by the Michigan Department of Transportation, with Roger Till, National Science Foundation Grant No. CMS-9730988 with Vijaya Gopu, and NCHRP/IDEA Project with Inam Jawed, which is gratefully acknowledged. Thanks are due to the former and current graduate students and post-doctoral fellows at the University of Michigan, Sanjin Kim, Ahmet Sanli, Vijay Saraf, Chan-Hee Park, Taejun Cho, Maria Szerszen, David Ferrand, Gerard Gaal, and Karol Szerszen, for their help in field measurements.

References

AASHTO Standard Specifications for Highway Bridges (1996), American Association of State and Transportation Officials, Washington, DC.

AASHTO LRFD Bridge Design Specifications (1998). American Association of State Highway and Transportation Officials, Washington, D.C.

ABAQUS User's Manual-version 5.6. (1996), Hibbit, Karlsson & Sorenson, Inc., Pawtucket, Rhode Island.

Bakht, B., and Jaeger, L. G. (1988). "Bearing Restraint in Slab-on-Girder Bridges." Journal of Structural Engineering, ASCE, Vol. 114, No. 12, pp. 2724-2740.

Bakht, B., and Jaeger, L. G. (1990). "Bridge Testing-A Surprise Every time." Journal of Structural Engineering, ASCE, Vol. 116, No. 5, pp. 1370-1383.

Kim, S-J. and Nowak, A.S (1997)., "Load Distribution and Impact Factors for I-Girder Bridges", Journal of Bridge Engineering, ASCE, Vol. 2, No. 3, pp. 97-104.

Nowak, A.S., Eom, J. and Sanli, A., "Control of Live Load on Bridges", Transportation Research Record, No. 1696, Vol. 2, 2000, pp. 136-143.

Nowak, A.S., Eom, J. and Sanli, A. and Till, R., "Verification of Girder Distribution Factors for Short-Span Steel Girder Bridges by Field Testing", Transp. Research Record, No. 1688, 1999, pp. 62-75.

Schultz, J.L., Commander, B., Goble, G.G., Frangopol, D.M. (1995), "Efficient Field Testing and Load Rating of Short and Medium Span Bridges" Structural Engineering Review, Vol. 7, No. 3 pp 181-194.

Tarhini K.M and Frederick, G. R. (1992), "Wheel load distribution in I-girder highway bridges" Journal of Structural Engineering, ASCE, Vol. 118, No 5, pp1285-1294.

Zokaie, T., Osterkamp, T.A., and Imbsen, R.A. (1991), "Distribution of Wheel Loads on Highway Bridges," National Cooperative Highway Research Program Report 12-26, Transportation Research Board, Washington, D.C.

SEISMIC RETROFITTING OF BRIDGES IN NEW YORK

Ayaz H. Malik, P.E.[1]

ABSTRACT

The N.Y.S.D.O.T. has established policies covering minimum requirements for resisting the seismic forces for new and existing structures. Based on the importance, location and other pertinent features, additional requirements are considered on a project-to-project basis. This paper presents an overview of the seismic design criteria for new structures and seismic retrofitting of conventional existing highway structures as part of the rehabilitation program of New York State. The scope of rehabilitation work considers the seismic vulnerability of the existing elements and aims at improving the seismic resistance and overall performance by retrofitting. Additionally, for larger and complex type of structures more sophisticated analysis techniques have been utilized for evaluating seismic vulnerability of these important structures.

INTRODUCTION

There are currently about 20,000 bridges in New York State under the jurisdiction of State, Bridge Authorities and Local governments. These bridges vary in structural types and materials. Many of these structures are more than fifty years old and the majority of these structures were constructed without any consideration for earthquake loadings. Any moderate to major seismic event can cause severe damage to the structures, endangering public safety and interrupting vital life lines. The New York State Department of Transportation has adopted measures to minimize the potential for catastrophic failure of its bridges (5). It is intended to maintain the retrofitted structures functional after the event, requiring a certain degree of repairs.

SEISMIC ACTIVITY

With the increased frequency of earthquake occurrences elsewhere, the probability of a significant seismic event occurring in the northeastern part of the United States is being given careful consideration. Over the past three centuries, the eastern United States experienced several major earthquakes. Historical evidence indicates that the largest earthquake ever occurred in the United States of America was in the eastern United States, at New Madrid, Missouri, in 1811, equivalent to 8.0 on the Richter scale.

The biggest earthquake ever recorded in New York State was in Messina, New York, two hundred miles northwest of Albany. It occurred on September 5, 1944, measuring 6.0 on the Richter scale. Since 1884, New York State has experienced four earthquakes with Richter Magnitudes greater than 5.0. The neighboring Canadian Provinces have experienced earthquakes of magnitude up to 7.0, which also affected New York State and adjacent areas.

Certain parts of the United States have traditionally ignored the potential for seismic damage, since evidence of past earthquakes of significance was relatively rare. The Northeast, including New York, was among those areas. Continuing advances in seismology and the increased public attention have caused scientists and engineers to look more closely at the threat that earthquakes pose to this area.

---

[1] Project Engineer, New York State Department of Transportation, Albany, New York, USA 12232.

NEW YORK STATE DEPARTMENT OF TRANSPORTATION SEISMIC CRITERIA

New York State Department of Transportation has developed "guidelines" to assist in the design of new structures and evaluate vulnerability of existing structures. They are summarized below.

Seismic Criteria for Design of New Structures

The seismic design requirements of Division 1A of the 1994 AASHTO Specifications are incorporated into the New York State Standard Specifications for Highway Bridges (1), with the following modifications:

- Although New York State comes under Seismic Performance Category 'A' (SPC 'A') and SPC 'B' on the AASHTO rock acceleration map, designers shall use a rock acceleration coefficient equal to 0.19 for the design of all new highway bridges.

- Seismic analysis shall consider the gravity load as the sum of the dead load plus live load.

- Continuous structures on a multiple span bridge shall be used rather than a series of simply-supported single spans.

- Steel rocker or steel sliding bearings shall not be used. Elastomeric pads (plain, or laminated with or without load plates) or multi-rotational (pot type or disc type) shall be used. Multi-rotational expansion bearings shall always be provided with guide bars, to allow limited lateral movement.

- Bridge bearings shall be anchored on the bridge seat. Anchor bolts shall be provided with a minimum of 200 mm (8 inch) edge cover with reinforced concrete substructures.

- Abutments shall be provided with a continuous bridge seat. Stub type abutments with isolated bearing seats shall not be used.

- Skewed layout of substructures shall be minimized as much as possible.

Superstructures and substructures are first designed for non-seismic group loadings specified by AASHTO. Seismic analysis is conducted on the "designed" structure. Designers have access to the SEISAB bridge seismic analysis program. Conventional structures are analyzed by modeling fixity condition at the substructure foundations. Multiple span bridges are analyzed by Response Spectrum Method, using the AASHTO Ground Motion Spectra and an acceleration coefficient of 0.19. Capacities of the designed structure are checked against the seismic forces (demand) and adjustments are made to the design as warranted. In all cases (whether seismic forces control or not), standard seismic details for: reinforcement, connections between superstructure and substructure, support length, column confinement, and footing anchorage details conforming to the AASHTO Division 1A are utilized.

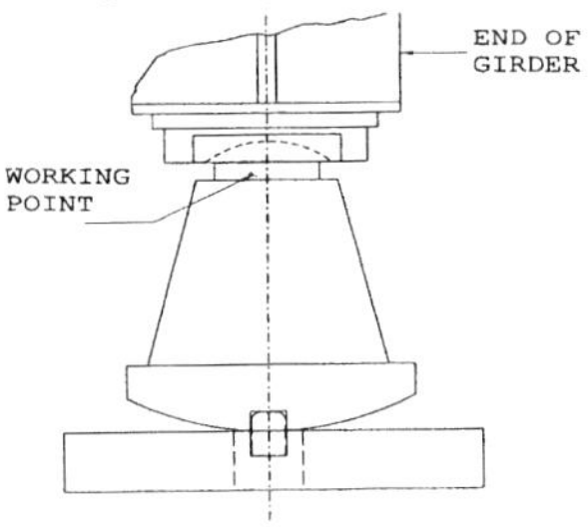

Figure 1 Steel Rocker Bearing

Seismic Criteria for Design of Existing Structures

- Existing bridges programmed for rehabilitation are evaluated for seismic failure vulnerability and retrofitting measures are incorporated into the rehabilitation contract.

- Initial seismic evaluations and seismic element analysis are based on the actual rock acceleration coefficient from AASHTO's rock acceleration map, attributable to the bridge site. In New York, AASHTO rock acceleration coefficients vary from .09 to .15.

- If retrofitting is deemed necessary, then all retrofitted elements are designed to meet NYSDOT 'new bridge' specifications.

- Capacity Demand ratios are computed for all elements of the bridge as outlined in FHWA's Seismic Retrofitting Manual for Highway Bridges (7).

Bearing Retrofitting

As observed in the previous earthquakes, as well as in the recent Kobe (Hanshin-Awaji, Japan) earthquake, steel bearings performed very poorly and bearing failure was the cause of the many structural failures (3, 9). During an earthquake, bearings are subjected to displacements, rotation and lateral forces in various directions, resulting in brittle failure of the unidirectional steel high rocker and low sliding bearings (Figures 1 and 2).

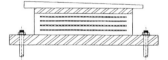

Figure 2 Steel Sliding Bearing

Replacing these bearings with ductile, multi-rotational and multi-directional bearings provide safety against potential unseating of the superstructure (Figures 3 and 4).

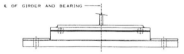

Figure 3 Elastomeric Bearing with Laminated Load Plates

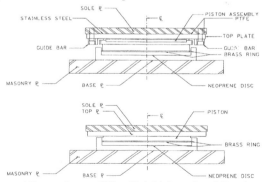

Figure 4 Expansion and Fixed Multi-Rotational Bearings

During the rehabilitation work, existing structures are jacked to remove the existing steel rocker or low sliding bearings. Due to the height difference between the elastomeric bearings and the existing rocker bearings, existing pedestals are built up to higher elevation, as recommended in the FHWA Retrofitting Manual (7). Alternately, steel extensions bolted to the bottom flange are used to adjust the height difference. Fixed steel bearings are replaced in a similar manner with fixed elastomeric bearings. New anchor bolts are set into the built up pedestals or drilled into the existing pedestals.

Depending upon the capacity of the existing substructure, various options have been considered to reduce the seismic design demand by adjusting the bearing configurations in one of the following manners:

♦ Providing fixed bearings at one of the abutments, and expansion bearings at the piers and the other abutment. This will help to reduce the seismic demand at the piers (Figure 6a).

♦ Providing all expansion bearings with lateral restrainers (Figure 5), thus reducing the transverse force demand by distributing it to all the substructures through expansion and fixed bearings (Figure 6b). Additionally, to further improve the longitudinal resistance, one abutment can be provided with fixed bearings and the other abutment can be provided with expansion bearings along with lockup devices (Figure 6b).

♦ Providing conventional laminated elastomeric bearings at the expansion supports and using a lead core base-isolation bearing at the fixed support. This will reduce the seismic demand, even at the fixed substructure (Figures 6c). For thermal expansion and contraction, the lead core bearing acts as a fixed bearing.

♦ Providing guided pot bearings at the expansion supports and a lead core base-isolation bearing at the fixed support. Since the coefficient of friction for expansion pot bearing is less than the elastomeric bearing, it will further cut down the demand in the longitudinal direction on the substructures (Figure 6d).

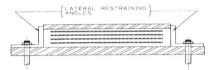

Figure 5 Elastomeric Bearing with Lateral Restraining Angles

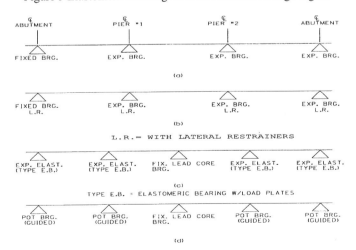

Figure 6 Various Bearing Configurations

RETROFITTING FOR CONTINUITY

Simply supported spans are made continuous, when feasible to provide redundancy (8). Continuity enhances the seismic response by distributing the in-plane forces to the piers and abutments and prevents loss of end support at piers due to longitudinal movement. When connecting the unrestrained ends of the two girder spans it is important to provide a complete splice between the flanges and the webs. Bolted splices are used since they provide ductility to the connection (Figure 7).

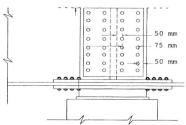

Figure 7 Typical Elevation at Piers

The use of slotted holes to allow any movement in the connection may cause brittle failure of the connection. Since continuity will introduce stress reversal, fatigue critical details at the end of bottom flange partial cover plates (if any) are retrofitted by a bolted splice (Figure 8) as per the AASHTO Standard Specification (1).

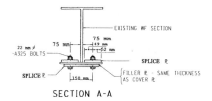

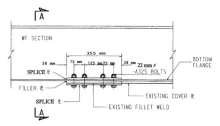

Figure 8 Bottom Flange Cover Plate Retrofit

Over stressing due to the continuity is controlled by using a lightweight deck. The two lines of bearings at the piers are replaced with a single elastomeric bearing, thus eliminating the dead load and/or live load eccentricity on the pier column(s) (Figure 9).

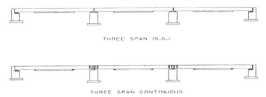

Figure 9 Continuity

For structures with pin and hanger systems, the suspended spans are fully-spliced with the cantilever to eliminate the vulnerable pin and hanger system (Figure 10).

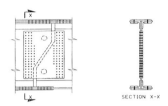

Figure 10 Splice Details

Where continuity is not feasible, restrainers and shear blocks (Figure 11) as recommended in the FHWA Retrofitting Manual (7) are used to prevent unseating of the superstructure girders.

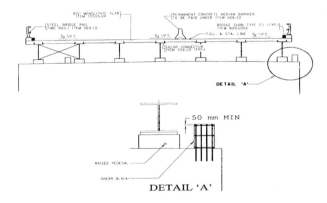

Figure 11 Shear Restrainers

Column Retrofitting

The current practice for earthquake resistance design of columns for bridge piers is to provide sufficient confinement of the potential plastic hinge locations by ties or spirals. Almost 90% of the existing pier columns are not provided with sufficient confinement necessary to improve the compressive strain and provide proper lateral support to the primary reinforcement. Hence, the failure of the column due to flexure or shear (in case of a short column) can be avoided by providing adequate confinement. In New York, this is typically done by retrofitting columns with steel or concrete jacketing and adding ties.

Steel Jacketing

By analyzing the existing columns for the design earthquake, the columns are checked for the flexural strength and the shear strength demands. Circular columns have been retrofitted with 10 mm (3/8") thick steel casing to provide passive confinement. To avoid excessive moment demand on the adjacent cap beam/footing a 10 mm (3/8") rubber sheet is placed between the in-fill grout and the steel casing around the column top/bottom. Figure 12 shows the details of the steel jacket retrofitting.

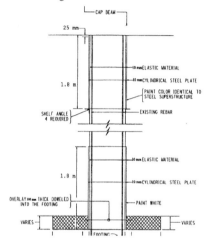

Figure 12 Seismic Retrofit for Pier Columns

Reinforced Concrete Jacketing

For solid (wall type) piers, a 305 mm (12 inch) thick layer of reinforced concrete is used as a jacket to provide flexural (longitudinal) as well as confinement reinforcement for the existing pier. A 1.2 m (4 foot) by 1.2 m (4 foot) grid pattern of drilling and grouting is used to dowel the concrete jacket to the existing pier. Figure 13 shows details of the retrofitted pier.

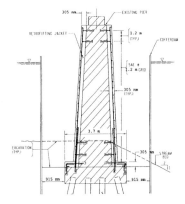

Figure 13 Reinforced Concrete Jacketing

Adding Ties

Adding of ties increases the shear strength of the column section along with the confinement for the column. This causes the flexural failure to occur before the shear failure, thus premature shear failure can be avoided. At the bottom of flared columns, adequate confinement is provided by adding ties (hoops) spaced equally between the existing 305 mm (12") centered ties. First the concrete cover is removed from the plastic hinge area and #5 hoops are tied around the existing flexural (longitudinal) reinforcement, alternating with the existing ties (Figure 14).

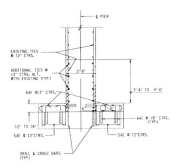

Figure 14 Footing Retrofit

Additionally, a 460 mm (18") thick reinforced concrete overlay is doweled into the existing footing. The retrofitted column has ties going 460 mm (18") into the retrofitted footing. Concrete cover is placed to match the existing column dimensions.

Foundation Retrofitting

Adding Piles

To satisfy the seismic demand for the design earthquake, piles are added to the existing foundation along with extension of the pile cap doweled into the existing cap. Additional piles with twice the capacity (soil and structural) are driven at the corners. By simple analysis of the pile group under the seismic load (plus any new load due to rehabilitation work) pile loads are computed by the following equation: Seismic load on pile = $P_s/n \pm M_{yy}/S_{xx} \pm M_{xx}/S_{yy}$; where P_s is the seismic axial load , n is the number of piles, M_{yy} and M_{xx} are the seismic longitudinal and transverse moment combinations, respectively, S_{xx} and S_{yy} being the corresponding section modulii. Existing pile loads plus the additional load due to earthquake, computed above, should not exceed the ultimate capacity of the existing piles. (Figure 15).

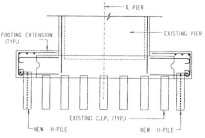

Figure 15 Foundation Retrofit

Bridging the Liquefiable Layer

To determine the liquefaction potential of cohesionless soils NYSDOT's Geotechnical Engineering Bureau uses two procedures (6). They are: 1) The empirical correlation based on the SPT blow count and field performance; and 2) a simplified steady-state strength procedure for analyzing stability of embankments and slopes against flow failures.

Depending on the thickness of the potential liquefiable layer at the existing pile foundation, additional steel piles are driven to rock to sustain the seismic loads under liquefiable conditions. The steel piles are over designed to carry the design loads below their buckling load capacity (Figure 16). This is determined by approximating the end conditions of the pile, to figure out the effective length factor,'K'. For example, for steel piles with a reinforced concrete footing and driven to bearing on rock, 'K' value of 1.0 will be used to determine $P_{ultimate}$ equal to $\Pi^2 AE/(Kl/r)^2$, where A is defined as the cross sectional area, E is the elastic modulus, Kl is the effective length, and r is the radius of gyration. Once the size is selected, Geotechnical Bureau performs the lateral analysis of the pile using FHWA's 'COM624' analysis program and its proprietary versions 'L-Pile' and 'Group' (distributed by Ensoft Products) to verify the adequacy of the selected pile(s).

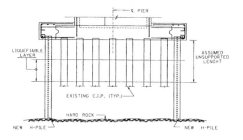

Figure 16 Bridging the Liquefiable Layer

Special Hazard study for New York City area Bridges

An expert panel, headed by Dr. Robin K. McGuire of Risk Engineering, Inc., Boulder, Colorado, was assigned the task of developing rock motions for the New York City area, which can be applied uniformly for the design of new structures and retrofitting of existing structures.

The recommended response spectra established by Dr. McGuire's report (4) has been implemented for the New York City and the surrounding areas. However, selection of Return Period, one or two level performance criteria, as well as importance classification and related site specific should be the responsibility of the Agency having the project jurisdiction.

These provisions are intended to be incorporated, as NYSDOT modifications to the current AASHTO seismic specifications for seismic design and analysis of bridges in New York City and the surrounding area.

Performance Criteria And Seismic Hazard

Bridges in the New York City and surrounding areas shall be designed to meet the performance criteria outlined in Table 1. Bridges shall be classified by the agency having the jurisdiction, as' critical',' essential' or 'other' meeting the requirements shown in Table 1.

TABLE- 1 PERFORMANCE CRITERIA AND SEISMIC HAZARD LEVEL FOR NEW YORK CITY AND SURROUNDING AREAS

Importance Categories	Return Period	Probability of Exceedance	Performance Criteria
Critical Bridges	2500 Yrs	2% in 50 Yrs	No collapse, limited access for emergency traffic in 48 hrs., full service within month(s)
	500 Yrs	10% in 50 Yrs	No collapse, no damage to primary structural elements, minimal repairable damage, full access to normal traffic available immediately (allow few hours for inspection)
Essential Bridges		2/3 (2% in 50 Yrs)	No collapse, repairable damage, one or two lanes available within 3 days, full service within month(s)
Other Bridges		2/3 (2% in 50 Yrs)	No collapse, significant but repairable damage in visible areas.. Traffic interruption acceptable

Two Level and One Level Design Approach

The probabilistic return periods and the corresponding performance criteria for functional evaluation and safety evaluation are shown in Table 1. In all cases, collapse shall be prevented; repairable significant damage may occur for one-level design approach and lower level of the two-level approach. For the upper level earthquake of the two-level approach the damage may occur but limited access for emergency traffic should be possible within 48 hours and full service within months.

Design and Analysis

Multimode Response Spectrum analysis should be performed for the two-level and the one-level design approach. However, for the upper level event, a non-linear static analysis, to assess the displacement and resulting damage, shall be performed. The critical primary load carrying members shall be evaluated to prevent brittle modes of failure. Different analysis procedures for different seismic input levels in a two-level approach should be considered Gravity loads shall include live loads, where they are likely to be critical in the seismic analysis. Effect of vertical ground motions should be considered, especially for long span structures.
Geotechnical analysis should be completed for both levels of design, rather than just the upper level.

The analysis should include assessments of the potential for liquefaction, lateral spreading, soil-structure interaction and uplift. In all cases, soil behavior that degrades the structural capacity of the foundations must be prevented at the lower level event, and that movement should be less than a maximum acceptable amount during the upper level event.

Site Specific Studies

'Critical','Essential' and 'Other' bridges with Site Class F (special soils, such as liquefiable soils) shall require site specific studies using as input the time histories provided by Dr. McGuire's Report (4). In a similar manner, site specific studies shall be conducted for all critical bridges with Site Class E (shear wave velocity<180m/sec).

SUMMARY

NYSDOT's seismic design performance goals are similar to AASHTO Specification requiring the earthquake forces to be resisted within the elastic range of the members without significant damage during a moderate earthquake. Under a severe earthquake, catastrophic failure should not occur and any damage should be detectable and can be repaired. For the new structures, design criteria requires elastic forces to be determined from a single mode spectral method of analysis as a minimum. In many cases, single mode method gives satisfactory results but it uses various approximations and it may compute some questionable results. Standard seismic details for reinforcement, connections, support length, column confinement and anchorages are provided, conforming to AASHTO Division 1A, for all structures.

A comparison of the recommended hazard for New York City (4) with the AASHTO Specifications (1) indicated that a significant reduction in seismic demand occurs with these recommendations for structures on rock. Whereas, an increase in seismic forces occurs at short periods (< 0.4 second) for structures on soil (2).

ACKNOWLEDGMENTS

The cooperation and assistance given by Lisa Maguire (NYSDOT) in preparing this paper is greatly appreciated.

REFERENCES

1. AASHTO Standard Specifications for Highway Bridges, Division 1A Seismic Design (1994) with NYSDOT "Blue Pages"- New York State Department of Transportation's Engineering Instructions 94-026, (July 1994)
2. Final Report: New York City DOT Seismic Design Criteria Guidelines, Dec.30, 1998 Weidlinger Associates, New York, N.Y.
3. Kawashima K., Damage of Highway Bridges by Hashin/Awaji, Japan, Earthquake and Seismic Design and Seismic Strengthening, proceedings of the International Conference on Retrofitting of Structures, Columbia University, New York City, NY March 11-13, 1996 (pages 187-207).
4. McGuire, Robin, Seismic Hazard for New York City, Risk Engineering, Inc.Boulder, Colorado, Jan.14, 1998
5. New York State Department of Transportation's Engineering Instruction "SEISMIC CRITERIA- Bridge Rehabilitation Projects, # E.I. 92-046 issued October 14, 1992.
6. New York State Department of Transportation- Geotechnical Engineering Bureau Manual- Liquefaction

Potential of Cohesionless Soils' Geotechnical Design Procedure GDP-9 January 1995.
7. Seismic Retrofitting Manual for Highway Bridges- F.H.W.A. RD-94-052, May 1995
8. Shirole, A. M. and Malik, A. H. Increasing the Seismic Failure Resistance of Highway Structures presented at the Second U.S.- Japan Workshop on Retrofit of Bridges, Berkeley, California, January 18-20,1994
9. Yashinsky, M., Lessons Learned from the January 17, 1995 Kobe Earthquake, proceedings of the National Seismic Conference on Bridges and Highways, sponsored by FHWA and CALTRANS, San Diego, California December 10-13, 1995.

Assessment

Current and Future Trends in the Heavy Haulage Bridge Assessment Process

Dr. S. N. SERGEEV
Lead Bridge Engineer, Halpern Glick Maunsell Pty Ltd and Senior Lecturer, University of Western Australia, Perth, Western Australia

G. SOBOL
Design Engineer, Transport SA, South Australia

C. C. CANDY
Director Structural Division, Halpern Glick Maunsell Pty Ltd

SYNOPSIS
The availability of a system for the structural assessment of bridges on heavy load routes is of paramount importance in any country with a developed road network. As such assessments generally constitute one of the core business activities of state and local Road Authorities, a system that is reliable and easy to operate, manage and maintain is of significant economic benefit.

A typical system generally requires collection and processing of bridge structure and load rating data and the subsequent assessment of the passage of specific heavy load vehicles over all (or nominated) bridge structures on the heavy haulage network. This is often complicated by the large variety of routes, destinations and heavy haulage vehicles as well as the requirement for various types of permit that may be issued by the Authority performing the assessment. Current assessment systems, particularly those used in Australia, are either purely manual or based on various empirical techniques. Both systems suffer significant problems when the asset base is large, the job load is very high, and experienced personnel are either non-existent or in short supply.

This paper presents an innovative, state-of-the-art process for assessing bridge structures on heavy haulage routes and highlights the fact that this process now forms a key component of a modern bridge management system. The cornerstone of the process is a user-friendly Windows-based bridge analysis program combined with powerful database facilities, that may be operated either as a stand-alone system or linked to a national road and bridge database.

1. INTRODUCTION
A number of different systems are currently used in Australia for assessing the effect of heavy load vehicles travelling over bridge structures on the national road network. Some are purely manual, generally relying on the accumulated knowledge and expertise of a single individual who has had a long-standing association with most structures on the road network and who possesses an in-depth knowledge of the heavy vehicle traffic history over them.

While this can be very effective in quickly isolating critical structures along a route, it suffers from the very significant disadvantage that the specialised knowledge disappears during the person's absence.

Other systems are empirical and generally based on comparison of the axle configuration and mass of the heavy vehicle with allowable limits derived from charts and tables. Although in many instances this is an effective method, problems are often encountered by inexperienced staff in identifying potentially critical structures, particularly for long routes with many bridges. The problem is complicated further by the fact that an increasing number of heavy haulage vehicles have varying axle widths and configurations.

Manual calculations in both of the above cases often prove time consuming, error-prone and tedious. The situation is being exacerbated by a dramatic increase in heavy load haulages and the perception within the transport industry that the assessment service must be quick, accurate and timely. However, while the demand for increased levels of service is rising, many authorities are faced with significant reductions in personnel sufficiently experienced in the assessment process. Manual and/or empirical systems are further disadvantaged by the lack of a method for rapidly and efficiently updating structure data associated with the construction of new bridges or the repair and refurbishment works carried out in accordance with an existing bridge maintenance plan.

A number of fully automated and computerised systems incorporating automatic route selection and minimal user intervention exist in Europe. Because of their complexity and need for large amounts of detailed structure data, these systems are generally very expensive to establish, operate and maintain. Most importantly, they are potentially vulnerable as they may either needlessly jeopodise the approval of the passage of the specified heavy vehicle along the nominated route or compromise the integrity of structures located on the route. This is due to the fact that any review facility provided by a fully automated system will, by its very nature, be simple, superficial and lacking in sufficient detail to allow well-founded and not overconservative decisions to be made.

Semi-automated systems that utilise a balanced mix of computer processing and manual post-analysis review have the benefit of allowing the numerically-intensive and repetitive processes of heavy load assessment to be automated while still relying on engineering judgement to determine the final outcome.

Several years ago a semi-automatic heavy load assessment system based on a restricted set of empirical rules was developed by Transport SA, South Australia. Recently, the underlying structural assessment methodology on which the original system was based, was modernised and the associated computer program enhanced and updated to the Windows environment, placing it at the forefront of heavy load route analysis systems in Australia today. This paper presents, in some detail, the entire heavy haulage bridge assessment process embodied in this innovative, state-of-the-art, semi-automatic system. It highlights the features that make it a key component of modern bridge management systems and discusses improvements that are planned for the near future.

2. HEAVY HAULAGE ASSESSMENT PROCESS
The heavy haulage assessment process is an integral and essential part of an overall bridge management system. It ensures that oversize and overweight vehicle movements across

bridges are systematically assessed and controlled, thereby reducing the possibility of serious structural damage.

The process encompasses three equally important tasks, namely:

1. collection, storage and updating of bridge structural data (generally referred to as *Bridge Data Management*);
2. bridge load rating and load posting; and
3. heavy and oversize vehicle load assessment.

A description of these three tasks is presented below.

2.1 Bridge Data Management

In any bridge data management system it is essential to develop and properly maintain a bridge database which, ideally, should be part of a broader, organisation-wide road database system. Bridge data collection is a continuing process of building knowledge about a bridge structure, from the time it is constructed until the moment it ceases to function as a bridge.

Within any heavy load assessment system, bridge data can be effectively divided into three progressively more complex information levels (refer to Table 1 below). The type of data available within each level influences the complexity and degree of accuracy of the heavy load assessment process.

Table 1

Bridge Information Levels

Information Level	Available Data
Level 1 Geometry	- design vehicle(s) - number of spans - span length(s) - structural articulation - minimum width - overhead clearance (if applicable) - underbridge clearance (if applicable)
Level 2 Stiffness	- as per data in Level 1 together with - *structural materials and dimensions* - *longitudinal girder spacings*
Level 3 Strength	- as per data in Levels 1 and 2 together with - *bridge structural capacities*

2.2 Bridge Load Rating and Load Posting

Bridge load rating is the process of determining the live load capacity of the structure. Typically, bridges are load rated for one of the following reasons:

- structural capacity data does not exist;

- the bridge structure has deteriorated to such a degree that its structural capacity may have been affected.

The load rating analysis may be carried out with various degrees of accuracy. It must be pointed out, however, that the additional time invested in performing a detailed analysis will ensure that heavy load permits will be processed more accurately and efficiently. The following data is generally required when undertaking a load rating analysis:

- the load rating vehicle(s);
- a detailed inspection report of the bridge structural condition;
- size and spacing of structural members;
- structural materials and their strength; and
- details of the structural connections.

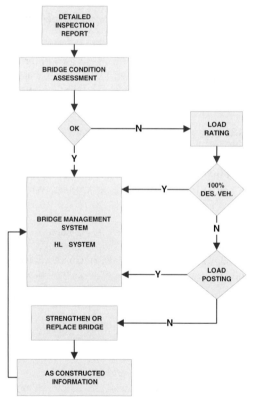

In most cases load rating of the bridge superstructure may be sufficient to determine its residual live load capacity.

The bridge load rating procedure forms a vital part of a Road Authority's global bridge management system (Figure 1). However, its success is very much dependent on visual and detailed inspections of the bridge stock being carried out on a regular basis and the information accurately stored and maintained in the road and bridge database.

Inspection reports must be reviewed by an experienced engineer who will generally determine whether a load rating analysis is warranted. This engineer must also be authorised to impose a live load limit on the bridge if the results of the load rating analysis indicate that the structure's live load capacity is below the legal limits set by the appropriate State Regulatory Authority.

Figure 1 – Bridge Assessment Process

2.3 Heavy and Oversize Vehicle Load Assessments

In Australia, heavy and oversize vehicle load assessments are carried out on a permit basis. Any vehicle that exceeds the legally prescribed limits is required to operate under either a *Period* or a *Single Trip Permit*. Each is different and therefore requires a different approach and methodology.

A *Single Trip Permit* is issued for a single movement of a vehicle on a specific route. This type of permit makes up the bulk of permits assessed by an Authority's heavy loads personnel.

Period Permits are issued by the Authority for the movement of a number of different types of licensed standard width vehicles, including road trains, B-doubles, mobile cranes and so on, with axle loads exceeding the legal limits. These permits are issued for a set period of time and are generally for specific routes or for a limited geographic area of operation.

In any country with an extensive and well-developed road network, processing of heavy load permits may ultimately become overwhelming unless a powerful, user friendly and inexpensive computerised aid can be found to assist with the engineering aspects of the process. In Australia this tool has been ***HLR*** – a PC based heavy load analysis system for bridges. It combines the benefits of using computers for the rapid analysis of structures and storage of data, with the option of exercising engineering judgement when determining the adequacy and sufficiency of the structure to carry the imposed heavy loads.

3. HLR – A PC BASED HEAVY LOAD BRIDGE ANALYSIS SYSTEM

HLR had its genesis over a decade ago, when the first simple DOS program based on working stress techniques was developed by the Bridge Section of the South Australian Department of Transport. With the advent of limit state philosophy and its inclusion into the Austroads 1992 Bridge Design Code [1], the shortcomings of this simplistic approach were recognised and a decision was made to expand and modernise the entire system.

The Windows 95/98/NT platform was chosen due to the low capital cost of PC-based computing hardware and operating systems, the ease of program development, testing, debugging and modification, and the universal familiarity with the PC desktop environment. The program itself consists of a number of fully interactive and integrated modules that are described in some detail below.

3.1 Route Creation

Prior to entering heavy load route data into HLR, the road network must first be described in terms of a series of links, or road sections, joined by nodes. Links are created by specifying a node number at each end of the required section of road. Link data contains the number of structures in the link and the identification number of each structure in the link. This identification number is unique and is given to the bridge by the relevant Authority. Heavy load routes may be created in one of two ways - either by specifying the individual links making up the route; or by specifying whole or part sections of routes that have previously been defined and saved to the route data base.

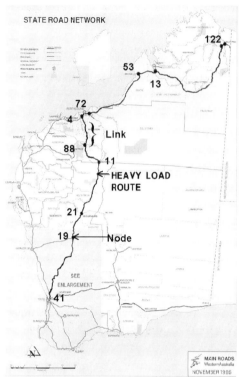

Figure 2 – Typical Example of Heavy Load Route

HLR has been designed in a manner which allows the heavy load route network to be built up over time. The network may initially consist of a single link with a single structure on that link. It can then be expanded as additional structure data becomes available.

Importantly, links need not initially have any structures assigned to them. This facility allows the framework for a heavy load network to be designed in terms of its component links and nodes, and structure data added later, as and when it becomes available.

For example, a substantial part of Western Australia's main road network is currently covered by the HLR system – approximately 500 road links containing more than 1250 bridges (and continually expanding). The system has become a powerful tool in the Main Roads WA heavy load assessment process. A typical example of one of Western Australia's heavy load routes is shown in Figure 2. It is approximately 3000 km long and contains more than a hundred bridges of various structural types.

3.2 Structure Data

Data requirements for each structure may be divided into three broad categories: information relating to identification and analysis control; geometry and structural data; and capacities (Level 1, 2 and 3 as defined in Table 1). Identification data includes the structure name/number, location information, the type of carriageway supported, the node associated with the "forward" direction of the link on which the structure lies and the type of restriction, (if any), to be imposed on the structure. This last item acts as a general purpose control filter, allowing the user to define a number of widely different conditions to be applied during the analysis phase e.g., whether HLR should:

- bypass the analysis of the structure altogether (e.g. the structure is undergoing repairs);
- perform only height/width restriction checks (e.g. for overpass structures);
- force the vehicle to travel at a predetermined speed; or
- prohibit access onto the structure altogether.

Geometry and structural data include structure type (bridge, culvert, pipe etc); girder or deck type (beam, truss, slab, etc); structural articulation (simply supported or continuous); design vehicle (e.g. HS20, T44 etc); span configuration, span lengths and girder spacings (if any); and moment of inertia details. Span data is entered in terms of span "groups". This reduces the amount of input information for simply supported structures consisting of multiple spans with identical properties.

Live load capacity data need only be provided if available. If girder spacings are specified then capacities of individual girders must be provided. If girder spacing is not given, capacities are assumed to apply to the whole cross-section. Moment capacities may be entered as required and, if available, for the:

- "working stress" level;
- 40% overstress level; and
- ultimate level.

This constitutes one the major advantages of HLR - its ability to operate within both Working Stress and Limit State methodologies.

3.3 Heavy Load Vehicle Data
Heavy load vehicles are specified in terms of their axle spacings, axle loads and overall axle width. Any number of axles may be specified and each axle can have a different load. The overall vehicle width and height may also be entered for clearance checking purposes. Once all data has been entered a scaled representation of the vehicle axle layout can be displayed for visual verification.

If required, the vehicle may be saved to the data base for future use. When next loaded from the data base, vehicle data can be edited by adding, deleting or changing axle information.

3.4 Overview of the Data Entry & Analysis Process
The normal procedure for processing a standard heavy load application typically involves the following steps (generally in the order given):

- enter relevant job identification data;
- enter vehicle data or select a vehicle from the data base and modify it if necessary;
- select a standard heavy load route or create a new route;
- select required analysis and output control options;
- perform the analysis; and
- review the results and print reports.

Prior to beginning the analysis the program allows a number of options to be specified:

- the types of bridges to be analysed (continuous, simply supported or all structures);
- the direction of travel along the route (in the forward or reverse directions); and
- the ultimate limit state strength factor (LS_f) and/or the working stress factor (F_w).

Since the ultimate strength factor is prescribed by the user prior to the analysis, scope exists to use a reduced factor for non-load carrying vehicles, (eg. cranes), where the possibility of overloading is reduced. Similarly, reduced ultimate strength load factors can be used when the field-weighing of heavy load platforms accurately confirms axle loads.

3.5 Details of the Analysis Process
A flow chart of the analysis process is given in Appendix A. Each structure along the designated route is assessed in turn and the results written to a log file. A number of checks are first performed to detect major non-structural restrictions. These include height or width

restrictions, absolute restrictions imposed by the asset manager and any other condition that may preclude the specified vehicle from crossing a particular structure, thereby obviating the need for further analysis. Warnings are also issued if link data does not exist or if vital structure data is incomplete or missing.

If the structure passes the above checks, the maximum moments induced by the heavy load vehicle, M_{hl}, are then calculated and compared with the live load capacities, M_c, as stored in the structure data base. The comparison is done for three different travel conditions viz:

- unrestricted vehicle travel speed and position on the bridge deck;
- travel speed only restricted (e.g. to 10 kph);
- both travel speed and vehicle position restricted (e.g. to 5 kph down the centreline of the bridge).

Shear force and support reaction capacity checks are only performed if these capacities are present in the structure data file. They are treated in the same way as moment capacities i.e. shears and reactions induced by the heavy vehicle are factored with the appropriate impact, distribution, ultimate and/or working stress factors to give a set of equivalent beam shears or reactions that are then compared to corresponding capacities from the structure data base.

The analysis method and the factors actually used in deriving the heavy load vehicle effects are dependent on whether moment capacities for the structure are known and whether the structure is simply supported or continuous.

3.5.1 Simply Supported Structures
Simply supported structures are checked for sufficiency by comparing moments at a single section within the maximum midspan region. All span groups are checked if more than one group has been specified for the structure. If moment capacities are not available, an empirical moment ratio method based on the original design standard for the structure is used (refer to Clause 3.5.4 and Appendix B for a description of the Ratio Method). Shear and reaction checks are performed in the same way.

3.5.2 Continuous Structures
During the analysis of a continuous structure an envelope is created of maximum sagging and hogging moments and shears at tenth points along each span of the entire bridge. Moments at every section are factored in accordance with the equations specified in Clause 3.5, then compared to the capacities at those sections for which non-zero values exist. All spans in the structure are analysed, irrespective of the nature and type of symmetry. Although results are calculated at tenth points along each span the heavy load vehicle is moved along the structure in increments of 1/20 of the span to ensure that an accurate envelope is obtained.

3.5.3 Moment Capacities are Available
Two cases are differentiated, depending on the type of superstructure being analysed. If girders are present, (effective girder spacing $G_{spacing} > 0$), HLR assumes that the capacity check is based on individual girder moments. Otherwise the check is performed on the full bridge deck cross-section ($G_{spacing} = 0$).

(a) **Girder Spacing Specified ($G_{spacing} > 0$)**

The maximum heavy load moment, M_{hl}, is calculated for each span then factored to produce a set of equivalent girder moments, (M_1, M_2, M_3) representing three possible travel conditions viz:

$$M_1 = M_{hl} * (1+DLA_1) * LS_f * DF_1$$
$$M_2 = M_{hl} * (1+DLA_2) * LS_f * DF_2$$
$$M_3 = M_{hl} * (1+DLA_3) * LS_f * DF_3$$

LS_f represents the ultimate strength (Limit State) factor. It is entered by the user prior to the analysis and is set to *1.0* for working stress checks. It may be adjusted to reflect various levels of control the user wishes to impose on specific heavy load movements.

DF_1, DF_2 and DF_3 are distribution factors that reflect various vehicle positions on the bridge deck while DLA_1, DLA_2, DLA_3 represent dynamic load allowances (impact factors) at the three possible travel speeds (refer also to Section 3.7 for details). Default values for all of these factors may be changed by the user to suit local requirements.

M_1 assumes that neither speed nor positional restrictions will be imposed on the vehicle. The girder distribution factor, DF_1, is based on two or more design lanes, each three metres wide. DLA_1 is a dynamic load allowance to account for the dynamic effect of vehicular loading at unrestricted travel speed. Typical values for DF_1 used by the South Australian Department of Transport (TSA), for example, are based on the empirical relationship $0.5*G_{spacing}/1.7$, while DLA_1 is based on the NAASRA, Bridge Design Specification (1976)[2].

M_2 assumes the vehicle speed is restricted to reduce the dynamic load allowance, DLA_2, to a minimum. A typical value for DLA_2 used by TSA is 0.05 (with the proviso that vehicle speed is reduced to 10 kph). Restrictions on the vehicle's position on the deck can be controlled by using an appropriate value for DF_2. (The value adopted by TSA, for example, is identical to DF_1).

M_3 assumes the vehicle speed is restricted to some practical minimum and the vehicle is required to travel in the most favourable position on the bridge deck. At this speed the dynamic effect, DLA_3, is generally considered to be negligible and is usually set to zero. Typical values for DF_3 used by TSA are based on the empirical relationship[2] $0.5*G_{spacing}/2.1$.

The calculated girder moments M_1, M_2, M_3 are compared to the girder moment capacity, M_c, and the appropriate travel restriction is determined in accordance with the following criteria:

- If $Mc > M_1$ No travel restrictions apply to the vehicle
- If $M_2 < Mc < M_1$ Vehicle travel speed restricted (e.g. to 10 kph)
- If $M_3 < Mc < M_2$ Vehicle travel speed restricted. Vehicle must travel in the most favourable position on the deck (e.g. 5 kph down the centreline)
- If $Mc < M_3$ The structure is assumed to be overloaded - no travel is permitted

HLR is able to produce a comprehensive series of reports and contains a preview feature that allows all restricted structures to be examined and changes to be made to any restrictions flagged by the program.

(b) No Girder Spacing ($G_{spacing} = 0$)

This case normally applies to structures that cannot be modelled as girders (such as wide box girders with only a few cells), or for structures where girder data is not yet available. The analytical procedure is essentially the same as that described in Clause 3.5.3(a), the only difference being that the calculation of the maximum moment (M_{hl}) is based on the entire deck cross-section, i.e. all distribution factors DF_1, DF_2 and DF_3 are set to unity.

3.5.4 Moment Capacities Not Available

If capacities are not present in the structure data file, the analysis is based on the comparison of empirically derived moment ratios. The load ratio, R_i, compares the maximum moment generated by the heavy load vehicle with that of the original design vehicle. A number of standard design vehicles have been incorporated into HLR and others can easily be added. Ratios R_1, R_2, R_3 are calculated for the three basic travel conditions described in clause 3.5.3(a) then compared to the specified working stress factor. A detailed explanation of the ratio method is given in Appendix B.

3.6 Analysis Based on Moment Multiplier Factors

A facility has also been provided withing the program to perform the analysis of non-standard structures, such as culverts and portal frames. The maximum moment due to the heavy load vehicle is calculated assuming the structure is a simply supported beam. The factored moments M_1, M_2, M_3 are then multiplied by a factor, K, and the resulting values compared to the structure capacity, M_c. Two *K-factors* may be specified for each span group in the structure - to check both positive and negative moments. In the case of a culvert, for example, this would permit a check to be made of the positive top-slab moment as well as the end corner moment.

3.7 Dynamic Load Allowance (*DLA*)

HLR allows a DLA factor to be specified for each span in the structure. If present, this value will be used to determine the heavy load vehicle effects as described in clause 3.5 above. If a DLA factor is not given, HLR will calculate one using an empirical method. A lower-bound structure frequency value of 0.9*120/Span and an upper value of 1.1*120/Span are used to determine two DLA factors (based on a user-specified frequency-DLA relationship). The largest of the two values is assumed to be DLA_1, the factor to be applied to the heavy load vehicle. DLA_2 and DLA_3 are selected by the user to control the travel speed of the vehicle.

4. BENEFITS OF THE SEMI-AUTOMATED ASSESSMENT SYSTEM

To date the experience of Australian Road Authorities using the HLR semi-automated assessment system has been extremely positive. Significant productivity increases have been achieved with a commensurate decrease in errors and an increase in the consistency and reliability of assessments. It is now possible to assess a heavy load movement over a moderately long route in an hour, a task that would otherwise have taken 2-3 days to perform if the route was checked manually, or 5-6 hours at best if the check was performed using a combination of computer-based calculations, spread-sheets, charts and an experienced analyst.

Another very useful benefit flowing from the system has been the identification of structures that may, in fact, be "weak links" in the heavy haulage road network. This enables the Authorities' relatively scarce engineering resources (and funds) to be focussed on these high-risk bridges, generally by undertaking a more detailed analysis of the structures or by load testing them to accurately determine their capacity.

Other, less tangible benefits, have included a considerable increase in client / operator satisfaction (due to much quicker assessment turn-around times) and the flexibility the system offers to the Authorities in using less experienced personnel to perform the initial computerised check. Only structures that have been assigned a restriction by the system may need to be assessed in more detail by an experienced engineer.

5. FUTURE DEVELOPMENT

The road network and bridge stock are continually changing in all countries. New roads and bridges are constantly being added and existing bridges are either decommissioned or upgraded. One of the easiest ways to manage and track this vast amount of information is via a GIS based electronic road network mapping system. Most modern road authorities now have access to GIS facilities, a situation that creates an opportunity of making HLR even more efficient by incorporating it into these systems.

Linking the two would allow structure data within HLR to be automatically compared to bridge and road data within the GIS database and update it where necessary. Being a graphics-oriented medium, the GIS system would provide an excellent graphical facility for creating heavy haulage routes and for performing overall maintenance of the bridge database. It would allow bridges to be picked directly from on-screen road network maps and allow all HLR-specific data to be edited and updated.

Incorporating HLR into a GIS system may ultimately allow centralisation of the heavy haulage permit processing system, either within existing government agencies or as an out-sourced business entity.

6. CONCLUSION

The process described in this paper represents the most advanced heavy haulage bridge assessment system used in Australia today. It is founded on a balanced combination of engineering expertise and judgment and a powerful computer-aided analysis tool. The process allows a heavy load route assessment system to be quickly, easily and cost effectively developed from a zero base level to a state where it can bring substantial economic benefits to any country with a developed, or developing, road network.

ACKNOWLEDGEMENTS

The authors would like to thank Transport South Australia (TSA), Main Roads Western Australia (MRWA) and Halpern Glick Maunsell Pty Ltd for the support provided in the course of preparation of this paper. Assistance provided by the TSA and MRWA Heavy Load Section staff is also greatly appreciated.

REFERENCES
1. AUSTROADS, Bridge Design Code (1992, 1998), Association of Australian and New Zealand Road Transport and Traffic Authorities
2. NAASRA, Bridge Design Specification (1976). National Association Of State Road Authorities

450 ASSESSMENT

APPENDIX A: HLR Assessment Methodology

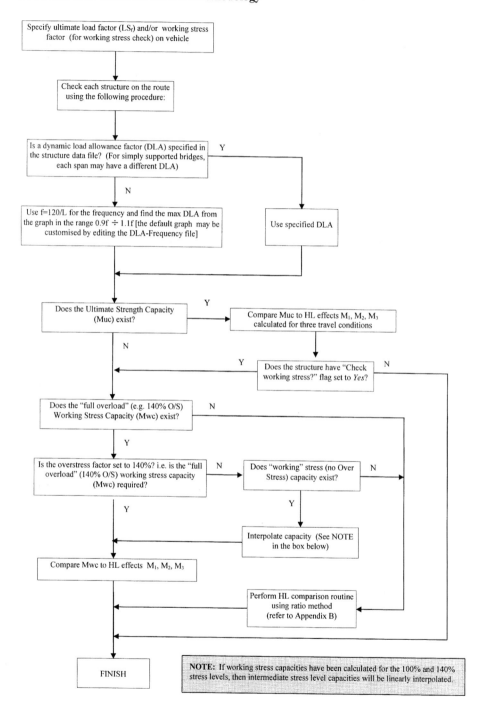

APPENDIX B: Moment Ratio Method

The conventional moment ratio method is an attempt to cater for situations where bridge capacity data is unavailable. It is an empirical technique that allows the maximum moment generated by a heavy load vehicle to be compared with the moment generated by the original design vehicle. Inherent in this method is the basic assumption that this latter moment notionally represents the section capacity. However, for older bridges in particular, this simple live load ratio approach takes no account of the effect of dead load (DL) and consequently can be unnecessarily conservative.

The methodology has therefore been modified in HLR to take DL into account. It may be summarised as follows:

(1) When determining the heavy load vehicle effect, (LL_{hl}), use the Dynamic Load Allowance factor (DLA_{LL}) specified in the structure data file or, if none exists, the method described in clause 3.7 of this paper. For the design vehicle effect, (LL_{design}), use the design Dynamic Load Allowance, DLA_{design}, described in clause 3.7 or, for pre-1992 structures, a default design DLA of $16/(L+40)$ [see reference 2 for details].

(2) In order to obtain a more accurate indication of overloading, modify the conventional LL_{hl}/LL_{design} ratio to incorporate the effect of dead load (DL). The modified ratio is given by:

$$\frac{DL + LL_{hl} \times (1 + DLA_{LL})}{DL + LL_{design} \times (1 + DLA_{design})}$$

(3) A recent study of the load capacity of South Australian bridges carried out by Transport SA showed that for pre-1976 bridges the relationship between the dead load effect (DL) in a structure and its design live load effect (LL_{design}) can be approximated using the expression:

$$\frac{DL}{LL_{design} \times (1 + DLA_{design})} = \frac{Span}{\Phi}$$

where Φ is a statistically derived 95% probability factor determined for different design standards and structure types. For example, Φ factors compiled by TSA for South Australian beam or girder bridges are:

Design Vehicle	Φ	Comment
HLGD	27	Design vehicle for pre 1950 structures
HS20	42	Design vehicle for 1950 - 1976 structures
T44	200	Current Australian design vehicle (this value is conservative and has not yet been verified)

(4) Transposing the expression in clause (3) above gives:

$$DL = \frac{Span \times LL_{design} \times (1 + DLA_{design})}{\Phi}$$

(5) Substituting for DL in expression (2) produces the following relationship for *unrestricted* travel:

$$R_1 = \frac{\frac{Span}{\Phi} \times \{LL_{design} \times (1 + DLA_{design})\} + \{LL_{hl} \times (1 + DLA_{LL})\}}{(\frac{Span}{\Phi} + 1) \times \{LL_{design} \times (1 + DLA_{design})\}}$$

(6) For travel at restricted speed, but no restriction imposed on the vehicle position on the deck, the expression becomes:

$$R_2 = \frac{\frac{Span}{\Phi} \times \{LL_{design} \times (1 + DLA_{design})\} + \{LL_{hl} \times (1 + DLA_2)\}}{(\frac{Span}{\Phi} + 1) \times \{LL_{design} \times (1 + DLA_{design})\}}$$

where DLA_2 is a reduced dynamic load allowance (refer to Clause 3.5.3(a))

(7) For travel restricted to minimum speed (5kph, for example) and the vehicle required to travel in its most favourable position on the bridge deck, the expression becomes:

$$R_3 = \frac{\frac{Span}{\Phi} \times \{LL_{design} \times (1 + DLA_{design})\} + [\{LL_{hl} \times (1 + DLA_3)\} \times DF_2 / DF_1]}{(\frac{Span}{\Phi} + 1) \times \{LL_{design} \times (1 + DLA_{design})\}}$$

At this highly restricted speed the dynamic load allowance, DLA_3, may be set to zero. For structures with beams and girders the general expression is modified by the ratio DF_2/DF_1 to account for dedicated single-lane travel, where:

DF_1 = *Empirically derived girder factor for single lane analysis (e.g 2.1)²*
DF_2 = *Empirically derived girder factor for multiple lane analysis (e.g 1.7)²*

For structures with no girders, (such as slabs and box girders), DF_1 and DF_2 are set to unity.

(8) Finally, compare R_1, R_2 and R_3 to the specified working stress factor, F_w, and determine the governing travel condition viz:

- If $R_1 < F_w$ No travel restrictions apply
- If $R_2 < F_w < R_1$ Vehicle speed is restricted, but no restriction is applied to the vehicle position on the deck
- If $R_3 < F_w < R_2$ Vehicle travel speed and position are both restricted
- If $R_3 > F_w$ The structure may be overstressed - no travel is permitted

The use of reliability-based assessment techniques for bridge management

DR. ROBERT J. LARK,
Cardiff School of Engineering,
Cardiff University, U.K.

DR. KATJA D. FLAIG,
WS Atkins Consultants Ltd.,
Epsom, U.K.

ABSTRACT

This paper reviews the use of reliability techniques for assessing the structural adequacy of bridges and highlights how the output of such an analysis can both inform and guide the user in their investigation of the structure under consideration. Three different ways of interpreting the value of reliability index that is obtained are identified and the feasibility of using each of these to prioritise repair and strengthening works is examined. An average acceptable value (AAV) of reliability index for different bridge types and failure modes is proposed and it is shown that in some cases priorities can be identified directly by comparing reliability indices, sometimes a measure of the significance of the reliability index is needed, while in other cases priorities are still not obvious and further investigation is required. To achieve this, the principles of risk management are explored and it is shown that where two or more structures have similar reliabilities such a procedure does offer a feasible approach.

It is concluded that, while there may be some dispute as to the absolute value of reliability based assessment procedures, when combined with other information that is a measure of the importance, value and criticality of the structure or element under consideration, such procedures can help bridge managers prioritise repair and strengthening works. The advantage of the proposed procedures is that they are able to identify priorities in a more rational way and one that explicitly reflects the safety of the structure.

INTRODUCTION

The process of assessment is of crucial importance for maintaining highway bridges in a safe and serviceable condition. In the past this process has used a deterministic approach based on the rules that were originally developed for design together with appropriate partial safety factors to ensure a reasonably consistent level of safety across a given bridge stock. The techniques employed by such an approach were, by implication, conservative, since 'they were developed for situations where a safe answer is more important than a realistic one and where economy comes more from ease of construction than from realistic strength assessment' (Jackson, 1993). Their objective was to provide rules such that if a structure complies with them it can be considered both safe and serviceable. However, the reverse of this is not true. A structure that does not comply with such rules is not necessarily unsafe or unserviceable. These deterministic rules provide only a lower bound to the risk of failure.

In order to ensure a reasonably consistent level of safety across a bridge stock these rules also have to be applicable to a wide range of structures and failure modes and be adequate throughout the life of the structure to which they are applied. Again this means that they are inevitably conservative and because of the natural variability in material, structural and system behaviour, this level of conservatism is also found to vary considerably from bridge to bridge. Using a deterministic approach to assessment does not allow this level of conservatism to be evaluated. The only result that can be obtained is that the structure either passes or fails the assessment.

This lack of a quantitative measure of the likelihood of failure of a structure is also a problem when it comes to trying to ensure that the limited funds that are available for repairing and refurbishing structures are targeted at those that are in most need. When managing a bridge stock, because of such budgetary constraints, a decision often has to be reached as to what work should be tackled first when the structural capacity of more than one of the stock is considered to be substandard. A key concern of any bridge manager is safety, however deterministic assessments do not enable the safety of a structure to be quantified and therefore it is not possible to differentiate between the risk of failure of different structures. Seen from the assessment engineer's point of view, condemning a structure that has failed a deterministic assessment, even if only by a very small margin, is the easy way out. For the bridge manager, however, the decision is much less straightforward. Once condemned, the bridge has to be either repaired or strengthened for which funds have to be obtained, or it has to be taken out of service, the consequences of which are likely to be either very unpalatable or totally unacceptable. What the bridge manager requires is a more rational and reliable evaluation of structural adequacy and safety on which to base his decisions. Support is required to highlight those structures that are in most need of attention and to provide confidence in the continuing use of the rest.

To address these issues assessment techniques based on the direct application of structural reliability theory have been developed (Shetty et. al., 2000). Such procedures enable both individual assessments to be more bridge specific and the levels of safety of different structures within a bridge stock to be compared on a rational basis. This is achieved by recognising that, other than due to gross errors, the variability in the perceived safety of structures is due to the uncertainty in the design or assessment procedures that are used and the 'normal', and therefore quantifiable, variability in materials and workmanship etc. In a reliability-based, probabilistic assessment procedure the uncertainty of each parameter that influences the failure mode under consideration is therefore modelled using an appropriate probability distribution function. The reliability analysis then gives an estimate of the probability of failure or reliability index of the element or structure under consideration.

This process is described in more detail elsewhere (Flaig, 1999). What is of interest here is how the output of such an analysis can both inform and guide the user in their investigation of the structure under consideration. Three different ways of interpreting the value of reliability index that is obtained are identified and the feasibility of using each of these to prioritise repair and strengthening works is examined.

PROBABILISTIC ASSESSMENTS

A variety of outputs can be obtained from a reliability-based, probabilistic assessment. The main outcome is the probability of failure, or an equivalent reliability index, but data is also available on information relating to the reliability analysis itself, such as the sensitivities and elasticities of the analysis. Figure 1 identifies three primary ways in which these results obtained from the assessment of a structure using a reliability analysis can be used.

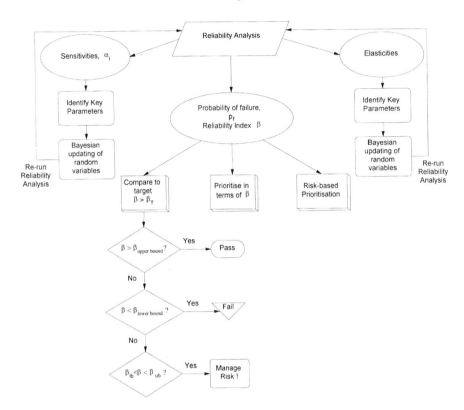

Figure 1 : Use of reliability analysis results in support of bridge management decisions

Having obtained the reliability index, there are also three secondary ways in which this value can be used. When considering the adequacy of a single structure it can be compared directly with some target value for the structural type and failure mode under consideration. Alternatively if a number of structures are being considered and the aim is to establish the relative priority of each structure, they can be ranked in terms of their respective reliabilities, using the ratio between the calculated reliability and the target value to compare different material types and failure modes. Finally, where there is more than one structure involved, they could also be ranked in order of risk, which can be established by using the probability of failure of each structure instead of the reliability index. These three approaches will be considered in more detail later in the paper.

If the calculated reliability, β is below the target level, β_T, provided that a standard procedure has been adopted and that default distributions were used to model the random variables, the sensitivities and elasticities can be used to determine which variables most influence the structure's reliability. It is these variables for which it would then be most appropriate to gather additional, structure specific information and, although in some cases structure specific values might be below the default values in which case the reliability would be reduced, an improvement in the confidence level of the data that is used can improve the reliability of the result that is obtained. Sensitivities are a measure of how important the coefficient of variation of a particular parameter is to the reliability of the structure. Additional structure specific data will often help in reducing the coefficient of variation associated with the default distributions. Elasticities on the other hand, indicate the significance of the mean value to the reliability. That is to say, if the structure specific mean value of a certain variable is higher or lower than the default value of that variable, the reliability will go up or down by a given amount which can be calculated from the respective elasticity.

AVERAGE ACCEPTABLE RELIABILITY

By accounting for structure specific information and modelling all the related uncertainties, the reliability of a particular structure can be calculated. When this reliability is then compared to acceptable levels conclusions can be made about the level of safety of the structure under consideration. The problem is that, currently, such acceptable levels are not explicitly defined. For establishing these acceptable levels three basic approaches can be found in the literature (Ditlevsen and Madsen, 1996):

i) Socially acceptable levels of risk derived from historical data
ii) Economic optimisation, in which risk is balanced against the costs incurred through either excessive or insufficient safety
iii) Calibration of acceptable values against existing codes

The major shortcoming of the first approach is that historical failure rates are so limited that a direct calibration is not possible and because they are not directly related to the 'nominal' failure probabilities computed from reliability analyses. Likewise, although economic optimisation can be used to establish an economically feasible range of reliability, the optimal level of reliability of a structure is highly dependent on a number of issues, many of which are not easily expressed in monetary terms. Economic optimisation is therefore not suitable for the derivation of structure specific acceptable levels of reliability.

Calibration to existing codes, although not free from difficulties, seems to be the most appropriate approach for establishing acceptable reliability levels. To date this approach has been used for the derivation of partial safety factors for limit-state design codes in order to make them compatible with the older permissible stress codes that they replaced. In this approach the reliability level implicit in existing codes is first quantified through reliability analysis of sections that exactly fulfil the existing code requirements. This code implicit level can then be taken as an acceptable level for the reliability based design or assessment of structures in the future. Adopting this approach, code implicit reliability levels for a large number of different bridge types and failure modes have been obtained and from the results average acceptable values of reliability index are identified as shown in Table 1.

The scatter in the values that have been obtained from the analysis of sections designed in accordance with current assessment codes substantiates the view that these codes do not ensure a consistent level of reliability across the bridge stock. The results also confirm that when an element fails a deterministic assessment it does not necessarily mean that the element is unsafe. The justification for adopting an average value is that it represents the most likely value for a given bridge type and failure mode, and thus, when combined with a measure of the variability of the code implicit reliability (i.e. the range), it best represents that which might be expected if a structure was designed exactly in accordance with the code. It is also anticipated that although further examples of the bridge types and failure modes reported here could be studied, they would not significantly alter the average values that have been adopted.

Table 1 : Proposed Average Acceptable Annual Reliabilities and Variability Ranges

Bridge Type	FAILURE MODE	Code implicit Reliability	Average Acceptable Value β_a	Variability	Variability Range
RC Slab Bridge	Bending: simply-supported continuous-sagging continuous-hogging	5.7 - 6.6 5.3 - 6.2 5.9 - 6.3	5.9	0.4	5.5 - 6.3
	Shear: without stirrups with stirrups	5.3 - 6.9 5.8 - 7.1	6.2	0.4	5.8 - 6.6
Pre-stressed Concrete Bridge	Bending: beam & in-fill slab beam & comp. Slab	7.5 - 8.0 6.7 - 7.1	7.3	0.4	6.9 - 7.7
	Shear: beam & in-fill slab beam & comp. Slab	5.9 - 6.6 7.0 - 7.2	6.6	0.6	6.0 - 7.2
Steel / RC Composite Bridge	Bending: simply-supported continuous-sagging max. bending + shear	4.6 - 5.6 3.9 - 4.5 5.5 – 5.8	4.9	0.7	4.2 - 5.6
	Shear: pure shear max. shear + bending	3.7 - 5.3 4.5 - 5.4	4.7	0.6	4.1 - 5.3

TARGET RELIABILITY

As identified above, having obtained the reliability index, there are then three ways in which this value can be used. When considering the adequacy of a single structure it can be compared directly with some target value for the structural type and failure mode under consideration.

One problem with this approach is that there is still some unease in the engineering community concerning the interpretation of results of reliability analyses, in particular in relation to the probability of failure, which is typically expressed as the reliability index. This is due in part to the difficulty in appreciating the significance of these parameters. They are not absolute values (i.e. a failure probability of 10^{-5} does not mean that one bridge in every 100,000 is going to fail every year) and they can only be compared with target values that have been calculated in exactly the same way, using the same resistance and load models (Flaig and Lark, 1999).

Nevertheless, assuming that this is the case, the calculated reliability of a bridge element can be compared with a target value such as the average acceptable value for the specific bridge type and failure mode under consideration that is given in Table 1. While it is clear that there is no such thing as absolute safety, this comparison does allow a judgement of the relative safety of the bridge in relation to the code inherent safety level to be made and, if expressed as a ratio, can be defined as a relative safety ratio (S_R). The underlying assumption being that the code inherent safety level, as a result of past experience and good practice, is widely accepted as being safe enough.

If this ratio is clearly well above or below unity then it is clear that the structure under consideration is either significantly safer or less safe than a structure designed exactly in accordance with the current codes of practice. The problem that arises, particularly given the variability that is inherent in the target values, is the interpretation of the significance of a ratio that is only just above or below unity.

One approach to dealing with this problem could be to try and refine the calculation of the structure's reliability. As noted earlier, one way of identifying the parameters that have the greatest influence on the structure's reliability is to examine the parameter sensitivities that are obtained from the reliability analysis. By collecting further data relating to these key parameters and applying Bayesian updating to the default distributions, structure specific probabilistic models can be obtained and the reliability analysis re-run. The use of Bayesian updating techniques is well documented in the literature (Melchers, 1999; Ang and Tang, 1975) and the use of Bayesian updating in bridge assessment has recently been demonstrated by Sterritt and Chryssanthopoulos (1999) and Das and Chryssanthopoulos (1999).

However, while such an approach can improve the nominal reliability of a structure and, with the help of structure specific data and a knowledge of its past performance, may be sufficient to provide the necessary confidence in its safety, it is possible that this reliability is still within the variability of the target value. A further option is therefore to simply acknowledge that the reliability is borderline and to assess and manage the risk associated with it. To do this the following criteria should be considered:

i) The reason for the inadequacy. For example, if the element is showing signs of deterioration and the process is still ongoing, action will have to be taken and it is simply a question of how long this action can be delayed. However, if the inadequacy is due to a change in code requirements, test data from the structure or similar structures may be sufficient to prove its adequacy, as described above.
ii) The type of failure mechanism (i.e. ductile or brittle).
iii) Warning of failure. For many failures early warning signs can be identified during routine inspections or by continuous monitoring, although care must be taken to ensure that these signs can reliably be detected in time.
iv) Redundancy and reserve strength of the structure. If a structure is redundant, the failure of a single element is likely to result only in local damage because the loads in the failed element can be redistributed to the rest of the structure.
v) Consequences of failure.

At present the only way of incorporating these issues is by the application of engineering judgement but, as will be shown later, a framework does exist for assessing risk and research is ongoing to identify how these issues can be quantified so that they can be explicitly accounted for in such an approach.

To summarise, when using a standard approach to the calculation of the reliability of a structure the result can be compared to a target value calculated in the same way. If this reliability is then below the lower limit of the variability range of the target value it is reasonable to conclude that the structure is less safe than expected of a structure designed in accordance with the codes of practice for that bridge stock. Likewise, if it is above the upper limit of the variability range of the target value the structure can be said to be safe. If, however, the calculated reliability is within the variability range of the target value no such clear-cut observation can be made and all that can be done is that the risk of failure is critically reviewed and managed.

RELIABILITY BASED PRIORITISATION

As identified earlier, one of the main problems with adopting reliability based assessment procedures is the unease that exists over the interpretation of the results that are obtained. However, by adopting a consistent and rational approach to dealing with the uncertainties that exist in the assessment process, it is surely reasonable to assume that this consistency will allow the reliability of one structure to be compared to that of another.

For structures of the same material type and failure mode this would seem to be no problem, but for different material types and failure modes the code implicit reliabilities are themselves very different and so a direct comparison is not immediately possible. To overcome this, the use of a relative safety ratio (S_R = bridge specific reliability / average acceptable reliability) reduces all values to a common base and by further quantifying the natural variability of the reliability of each material type and failure mode, for example as the ratio of the lower bound reliability or the variability range to the average acceptable value for that given material type and failure mode, a measure of the significance of the relative safety ratio can be obtained.

In this way it is possible to compare the reliability of one structure and failure mode with that of another and therefore, for bridge management purposes, to prioritise the work that is required to be undertaken on a number of bridges when they all have reliabilities that are less than ideal. Sometimes these priorities can be established directly and sometimes a measure of the significance of the safety ratio is needed, but in both cases the fact that this can be achieved confirms that reliability based assessment procedures are capable of revealing additional safety information about a structure that is beyond that which could be obtained from a deterministic assessment procedure.

RISK BASED PRIORITISATION

The last way in which the calculated reliability indices can be used is to compare and prioritise a number of structures based on risk rather than reliability. In this context risk is defined as the probability of failure multiplied by the cost of the possible consequences of a failure, i.e.:

$$Risk = Probability\ of\ failure\ *\ Consequences$$

In this case the underlying aim is to provide a constant risk across the whole bridge stock as opposed to a constant reliability. This is appropriate because the latter can be criticised as not being logical in that it provides the same reliability for a major and heavily trafficked motorway bridge as for a local road bridge which carries much less traffic. There are however two sides to this argument. Firstly the user will always expect the same reliability whether it is a small country road bridge that is being crossed or a major motorway. In other words no one wants to feel less safe on smaller bridges. On the other hand, seen from the highway authority's point of view, the consequences of a motorway bridge failing exceed those of a minor road failure by several orders of magnitude. Consider a hypothetical example given by Das (1999). The cost of total collapse of a particular bridge, including road user delay costs, is taken to be £2 million and that resulting from the failure of a major element is £150,000. It is also assumed that the probability of total collapse occurring is 10^{-8} and that for element failure 10^{-3}. The respective risks can therefore be calculated as 0.02 for total collapse and 150 for element failure. The conclusion drawn by Das (1999) is that "structures management must therefore be aimed at eliminating as far as practicable, not only the risk to life safety but also the risk of element failures" that have the potential to cause traffic disruption and other costs. It is agreed that it is a valid exercise to balance the risk of failure of a key element on an important bridge against the total collapse of a minor bridge when considering strengthening priorities. Taking the concept of constant risk further however, it has never been explicitly examined how feasible it really is to implement risk-based prioritisation.

The problem that arises is that, because the probability of failure of most bridge types are relatively similar whereas the consequences associated with a high risk structure such as a motorway bridge are several orders of magnitude greater than those of a minor road bridge, the risk associated with the former will always be greater than that of the latter. Further work is therefore required to identify how the definition of risk can be refined to make the approach more worthwhile.

As a first attempt, it may again be more appropriate to express both the probability of failure and the consequences as ratios instead of as absolute numbers. For the former this can be done by using the relative safety ratio (S_R) defined earlier, while for the consequences it could be taken, for example, as the ratio of the consequences associated with the bridge under consideration to the consequences of the bridge with the 'worst' consequences in the bridge stock. This way the risk score given by:

$$Risk = Relative\ Safety\ Ratio\ *\ Consequence\ Ratio$$

would be calculated from two relative measures and this should make it more robust to differences in the consequences associated with different bridge stocks and more amenable to calibration with respect to the bridge stock under consideration.

CONCLUSIONS

This paper reviews the use of reliability techniques for assessing the structural adequacy of bridges and highlights how the output of such an analysis can both inform and guide the user in their investigation of the structure under consideration.

It is identified that in order to judge whether the calculated reliability of a structure is adequate, benchmark reliabilities are needed. The best way of establishing these is by calibrating them such that they correspond to the reliability that is inherent in structures that have been designed so as to exactly fulfil the existing code requirements. The problem with this approach is that the code-inherent reliability varies not only between different material types but also, for structures of the same type, between structural forms and across spans. This natural variability has to be accounted for when establishing acceptance levels and so an average acceptable reliability is proposed for different bridge types and failure modes, which, together with its anticipated variability, provide an appropriate benchmark.

Having calculated the reliability of an element of a bridge there are various ways in which the value can be used as part of the bridge management process. Firstly, this reliability can be compared with the average acceptable reliability for the bridge type and failure mode under consideration. If the calculated reliability is below the lower limit of the variability range of the average acceptable value it is reasonable to conclude that the structure is less safe than might be expected of a structure designed in accordance with the codes of practice for that bridge stock. Conversely, if it is above the upper limit of the variability range of the average acceptable value the structure can be said to be safe. If, however, the calculated reliability is within the variability range of the average acceptable value the outcome is less clear-cut and the risk of failure must be critically reviewed and then managed.

The second way in which the calculated reliability of a bridge can be used is that a number of structures can be prioritised in terms of their respective reliabilities. For structures of different material types, which consequently have different average acceptable values, a safety ratio that relates the calculated reliability to the average acceptable value of the bridge type under consideration, allows rational comparisons to be made. Using this approach it is found that sometimes priorities can be established directly, sometimes a measure of the significance of the safety ratio is needed while in other cases priorities are still not obvious and further investigation is required.

The third option is therefore to rank structures in terms of risk rather than reliability. This approach takes the probability of failure as well as the consequences of failure into account. However at present, because of a lack of data for the latter, this approach is still found wanting. Further work is still required to identify how the definition of risk can be refined to make the approach more worthwhile and one way in which way this can be done has been suggested.

It is concluded that, while there may be some dispute as to the absolute value of reliability based assessment procedures, when combined with other information that is indicative of the importance, value and criticality of the structure or element under consideration, such procedures can help bridge managers prioritise repair and strengthening works. The advantage of the proposed procedures is that they are able to identify priorities in a more rational way and one that explicitly reflects the safety of the structure.

ACKNOWLEDGEMENTS

The authors would like to thank Dr. Nigel Knowles and Dr. Navil Shetty of W.S. Atkins and Dr. Parag Das from the Highways Agency for their support of the research project of which the work described in this paper formed part. The results presented in Table 1 were obtained with the assistance of the Transportation Engineering Division of W.S. Atkins, whose help is also gratefully acknowledged. The interpretation of these results and the views expressed in this paper are however only those of the authors and do not necessarily represent the views of either W.S. Atkins or the Highways Agency.

REFERENCES

Ang, A.H.S. and Tang, W.H. (1975). *Probability concepts in engineering planning and design, Volume 1 – Basic Principles.* John Wiley, USA.

Das, P.C. (1999). *Development of a comprehensive structures management methodology for the Highways Agency,* in 'Management of highway structures', ed. Das P.C., Thomas Telford Publishing, London, UK.

Das, P.C. and Chryssanthopoulos, M.K. (1999). *Uncertainty analysis of the bridge management process and the significance of updating,* in 'Current and future trends in bridge design, construction and maintenance', ed. Das P.C., Frangopol D.M. and Nowak A.S., Thomas Telford Publishing, London, UK.

Ditlevsen, O. and Madsen, H.O. (1996). *Structural Reliability Methods.* John Wiley, UK.

Flaig, K.D. (1999). *Risk and reliability-based bridge assessment and management techniques,* PhD Thesis, Cardiff University, December 1999, U.K.

Flaig, K.D. and Lark, R.J. (1999). *Integration of reliability-based assessment techniques into an advanced BMS.* Proceedings of International Bridge Management Systems Conference, April 1999, Denver, Colorado, USA.

Jackson, P.A. (1993). *Strength assessment to keep bridges in service,* in 'Bridge Management 2 - Inspection, Maintenance Assessment and Repair', ed. Harding J.E., Parke G.A.R., and Ryall M.J., Thomas Telford Publishing, London, UK.

Melchers, R.E. (1999). *Structural reliability analysis and prediction.* 2nd edition, John Wiley, UK.

Shetty, N.K., Flaig, K.D., Rubakantha, S. and Gibbin, N. (2000). *Levels 3, 4 and 5 methods of assessment for bridges,* Symposium on the management of highway structures, Institution of Civil Engineers, October 2000, London, U.K.

Sterritt, G. and Chryssanthopoulos, M.K. (1999). *Probabilistic limit state modelling of deteriorating RC bridges using a spatial approach,* in 'Current and future trends in bridge design, construction and maintenance', ed. Das P.C., Frangopol D.M. and Nowak A.S., Thomas Telford Publishing, London, UK.

# ASSESSMENT OF FATIGUE DAMAGE IN THE TSING MA BRIDGE UNDER TRAFFIC LOADINGS BY FINITE ELEMENT METHOD

T. H. T. CHAN [1], L. GUO [2], Z. X. LI [2]
[1] Department of Civil and Structural Engineering, The Hong Kong Polytechnic University, Kowloon, Hong Kong
[2] College of Civil Engineering, Southeast University, Nanjing, 210096, PR China

ABSTRACT Steel bridges are vulnerable to fatigue damage caused by traffic loadings. The Tsing Ma Bridge (TMB) in Hong Kong, of 2.2 km total length and a main span of 1377 m, is the longest steel suspension bridge in the world carrying both road and rail traffic. This paper is aimed to study the fatigue damage of the TMB using the Finite Element Method. An effective finite element (FE) model of TMB embodying the properties of almost all the structural members was developed. The verification of the model was carried out with the help of online data. The dynamic response of the TMB under traffic loading was analyzed using the developed FE model. Considering the bridge fatigue damage is local damage, a local damage model is applied in the FE model of the TMB with possible fatigue damage degree and its evolution equation. As the TMB is at its virgin state and the traffic volume on the bridge is relatively light, an update ratio of the traffic volume is also considered in the proposed FE model. With the established FE model, the future response of the Bridge could be simulated and, the damage degree in the Bridge under traffic loading could be assessed too. All of these results provide a basis of estimating the remaining service life of the TMB.

1. Introduction

Fatigue is an important failure mode for steel structures. In fact, 80-90% of failures in steel structures are related to fatigue and fracture [1,2]. The TMB (shown in Figure 1), of 2.2 km total length and a main span of 1377 m, is the longest steel suspension bridge in the world carrying both the road and rail traffic. It is also vulnerable to fatigue damage. Therefore, it is significant to study the fatigue damages in the bridge for its safety assessment.

Fatigue analysis for an existing bridge is dominantly based on the stress analysis to get the distribution of stress in structures. Recently, some important works on fatigue analysis for bridges are found [3, 4]. In these works, the short or medium span bridges were the focus and, the finite element method (FEM) was used to decide the critical locations due to fatigue damage. Field tests were then conducted for assessment of the stress range in these critical locations. However, a long suspension bridge has a sheer size and is often over water that makes access for inspection, instrumentation and testing of them very difficult. Many experimental techniques that have been shown as successful for structural identification of short and medium span bridges cannot simply be scaled-up to long span bridges. All of these make it very difficult to study the fatigue damage in long-suspension bridges with experimental measurement. Correspondingly, numerical simulation is a feasible way to study the fatigue damage in a large practical bridge.

Finite element method (FEM) was recommended by the British standards as a rigorous method for structural fatigue stress analysis. The studies [5, 6] show that FEM has become a highly valuable tool as a basis for an evaluation of the fatigue behavior. The latest work by Lin and Smith [7, 8, 9] described finite element modeling of fatigue crack growth of surface cracked plates and provided an accurate evaluation of fatigue life of the structures. Unfortunately, all of these works are just on the fatigue analysis of some of the elements and/or parts of the structures. However, a long suspension bridge is a very flexible and lightly damped structure. Heavy trucks and railway vehicles running on a long suspension bridge may significantly change the local dynamic behavior and affect the fatigue life of the bridge. A fatigue analysis of a whole bridge is necessary, but there is not any related works found, which therefore becomes a significant problem to be studied.

Conventionally, finite element method is used to conduct the modal analysis of a practical bridge. In this connection, the single-girder beam element model, the double-girder beam model, the triple-girder model, the shell element model and the thin-walled element model have been developed to model a bridge deck [10-14]. All of the finite element models in these works are equivalent models, and could not be used to analyze the stress distribution in the structure.

To analyze the fatigue damage in the TMB, this paper presents a complicated FE model embodying the properties of almost all the structural elements in the bridge. Moreover, in considering that fatigue damage is a local failure mode and is often occurred in the welded regions, the weld details should also be included in the FE model. The verification of the model was carried out with the help of online data measured by the structural health monitoring system permanently set on the bridge. The constructed model could be used to analyze the dynamic response of the bridge under traffic loadings. In thinking of the TMB is at its virgin state and the traffic volume on the bridge is relatively light, an update ratio of the traffic volume is also considered in the proposed FE model. With the established FE model, the future response of the Bridge was simulated and, the damage degree in the Bridge under traffic loading was assessed too. All of these results provide a basis of estimating the remaining service life of the TMB.

2. FE model of the TMB

The TMB is a double deck suspension bridge. It contains about twenty thousand structural members, including longitudinal trusses, cross frames, deck plates, tower beams, main cables and hangers, etc. In recognizing that the conventional modeling procedure for cable supported bridges by approximating the bridge deck as analogous beams or grids is not applicable for accurate fatigue stress analysis, a precise finite element model of the TMB was constructed using the commercial software package ABAQUS in the Department of Civil & Structural Engineering, the Hong Kong Polytechnic University. It is very complicated to establish a FE model of such a large bridge. Not only the structural elements should be modeled as related appropriate element type, e.g. the longitudinal trusses in the bridge deck simulated as space beam elements with section details, such as box-section, I-section, the pavement represented as shell elements, the hangers modeled as space beam elements with circular cross section, but the connected conditions between the structural elements and initial conditions of relative support should also be considered carefully. Welded regions should also be given stress in the FE model as fatigue damage frequently occurs at these locations. A relative mesh method was used to consider the weld connections. The spatial configurations of the original structure are remained in the model. Fig. 2 shows the FE model of a typical deck unit. More than 7300 nodes and 19000 elements are included in the TMB FE model. More details of the model could be found in the reference [15].

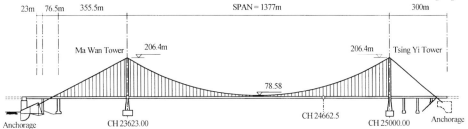

Figure 1: Sketch of Tsing Ma Bridge

3. Verification of the model

Dynamic characteristics and dynamic responses of the Bridge under traffic loadings were analyzed using the developed model. Comparison of the computed results with the related measured data was carried out to verify the efficiency of the model.

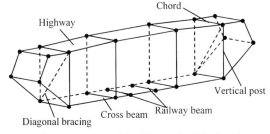

Figure 2: FE model of the typical deck unit

Modal characteristics of the TMB The dynamic analysis of a bridge subject to ambient loading depends largely on knowledge of the bridge modal properties. Modal analysis is therefore a most important step towards a successful dynamic analysis. The modal properties of the TMB were analyzed with the developed FE model. In the eigenvalue calculation, the initial stresses in the hangers and main cables will affect the structural stiffness significantly. The initial stresses were tested when the bridge was just constructed and adopted as stress initial conditions in the FE model.

The periods of the first 80 modes of free vibration computed with FEM range from 2.305 to 0.134s. Some free-vibration frequencies were measured separately by the Tsing Hua University (TH U) in China, the Hong Kong Polytechnic University (HK PolyU) and the Hong Kong Highways Department (HKHD). The designer Mott MacDonald Hong Kong Ltd (MMHK) and the checker Flint & Neill Partnership of UN (FNP) of TMB also analyzed theoretically the first few frequencies [16]. Table 1 gives a comparison between the first few measured and analytical natural frequencies. In Table 1, the "Mean" data in the 7th column are the mean values of measured and theoretically analyzed values of frequencies of the 2nd to 6th columns. The deviation = (FEM-Mean)/FEM ×100%. The maximum relative deviation is 9.7%, which shows the calculated frequencies by the constructed FE model compare well with the measured and the analyzed ones. The main dynamic response properties are included in the FE model. It is efficient to study the modal property of the TMB using the FE model.

Table 1: The few free-vibration frequencies of TMB, unit (Hz)

MODE	MMHK	FNP	TH U	HK POLYU	HKHD	MEAN	FEM	Deviation (%)
Lateral								
First	0.065	0.064	0.069	0.069	0.070	0.0674	0.069	+2.3
Second	0.164	0.149	0.161	0.164	0.170	0.1616	0.161	-0.4
Vertical								
First	0.112	0.112	0.114	0.113	0.114	0.1130	0.117	+3.4
Second	0.141	0.133	0.137	0.139	0.133	0.1366	0.144	+5.1
Torsion								
First	0.259	0.253	0.265	0.267	0.270	0.2592	0.262	+1.1
Second	0.276	0.268	0.320	0.320	0.324	0.3016	0.332	+9.2

Stress response of the TMB under train loading As aforementioned, the fatigue damage analysis depends on the research on stress distribution in the structure. The effectiveness of the model for dynamic stress analysis is therefore the primary concern. The TMB carries both highway and railway traffic. Relatively, its dynamic response was caused by the combination of the railway loading together with the highway loading. The highway loading is very complex as it varies with the traffic flow, the vehicle type, the road roughness and the vehicle positions. There are many uncertainties in the highway loading model. Compared with the highway loading, a train has fixed axle spacings and known axle loads. It is more reliable for a railway loading model than that for the highway loading model. The dynamic response of the bridge under a running train was therefore studied to investigate the validity the FE model.

While a train passing through a bridge, periodic excitations on the bridge could be induced and bridge fatigue damage could then be produced. For a long span bridge-train system, Diana and Cheli [17] pointed out that the interaction between the train and bridge should be considered while studying the bridge deformation. Recently, Xia et al [18] established an accurate train-bridge model to study the dynamic interaction of long suspension bridges with running trains. Their results indicated that the interaction between the train and long span bridge is insignificant. To avoid the model being more complicated, the running train is simulated as moving loads in the established model. When the speed of the moving load is so slow, the crawling speed load could be considered as static load and the corresponding response could be considered as the influence line of the response. The works by Moses et al [19] indicated that the difference of long suspension bridge effect under static load and under dynamic load is so small that the impact factor would not be over 10%. Therefore, the moving load model is accurate enough to study the bridge response under running trains.

A typical MTR train running on the TMB [18] was adopted to model the train loading, where each axle loading of the train is about 170kN when the train is fully loaded. The train with 10 cars with full passengers was adopted in this model. Equivalent nodal forces were added to the FE model, where the train speed considered as 30m/s. The modal superposition method was adopted to analyze the dynamic response. As long suspension bridges bear global deformation at very low frequency and local deformation at relatively high frequency, assuming the bridge local deformation being included in the response, the first 80 modes were selected in the analysis.

Considering that long suspension bridges are lightly damped structures, the modal damping was chosen as 0.5% in the FEM.

In order to ensure user comfort and safety of the TMB, a structural health monitoring system has been devised and installed to monitor the integrity, durability and reliability of the bridge. This monitoring system comprises a total of approximately 350 sensors, including accelerometers, strain gauges, displacement transducers, level sensors, anemometers, temperature sensors and weigh-in-motion sensors, installed permanently on the bridge. The strain gauges were installed to measure stresses at bridge-deck sections. Comparison of the computed results with the measured online data was taken to validate the efficiency of the FEM. Figure 3 shows the typical locations of the installed strain gauges. The strain-time history recorded by the strain gauges located at detail "D" and detail "J" were considered, where the strain gauge "SR-TLN-01" is a strain rosette, and the corresponding principle stresses in the location were computed to compare.

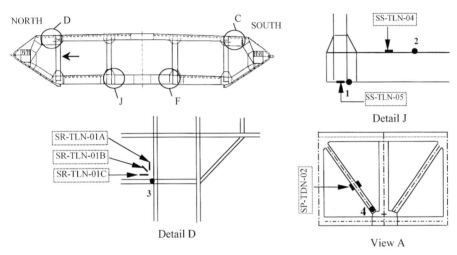

Figure 3: Typical locations of strain gauges in bridge deck

Because the TMB is symmetrical about the bridge centerline for each section, the strain-time histories recorded by the strain gauges in details "C" and "F" are similar to that in details "D" and "J", and would not be studied separately. The stress history in the diagonal bracing recorded by "SP-TDN-02" (from View A) was also considered. The points "1", "2", "3", and "4" shown here are the relative stress output points in the FE analysis.

With the strain-time history data, the stresses versus time in the locations could be obtained easily with the Hooke's law (where the Young's modulus is 200GPa). A typical stress history is shown in Figure 4. Each sudden change on the curves is

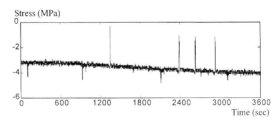

Figure 4: Typical stress-time history

corresponding to the passage of a train. Obviously there were eight trains passing through the TMB during the hour of interest, four of which ran out bound from the Airport and vice versa. When a train passed through the bridge in different directions, the differences of the responses were marked. In the constructed FE model, the bridge responses under a train running in bound towards the Airport were simulated and correspondingly compared to the measured data under the same train direction.

To be clearer, the stresses recorded by strain gauges should be calibrated to compare with the computed data. The comparison of output stresses in point "1", "2", "3" and "4" with the recorded data by "SS-TLN-05", "SS-TLN-04", "SR-TLN-01" and "SP-TDN-02" (shown in Figure3) are depicted in Figure 5. It could be seen that there is fine agreement between the computed results and measured ones. The model is efficient to calculate the dynamic stress response in the TMB.

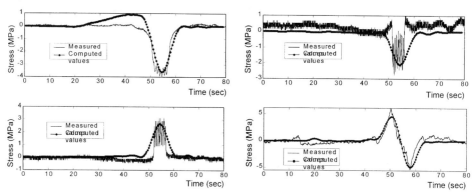

Figure 5 Comparisons of the Computed Stress Histories with Measured Ones at (a) "SR-TLN-01", (b) "SR-TLN-05", (c) "SR-TLN-04", and (d) "SP-TDN-02"

From the aforementioned comparison of the computed dynamic characteristics and dynamic stress of the bridge with the measured results, the constructed model could be used to study the bridge fatigue damage under traffic loadings.

4. Dynamic response of the bridge under traffic loading

The dynamic response of the bridge under ambient loading and traffic loading could be measured by the health monitoring system installed on the bridge. Using the online data, some important works of Chan and Li [20, 21] on fatigue damage analysis of the bridge deck section were carried out. The works of Chan and Li indicate that traffic loading mainly effected the stress oscillation or stress spectra in the bridge deck. However, the locations of the monitoring sensors installed on the bridge were selected for general purposes and they might not be critical to fatigue damage. The analysis of the TMB global dynamic response and the stress time history under traffic loading in the members are necessary.

In the case of bridges carrying both highway and railway loadings, the total fatigue damage in the bridges should be determined for each loading condition separately. As there is a single rail track only passing through the TMB in each direction, the additional combined stress history by more

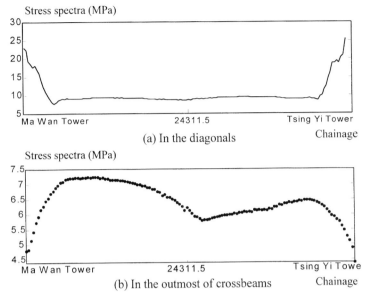

(a) In the diagonals

(b) In the outmost of crossbeams

Figure 6: stress spectra along the main pan

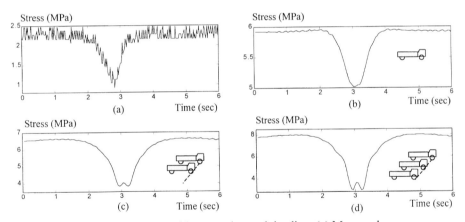

Figure 7: stress history under truck loading, (a) Measured; (b) under single truck; (c) under double trucks; (d) under triple truck

than two trains could be neglected. The computed stress history depicted in Figure 5 should be used directly to analyze the fatigue damage in related locations. With the stress history, the stress spectrum in relative locations could be obtained and, the fatigue damage could be computed using the Palmgren-Miner rule. Figure 6 gives the stress spectra in the related locations along the main span. Correspondingly the fatigue damages in the relative locations along the main span are determined.

From Figure 6, it could be seen that the stress spectrum in the diagonals or in the outmost of the cross beams are varying along the main span. Correspondingly the fatigue damages in the relative locations along the main span are changeable.

The bridge dynamic response under the highway loading is also analyzed using the FE model. The standard fatigue vehicle specified in BS 5400 [2] was adopted in the calculation. The bridge responses under a single truck, double trucks and triple trucks were computed separately. The measured and computed stress history under various truck loadings are depicted in Figure 7. Comparison of Figure 7 with Figure 5 shows that the stress spectra caused by truck loadings are much smaller than the ones caused by train loadings. However the passages of truck are much heavier than of the train in the same period, and the distribution of the trucks on the bridge could obviously affect the stress distribution in the bridge deck. The stress level in each element under heavy traffic was computed using the FE model. In the calculation of the fatigue damage, the mean stress caused heavy truck traffics was considered carefully.

5. Updating model with considering fatigue damage

Generally the highway loading and railway loading on a bridge vary with time. The fatigue damage is therefore caused by variable amplitude loads. Li and Chan [20, 21] presented an effective method to calculate the fatigue damage under variable amplitude loadings. With the computed stress spectra, the damage degree in the relative locations could be calculated using their method. The WIM system set on the bridge could identify the vehicle types and, the related stress cycle numbers in a block time could be obtained. Considering that the TMB is at virgin state and the traffic volume is relatively light, an updating ratio of traffic volume is included in the calculation of the stress cycles. The damage degrees in the outmost of the cross beams along the main span after about 50 years service are shown in the Figure 8.

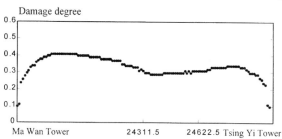

Figure 8: Damage degree along the main span

Figure 9: Procedure for updating model

Fatigue damage will be occurred in the structural elements after using some years. Correspondingly the stress or stress spectra in the elements will be changed even under the same loading condition. To compute the stress or stress spectra in the **damaged** structure, the relative FE model should be updated. A procedure of updating FE model with considering damage effect is depicted in Figure 9. Here the D_0 is the initial damage degree in an element and could be considered as very small in the virgin state; $\Delta\sigma$ is the stress spectrum in the related element, which could be calculated using the FE model; N is the stress cycles in the block time (selected as 10 years), which could be obtained from the WIM data using the method presented by Moses [19]; B is the element stiffness and could be updated with the formula presented by Li [23]. With the procedure, the stress or stress spectrum level in the related element could be decided with the considered damage effects. Figure 10 shows the stress spectra span in the outmost of crossbeams along the main in about 50 years. From the comparison of Figure 10 with Figure 6(b), it was shown that the stress spectra in the elements were reduced when considering damage effects and the differences of the stress spectra in the elements also became smaller. This is the most important effect caused by damage. There are not any reports found before on the phenomena.

In BS5400 [2], the stress level or stress spectra in the elements were considered as invariable in its whole service time. However, from the updating model calculation, the variation of tress spectra in the elements is significant (about 30% here). This variation of stress spectra in the structure whole service life is suggested to be considered in the structural fatigue design.

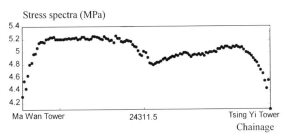

Figure 10: Stress Spectra along the main span in 50 yeas

6. Conclusions

The finite element method is found to be efficient for large suspension bridge fatigue analysis. The TMB, as a case study, was simulated as a large finite element model embodying the properties of almost all the practical structural elements using commercial software ABAQUS. Verification of the model and its efficiency for fatigue damage analysis were presented with the help of the online data measured by the health monitoring system. Dynamic response of the bridge under traffic loadings was studied using the developed FE model. The results indicate that the stress spectra in the relative elements are mainly caused by the railway loading and, the highway loading could take an effect on the mean stress level in the elements. An updating model considering the damage effects was also presented and, a phenomenon was first reported here that the stress level in the damaged elements could be reduced with the increasing of the damage degree. This effect is suggested to be considered in the structural fatigue design.

Acknowledgements

Funding support to this project by the Hong Kong Polytechnic University is gratefully acknowledged. Appreciation is extended to the Highways Department of the Hong Kong SAR Government who provided the data measured by the Health Monitoring System on the TMB and relevant documents.

References

1 ASCE, 1982 *Journal of structural engineering* **108**, 3-88. Committee on fatigue and fracture reliability of the committee on structural safety and reliability of the structural division, fatigue reliability 1-4.
2 BSI, 1982 *BS5400: Part 10*, Code of Practice for Fatigue.
3 R. Pullin, D.C. Carter and K. M. Holford 1999 *Key Engineering Materials* **167-168**, 335-342. Damage assessment in steel bridges.
4 W. R Charles, M. Gregory, C. Paul, A. Kayoko and W. Scott 2000 *Journal of Bridge Engineering* **5**, 14-21. Dynamic response and fatigue of steel Tied-Arch Bridge.

5. H. Nowack, and U. Schulz 1996 *Fatigue 96*, Pergamon, Oxford **2**, 1057-1068. Significance of finite element methods in fatigue analysis.
6. I. V. Putchkov 1995 *International Journal of Fatigue* **17**, 385-398. Development of a finite element based strain accumulation model for the prediction of fatigue lives in highly stresses Ti components.
7. X. B. Lin and R. A. Smith 1999 *Engineering Fracture Mechanics* **63**, 503-522. Finite element modeling of fatigue crack growth of surface cracked plates Part I: The numerical technique.
8. X. B. Lin and R. A. Smith 1999 *Engineering Fracture Mechanics* **63,** 523-540. Finite element modeling of fatigue crack growth of surface cracked plates Part II: Crack shape change. Engineering Fracture Mechanics.
9. X. B. Lin and R. A. Smith 1999 *Engineering Fracture Mechanics* **63**, 541-556. Finite element modeling of fatigue crack growth of surface cracked plates Part III: Stress intensity factor and fatigue crack growth life. Engineering Fracture Mechanics.
10. E A Branco, J Azevedo, M Ritto-Corretia 1993 *Structural Engineering* 4, 240-244. Dynamic analysis of the international Guadiana Bridge.
11. Y B Yang, W McGuire 1986 *ASCE Journal of Structural Engineering* 112 (ST4) 853-877 A stiffness matrix for geometric nonlinear analysis.
12. Y B Yang, W McGuire 1986 *ASCE Journal of Structural Engineering* 112 (ST4) 879-905 Joint rotations and geometric nonlinear analysis.
13. V Boonyapinyo, H Yamada, T Miyata 1994 *ASCE Journal of Structural Engineering* 120 (2) 486-506 Wind-induced nonlinear analysis
14. L D zhu, H F Xiang, Y L Xu 2000 *Engineering Structures* 22 1313-1323 Triple-girder model for modal analysis of cable-stayed bridges with warping effect
15. T H T Chan, L Guo, Z X Li, *summit to Journal of sound and vibration* Finite element method for fatigue stress analysis of long suspension bridge
16. C.K. LAU, W.P. MAK, W.Y., CHAN, K.L. MAN, and K.F. WONG 1999 *Advances in Steel Structures ICASS'99, Hong Kong,* 487-496. Structural Performance Measurements and Design Parameter Validation for Tsing Ma Suspension Bridge.
17. G. Diana and F. Cheli 1989 *Vehicle System Dynamics*, **18**, 71-106. Dynamic interaction of railway systems with large bridges.
18. H. Xia, Y.L. Xu, and T.H.T. Chan 2000 *Journal of Sound and Vibration* **237**, 263-280 Dynamic Interaction of Long Suspension Bridges with running Trains.
19. F. Moses, C.G. Schilling, and K. S. Raju 1987 *National Cooperative highway Research Program Report* **299**, Fatigue evaluation procedures for steel bridges.
20. T. H. T. Chan, Z. X. Li and J. M. Ko 2001 *International Journal of Fatigue* 23, 55-64 Analysis and Life Prediction of Bridges with Structural Health Monitoring Data- Part II: Application.
21. Z. X. Li, T. H. T. Chan and J. M. Ko 2001 *International Journal of Fatigue* 23 45-53. Fatigue Analysis and Life Prediction of Bridges with Structural Health Monitoring Data- Part I: Methodology and strategy.
22. J.Y. WANG, J.M. KO, and Y.Q. NI 2000 *Nondestructive Evaluation of Highways, Utilities, and Pipeplines IV, Proceedings of SPIE,* **3995**, 300-311. Modal sensitivity analysis of Tsing Ma Bridge for structural damage detection.
23. Z X Li 1995 *Theoretical and Applied Fracture Mechanics*, 22, 165-170 Viscoplastic damage model applied to cracking gravity dam.

Some Outcomes from Load Testing of Small Span Bridges in Western Australia

Dr I CHANDLER
Curtin University of Technology, Perth, Australia

INTRODUCTION

In the rural areas of the south west of Western Australia the countryside is relatively flat and there are numerous small creeks and rivers, many of which only flow in the winter months. To cross these creeks numerous short span bridges of timber or concrete construction have been built.

Main Roads Western Australia (MRWA) is responsible for the management and technical guidance of the states road network system of which the 2400 bridges currently in service form a significant part. The bridge section of MRWA has invested resources into developing methods of determining the structural integrity of the bridges under their control and supervision.

Since 1992, a bridge load testing program has been developed to further understanding of the real structural performance of timber and concrete bridges. Initial investigations were with modest static loads applied by a truck and the work concentrated on the measurement of longitudinal strains in timber stringers (Putt et al, 1992). This was extended to include measurement of deflections and in 1994 deflection measurements under dynamic loads were also recorded (Chandler & van Kleef, 1997). In a recent investigation (Chandler & Haritos, 2000) measurement of accelerations under traffic loading were also used to supplement data collected from the strain and deflection measurements.

Two bridges that have been recently tested are a multispan timber stringer bridge - Bridge 631 and a very short span rail reinforced concrete bridge – Bridge 6028. In this paper, the testing of these bridges will be briefly described and some of the outcomes from the testing programs will be presented and discussed.

BACKGROUND DETAILS OF BRIDGE 631

Bridge 631 is a 190m long timber bridge with 31 spans (each approximately 6 metres) over the Avon River in Toodyay (a small town approximately 100 km from Perth the capital city of Western Australia) - see Figure 1. The bridge has round log stringers supporting sawn timber bearers and a longitudinal timber deck overlain with 100 mm thick concrete deck. It was constructed from local hardwood timbers (Jarrah and Wandoo) in 1950 and has had a series of repairs in 1965, 1980, 1994 and 1998. However as the width between kerbs is only 5.5 metres and it does not have a footpath it has been decided to replace the bridge with a new wider concrete structure during 2001.

The proposed removal of the existing timber bridge has provided an opportunity to pursue a concentrated program of research into aspects of bridge inspection, timber material properties, structural failure and modelling and improved repair techniques and strengthening methods.

Non destructive evaluation of this bridge prior to construction of the new bridge was conducted during May 2000 using trucks to provide both static and dynamic loading. It is planned to conduct destructive tests in 2002 once the replacement bridge is completed.

Figure 1:   General View of Bridge 631

DETAILS OF THE LOAD TESTS OF BRIDGE 631

Static and dynamic load testing of the bridge with two trucks took place over two weekends and allowed strain and deflection measurements to be recorded for 17 of the 31 spans. Over 600 Mbytes of data was collected from the tests.

For the static tests the trucks were positioned in 3 lateral positions across the width of the bridge and multiple positions along the bridge with the rear axle group over a pier or midspan. Dynamic tests were also conducted with the trucks crossing the bridge at designated speeds of walking pace, 25, 40, 50, 60, 70 and 80 km/hr and in one of the three lateral positions.

The trucks were chosen to represent the extremes of the Austroads T44 axle spacings with the maximum load limited by MRWA concerns about the load rating (capacity) of some bridge elements. The shorter of the trucks had axle spacings of 3.05, 1.36, 3.42, 1.30 metres and an overall length between axles of 9.13 metres. The dual axles on the trailer were loaded to 16.4 tonnes and the gross vehicle mass (GVM) was 35.6 tonnes. The longer truck had axle spacings of 3.36, 1.32, 8.70, 1.37, 1.36 metres and an overall length between axles of 16.11 metres. The triaxles on the trailer were loaded to 21.9 tonnes and the GVM was 41.2 tonnes.

Measurements were recorded electronically via a purpose built data logging system with 32 channels. Linearly Variable Displacement Transducers (LVDT's) were used to measure deflections at midspans of stringers, ends of corbels and various locations on halfcaps and piles. Strains were recorded at midspan of stringers and various locations on the substructure.

It is difficult to install instruments and move them between set-ups with the bridge deck 4 to 6 metres above the ground. Therefore, a series of aluminium tube descenders were attached to the bridge elements transferring all vertical movements closer to the ground and allowing instruments to be installed at a convenient height on standard scaffolding reference frames. An example of a typical setup for midspan of a set of stringers is shown in Figure 2.

Strain readings were obtained by measurement of the movement between two 6mm diameter steel pins inserted in drilled holes in the timber at 500 mm centres. A purpose built extensometer was attached to the pins and movement recorded via a short travel LVDT. These instruments have been successfully used on a number of tests, but it is time consuming installing and shifting the instruments.

Figure 2: Underside Of A Span Of Bridge 631 Ready For Testing

RESULTS FROM LOAD TESTS OF BRIDGE 631

The maximum measured deflections under static loads were from 6 mm to 12 mm and the maximum measured strains were 150 to 200 microstrain. These values were usually produced by the triaxle group of the longer wheelbase truck and when the truck was located in either the North or South loading positions near the outside stringers.

A number of factors contribute to the large variation in deflection response of the bridge including variations in stringer stiffness and variations in repairs to the spans with up to three additional stringers in a few spans. However the factor that was seen to be particularly significant in this bridge was the variations in stiffness of the corbel and halfcap interface with some corbels rotating and bending so that the end deflected as much as 30% of the midspan deflection. This has a significant influence on load sharing between the stringers and warrants further investigation.

The bridge has very visible undulations in the deck surface at approximately 6 metre intervals. It shook and vibrated noticeably under traffic and all vehicles bounced up and down due to the uneven surface and flexible deck of the bridge. The length of the bridge and the repeated undulations increase this motion as vehicles travel along the bridge and make vertical bounce quite dramatic for trucks around 60 km per hour – the legal speed limit on the bridge. As a result, maximum dynamic effects can occur for short vehicles and lighter loads. Figure 3 shows that the drive axles of the short truck with only 14.1 tonnes produced a larger deflection than any of the other axle groups, even more than the triaxle with 21.9 tonnes.

The dynamic load allowance varied considerably with vehicle speed and was affected by the matching of vehicle motion with surface undulation. It was greater for the short truck and values of 45% to 50% were recorded. The natural frequency for this bridge was in the range 9 to 9.5 Hz giving an Austroads code value for dynamic magnification of 25% - significantly less than that observed.

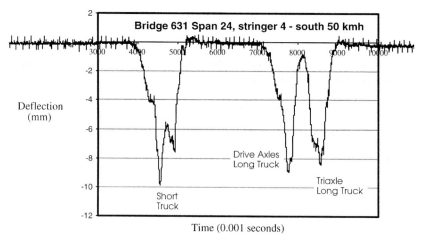

Figure 3: Typical Dynamic Deflection Response for Bridge 631 – Trucks Travelling in Convoy

ANALYTICAL MODELS OF BRIDGE 631

Spans 2 to 11 of the bridge were modelled using the software package Strand 7. The model consisted of beam elements with user defined cross-sections to match the individual stringers, beams for the bearers and corbels, plate elements for the deck and three dimensional brick elements for the half caps. Figure 4 gives an idea of the mesh for a portion of this model. Overall 18500 elements were used in the model.

Data for the model was obtained from site surveys with a total station giving cartesian coordinates of key points throughout the structure. Visual inspection data was available giving the vertical and horizontal dimensions of each stringer at both ends and at midspan. However to simplify the model it was decided that the section at midspan would be used throughout the stringer length.

The real deck was constructed of longitudinal timber planks 250 mm wide by 125 mm thick overlayed with concrete of varying thickness. This was represented by a single layer of plate elements approximating the combined stiffness. Othotropic properties were used to reduce the lateral stiffness due to the discontinuities in the deck timbers. The deck timbers were joined longitudinally in a butt joint at the middle of each span and this was modelled by a reduced thickness of elements at that point as seen in the lower diagram of Figure 4.

A series of link elements were used to join the various components together without over constraining the connections. Master-slave links at the end of the corbels were used to connect stringers to corbels and also to connect the bearers to the stringers. It was found that the rotational degree of freedom about the longitudinal axis of the model had to be released to match bearer and deck displacements with measured results.

Results obtained from this model generally matched the measured deflections to within 5% although there were some local anomalies. The lateral stiffness of the deck and the method of linking different parts of the model were significant influences on the results. A better match could have been obtained by separate modelling of the concrete and timber portions of the deck at the expense of increased model complexity and increased solution time.

The model indicated maximum stresses in the timber stringers of 5 MPa and maximum stresses in the steel beams of 60 MPa. These results were in general agreement with the strain measurements on steel and timber stringers.

The longitudinal deck stiffness was varied to use the model to test predictions about the lateral load sharing and relative deflection of the stringers. The model indicated that the relative deflections remained unchanged with significant changes in the deck stiffness.

DISCUSSION OF BRIDGE 631

The preliminary load testing and subsequent analysis have both proved useful in identifying key parameters influencing the structural response. Further analysis of the test data is required and the finite element model applied to the eastern end of the bridge to verify its consistency with the observed behaviour.

The information gained from both will be used to direct further investigation into this bridges structural behaviour and to assist in planning for the destructive testing of selected spans and components.

478 ASSESSMENT

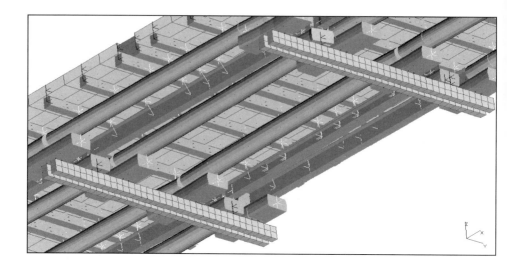

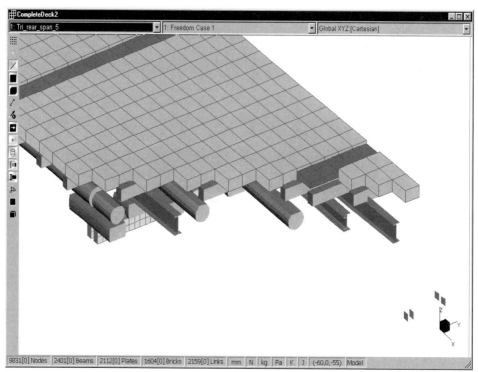

Figure 4:   Two Views of the Finite Element Model of Bridge 631 Showing Deck, Steel and Timber Stringers, Corbels and Half Cap Beams

BACKGROUND DETAILS OF BRIDGE 6028

Bridge #6028 is a two span rail reinforced concrete bridge. It has clear spans of 3.05 metres with abutment and pier walls of 300 mm nominal thickness. The overall concrete depth is 240 mm with inverted rail (45 lb/yd) at nominal 600 mm centres as main reinforcement (see Figure 5). The concrete for this bridge was of poor quality and 100 mm diameter cores taken from the slab had compressive strengths between 7 MPa and 11 MPa.

A typical section through the slab showing the inverted rails can be seen in Figure 5. The delamination of the slab just above the rails and at the plane of the mesh reinforcement is evident and extended across much of the slab. Construction of the slab involved placing premixed concrete to cover the rails, placing the mesh reinforcement and placing additional concrete to complete the slab. Due to remoteness of the site and the use of a small capacity concrete truck it was inevitable that a cold joint would form at half depth of the slab.

After removal of the gravel road surface and prior to testing there were several readily visible cracks on the upper surface including significant transverse cracks directly above the central pier. There were also extensive cracks on the underside of both spans of the bridge with longitudinal cracks coinciding with the rail locations and transverse flexural cracks in several locations.

Figure 5: View Of A Cut Section Of Bridge 6028 Showing The Inverted Rail Reinforcement

DETAILS OF THE LOAD TESTS OF BRIDGE 6028

The bridge was loaded through 4 points (each 400 mm by 200 mm in area) at the corners of an 1800 mm by 1200 mm rectangle – representing a dual axle configuration central on the span. Thus the bridge was loaded through 4 steel plates (each 400 mm by 200 mm in area) whose centres were at the corners of an 1800 mm by 1200 mm. Fifty five tonne hydraulic jacks were placed centrally on these plates and connected via a common manifold to an electric pump and a hand actuated pump. The hand actuated pump was used for small increments of load and fine control with the electric pump available for large displacements. This system had been calibrated with a large dial pressure gauge and readings were recorded in MPa of line pressure.

The reaction for the jacks was provided by a series of masses stacked onto a set of eight 690UB125 steel beams which in turn were supported by transverse pairs of 690UB125 beams sitting on concrete blocks placed in a trench created alongside the abutments. The arrangement with 120 tonnes of reaction in place is shown in Figure 6 and was setup with the assistance of a 130 tonne capacity mobile crane.

Deflection measurements were taken at up to 20 locations on the bridge for each test. Linearly Variable Displacement Transducers and a data logging system were used to record displacements at a sampling rate of 5 times per second. Key deflection points were also monitored by using dial indicators with 50 mm travel.

Four tests were conducted, a proof load test of span 2 on the east side followed by three destructive tests as follows:-
- Span 1 East Side with 4 point loads
- Span 2 East Side with a single point load central between the widest spaced rails
- Span 1 West Side with 4 point loads on a cut section of the slab acting as a wide beam.

Figure 6: Bridge 6028 Ready For Testing With 120 Tonne Reaction In Place Over The Bridge

RESULTS FROM LOAD TESTS OF BRIDGE 6028

For the proof load the load was applied in small increments up to approximately 190 kN and unloaded and reapplied. On the third load application the load was continued in increments until 380 kN when the maximum deflection was 3.5 mm. There was no evidence of cracking and the load deflection plot remained linear.

The destructive tests all showed the slabs weakness in punching shear. Failure for Test 1 with four point loads occurred when one load point punched through at a total load of 620 kN (155 kN at each load point). Maximum deflection prior to failure was 32 mm near the outer edge load point. The load deflection plots for two points on the centreline of the span are shown in Figure 7.

At the time of failure, a loud noise was heard and a large section of concrete approximately 600 mm wide and 2000 mm long fell out from the underside of the slab, directly below the punching failure.

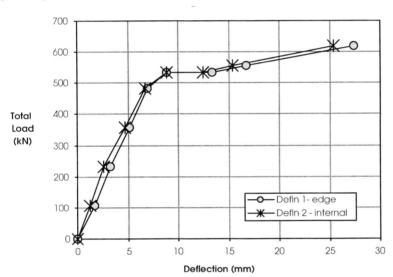

Figure 7: Graph of Load-Deflection Measurements for Test 1 Span 1 East side for Bridge 6028

In test 2 the load was placed on one patch 400 mm by 200 mm located between the widest spaced rails in span 2. The steel loading plate began to punch through the concrete surface at a load of 300kN and failure occurred at a load of 340 kN. Maximum deflection prior to failure was approximately 14 mm. Once again a large portion of the lower half of the slab (below the mesh and between the rails adjacent to the load) was pushed off at the time of the failure.

ANALYTICAL MODELS OF BRIDGE 6028

The bridge was modelled using a grillage analysis with 15 rows of longitudinal members and 21 rows of transverse members. Although several assumptions had to be made regarding concrete properties and the extent of delamination of the deck, agreement within 20% was obtained between the elastic deflections of the model and those measured during the linear portions of the loading.

Punching shear capacities were calculated using the Austroads code approach and predicted failure loads to within ±16% of the measured failured loads. However at these concrete strengths small variations in strength had a large change in calculated shear capacity – a 1 MPa increase in f'c produced a 28% increase in capacity.

DISCUSSION OF BRIDGE 6028

Bridges such as 6028 represent a low-cost solution for roads with small traffic volumes and could be expected to have a useful life of 30 years or more. Thus it may be as effective to build such bridges and regularly replace them as would providing more expensive solutions to meet a 100 year design life. Whole of life costing including risk management and assessment should be applied to the maintenance and replacement of bridge stock such as this bridge, particularly for bridges on minor roads in rural areas with small traffic counts.

SUMMARY OF THE OUTCOMES FROM TESTING

In the short length of this paper it has only been possible to briefly mention some of the outcomes from the testing program and subsequent analysis. Further analysis and testing is needed to clarify the behaviour of the complex structural system represented by timber stringer bridges.

The testing did identify that steel stringer inserts were carrying significant proportions of the vehicle loads and showed that the corbel movement was larger and more significant than previously identified.

Finite Element modelling proved valuable in matching analytical stresses and deflections to the observed behaviour and identifying effects, such as the bearer location, that had a large influence on the modelled behaviour.

The load testing of Bridge #6028 highlighted the weakness in punching shear of this bridge due to use of poor quality concrete and a wide spacing of the rail reinforcement. Improvements in quality of concrete and better control on construction methods would enable low-cost solutions such as these bridges to form a useful part of the road network.

The research work presented in this paper offers significant benefits to the community, including local government, road users, the road freight industry and tertiary institutions.

ACKNOWLEDGEMENT

The author gratefully acknowledges the support and sponsorship of this work by Main Roads, Western Australia. The author also thanks the many others who assisted with these projects and in particular, Matthew Old, Daniel Chatley and Peter Truphet. The experimental and analytical work on bridge #631 was part of undergraduate project work by Matthew Old and

Daniel Chatley. The testing of bridge #6028 was conducted with the assistance of BGE Consulting Engineers, and most of the analytical work was done by Peter Truphet.

REFERENCES

Chandler, I and Haritos, N, 2000 *Evaluation of the Strengthening of MRWA Bridge #617* Austroads 4th Bridge Engineering Conference, 29 Nov – 1 Dec, Adelaide, Australia

Chandler, I. and Van Kleef, H., 1997 *Strength Assessment Of Timber Bridges - Field Load Testing And Laboratory Investigations,* Proc Austroads 1997 Bridge Conference, Dec., Sydney, Australia, Vol 3:229-239.

Putt, I, Chandler, I and Margetts, L 1992, *A Strength Assessment Method For Timber Bridges* Timber Bridge Conference, Melbourne, Australia.

SENSITIVITY OF STEEL BRIDGE FATIGUE LIFE ESTIMATES TO FATIGUE CRACK MODELLING

T.D. Righiniotis
Dept of Civil and Environmental Engineering
Imperial College
London SW7 2BU
UK

M.K. Chryssanthopoulos
Dept of Civil Engineering
University of Surrey
Surrey GU2 7XH
UK

ABSTRACT

The objective of this paper is to present the results of a sensitivity analysis arising from the application of a probabilistic fracture mechanics procedure to steel bridges. The latter follows closely the principles and recommendations of BS 7910 [1] but also accounts for the uncertainties inherently present in fatigue crack growth and failure. Thus, a range of failure probability estimates is determined for a typical fatigue detail under more or less refined modelling assumptions. It is observed that accuracy in the adopted crack growth law is very important, since a significant part of the bridge stress spectrum is often associated with low amplitudes, whereas rigour in the fracture failure criterion, although desirable, seems to have less influence on the results.

1 INTRODUCTION

The actual fatigue life of steel bridge components can be considerably shorter than the original design life due to a number of factors, e.g. unfavourable increase in traffic loads (in frequency and/or in vehicle weights), poorer than specified workmanship, or unconservative estimation of anticipated load effects on particular details. For these reasons, quantification and assessment of fatigue effects is expected to play an increasingly important role in the future, as the stock of existing bridges gets older in many countries around the world.

Given the large number of fatigue details on any bridge, and the cost associated with inspection and (where necessary) repair of defects, it is important to have methodologies in hand that can furnish remaining life estimates and inspection intervals with an acceptable degree of confidence with due attention paid to the range of uncertainties that characterize fatigue crack growth and failure. In this context, the UK Highways Agency has initiated a project aimed at producing procedures for the assessment of steel bridge components with cracks using fracture mechanics principles. In other sectors such as the offshore industry [2] guidance in this area already exists but the application of probabilistic fracture mechanics for bridge components is still under development [3,4]. It is also important to note that in the UK the code of practice relevant to fracture mechanics assessment of welded structures has recently been updated and enhanced [1], in the light of research carried out in the last ten years or so.

Thus, the purpose of this paper is to present the results of a sensitivity analysis, using a probabilistic fracture mechanics based methodology developed by the authors [5] following the principles of BS7910. Given the current state of fracture mechanics applications in bridges, and the need to quantify the influence of a range of modelling assumptions, this is considered both timely and important.

2 GENERAL CRACK GROWTH CONSIDERATIONS

Fatigue crack growth may be modelled through the Paris-Erdogan equation [6]

$$\frac{da}{dN} = C_A (\Delta K_A)^m \tag{1}$$

where a is the crack size, N is the number of applied cycles, C and m are the crack growth (Paris) parameters and ΔK_A is the stress intensity range defined as

$$\Delta K_A = S_r Y_A \sqrt{\pi a} \tag{2}$$

In equation (2), S_r is the stress range and Y_A is the stress intensity magnification factor, a non-dimensional function of the cracked geometry and loading. Implicit in equations (1) and (2) is the assumption that the crack is fully described by a single dimension, a. This model is in the sequel termed the one degree-of-freedom model (1-DOF). Solution of the differential equation (1), in terms of the number of cycles required to propagate the crack from an initial crack size a_{in} to a final crack size a_f, may be obtained either by using a finite difference scheme [5] or numerical integration [7].

In general, cracks grow in a semi-elliptical shape, which is characterised by the crack depth a and the crack half-length c [8]. During fatigue, the crack may now grow in both directions as prescribed by their associated Paris-Erdogan law thus in addition to (1)

$$\frac{dc}{dN} = C_C (\Delta K_C)^m \tag{3}$$

where

$$\Delta K_C = S_r Y_C \sqrt{\pi a} \tag{4}$$

Since the crack can grow in both length and depth directions, it is now called a two degree-of-freedom crack (2-DOF). The subscripts A and C indicate that the Y factors as well as the Paris C parameters are different for the deepest (point A) and surface points (C). Based on the different conditions of constraint existing at the deepest and surface points, researchers have proposed different values for C_A and C_C. However, BS 7910 [1] proposes that $C_A = C_C$ which is conservative since, in general, the crack length will increase slower than the crack depth and this assumption is adopted in this paper.

It is generally accepted that fatigue crack growth displays a sigmoidal characteristic [8] (see Figure1), which is highly dependent on the stress ratio R ($=S_{min}/S_{max}$) and the environment. Crack growth below the threshold stress intensity range (ΔK_{thr}), which is material property, does not occur. ΔK_{thr} is also highly environment and R-dependent. These considerations somewhat limit the use of equations (1) and (3) within certain ΔK values as well as specific stress ratios. Attempts have been made in the past to condense all the fatigue crack growth data for a given environment into a single curve by introducing into the power law quantities such as R, ΔK_{thr} and fracture toughness K_{mat}. Recently, BS 7910 [1], following research work spurred by the offshore industry [9], has proposed a bi-linear crack growth law (see Figure 1) which describes crack growth in the near threshold region (Region I) and the Paris region (Region II) via two Paris-type equations thus

$$\frac{da}{dN} = C_1 (\Delta K_A)^{m_1} \qquad \Delta K_{thr} < \Delta K_A \leq \Delta K_{tr} \qquad (5)$$

$$\frac{da}{dN} = C_2 (\Delta K_A)^{m_2} \qquad \Delta K_{thr} < \Delta K_A \leq \Delta K_{tr} \qquad (6)$$

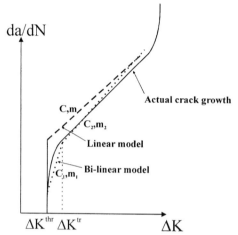

Fig 1. Schematic representation of actual crack growth and its linear and bi-linear approximations

with a similar set of equations for the crack length c. In BS 7910, different parameters C_1, C_2, m_1 and m_2 are specified for $R > 0.5$ and $R < 0.5$. In welded structures it is reasonable to assume that $R > 0.5$ due to the presence of residual stresses. For higher stress ratios, the rate of crack growth is higher and this is reflected in the crack growth parameters. BS 7910 proposes also the use of a linear crack growth law as a conservative alternative to the bi-linear crack growth law.

Crack growth terminates either when a physical limit is reached (e.g. the thickness of a plate), or, when, under the specified externally applied loads, the crack has grown to a size that cannot be sustained by the structure and fracture or plastic collapse occurs. The former is invariably adopted as a serviceability criterion in the offshore and nuclear industries to prevent leakage. With regard to the latter, plastic collapse or fracture may be specified according the Level 2A Failure Assessment Diagram (FAD) of BS 7910 [1]. According to this FAD, for a given set of loads, fracture toughness K_{mat}, yield strength σ_y and ultimate tensile stress, acceptability of a certain flaw is determined based on whether the point associated with these quantities in K_r-L_r space lies inside the curve

$$K_r = \left(1 - 0.14 L_r^2\right)\left(0.3 + 0.7 e^{-0.65 L_r^6}\right) \qquad L_r \leq L_r^{max} \qquad (7)$$
$$K_r = 0 \qquad L_r > L_r^{max}$$

A description of the parameters appearing in equations (7) is given in BS7910 [1]. For a 2-DOF crack, each point is associated with its own FAD. Furthermore, under alternating loads the critical crack size as determined by equations (7) is varying. In the following, it is conservatively assumed that (7) pertain to the maximum stress occurring in the entire stress history. As the crack grows, different points are traced inside the FAD and the critical crack size corresponds to the intersection of the trace with the line described by equations (7).

From the previous discussion, it becomes quite apparent that the results obtained from a fracture mechanics-based fatigue analysis, either deterministic or probabilistic, will depend on the modelling assumptions made. Clearly, application of a linear crack growth law over the entire range of ΔK values will yield different results to a bi-linear law over the same ΔK values. Likewise, crack growth termination based on geometric considerations will result in a different fatigue life to that obtained by using a fracture-based criterion. Finally, modelling of a crack that grows in a plane as a 1-DOF crack, may be in some cases unrealistic. These modelling assumptions are the subject of this paper. Modelling of the crack initiation period is not considered here since it is generally accepted that in welded structures this part of the component's life is negligible compared to the crack propagation life [8].

3 LOAD MODELLING

The differential equations (1), (3), (5) and (6) are applicable for constant amplitude loading.

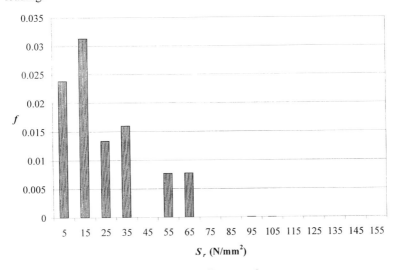

Fig 2. Frequency diagram of stress ranges

They can also be applied for variable amplitude loading once the stress spectrum has been converted to blocks of constant stress range. This may be accomplished by way of the reservoir, rainflow and other range-counting procedures. For the purposes of the following analyses, stress spectra were derived from the bending moment influence line at mid-span of a 10m simply supported beam [10] while traversing a specified garage of vehicles typical of conditions on UK roads [11] over the span. The stress ranges thus obtained were placed on a histogram with their associated frequencies determined from the annual number of vehicles reported in the same reference. The histogram is depicted in Figure (2).

4 PROBABILISTIC ANALYSIS

The associated fatigue reliability problem may be solved using simulation. Values for the random variables are randomly selected based on their frequency of occurrence as this is determined by the assumed probability density function. Cycles to failure can then be evaluated by solving the Paris-Erdogan equation for each of the selected samples and the cumulative probability of failure may obtained from the condition

$$P_f(T) = P[a_{fail} \leq a_f(T)] \quad (8)$$

where a_{fail} is the crack size at failure as this is specified by a geometry or fracture criterion and $a_f(T)$ is the final crack size at time T. Strictly, a_{fail}, when defined in terms of fracture or plastic collapse, is also time-dependent since the maximum loads are random but is here treated as being constant over time and determined by the maximum stress in the spectrum. Therefore, in the following, a_{fail} is randomised via the fracture toughness, the yield stress and the ultimate tensile stress. On the other hand, $a_f(T)$ is randomised via the Paris C parameter(s), ΔK_{thr}, ΔK_{tr} (for a bi-linear model), a_{in} and, for a 2-DOF model, the initial crack shape $(a/c)_0$. The loading is considered deterministic, as discussed in the previous section, and its statistical characteristics are used for crack growth as described in Reference [7].

5 SENSITIVITY ANALYSIS

For the purposes of the sensitivity analysis, the butt weld containing a toe crack is considered. It is assumed that $t = 20$mm and $l = 10$mm. This geometry is a class E fatigue detail [12] and the section modulus used to convert bending moments into nominal stresses was selected so that the detail would satisfy the 120-year design criterion. The random variables used in these analyses are given in Table (1).

VARIABLE	TYPE	MEAN	COV	MODEL APPLICABILITY
C	LOGNORMAL	2.5×10^{-13}	0.5	Linear
C_1	LOGNORMAL	4.8×10^{-18}	1.7	Bi-linear
C_2	LOGNORMAL	5.86×10^{-13}	0.6	Bi-linear
ΔK_{thr}	LOGNORMAL	147	0.4	All
a_{in}	LOGNORMAL	0.2	0.2	All
$(a/c)_0$	LOGNORMAL	0.01	0.2	2-DOF
K_{mat}	WEIBULL	2250	0.25	FAD
σ_y	LOGNORMAL	380	0.07	FAD

Table 1. Random variables

Table (1) describes the type of distribution used, its mean and coefficient of variation (COV) and finally the appropriate model(s) for each variable. The parameters C_1 and C_2 were assumed to be uncorrelated. The ultimate tensile stress was assumed to be fully correlated with the yield stress, thus $\sigma_{UTS} = 1.5\ \sigma_y$. The stress intensity range ΔK_{tr} was also assumed to be random and determined from the intersection of the two linear segments. All analyses were run for the bridge spectrum described in Section (3). The Paris m parameters were assumed to be deterministic and are given in Table (2). The factors Y_A and Y_C used for this geometry are given in the Appendix. The differential equation (1) was solved using numerical integration while, for a 2-DOF model in addition to (1), a modified equation (3) was solved using the fourth order Runge-Kutta method [13].

	VALUE	MODEL APPLICABILITY
m	3.0	Linear
m_1	5.10	Bi-linear
m_2	2.88	Bi-linear

Table 2. Paris' m parameters

5.1 One and two-dimensional modelling of the toe crack

Typically toe cracks are two-dimensional i.e. planar. Although several crack-like defects may initiate at various locations along the weld toe, these quickly coalesce to form a dominant crack, which is roughly of semi-elliptical shape [8]. Fatigue crack growth modelling of this crack may be accomplished by (i) approximating the geometry by an edge crack in a finite width strip [5,7] (see Figure 3a), (ii) approximating the geometry by a semi-elliptical surface crack and assuming a fixed crack shape which remains constant while the crack depth increases [4] (see Figure 3b) and (iii) as in (ii) but allowing both crack depth and crack length to evolve as prescribed by the Paris-Erdogan equations [13,14]. Differences in these three approaches arise purely as a result of the different Y factors. Approaches (i) and (ii) require the same computational time because, from a numerical viewpoint, the model is essentially 1-DOF. On the other hand, approach (iii) requires significantly greater computational time than the previous two due to the additional differential equation (3). Some of the numerical intricacies of approach (iii) are described in more detail in Reference [13]. Alternatively, this type of crack may be modelled as a through-thickness crack, which grows in the length direction [3]. This last assumption is rather crude but could be used once the depth of the semi-elliptical surface crack in (ii) and (iii) has reached the thickness of the plate. Failure probabilities derived from approaches (i) and (iii) are presented in Figure (4) for the 10m-bridge spectrum. The failure criterion used here to evaluate the probabilities in equation (8) is the crack depth reaching the thickness of the plate i.e. $a_{fail} = t$. Finite width effects on the stress intensity factors and hence the possibility of the length reaching the width of the plate are ignored. In other words, the plate is assumed to be infinitely wide. This assumption is also followed through in subsequent analyses. Both analyses were run for the bi-linear crack growth law. Note that for the one degree-of-freedom crack of Figure (3a), this signifies that the plate is severed. As can be seen in Figure (4), significant differences arise as a result of the different modelling of the cracked geometry. As mentioned previously, these discrepancies are caused by the different Y factors as these are raised to the mth power. The Y factors pertaining to the 2-DOF model are highly dependent on the crack shape (a/c). As the crack shape becomes more half-circular, their Y factors become considerably less than their 1-DOF counterparts for the same crack depth. The computational time required to carry out a 2-DOF analysis is significant but there is another advantage associated with it.

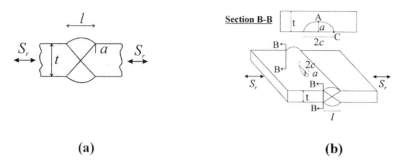

Fig 3. Schematic representation of a toe crack as (a) 1-DOF model and (b) 2-DOF model

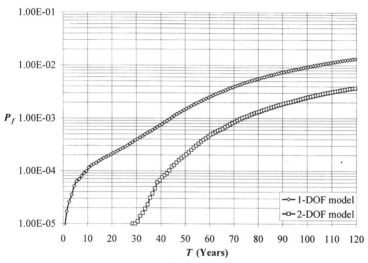

Fig 4. Cumulative failure probabilities for a bi-linear crack growth law, a thickness failure criterion $a_{fail} = t$ and the crack models depicted in Figure (3a) and (3b).

The 2-DOF methodology can be applied for a wider range of cracked geometries and can yield the statistics of not only the crack depth but also the crack length. This becomes important when inspection results must be considered. However, it has to be emphasised that 2-DOF crack modelling is not always appropriate. Cracks growing in webs, for example, may be more accurately modelled based on a 1-DOF assumption.

5.2 Fatigue and fracture modelling

The results based on $a_{fail} = t$ and a_{fail} determined by the Level 2A FAD of BS 7910 are depicted in Figure (5) for the 10m-bridge spectrum. These analyses are related to the 2-DOF crack (see Figure 3b) and the bi-linear crack growth law. The failure probabilities associated with the FAD are greater since critical crack sizes derived from the FAD are by definition smaller than the thickness of the plate. However, it has to be emphasized that these crack sizes are highly dependent on the fracture toughness.

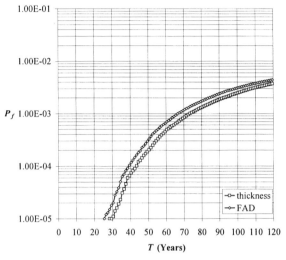

Fig 5. Cumulative failure probabilities for a 2-DOF crack, a bi-linear crack growth law and two failure criteria

If the mean of K_{mat} was less, the failure probabilities shown in Figure (4) for FAD would be even greater. Although both the deepest and the surface points of the crack are checked for failure, use of a FAD-based criterion does not require significant computational time. The disadvantage of using this criterion is that in the absence of any experimental data for K_{mat}, caution should be exercised when selecting the Weibull parameters and account should be taken of the temperature ranges within which the component is operating. Failing to do so, the probability curve may turn out to be unconservative.

5.3 Linear and bi-linear crack growth modelling

The results for a FAD criterion and a 2-DOF crack model subjected to the 10m-bridge spectrum are depicted in Figure (6) for the two crack growth laws described in Section (2). The figure clearly demonstrates the large differences arising from the application of these two laws. These differences are attributed to the spectrum content in low stress ranges. For a linear crack growth law, which is consistent with a straight S-N line, the fatigue lives are severely underestimated and the model becomes very conservative [7]. On the other hand, the bi-linear model has the ability to capture the kink in the S-N curve, which appears in most fatigue design codes. This phenomenon is also in line with experimental observations where at low stress ranges fatigue lives deviate from the straight line considerably. For a spectrum that contains many high stress ranges the differences between the two crack growth models would be significantly less.

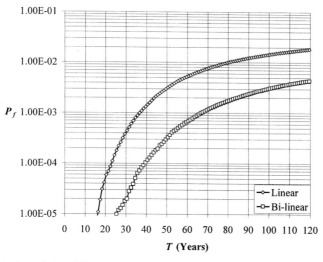

Fig 6. Cumulative failure probabilities for a 2-DOF crack model, a FAD criterion and two crack growth laws

6 CONCLUSIONS

This paper has quantified the effects of fracture mechanics modelling assumptions on fatigue life under bridge loading within a reliability framework, which also follows the principles of BS 7910. Factors investigated included the crack growth law, the crack geometry and the failure criterion, all pertaining to a butt-welded plate containing a toe crack. It was confirmed that 1-DOF modelling of crack geometry and the adoption of a simple linear crack growth law lead to conservative failure probability estimates. However, on the basis of the particular results obtained, both these assumptions may be over-conservative for bridge applications. Use of a FAD-based failure criterion is certainly more realistic for fatigue details but caution should be exercised when selecting the appropriate material properties, especially in the case of bridge steels where data is generally scant. For the specific case examined, the observed differences between a FAD-based and a thickness criterion were considerably smaller than those observed as a result of the modelling assumptions discussed earlier.

ACKNOWLEDGEMENTS

The work in this paper forms part of a project sponsored by the Highways Agency and undertaken in co-operation with Flint and Neill Partnership and TWI. The authors would like to thank the Highways Agency for their financial support. The opinions expressed are those of the authors and do not necessarily reflect those of the sponsors.

REFERENCES

[1] BS 7910 (1999). Guide on methods for assessing the acceptability of flaws in fusion welded structures, BSI, London.

[2] MTD Ltd (1992). *Probability based fatigue inspection planning*, The Marine Technology Directorate, London.
[3] Zhao Z., Haldar A. (1996). Bridge fatigue damage evaluation and updating using non-destructive inspections, *Eng. Fract. Mechs.*, **53**(5), 775-788.
[4] Lukić M., Cremona C. (2001). Probabilistic assessment of welded joints versus fatigue and fracture, *J. Str. Engng* (ASCE), **127**(2), 211-218.
[5] Righiniotis T.D. and Chryssanthopoulos M.K. (2000). Fracture mechanics based fatigue assessment of steel highway bridges, in *Conf. on Bridge Rehabilitation in the UK*, Inst. of Civil Engs, London.
[6] Paris P.C., Erdogan F. (1963). A critical analysis of crack propagation laws, *J. Basic Engng* (ASME), 528-534.
[7] Righiniotis T.D., Chryssanthopoulos M.K. (2001). Probabilistic estimation of residual fatigue life for steel bridge components, in *8th International Conference on Structural Safety and Reliability*, accepted for presentation.
[8] Maddox S.J. (1994) *Fatigue strength of welded structures*. Abington Publishing.
[9] King R.N., Stacey A. and Sharp J.V. (1996). A review of fatigue crack growth rates for offshore steels in air and seawater environments, in *15th Int. Conf. on Offshore Mech. and Arctic Eng.*, OMAE-96, 341-348.
[10] Cooper D.I. (2000). Unpublished communication on traffic load spectra for fatigue analysis, Flint and Neill Partnership, London.
[11] BD 77/98 (1999). Assessment of steel highway bridges for residual fatigue life, Draft Highways Agency Standard, London.
[12] BS 5400: Part 10 (1980). Steel, concrete and composite bridges, Code of practice for fatigue, BSI, London.
[13] Righiniotis T.D. (2001). Error estimation of shape changes during fatigue crack growth, in *10th International Congress of Fracture*, accepted for presentation.
[14] Shetty N.K. and Baker M.J. (1990). Fatigue reliability of tubular joints in offshore structures: Crack propagation model, in *9th Int. Conf. on Offshore Mech. and Arctic Eng.*, OMAE-90, 223-230.

APPENDIX

For the 2-DOF crack geometry, the Y factors used in this study are given by

$$Y_A = M_{km} M_m^A / \sqrt{Q} \qquad (A1)$$
$$Y_C = M_{km} M_m^C / \sqrt{Q} \qquad (A2)$$

The factors M_m^A and M_m^C are the Newman-Raju solutions for points A and C of a surface crack in an infinitely wide plate under tension [1] while M_{km} is the magnification factor associated with the weld [1]. Here it has been assumed that the latter factor is the same for points A and C. The variable Q in (A1) and (A2) is the crack's shape factor defined as [1]

$$Q = 1 + 1.464 (a/c)^{1.65} \qquad (A3)$$

For the 1-DOF crack model, the Y_A factor is reported in Reference [5].

Research

FULL STRENGTH JOINTS FOR PRECAST REINFORCED CONCRETE UNITS IN BRIDGE DECKS

STUART R. GORDON
Chartered Engineer, Edinburgh, Scotland
IAN M. MAY
Professor of Civil Engineering, Heriot-Watt University, Edinburgh, Scotland.

Synopsis

This paper describes a series of tests carried out to develop *in situ* joints suitable for precast reinforced concrete deck units, to be used for the decks of steel concrete composite bridges, in negative moment areas. The results of the tests are given together with comparisons to the theoretical strengths. The behaviour of the joints at serviceability and ultimate loads and beyond, are discussed. A technique for estimating the bearing stresses in the joints at ultimate load is described.

Introduction

Composite steel and concrete bridges have been designed and built in the United Kingdom for many years. The traditional form of construction has relied on the use of *in situ* systems of formwork and falsework to form *in situ* concrete decks. More recently, with the advent of motorway widening and design and build forms of tender procurement, moves have been made towards forming the concrete deck in precast construction. The advantages of the use of precast units are, speed of construction, avoidance of falsework and formwork for the deck and, since the units can be cast under factory conditions, a high standard of workmanship.

Several bridges have been built using partial depth precast units, which effectively rely on units to act as participating formwork. This form of construction requires a partial depth unit to be placed on the steel superstructure with pockets included to allow full interaction with shear connectors attached to the top flanges of the steel beams. Recent examples in Scotland include the A828 Creagan Diversion Bridge at Loch Creran in the West Highlands, designed by Carl Bro, the Muckle Roe Bridge in Shetland and the Hutton Mill Bridge in Berwickshire, both designed by The Babtie Group. These structures are all of similar form. However, the sizes of the units vary, those at Creagan being significantly larger when compared to the others at 11m wide and 3.5m long, Figure 1. The partial depth units provide one solution to the problem of precast deck units but suffer from several inherent disadvantages. The disadvantages include the large amount of reinforcement required in the units for the temporary erection conditions, the inherent flexibility of the units and the need for an *in situ* slab to be cast over the units to form the composite deck.

It would clearly be beneficial to adopt full depth precast units for the deck construction to overcome these inherent weaknesses. There is considerable interest amongst both designers and contractors alike to develop such units. This paper describes preliminary results from a programme of research being undertaken in association with the Steel Construction Institute to develop a suitable unit for full depth unit construction. The primary task is to provide a unit, which, when used in a bridge deck, will give at the least similar performance to that of a traditional *in situ* reinforced concrete deck.

The advantages of the use of full depth precast units can be summarised as:
- Once the units are in place they would require little further reinforcement and the amount of *in situ* concreting would be reduced dramatically to a minimal amount;
- The amount of reinforcement required in the full depth units can be significantly reduced in comparison to the half depth alternative;
- Construction traffic can be accommodated on the bridge deck at an earlier stage;
- The speed of construction would be improved leading to cost savings.

Figure 1 Construction of Loch Creran Bridge on the A828, showing placing of precast units.

However, one problem that remains is the development of suitable details to give continuity in negative, hogging, moment regions. This paper will describes research aimed at addressing this problem and describes the development of joint details necessary to provide longitudinal continuity in the negative moment areas of composite bridges.

In a negative moment area the deck slab is predominantly loaded in longitudinal global tension. At the joint a suitable reinforcement detail is required to ensure that the tension forces can be transferred through the joint and that the overall size of the joint can be minimised to aid construction. Full length lapped joints using straight bars are considered as an option but these produce excessive joint lengths. Details to minimise the joints are investigated in the test programme.

Positive, or sagging, moments are not considered to present any significant problems in this form of construction. In such areas the slab is in global compression and provided the joints are formed with a material that has low shrinkage, transfer of global compression through the deck slab joint should be ensured. The joint will still be required to resist the load effects from local wheel loadings. In both positive and negative moment regions the joint must be sufficiently robust to satisfy the serviceability requirements. The joint must also satisfy the fatigue loading requirements. However, this aspect is not addressed in this paper.

Test Programme

A series of full-scale specimens, which represent a section of the joint between two precast units, has been tested. Various arrangements of reinforcement and various geometrical arrangements of the joint have been investigated. In the tests the units have been loaded incrementally in tension up to and beyond the peak load. To date 12 specimens have been constructed and tested.

Details of typical specimens are shown in Figures 2 to 5. The unshaded parts of the specimens shown in the figures represent the precast units and the shaded part the *in situ* joints. The specimens were 1250mm long, 525mm wide and 265mm deep when the units incorporated precast downstands, Figures 4 and 5. These are required in practice to allow the *in situ* section to be formed without the use of on-site formwork. In the specimens shown in Figures 2 and 3 the downstand sections were removed and a slab depth of 190mm was adopted to represent the core of the *in situ* section. The specimens were cast with stiffened 203x203x86 Universal Column (UC) sections at the two ends in order that the specimens could be loaded by a 250 tonne capacity Losenhausen testing machine, Figure 6. The UC sections ensured that the loading in the specimens was uniform across the width. Fin plates were attached to the UC sections and the reinforcement, in the precast units, was welded to the plates. The specimens were constructed in 3 sections, the 2 precast units were constructed first and the *in situ* part was cast, usually, about a week later. For testing the specimens were mounted vertically in the rig and loaded incrementally in tension, Figure 6.

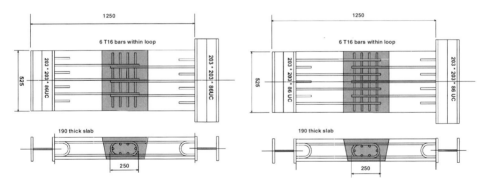

Figure 2. Test specimen for full depth units – Unit T11

Figure 3. Test specimen for full depth units – Units T10, T12

The tests were carried out in 4 series. Series 1 (T1-T3) was used to identify problems associated with the unit fabrication, curing, and handling and to develop the testing technique. Subsequent series investigated various arrangements of the joint. Series 2 (T4-T6) investigated arrangements of symmetric and non-symmetric joints with downstands. In one case, the downstand was found to enhance the strength of the *in situ* section. Therefore in Series 3 (T7-T9) a set of non-symmetric units without downstands was tested to investigate this effect. The *in situ* part of the Series 3 units tended to rotate in-plane under loading and failed at a load less than the predicted failure load. Hence, Series 4 (T10-T12) was carried out using symmetric units without downstands to overcome the deficiencies of the previous series.

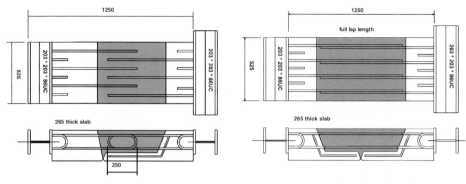

Figure 4. Test specimen for full depth units – Units T3, T6 (T7, T8, T9 excl. d/s) transverse bars omitted for clarity

Figure 5. Test specimen for full depth units – Units T1, T5 transverse bars omitted for clarity

In the early tests, the full theoretical strengths of the units were not attained, which lead to changes in the details of later specimens. In some of the later tests specimens were loaded beyond the peak load in order to investigate their ductility. Throughout all the tests surface strain readings were recorded using Demec gauges and crack patterns were noted. The behaviour of the specimens throughout the loading was recorded; in particular the response at the serviceability load.

The units can be divided into 4 distinct groups when considering the behaviour at loads exceeding the service load. Each unit contains transverse bars unless otherwise noted. These groups are:

- Non-symmetrical arrangements of looped bars with a central confined core (tests T6, T7, T8)
- Non-symmetrical arrangements of looped bars with a central confined core without transverse bars (test T9)
- Symmetrical arrangements of looped bars with a central confined core (tests T4, T10, T11, T12)
- Symmetrical arrangements of standard straight lapped bars (test T5)

Details of the units tested are given in Table 1.

Terms Adopted

The following terms are used to describe the behaviour of the specimens. The predicted yield load, P_y is the product of the area of bars and the measured 0.2% proof stress for the reinforcement; the service load has been taken as 50% of the predicted yield load, P_y; the ultimate failure load, P_u is the product of the area of bars and the measured ultimate tensile strength of the reinforcement and the cracking load, P_c as the product of the cross sectional area of the *in situ* concrete and the tensile capacity of the concrete, f_t, where $f_t = 0.33(0.8f_{cu})^{1/2}$. P_s is the maximum load carried by the specimen under test.

Table 1 Details of Test Specimens

Unit & Figure No	Reinforcement arrangement	P_y (kN)	P_u (kN)	P_c (kN)	P_s (kN)	P_s/P_u	Failure/ Remarks
T4 Fig 2, 8 bars + d/s	8 L – sym 125, 400 core (d/s) 4 no.T16 T&B, 6T16 transv	895.0	1023.4	---	900	0.88	Y R
T5 Fig 5	8 S – sym 125, lap (d/s) 4 no.T16 T&B, T16 transv T&B	895.0 1023.4*	1105.8 1264.4*	---	1075	0.85-0.97	Y NSD B
T6 Fig 4	6 L – n/sym 150, 250 core (d/s) 3 no.T16 T&B, 4T16 transv	671.3	767.5	---	620	0.81	Y 75mm R
T7 Fig 4 excl d/s	6 L – n/sym 150, 250 core 3 no.T16 T&B, 6T16 transv	612.5	723.6	207.8	660	0.91	Y 66mm CR
T8 Fig 4 excl d/s	6 L – n/sym 150, 250 core 3 no.T16 T&B, 8T8 transv	612.5	723.6	207.8	640	0.88	Y 100mm CR
T9 Fig 4 excl d/s	6 L – n/sym 150, 250 core 3 no.T16 T&B, no transv/ fibres	612.5	723.6	211.4	470	0.65	S LD/ CR
T10 Fig 3	7 L – sym 125, 250 core 4 no.T16 U bars T 3 no.T16 U bars B, 8T8 transv	669.3	768.0	218.9	730	0.95	F
T11 Fig 2	6 L – sym 150, 250 core 3 no.T16 U bars T, 3 no.T16 U bars B, 6T16 transv	669.3	768.0	218.9	770	1.00	Y
T12 Fig 3	7 L – sym 125, 250 core 4 no.T16 U bars T 3 no.T16 U bars B, 6T16 transv	669.3	768.0	218.9	740	0.96	Y Nom CR

* denotes strength includes additional T12 bars in the downstand

Key to Table 1

Reinforcement arrangement			
eg. T4	8 L – sym 125, 400 core (d/s) 4 no.T16 T&B, 6T16 transv		
key:	no.of Looped/Straight bars L/S – symmetrical/non sym about CL, bar spacing, core length (downstand) no. of main bars, diameter, positions top and bottom (T&B), no. of transverse bars		
Failure types and remarks			
Y	Yielding of reinforcement	B	Benchmark test using standard straight bars
F	Fracture of reinforcement	CR	Core rotation – in-plane
S	Sudden failure of *in situ* concrete section	LD	Low ductility
NSD	No separation of downstand	R	Rotation of downstand away from *in situ* section

Materials

Material strengths were measured for the concrete and reinforcing bars used in the units and were used to determine the various parameters in Table 1 and in the bearing calculations.

Test Results

The results from first Series, T1-T3, have not been presented as these tests were used to develop the test techniques used in subsequent tests.

Figure 6 Testing of units in Losenhausen Rig

Behaviour at the service load, tests T4-T12

In all specimens transverse cracks developed throughout the units, mostly within the surface concrete of the *in situ* section, but to a lesser extent in the precast sections. At the service load the cracking was generally minor and evenly spaced across the units. In units T8, T9, T11 and T12 cracks developed primarily on one side of the units indicating a slight misalignment during loading. However, following the cracking of the concrete, some redistribution of loading occurred beyond this point and the cracks formed more evenly. As expected, cracks developed along the interface between the precast and *in situ* sections in all the units tested. The interface surfaces were not roughened to form a proper construction joint, as would happen in practice, as this was considered likely to give lower cracking loads. Full depth cracks developed through all of the units at the service load. This was to be expected as the service load exceeds the cracking load by approximately 50%. The cracks appeared in both the precast and *in situ* sections. Minor splitting cracks occurred in most of the units over the positions of the main reinforcing bars. Minor diagonal cracking was noted particularly in the units with staggered arrangements of reinforcement (T6, T7, T8 and T9) predominantly in the *in situ* concrete. Strain levels were recorded during loading using a series of Demec points on the units to ensure that the loading was uniform across the width of the unit.

Behaviour at loads exceeding service load

The units were progressively loaded beyond the service load, initially in increments of 70kN, which were reduced to 50kN increments near failure. After the peak loads displacement control was used to continue some of the tests in order to investigate the ductility of the specimens. The extent of the cracking increased significantly near the failure loads. In all cases, on removing concrete from the top of the unit after failure, it could be seen that the bars had remained attached to the fin plates on the UC sections indicating that failure had been due to the reinforcement bars yielding or fracturing within their length. The predicted yield and failure loads for the units and the maximum test loads sustained are given in Table 1.

Non symmetrical units T6, T7, T8 - looped bars

The units failed by yielding of the main reinforcement. Because of the non-symmetric arrangement of the reinforcing bars there was a tendency for local in-plane rotations in the *in situ* sections (relative to the precast ends) to occur. The rotations were observed at loads above 550kN. Part of the area at the ends of the *in situ* section contained no longitudinal looped bar reinforcement, point A, Figure 7. Wide transverse cracks propagated across the insitu concrete at these unreinforced areas. The concrete at the outer bars spalled off due to cracking within the unreinforced areas and the *in situ* concrete moved away from the precast sections. The strains increased and major cracking developed throughout the precast sections and the concrete near the surface of the *in situ* sections. Two major transverse cracks opened on diagonally opposite corners, point B, Figure 7. A parallel transverse crack formed at the interface of the *in situ* and precast concrete, point C, Figure 7. The cracks in the *in situ* section were closely spaced and comprised many diagonal and splitting cracks. The spalled surface concrete was removed after the failure load had been reached for each of the specimens to allow the core concrete to be inspected. The central portions of the core were seen to be undamaged with no cracking present. The 16mm diameter lacer bars had not deformed except near their ends. Local crushing of the concrete in the core had occurred at the unrestrained ends.

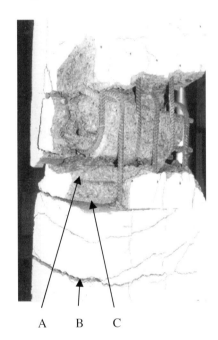

Figure 7 Testing of unit T6 showing extension at failure load

In unit T8, the 8mm diameter lacer bars were restrained tightly within the looped bars and bulged out between the looped bars at the top and bottom edges of the concrete core. One of the lacer bars had sheared through completely adjacent to one of the looped bars. At an overall extension of 100mm of the unit, the central core developed a diagonal crack, which propagated between the ends of two adjacent looped bar ends. Some of the tests were carried out until the final extension of the units was 75 to 100mm with only a modest reduction of load.

Non-symmetrical unit T9 – looped bars without transverse bars

Unit T9 had no transverse reinforcement present and the *in situ* concrete was fibre reinforced only. Above the service load the load was increased in increments of 70kN, but a sudden failure of the unit occurred at 470kN. The central core concrete parted along diagonal lines joining the looped ends. The looped bars were visible and the core showed distinct failure planes radiating between the ends of the loops.

Symmetrical units T4, T10, T11, T12 - looped bars

The units failed by yielding or fracture of the main reinforcement. Diagonal and splitting cracking was principally concentrated around the centre of the *in situ* sections with wide transverse cracks at the top and bottom edges of the cores. Transverse cracking was noted throughout the units with the *in situ* major cracks being spaced at approximately 150mm. In units T10 and T12, the central concrete core was intact after failure had occurred except at the ends where the concrete core had failed locally and the looped bar had slipped off the concrete core. Where local crushing of the concrete and slip of the outer looped bars occurred slight rotation of the core was noted but no cracking was observed within the core. In the units in which 8mm diameter lacer bars were used they behaved as unit T8. In unit T10 the bottom central looped bar fractured, at the ultimate load, beneath the central core and 'necked' at the corresponding location on the other side of the unit. In units T4, T11 and T12, the reinforcement had clearly yielded with crack widths of up to approximately 11mm but there was no evidence of the bars fracturing.

Symmetrical unit T5 - standard straight lapped bars

At failure, the main reinforcing bars yielded near the interface of the precast and *in situ* concrete. Extensive cracking was noted in the precast sections and at the interface. Large transverse cracks appeared across the precast downstand. It was clear that the bond between the downstand and the *in situ* concrete stitch was maintained throughout, thus contributing to the strength of the *in situ* section. Including the contribution of the reinforcement in the downstand increases the predicted yield load capacity by 23.6%, Table 1.

Discussion of results

In Series 2 and 3, because the central cores of specimens T6, T7, T8, and T9 were non-symmetrically loaded due to the non-symmetric distribution of the looped bars the outer looped bars tended to slip off the concrete core as the adjacent concrete split and spalled. Because of this the load capacity of the outer looped bars was limited and the units carried less than the predicted ultimate loads. It was anticipated that by using symmetric arrangements of bars the specimens would be able to carry the predicted failure load. Thus the units for Series 4 were designed to be symmetrical to avoid in-plane rotations and develop the full failure load. Two arrangements of units were used for Series 4, one having staggered symmetrical bars, Figure 3, and one having paired symmetrical bars, Figure 2. The staggered arrangement tended to force a failure at the more lightly reinforced end of the unit, the reinforcing bars carrying approximately 33% more load than those in the more heavily reinforced end. The core was hence, loaded unevenly. However, the core concrete showed no signs of distress under this form of loading. The Series 4 units carried the predicted failure loads. In Series 3, the unsymmetrically reinforced specimens the use of 16mm diameter lacer bars was preferred as it reduced the effects at the ends of the core. However, in the symmetrically reinforced units the smaller lacer bars were adequate.

To study the in-plane rotations observed in the tests, an elastic analysis was used. A non-symmetric frame model was set up using members with the properties of a single 16mm diameter bar and with a stiffer transverse beam to represent the central concrete core. The analysis showed clearly the core rotations observed in the tests. The central bars carried the correct proportion of the load applied, the intermediate bars were more heavily loaded than expected, and the outer bars carried less load. With this uneven distribution of loading there would be a likelihood of failure at loads below the predicted ultimate loads because of the tendency noted in the tests for the end bars to slip off the core. The limited width of the unit used in the test also influenced the degree of rotation. For a very narrow unit this effect would be significant as the non-symmetry would be a dominant factor. In comparison, in a unit as wide as a bridge deck with a large number of staggered reinforcing bars significant rotations would not occur.

The sudden brittle failure of unit T9 shows that the inclusion of the lacer bars was important. It is clear that the lacer bars provide restraint to avoid tensile failure of the core. Thus the use of fibre reinforced concrete without lacer bars is not suitable for providing sufficient strength in the *in situ* concrete core. Transverse reinforcement was omitted in unit T9 in order to investigate the use of fibres in the concrete to resist direct global tension. It is appreciated that local wheel load effects would require transverse reinforcement to be incorporated in the bridge deck joints to resist transverse bending effects. Hence in practice it would not be possible to omit this reinforcement.

Bearing stresses

The behaviour of the lacer bars depended on the diameter used. The 16mm diameter lacer bars remained straight, whereas the 8mm bars were observed to have deformed significantly when inspected after failure. The major contribution of the lacer bars is to confine the concrete in the central core. The bearing stresses calculated for the inside of the looped bars is significantly higher than that permitted in BS5400[1]. However, no evidence of crushing of the concrete was seen inside the loops.

Table 2 Details of Bearing Stresses in Looped Bars for Series 4

Unit	Ultimate load carried per loop (N)	Effective cruciform area (mm^2)	Average bearing stress, σ_b (N/mm^2)	Allowable bearing stress, σ_{all} (N/mm^2)*	ratio σ_b / σ_{all}	Transverse lacer bars and spacing
T10	243,333	2408	101.1	107.7	0.94	8T8 transv, 125
T11	256,667	3680	69.8	102.6	0.68	6T16 transv, 150
T12	246,667	3280	75.2	96.7	0.78	6T16 transv, 125

* partial factor of 1.5 removed and actual material strength used for concrete

The equation given for the bearing stress calculation in BS5400 has a partial safety factor of 1.5, this has been taken as 1 in the following calculations. In order to include a contribution from the lacer bars a cruciform has been considered, which leads to a larger area for the bearing stresses to be distributed over. For unit T12, using 16mm diameter bars at 125mm centres, bent to a minimum radius former of 48mm, the allowable stress of $96.7 N/mm^2$, taking f_{cu} as $54.0 N/mm^2$, is greater than applied stress. The effective cruciform area is $3280 mm^2$ compared to $1536 mm^2$ for the looped bar only. Adopting the above approach the bearing stresses at the ultimate loads are found to be acceptable.

Conclusions

The results of tests of 12 specimens have been presented. The specimens included symmetric and non-symmetric arrangements of straight and looped bars. The failure load in the non-symmetric units occurred at approximately 90% of the predicted value (units T6-T8). The failure load in the symmetric units occurred at approximately 95% of the predicted value (units T4, T10-T12). The specimens showed good ductility. The effect of rotation of the units due to non-symmetric arrangement of the reinforcement has been discussed. The tests show that providing a symmetric arrangement of the reinforcement avoids the possibility of the bars slipping off the concrete core and other edge effects, and leads to an increase in the strength of the unit.

The tests undertaken demonstrate that the ultimate strength of an arrangement of looped bars around a central confined core of concrete with lacer bars is capable of carrying the ultimate load based on the ultimate strength of the looped bars.

In the benchmark test using straight bars, unit T5, the failure load exceeded the predicted failure load based on the main reinforcing bars. However it was difficult to determine the contribution, if any, to the strength from the downstand reinforcement.

Bearing stresses have been discussed with respect to the requirements of BS5400. Adopting an approach to the determination of the bearing stress which includes a contribution from the lacer bars, shows that the stresses are within the limits proposed by BS5400. The tests show that the use of 8mm diameter lacer bars is adequate for the joint to carry the predicted failure load. The use of fibre reinforced concrete on its own as a means of providing adequate strength within the *in situ* concrete core is not sufficient.

Acknowledgements

The authors would like to acknowledge the financial support and the guidance given by the Steel Construction Institute and Corus throughout the test programme.

Reference

1   British Standards Institution, Code of Practice for Design of Concrete Bridges, Part 4:1990. Steel, Concrete and Composite Bridges. BS5400: London, 1990.

Vibration and Impact Studies of Multi-girder Steel Bridge in Laboratory

Ling YU and T. H. T. CHAN

Department of Civil and Structural Engineering
The Hong Kong Polytechnic University
Hung Hom, Kowloon
HONG KONG

ABSTRACT

The impact of highway bridges resulting from the passage of vehicles across spans is an important problem encountered in the design of bridges. Many parameters that may affect the impact factor include the velocity and dynamic properties of the moving vehicle, the dynamic properties and boundary conditions of the bridge, and the roughness in road pavement. According to the truck design loading specified by AASHTO and the structural similarity principle, a vehicle and a multi-girder steel bridge are designed and constructed in laboratory for impact studies and further identification of the moving axle loads. The 3-D finite element method in the professional software package, ABAQUS, is used to calculate the dynamic characteristics of the bridge. The experimental modal analysis technique is adopted to measure the modal properties of both vehicle and bridge. The influence of both the vehicle velocity and lateral lane position on the impact factor is investigated. Accurately estimation of the static responses of bridges due to moving vehicles is considered and compared with the filtered static responses. Results are useful for the bridge design and the further study on the development of impact formulae proposed by the bridge loading code.

1. INTRODUCTION

The dynamic response of a multi-girder bridge to crossing vehicles is a complex problem affected by the dynamic characteristics of both the bridge and the vehicle and by the bridge surface conditions. Many of parameters interact with one another, further complicating the issue, and consequently, many research studies have reported seemingly conflicting conclusions (Bahkt and Pinjarkar 1989, Hwang and Nowak 1991, Fafard and Savard 1993, Cantienti and Barella 1995 and Mabsout et al 1999). Nevertheless, any theoretical and experimental investigation will provide useful results and will enhance the knowledge base in this field.

The objectives of this paper is to evaluate the vibration and impact of multi-girder steel bridge made in laboratory based on the structural similarity of bridge. It describes the experimental testing and discusses the results and matters arising therefore. The results provided several conclusions. They are associated with the design of bridge-vehicle model in laboratory, the agreement of the modal characteristics of bridge theoretically and experimentally. Then with evaluation of equivalent static responses, effect of both vehicle speed and multilane on the impact factor, which vibration modes excited by the moving vehicle. And finally with the ability of code provision to account accurately for dynamic vehicle-bridge interaction.

2. EXPERIMENTAL MODEL

The experimental model is used to simulate the vehicle-bridge system that includes the interaction between the vehicle and the bridge. Based on the structural similarity theory, the experimental results of the model can be extended to the real cases. Here, the similarities main include the geometry similarity, load similarity, time similarity and dynamic behavior similarity. The ratio of the axle spacing and the bridge length, the ratio of the vehicle weight and the bridge weight, the frequency ratio of the vehicle and the bridge are also considered.

Bridge Model

The bridge deck is composed of a steel deck and five I-section or rectangular-section steel girders. According to a typical single span bridge [Fafard, 1993], the size of the bridge deck (R.C.C.slab) is assumed as 8 feet in length, 4 feet in width and with various thickness. Firstly, the steel deck without girders is considered. The natural frequencies are listed in Table 1.

Table 1. Natural frequencies of steel deck without girders

Dimensions			Natural frequencies (Hz)				
Length a (feet)	Width b (feet)	Thickness h (inch)	1	2	3	4	5
8	4	1/8	1.263	3.612	5.130	8.515	11.643
8	4	3/16	1.895	5.418	7.695	12.773	17.464
8	4	¼	2.526	7.223	10.260	17.031	23.285
8	4	½	3.789	10.835	15.390	25.546	34.928

Table 2. Natural frequencies of steel deck with girders

Dimensions of ribs		Natural frequencies (Hz)				
Width (inch)	High (inch)	1	2	3	4	5
1	0.5	3.212	12.900	20.473	29.064	41.519
		<u>3.306</u>	<u>12.165</u>	<u>22.094</u>	<u>27.385</u>	<u>44.636</u>
0.5	1	4.945	10.687	19.837	27.293	29.710
		<u>5.255</u>	<u>10.525</u>	<u>21.046</u>	<u>25.291</u>	<u>27.624</u>
1	1	6.314	16.804	25.287	38.772	39.543
1.5	1	7.268	27.032	29.085	57.625	58.757
1	1.5	10.627	22.959	42.519	47.661	58.208

Notes: Values underlined are for deck with 3/16b thickness

Then the girders are along the longitudinal direction and located in 1/10b, 3/10b, 1/2b, 7/10b, 9/10b respectively. If the length and width of deck is not changed but their thickness are 1/4b and 3/16b respectively, the corresponding deck natural frequencies of the bridge deck are listed in Table 2 when girders with different sections. Obviously, the frequencies increase in comparison with those in Table 1.

The structural model of multi-lane bridge deck is shown in Figure 1. The width of each lane can be determined as 0.3048 m. Since the lanes are symmetrical, only three tracks are glued on the top surface and they are located at 1/8b, 3/8b and 1/2b respectively. So the wheel spacing of the moving vehicle is less than 0.183 m by AASHTO.

Moving Car
Based on the AASHTO specifications (H20-44 or H15-44), the wheel spacing is 0.180 m. Then the axle spacing is 0.42m because the ratio of the wheel spacing and the axle spacing is 6/14 in the specifications (H20-44 or H15-44). The front axle weight is 0.2W, the rear 0.8W. Here, W is the total weight of the moving car. Referring to Fafard (1993), the ratio of the vehicle (H20-44) and the bridge weight is 0.1. The weight of the bridge model is 172.3 Kg for ¼ inch thickness plate (137.1 Kg for 3/16 inch thickness). Therefore the weight of moving car can be initial determined as 17.2 Kg (13.7 Kg for 3/16 one).

Regarding the strain induced if the strain responses are too small to measure accurately, the weight of the moving should be added. When the thickness of is ¼ inch and the weight of car 17.2 Kg, the calculated results show that the maximum strains are about 12 micro strain on the plate, 100, 105 and 110 micro strain at the middle span of the first, second and third girder respectively. This amount of responses is appropriate for measurement. Therefore, more than 17.2 Kg weight of moving car is suitable. In practice, in order to get a bigger amplitude of bridge response induced by the vehicle moving across the bridge, the actual car model was fabricated with a wheel spacing of 0.2 m and an axle spacing of 0.46 m. The total weight of the car is 19.4 Kg, in which the rear axle 14.2 Kg.

Measuring System
The responses of the model are measured by the strain gauges and accelerometers, which are mounted under the bottom surface of girders. Total 26 strain gauges and 6 accelerometers are mounted at 1/4a, 3/8a 1/2a, 3/4a and 7/8a as shown in Figure 1. The car speed is measured by two columns of triggers, which are installed on the top surface of the bridge deck. There are nine equally spaced triggers for each column as shown in Figure 1. Figure 2 shows the experimental setup in laboratory.

3. ANALYSIS OF 3-D MODEL USING FINITE ELEMENT METHOD

In order to analyze the dynamic characteristics of the model, the commercial software ABAQUS is adopted to model the slab-on-girder bridge. Usually, shell elements or brick elements are used to model the slab, and the beam elements or shell elements are used to model the girders. As the neutral axis of each girder is different from that of the slab, rigid links are generally used to connect the girder elements to the slab model. ABAQUS offers a wide variety of shell and beam elements. After comparing the analytical results with the experimental ones, the S8R5 type of thin shell elements are finally chosen for slab and B32 beam element for girders respectively. The slab model consists of 25 x 25 numbers of shell elements in both longitudinal direction and transverse direction. The girder model consists of 3 x 25 numbers of beam elements. The constraints support at the middle surface of the plate. One end of the model is simply supported. It restrains the movement along x, y and z directions, but free to rotate about the three axes. The other end is on a roller support. It allows additional movement in vertical direction.

The results are listed in Table 3. It shows that the analytical results are in good agreement with the experimental ones, especially for the first two lower modes. Although the relative percentage errors are larger for the higher modes, they are within acceptable region.

Table 3. Modal Frequencies of Model Bridge (Hz)

Mode No.	1	2	3	4	5	6	7	8
Analytical	9.52	12.6	28.9	37.8	40.9	54.6	65.6	83.8
Experimental	9.42	12.76	27.01	34.9	38.45	49.77	61.65	73.2
error (%)	1.05	-1.27	6.53	7.67	5.99	8.85	6.02	12.65

4. PRELIMINARY EXPERIMENTAL STUDIES

Static Tests

The static tests include calibration and the "equivalent" static experiments. Calibration process is needed to convert data from voltage to strain or moment format. Three locations were chosen to add the load. They are at the middle of the three lanes on which the dynamic test will be carried out. Several weights varying from 0 to 30 Kg are placed at one of these locations. And the corresponding voltage variation values are then found for different loads. Finally the calibration factors for each gauges can be determined by linear fitting method for each location. A good linearity of calibration factor can be found for the load location and the near locations. But the linearity is not so good for the far locations from the load location. The calibration factor should be verified and adjusted with the theoretical static calculation under a unit load. The final calibration factors at some typical locations are listed in Table 4.

Table 4. Calibration Factors at Some Particular Locations

Load Location	Calibration factor (micro strain / voltage)								
	3	4	5	13	14	15	18	19	20
#15, Girder 1	16.83	16.66	13.28	31.16	11.83	16.32	23.54	12.40	12.40
#13, Girder 3	12.36	10.21	17.56	13.54	20.01	19.12	12.41	9.57	16.85

The "equivalent" static experiments involved the test are traversing the spans along different lanes at a crawling speed, say about 0.5 m/s here, so that dynamic effects would not be induced in the superstructure (Bahkt and Pinjarkar 1989). It has been suggested convincing that, at this speed, dynamic responses are not present and a static response is obtained. In order to obtain the actual static responses accurately, a filtering technique is required to filter out the collected responses. Different kinds of filters can be chosen for the equivalent static responses in the signal processing toolbox of Matlab (Henselman and Littlefield 1997). The type of filter used in this study was a low-pass Butterworth digital filter. This type of filter, ideally, retains all frequency responses below the cutoff frequency and omits all frequency responses above the cutoff frequency (Thater et al 1998). Several Butterworth filter designs were tested by varying various filter orders as well as cutoff frequencies. By trial-and-error it was found the best approximation to the correct static curve was obtained using a fourth-order Butterworth filter at a cutoff frequency between [0.001, 5] Hz for the experimental model. A particular static response comparison after and before filtering is illustrated in Figure 3. The data collected during the equivalent static experiments will be used to calculate the impact factors. Under each experiment condition, the experiment repeated eight times so that the static response can be evaluated correctly. Finally, the corresponding maximum static response is chosen to be the mean value of the eight maximum static responses. Figure 4 shows a comparison of maximum static values. It shows that selection of the mean value of static responses is appropriate for calculation of the impact factors.

Dynamic Tests
The dynamic tests constitutes three series of tests where the vehicle traveled at various speeds ranging from 0.5 m/s to 1.55 m/s. Same as the previous static tests, the experiment repeated eight times at each speed. The three series of tests are as follows: the vehicle run along the lane 1, 2 and 3 respectively. The data collected during the dynamic experiments, together with the data collected earlier from the static experiments, were used to determine the impact factors. The lateral vehicle lane was varied to investigate the excitation of modes other than the longitudinal flexural mode, which is generally assumed to be the fundamental mode. The test speed was varied to investigate the effect of speed and the consequent effect of time taken to cross a span. A typical comparison of maximum dynamic response is illustrated in Figure 5.

Modal Analysis
A separate experiment was conducted to determine the dynamic properties of the bridge-vehicle model in laboratory. The experimental modal analysis technique is employed here. According to the features of the slab-on-girder bridge, total 45 acceleration response points are placed on the top surface of the slab, in which a column of 9 points are equally spaced put on the slab surface along each girder below. A single input and multi output method is used in the experimental modal analysis of the model. The identification of the modal parameters of the model is based on the measured frequency response function (FRF) of the structure. If sufficient FRFs are measured, the modal parameters can be estimated accordingly. The software package is the Data Auto Sample and Process system provided by the China Orient Institute of Noise & Vibration (INV-DASP). The data acquisition adopted the time-varying base method in order to improve the accuracy of the FRF calculation. The curve fitting method employed the multiple degree of freedoms model of the complex modes. The satisfactory identified modal frequencies are listed in Table 3 together with the FEM results.

5. IMPACT FACTOR (IF) STUDIES

Definition of Impact Factor
Bahkt and Pinjarkar (1989) have given a review of dynamic testing of highway bridges. They summarized eight definitions of impact factors, which are currently used in different countries. After comparing the eight definitions, they concluded that the definition 8 is apparently more logical as the dynamic and static values substituting into the other seven equations may not be taken from the same location. Therefore, the definition 8 is used in this study, which is the same as in the AASHTO (1994) LFRD code. It is as follows,

$$IF = \frac{\delta_{dyn} - \delta_{sta}}{\delta_{sta}} \quad (1)$$

where δ_{dyn} is the maximum response under the vehicle travelling at normal speed, δ_{sta} is the maximum response under the vehicle travelling at crawling speed.

Effects of Vehicle Speed
Several researchers have investigated the vehicle speed as a parameter influencing IF values. Hwang and Nowak (1991) investigated the relationship analytically with three different vehicles weighting 178, 356 and 534 kN. The lightest truck induced increasing IF with

increasing speed; however, the two heavier vehicles induced a smaller IF value as speed increased. Haywood (1995) found that values of IF are small for speeds <40 km/h on the one timber, and two concrete short span bridges tested with no other conclusive indication of a correlation found. Cantienti and Barella (1995) also discovered somewhat inconclusive results from tests at three prestressed box bridges of 24-70 m span. Two of the three bridges exhibited increasing IF with speed; one did not. It seems from the previous literatures that the relationship between IF and speed is different for each bridge.

Results of the present study of the relationship between IF and speed are shown in Figure 6. Plots of IF versus speed are shown for all three cases where the car ran along three different lanes. Obviously, the IF increases with increasing speed for all the three cases. Of course, there is a different increasing extent for each case. When the car ran at a speed less than 1.2 m/s (also say 10U), the IF increases gradually with the speed. After that, the IF increases sharply when the car ran at a higher speed 1.44 m/s (15U). In particular for car running along lane 1, the side lane, the IF increment is dramatic. This is probably due to vehicle excitation becoming more significant at higher speeds and leading to larger dynamic response. Since the static response is not influenced by the excitation, the IF is consequently larger.

It is also interesting to find that the smallest IF in each case always corresponds to the girder that are welded under the running lane. More distant the girder from the running lane, the bigger the IF is. This is mainly because the static live load is very small for the farther girder from the running lane.

Effects of Lateral Lane Position
Figure 7 illustrated the effects of lateral lane position on the impact factor. Values of impact factor are plotted for each girder. At one vehicle speed, the impact factor is largest when the car ran along the side lane 1. The smallest impact factor occurs in the case of car running along the central line lane 3. When the car runs along one lane, the impact factor under the corresponding girder is always smallest, but the largest impact factor occurs in the girder 5. For example, the car runs along the central symmetric lane 3, the interior girder 3 has a smallest impact factor but the exterior girders 1 and 5 have larger impact factors. This is because the static response in girder 3 is larger than one of the exterior girders when the car runs along central lane 3.

Table 5. Impact factors for each individual girder

Girder Number	5 U (0.76 m/s)			10 U (1.1 m/s)			15 U (1.44 m/s)		
	Lane 1	Lane 2	Lane 3	Lane 1	Lane 2	Lane 3	Lane 1	Lane 2	Lane 3
Girder 1	0.082	0.086	0.061	0.124	0.224	0.206	0.571	0.719	0.896
Girder 2	0.116	0.042	-0.019	0.149	0.121	0.131	0.842	0.512	0.565
Girder 3	0.198	0.095	0.037	0.264	0.198	0.167	1.321	0.728	0.480
Girder 4	0.293	0.071	0.028	0.309	0.234	0.202	2.055	0.898	0.579
Girder 5	0.804	0.134	0.196	1.059	0.364	0.349	4.979	1.349	1.022
Average	0.298	0.085	0.061	0.381	0.228	0.210	1.953	0.841	0.708

When the car ran along different lanes at three speeds, the impact factors for each individual girder are listed in Table 5. The average impact factors also shown in Table 5. Results together with Figure 7 show that different girders might have different values of the impact factor and therefore one value of the impact factor might not be good for evaluating the impact factor in bridge design. Results also show that many cases have impact factors greater than 0.25 that is specified in the current design code in Hong Kong (BS5400 1978). For instance, all impact factors at speed 15U, impact factor at girder 5 along lanes 2 and 3, and impact factors at girder 3, 4 and 5 along side lane 1. Therefore there is an interest of studying the cause of high impact factors.

Which Mode Excited by Moving Vehicle
Whether the vehicle runs along which lane, the impact factors are larger in girder 4 as shown in Figure 7. The power spectrum analysis of the corresponding strain responses at 12 point of the middle span is illustrated in Figure 8. It shows that the larger impact factor is mainly due to the fundamental frequency of about 9.5 Hz, i.e. the first flexural vibration mode of the plate in longitudinal direction. Some of the other higher modes, which have lateral and/or torsional components, are difficult to be found from these spectrum analyses. This means the vibration of the bridge with a larger impact factor does not associate the lateral vibration of the bridge.

6. CONCLUSIONS

The vibration and impact studies have been carried out in this paper. Some conclusions can be drawn as follows: 1) The FEM and experimental results have a good match, and it showed that the scaled bridge-vehicle model could be used in evaluation of the vibration and impact studies. 2) The fourth order Butterworth filter employed is a good filter for determination of the equivalent static responses and for filtering the dynamic responses. 3) The impact factor increases with the increase of the vehicle speed. 4) The larger impact factor is mainly due to the first flexural vibration mode in longitudinal direction of the bridge. The lateral modes do not contribute to the vibration of bridge.

ACKNOWLEDGEMENTS

The support provided by the Hong Kong Research Grants Council is gratefully acknowledged.

REFERENCES

1. AASHTO, (1996). *Standard Specification for Highway Bridges*, American Association of State Highway and Transportation Officials.
2. ABAQUS, (1998). *Standard User's Manual*, Volume II, Version 5.8, Hibbitt, Karlsson & Sorensen, Inc.
3. Bakht, B. and Pinjarkar, S. G., (1989). "Dynamic Testing of Highway Bridges-A Review". *Transportation Research Record 1223*, 93-100.
4. BS5400: Part 2. (1978). *Steel, Concrete and Composite Bridges. Specification of Load.* British Standards Institution, London, United Kingdom.
5. Cantienti, R. and Barella, S., (1995). "Swiss Testing of Medium Span Bridges." *Proceedings of the fourth international bridge engineering conference.*
6. Fafard M. and Savard M. (1993). "Dynamics of bridge-vehicle interaction". *Structural Dynamics --- EURODYN'93*, 951-960.
7. Henselman, D., and Littlefield, B. *The student edition of MATLAB*, Version 5. Printice-Hall, Upper Saddle River, N.J.
8. Heywood, R. J., (1995). "Are Road-Friendly Suspensions Bridge-Friendly?" OECD DIVINE, *Proceedings of the fourth international bridge engineering conference*, 281-295.

9. Hwang, E. S. and Nowak, A. S., (1991). "Simulation of Dynamic Load for Bridges". *Journal of Structural Engineering*, 117 (5), pp.1413-1434
10. Mabsout M. E., Tarhini K. M., Frederick G. R. and Kesserwan A. (1999). "Effect of multilanes load distribution in steel girder bridges". *Journal of Bridge Engineering*, 4(2), 99-106.
11. Nowak, A. S., (1995). "Calibration of LRFD code". *Journal of Structural Engineering*, 121 (8), 1245-1251
12. Thater, G., Chang, P. Schelling, D. R. and Fu, C. C., (1998). "Estimation of Bridge Static Response and Vehicle Weights by Frequency Response Analysis". *Can. J. Civ. Eng.* 25: 631-639

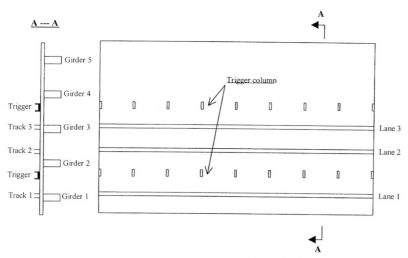

1.a) Top face of bridge deck

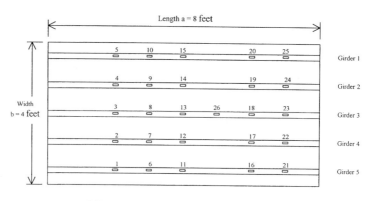

1.b) Bottom of bridge deck

Figure 1. Layout of experimental bridge deck

Figure 2. Setup in experiments

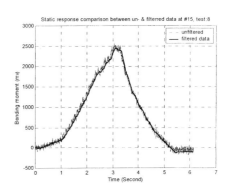

Figure 3. Typical static responses

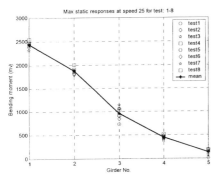

Figure 4. Maximum static response

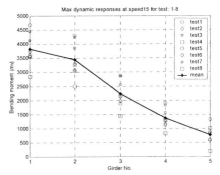

Figure 5. Maximum dynamic responses

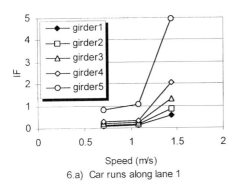

6.a) Car runs along lane 1

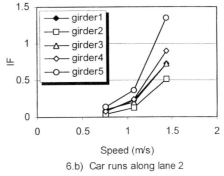

6.b) Car runs along lane 2

Figure 6. Impact factor vs. vehicle speed

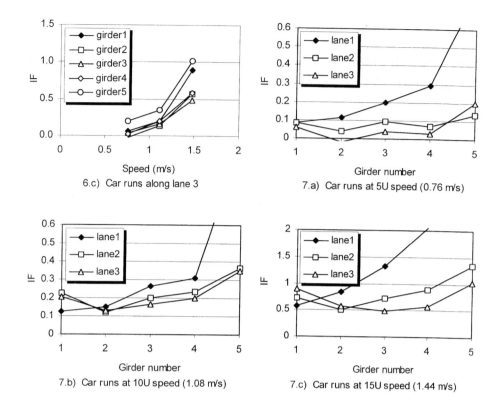

Figure 7. Impact factor vs. lateral location

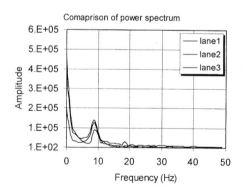

Figure 8. Comparison of spectral analysis

Fundamental study on application of carbon fiber reinforced polymer strips to a notched steel member

HIROYUKI SUZUKI, MEISEI University,
Hino-city, Tokyo, JAPAN

1. Introduction

It is reported in recent year that many failures occur in highway bridges[1]. The causes of the failures are an increase of a vehicle weight, an increase of traffic, and an increase of impact according to them. Bridges, which need strengthening to resist an increase of the load, are being more. When steel members are used in repair and/or strengthening of the existing steel bridges, high strength bolts or welding are used to connect of members. However, as working space is very narrow, workability is very bad in both cases. Therefore, it is considering that carbon fiber reinforced polymer strips, hereinafter called CFRP strips, are used in repair and/or strengthening of the existing steel bridges.

The CFRP strips are produced by solidifying carbon fiber with resin. The CFRP strips are very light and are only attached an existing member using an epoxy adhesive. So, procedure is very easy and workability is very good.

However, there are no studies on application of the CFRP strips to steel members, and only two studies on concrete beams strengthened by the CFRP strips[2,3]. Of course, the CFRP strips have never been used to steel members in Japan.

Table 1 Size of the CFRP strips

Specimen	$W=a/w$ [*]	b(mm) [**]	$B=b/a$	l(mm) [***]	$L=l/2a$
Tn	0.25	0	0	0	0
Tn0.4-2	0.25	10	0.4	100	2
Tn1.0-2	0.25	25	1.0	100	2
Tn2.0-2	0.25	50	2.0	100	2
Tn0.4-6	0.25	10	0.4	300	6
Tn1.0-6	0.25	25	1.0	300	6
Tn2.0-6	0.25	50	2.0	300	6

[*]: a is notch length(25mm), and w is notch width(100mm).
[**]: b is width of the CFRP strips.   [***]: l is length of the CFRP strips.

Table 2 Mechanical properties

Material	E^{*} (N/mm^2)	σ_Y^{**} (N/mm^2)	σ_B^{***} (N/mm^2)	El^{****}
SM400	2.1×10^5	265	402	0.180
CFRP strips	1.5×10^5	-	2400	0.014

[*]: Elastic modulus   [**]: Yield stress   [***]: Tensile strength   [****]: Elongation

The aim of this study is to obtain basic data to apply the CFRP strips to steel structures. So, in this paper, tensile tests are done using notched steel plates strengthened by the CFRP strips. And application of the CFRP strips as a strengthening member to notched steel members is discussed.

2. Experimental procedure

Specimen configuration is shown in Fig. 1. Size of the specimen is 100mm ×9mm×1,000mm. A notch, which length is 25mm and tip radius is 2mm, was machined in the center of the specimen. Gauge length is 600mm. Size of the CFRP strips attached to the steel plate is shown in Table 1. Tn

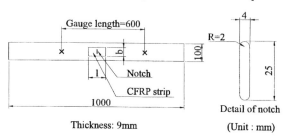

Fig.1 Specimen configuration

is a specimen without strengthening. The CFRP strips were attached to a steel plate by an epoxy adhesive. Mechanical properties of the steel and the CFRP strips are shown in Table 2. The mechanical properties of the steel were obtained from tensile test of specimen without notch of 100mm×9mm×1,000mm and the CFRP strips are data from a catalogue.

3. Results and discussion

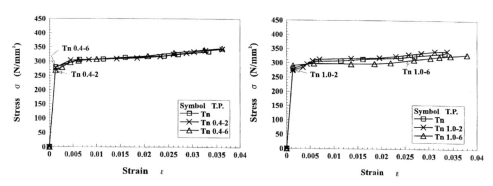

Fig. 2 Stress-strain curves (B=0.4)   **Fig. 3 Stress-strain curves (B=1.0)**

(1)   Stress-strain curves

Figure 2 is a stress-strain curves in the case of B=0.4. The vertical axis is a stress divided a load by the gross sectional area. The horizontal axis is a strain in the gauge length of 600mm. The stress-strain curve of Tn is also shown for reference in the figure. Arrows in the figure show a point that the CFRP strips detached. The CFRP strips of Tn0.4-2 and Tn0.4-6 detach when the stress of steel has reached close to the yield point. Therefore, the stress-strain curves after yielding of steel are the same as one of Tn. The stress-strain curves before the yield point are also the same. It is concluded from the above results that the CFRP strips do not have effect on a stress-strain curve of a member even if the CFRP strips of B=0.4 are attached to the notched region of a steel member with a notch of W=0.25.

Figure 3 is in the case of B=1.0. In Tn1.0-2, the CFRP strips detach when the stress of steel has reached close to the yield point. On the other hand, in Tn1.0-6, the CFRP strips detach at the stress of $310N/mm^2$ and the strain of 0.026. Therefore, it is said that length of the CFRP strips needs longer than L=6 not to be detach the CFRP strips before yielding of a steel in the case of strengthening a steel member with a notch of W=0.25 using the CFRP strips of B=1.0. There is no difference between the stress-strain curves of Tn1.0-2 and Tn1.0-6, and they are also good agreement with Tn. It is concluded from the above results that the CFRP strips do not have effect on a stress-strain curve of a member even if the CFRP strips of B=1.0 are attached to the notched region of a steel member with a notch of W=0.25.

Figure 4 is in the case of B=2.0. The CFRP strips of Tn2.0-2 and Tn2.0-6 detach near the stress of $310N/mm^2$ and the strain of 0.015. So, in the case of B=2.0, there is not a influence of the length of the CFRP strips on the detached load, and when length of the CFRP strips is more than L=2, it is known that the CFRP strips do not detach before yielding of a steel. There is not meaningful difference between the stress-strain curves of Tn2.0-2 and Tn2.0-6, and they are also good agreement with Tn. It is obvious from the results that the CFRP strips do not have effect on a stress-strain curve of a member even if the CFRP strips of B=2.0 are attached to the notched region of a steel member with a notch of W=0.25.

Let us examine an influence of width of the CFRP strips. Figure 5 is stress-strain curves of the specimens of L=2. There is no difference between the stress-strain curves in the figure. In Tn0.4-2 and Tn1.0-2, the CFRP strips detach when the stress of steel has reached close to the yield point. However, the CFRP strips of Tn2.0-2 do not detach till the stress of steel has reached $310N/mm^2$. Therefore, it is concluded that width of the CFRP strips needs wider than B=2.0 not to be detached the CFRP strips after yielding of a steel in the case of strengthening a steel member with a notch of W=0.25 using the CFRP strips of L=2.

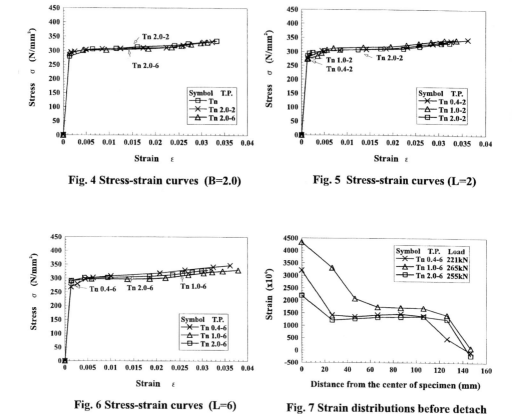

Fig. 4 Stress-strain curves (B=2.0)

Fig. 5 Stress-strain curves (L=2)

Fig. 6 Stress-strain curves (L=6)

Fig. 7 Strain distributions before detach of the CFRP strips (L=6)

Figure 6 is in the case of L=6. There is also no difference between the stress-strain curves in the figure. The difference between strain at detach of the CFRP strips of Tn1.0-6 and Tn2.0-6 is considered to be error in an attaching work of the CFRP strips. Therefore, in the case of strengthening a steel member with a notch of W=0.25 using the CFRP strips of L=6, it is concluded that width of the CFRP strips needs wider than B=1.0 not to be detached the CFRP strips after yielding of a steel.

(2) Strain distribution before detach of the CFRP strips
Strain distribution before detach of the CFRP strips is shown in Figure 7, which is the case of L=6. Strain increases between x=150mm and x=110mm, where x is distance from the center of the CFRP strips, is constant between x=100mm and x=60mm or x=25mm, and increases again between x=60mm or x=25mm and x=0mm.

(3) Shearing force between steel and the CFRP strips
Shearing force between steel and the CFRP strips is shown in Figure 8, which is the case of L=6 and at the load of 98kN. The shearing force at the load of 220kN, which is the load just before detach of the CFRP strips, is shown in Figure 9. The shearing force was calculated by the following equation and strains measured on the CFRP strips.

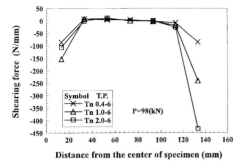

Fig. 8 Stress distributions on the notched section

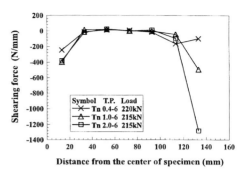

Fig. 9 Stress distributions on the notched section

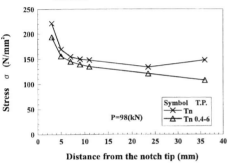

Fig. 10 Stress distributions on the notched section

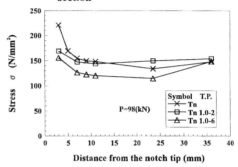

Fig. 11 Stress distributions on the notched section (B=1.0)

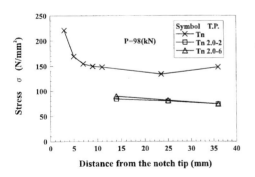

Fig. 12 Stress distributions on the notched section (B=2.0)

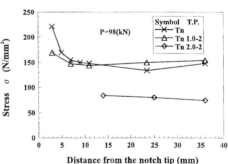

Fig. 13 Stress distributions on the notched section (L=2)

$$Q = AE(e_{x+dx} - e_x)/dx \quad \text{------------(1)}$$

Where Q is the shearing force per unit length (N/mm),
A is a sectional area of the CFRP strips (mm^2),
E is young's modulus of the CFRP strips ($1.5 \times 10^5 (N/mm^2)$),
e_x is strain at location of x,
e_{x+dx} is strain at location of (x+dx),
dx is distance (mm).

In Figures 8 and 9, the shearing force is greater in the region of the end and the center of the CFRP strips and is nearly equal to 0 from x=30mm to x=110mm. So, it is concluded easily that the CFRP strips detached from the end of or the center of the CFRP strips.

(4) Stress distribution on the notched section
Figure 10 shows stress distribution on the notched section of Tn0.4-6 at P=98kN with Tn not to be strengthened. It is found from the figure that the notch tip stress of Tn0.4-6 is lower of $30N/mm^2$ than Tn and the stress far from the notch is also lower of $20N/mm^2$. So, even though a width of the CFRP strips is shorter than a length of a notch, such as B=0.4, it is concluded in the case of L=6 that the CFRP strips has an effect to be decreased stress on the notched section.

Figure 11 is in the case of B=1.0. It is found from the figure that the notch tip stress of Tn1.0-2 and Tn1.0-6 is lower of $60N/mm^2$ than Tn. As to the stress far from the notch, Tn1.0-2 has not a meaningful difference from Tn and Tn1.0-6 shows decrease of $30N/mm^2$ in comparison with Tn. Therefore, it is known that Tn1.0-6 can decrease the greater stress on the notched section than Tn1.0-2.

Stress distribution in the case of B=2.0 is shown in Figure 12. It is found from the figure that the stress of Tn2.0-2 and Tn2.0-6 at 13.5mm from the notch tip is lower of $60N/mm^2$ than Tn. And an influence of length of the CFRP strips on the stress distribution is not recognized.

Figure 13 is stress distribution on the notched section at P=98kN in the case of L=2. Tn is also shown in the figure. The notch tip stress of Tn1.0-2 shows decrease of $60N/mm^2$ in comparison with Tn. So, it is known that strengthening using the CFRP strips can decrease the notch tip stress. As regards the stress over 13.5mm from the notch tip, Tn1.0-2 and Tn have not a meaningful difference, and Tn2.0-2 is lower of $60N/mm^2$ than Tn. Therefore, it is concluded that width of the CFRP strips needs B≥2.0 to be decreased the stress on the notched section in the case of L=2.

Figure 14 is in the case of L=6. It is found from the figure that the notch tip stress of Tn0.4-6 and Tn1.0-6 show decrease of $30N/mm^2$ and $70N/mm^2$ in comparison with Tn, respectively.

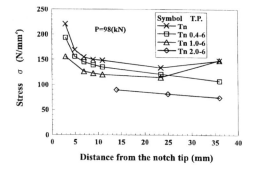

Fig. 14 Stress distributions on the notched section (L=6)

The stress in the region far from the notch tip also decreases according to wider of the CFRP strips.

Therefore, it is concluded in the case of L=6 that the effect of decreasing the notch tip stress and the stress in the region far from the notch tip become greater by being wider of the CFRP strips.

4. Conclusions

In this paper, static tensile tests of a notched steel plate strengthened by the CFRP strips were described. And application of the CFRP strips to strengthening of notched steel member was examined. The influence of a length and a width of the CFRP strips on the stress-strain curves, on the load when the CFRP strips detached from the steel member and on the stress distribution on the notched section were discussed on the basis of the data obtained from the experiments.

The following conclusions were obtained.
(1) In the case of b/a=0.4, where b was a width of the CFRP strips and a was a notch length, the notch tip stress at the load of 98kN decreased from $220N/mm^2$ of the specimen without strengthening to $190N/mm^2$. However, the CFRP strips had detached before yield of gross section of a steel member even in the longest CFRP strips l/2a=6, where l/2 was a half-length of the CFRP strips.
(2) In the case of b/a=1.0 and the longest CFRP strips l/2a=6, the notch tip stress decreased from $220N/mm^2$ to $160N/mm^2$ at the load of 98kN and the CFRP strips had not detached till yield of gross section of a steel member.
(3) In the case of b/a=2.0 and l/2a=2, the CFRP strips had effect to be decreased the stress on the notched section from $150N/mm^2$ to $90N/mm^2$ at the load of 98kN and the CFRP strips had not detached till yield of gross section of a steel member.
(4) It was considered that the CFRP strips had detached from the end of the CFRP strips or notched region.

References
1) Hanshin Expressway Administration and Technology Center: Maintenance of highway bridges --Case studies of failures and repairs--, 1993-3(In Japanese).
2) H. Suzuki et al: Bending tests of concrete beams strengthened using carbon fiber reinforced polymer strips, Proceedings of Annual Conference of Civil Engineers '96, JSCE Kansai Chapter, I-133-1~I-133-2, 1997-5(In Japanese).
3) H. Suzuki et al: Study on strengthening of concrete beams using carbon fiber reinforced polymer strips, Proceedings of the 52nd Annual Conference of the JSCE, 1-(A), I-A203, pp.404-405, 1997-9(In Japanese).

Three Dimensional Modelling of Masonry Arch Bridges

PAUL J. FANNING[1] and THOMAS E. BOOTHBY[2]
[1]Dept. of Civil Eng., University College Dublin, Dublin, Ireland
[2]Dept. of Architectural Eng., The Pennsylvania State University, Pennsylvania, USA

Abstract

Stone and masonry arch bridges are complex systems whose structural response is a function of the composite masonry and mortar material, the contained fill material and the interaction between these and the surrounding soil medium. The predictability of masonry arch bridges and their behaviour is widely considered doubtful.

Simulation of virtually all known effects governing the response of arch bridges, including cracking in the arch structure, plastic response of the fill material and the transfer of compressive and frictional stresses between the fill material is possible using modern finite element codes.

This paper compares finite element model predictions with service load test results for two arch bridges in Dublin Ireland and shows that provided sensible material parameters are used for the main bridge components that the modelling tools enable prediction of the test results. The implications for assessment ratings of both bridges are discussed in the light of the test and model results.

Introduction

The most widespread and authoritative document on the assessment of masonry arch bridges is the UK Department of Transport's Advice Note BA 16/97, which says; "The long term strength of a brick or masonry arch bridge is almost impossible to calculate accurately and recourse has, therefore, been made to an empirical formula based on the arch dimensions." In view of the confidence with which bridge engineers enter assessments of steel or concrete bridges, such statements regarding a bridge form that has persisted for over 2000 years, and has been the subject of scientific analysis for over 300 years are surprising. This outlook on masonry bridge assessment arises from two primary reasons, the lack of knowledge about the conditions of a given masonry arch bridge, and the lack of an accepted procedure for the analysis of masonry arch bridges.

The most widely used method of assessment for masonry arch bridges is the MEXE method. The MEXE method is a semi-empirical method, including multiple reduction factors based on conditions noted in a visual inspection. The method is crude at best, and the rational basis of the reduction factors has never been well established. It is also extremely conservative, and since the basic formula for allowable axle load is based on a parabolic arch, the method becomes grossly over-conservative for shapes of arch other than parabolic, especially segmental or elliptical arches, both of which are quite common in Ireland. The method is widely used primarily because it is extraordinarily simple to apply. The method is however generally unsuitable for the management of an important population of bridges, because it will result in unwarranted bridge postings, unwarranted rejections of heavy vehicle permits, and unwarranted strengthening of bridges.

In this paper MEXE assessments, following United Kingdom (UK) and Ireland guidelines, are undertaken for two masonry arch bridges. The three dimensional modelling techniques used for the analysis of both bridges are discussed. Service load testing, up to and above the maximum axle loads allowed by the MEXE method, and three dimensional finite element models are demonstrated to be in close agreement. It is argued that the allowable axle loads, determined by the MEXE method, are so unduly conservative that the recommendations made by the Irish Department of the Environment should at least be considered the norm for this type of assessment.

Sarah Bridge and Griffith Bridge

Sarah Bridge was constructed over the River Liffey on the outskirts of Dublin city centre in 1793. It is a granite ashlar masonry arch bridge spanning 32m with a rise of 6.58m, Figure 1. The barrel width is approximately 12.45m. The bridge surface is split into two carriageways for traffic and two pedestrian footpaths both 1.9m wide.

Figure 1 : Sarah Bridge

For the purposes of testing a single carriageway was closed and two vehicles, Figure 2, each weighing approximately 30 tonnes, were driven back to back across the bridge in the closed carriageway. Because of the size of Sarah Bridge, and its location over a tidal river, it was impractical to install a reference frame beneath the bridge. Instead, displacement measurements were taken with a digital level, with a sensitivity of 10μm, on the outside edge of the barrel at the crown. Vibrating wire strain gauges were also used at the crown to determine maximum live load strain differences in the structure. These gauges, mounted in the longitudinal and transverse directions, were located at midspan and approximately 5.25m in from the edge of the barrel beneath the closed lane.

Griffith Bridge, Figure 3, has a span of 9.49 m, a rise over the abutments of 2.67 m, a width of 7.85 m, and an arch ring thickness of 45 cm. The arch ring is constructed of granite on the face, while the remainder of the arch ring is limestone, with joints about 0.5 cm thick. The spandrel walls are also of ashlar limestone construction, with joint thicknesses of approximately 1 cm.

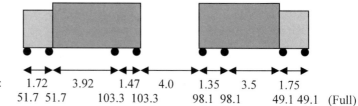

```
Axle Spacings (m):    1.72      3.92      1.47    4.0    1.35    3.5    1.75
Axle Loads (kN):      51.7 51.7        103.3 103.3      98.1 98.1      49.1 49.1   (Full)
```

Figure 2 : Truck Loads and Loading Arrangement

Figure 3 : Griffith Bridge

For Griffith Bridge, DCLVDT's having a linear range of ±1.27 mm, mounted on a professionally assembled scaffolding system underneath the structure, were used to measure displacements. Displacement measurements were taken at the centreline of the roadway at the crown, the two abutments, and at one of the haunches. Additional displacements measured at abutment, haunch, and crown on one edge of the bridge were also recorded. The bridge was loaded with a four axle truck, the left hand truck in Figure 2, loaded full, half-loaded, and empty. Griffith Bridge was closed for the purposes of testing and the truck was driven along the centreline of the bridge for each test.

The bridge parameters used in MEXE assessments of both bridges are listed in Table 1. The calculated assessment ratings are tabulated in Table 2. The MEXE (Ireland) ratings follow from a Department of Environment report issued in Ireland in May 1988 which allows upgrading of the modified axle loads for condition factors greater than 0.5. None of the three bridges had observable defects that would warrant a downgrading of the condition factor below the maximum of 0.9 and hence the MEXE (Ireland) ratings are higher than those arrived at by following BA16/97.

Bridge	Span (s)	Rise at crown (r_c)	Rise at 1/4 point (r_q)	Depth of key (d)	Depth of fill (h)	Condition Factor F_c
	m	m	m	m	m	----
Griffith	9.45	2.70	2.30	0.45	0.20	0.9
Sarah	31.8	6.58	5.28	0.80	0.40	0.9

Table 1. Bridge Parameters

Bridge	MEXE (UK)		MEXE (Ireland)	
	single	double	single	double
Griffith	126	81	227	146
Sarah	180*	103*	324	185

*MEXE generally considered excessively conservative for this span length.

Table 2. Bridge Ratings (load on individual axle in kN)

Three Dimensional Modelling of the Test Bridges

Current assessment methods for masonry arch bridges use two-dimensional modelling techniques. In this study three dimensional nonlinear models of both bridges were constructed using ANSYS V5.5.

Reliably modelling the geometry of an arch bridge poses several challenges. Although much of the structure may be visible, a significant part of the structure is usually concealed, Figure 1 or Figure 2. The main support at the abutment of any arch bridge is only rarely available for inspection. Moreover, the depth of the arch barrel visible on the face of the structure is not necessarily the same as the depth within the structure. The properties of the fill are generally unknown, and have been shown in testing to have a substantial influence on the ultimate strength of the structure (Royles and Hendry 1991).

The geometry and profiles of both bridges were determined from photographic surveys. The fill was assumed to be uniform through its depth and the barrel thickness was assumed to be equal to the thickness of the facing blocks. A two dimensional elevation of the bridge was constructed and meshed before being extruded through the transverse thickness to arrive at the three dimensional representation of the bridge.

The masonry stones and mortar joints were modelled as a continuum. Three-dimensional eight noded isoparametric elements, which can account for cracking and crushing, where employed for the masonry material. The element used includes a smeared crack model to allow the formation of cracks perpendicular to the direction of principal stresses, which exceed the specified tensile strength of the masonry material (ANSYS, 1999). The element behaves in a linear elastic manner until either of the specified tensile or compressive strengths are exceeded. The element is thus nonlinear and requires an iterative solver.

Overlaid on the arch barrel, and contained within the spandrel walls, three-dimensional solid elements are also used to model the fill material and any masonry or rubble backing to the arch barrel. The primary functions of the fill material are to lock compressive stresses into the arch ring under dead load, to distribute relatively concentrated loads over greater lengths and widths of the arch barrel and to provide longitudinal restraint to the arch by its interaction with the surrounding soil medium. The fill material is generally a soil material or unbonded masonry or rubble and is often very variable in its structural characteristics, although it is not uncommon for masonry arch bridges to have regularly arranged backing near the arch abutments. In constructing the numerical models for the bridges discussed in this paper no backing was considered and the fill material was modelled using a Drucker-Prager material law. The Drucker-Prager yield criterion requires three material parameters: the cohesion, c, the angle of internal friction, ϕ, and the angle of dilation.

The material properties used in model, based on recommendations for masonry arch analysis by Boothby (2001) and Boothby and Roberts (2001), are summarised in Table 3. Analyses were performed for the masonry/mortar continuum modelled using linear (LM analysis) and nonlinear (NLM analysis) material laws.

Masonry	Young's Modulus (GPa)	Poisson's Ratio	Density (kg/m^3)	Tensile Strength (MPa)	Compressive Strength (MPa)		
Griffith	10	0.3	2200	0.5	10		
Sarah	15	0.3	2200	0.7	10		
Fill	Young's Modulus (MPa)	Poisson's Ratio	Density (kg/m^3)	Cohesion (MPa)	Angle of Friction (degrees)	Angle of Dilatency (degrees)
Griffith	15	0.231	1700	1x10^{-3}	44.4	44.4
Sarah	15	0.231	1700	1x10^{-3}	64.1	64.1

Table 3 : FE Model Material Assignments

To facilitate sliding or movement of the fill material relative to the arch barrel and the spandrel walls, without generating significant tensile stresses at the interface between these materials, three dimensional frictional contact surfaces are included. The masonry elements of Sarah Bridge are shown in Figure 4, while the full model, including fill elements, for Griffith Bridge is shown in Figure 5. The symmetrical nature of the loading on Griffith Bridge resulted in only a half model of the bridge being necessary.

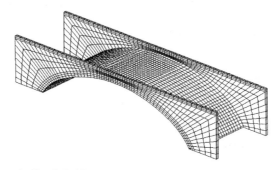

Figure 4 : Sarah Bridge – FE Mesh for Masonry/Mortar Continuum

The boundary conditions applied to the finite element mesh were the same for both models. Symmetry boundary conditions were applied along the centreline of the Griffith Bridge model. The fill was restrained in the span direction at opposite ends of the bridge. The underside of the fill material was restrained vertically. The arch barrel and spandrel walls were extended to a depth of 1.0m below the arch springing level in all models. The base of the arch barrel was restrained vertically and in the span direction at this depth. The base of the spandrel walls, which are generally embedded in the surrounding soil medium and are not usually visible for inspection, were restrained in the vertical and transverse directions.

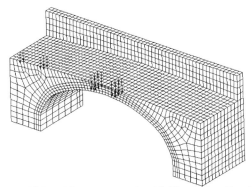

Figure 5 : Griffith Bridge – FE Mesh with Illustrative Truck Loading

The loading on a masonry arch bridge is a combination of self-weight loading and traffic loading. The proportion of loading due to the selfweight is significant and indeed much of the strength of these bridges is attributed to the stresses induced in the masonry material due to selfweight effects. In modelling this type of bridge an initial gravity-loading step is undertaken in an effort to generate the in-situ stresses, both longitudinal and transverse, to which the bridge is subjected. Subsequent loading events, such as the passing of a truck over the bridge, use the equilibrium solution from the gravitational loadstep as a set of initial conditions. Appropriately distributed nodal forces, applied to the roadway surface, were used in the finite element models to represent the test truck weight distributions, Figure 2. In each case simulation of the moving vehicle, which travelled typically at less than 20km/hr, was undertaken by a sequential series of static loadsteps with the equilibrium solution at the end of one loadstep resulting in a set of initial conditions for the subsequent load step.

Comparison of Test and 3D Model Results

Griffith Bridge

The deflected shape of the arch barrel and spandrel walls when test truck, fully loaded, has its rear axle at mid-span is plotted in Figure 6. The deflected shape of the bridge demonstrates clearly the important three-dimensional effects that contribute to the strength and stiffness of the bridge. The maximum deflection occurs at the centreline of the bridge with the spandrel walls stiffening the outer edges of the arch barrel. These transverse effects are evident in the both bridges analysed and are considered to be the source of longitudinal cracking that is evident in many arch bridges.

The numerical and experimental bridge responses at midpsan, at the centreline of the bridge, as the fully loaded truck passes over the bridge are plotted in Figure 7. The maximum deflection measured in the test was 0.43mm compared to 0.54mm predicted by the finite element model. The profile of the test and numerical data are also consistent, with the front and rear axle combinations being evident in both responses.

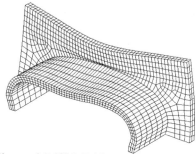

Figure 6 : Deformed Shape of Griffith Bridge Arch Barrel with Rear Axle at Crown

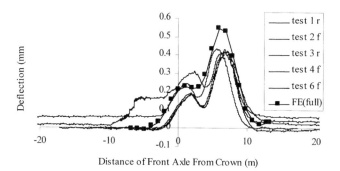

Figure 7 : Griffith Bridge – Deflection at Crown due to Fully Loaded Truck

The correlation between test and numerical data at midspan, at the edge of the bridge, is shown in Figure 8. The test response at the crown on the centreline is 0.54mm, Figure 7, compared to an edge test response of 0.15mm. Three-dimensional effects are thus significant and although a two-dimensional analysis may be able to accurately predict stresses and strains in the span direction of certain arch forms, namely those with relatively large barrel widths, an examination of transverse effects is also clearly desirable.

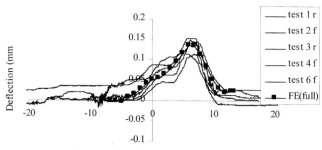

Figure 8 : Griffith Bridge –Deflections at Edge due to Fully Loaded Truck

Sarah Bridge

Longitudinal and transverse strains, for two successive passings of the two trucks (T1 and T2), for Sarah Bridge, are plotted in Figure 9. The successive tests demonstrate the repeatability of the bridge response. The transverse effects are significant with the transverse strains, maximum 17.8 microstrain, being larger than the longitudinal strains, maximum 2 microstrain. The model predictions of strain at the measurement points are also plotted in Figure 9. The strains at the measurement point in both finite element models (LM and NLM) are almost identical and only one set is illustrated for clarity. The models are seen to capture the dominant transverse strain in the bridge during the tests, with a maximum transverse strain of 14.2 microstrain.

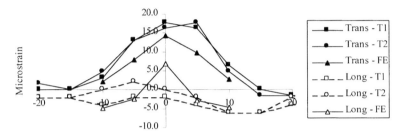

Distance Rear Axle of Front Truck from Centreline (m)

Figure 9 : Comparison of Test and FE Model Strains

The maximum test deflections for the two tests were 0.56mm and 0.38mm respectively. The FE model results were 0.84mm for NLM and 0.57mm for LM. Significant cracking of the arch system was however evident in the non-linear model. The crack distribution under self-weight loading conditions indicated partial separation between the barrel and the spandrel walls. This was not evident from bridge inspections. Contours of principal tensile stresses, which have a maximum value of 1.2MPa, for the linear material model are plotted in Figure 10, the more heavily shaded regions representing those that are stressed highest.

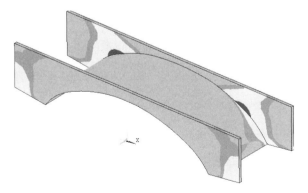

Figure 10 : Principal Stress Intensities Under Selfweight

The cracks in the non-linear model were formed initially at the location of maximum tensile stress, i.e. at approximately the quarter span points. The eventual spread of the cracked zone along the length of the arch/barrel interface is governed by the specified tensile strength of the masonry/mortar continuum, 0.7MPa, in the nonlinear model. In the non-linear material model the midspan deflections are larger due to this separation. The transverse strains in both however are similar due to the relatively low height of the spandrel walls above the arch barrel at midspan and also because the cracks induced in the non-linear material model did not extend along the full length of the arch barrel/spandrel wall interface.

In this instance, and mindful of the condition of the bridge system, it is concluded that the actual response of the stone arch is closer to the linear model results, suggesting that a higher value of tensile strength is required in the finite element models.

A conventional two-dimensional assessment is based on a unit width of the bridge. This assumption is considered conservative in so far as the spandrel walls are assumed to add to the strength of the system. However even in allowing for this eventuality the 3D models and the measured bridge response indicate that it is still the transverse strains that are more significant and in an overload situation the first indication of distress in Sarah Bridge would be longitudinal cracks initiated by transverse effects.

Implications for Assessment

The service load tests demonstrate a high degree of repeatability for both bridges with neither bridge showing any signs of distress during testing. Equally both bridges were seen to demonstrate significant transverse responses under loading.

The MEXE (UK) assessment resulted in a maximum axle load of 81kN for a double axle bogey on Griffith Bridge. Given the response of the bridge during testing, with each axle weighing 103.3kN, the MEXE (UK) assessment is over-conservative. An intervention, for example structural strengthening or a weight restriction on the bridge, based on a MEXE (UK) assessment alone would be unwarranted. This is true also for Sarah Bridge although it is recognised that the MEXE method is known to be grossly over conservative for such long spans.

The finite element model solutions demonstrate the predictability of response that can be achieved. In general the numerical model results are within 20% of those recorded during testing.

The MEXE assessment, modified following guidelines published by the Department of the Environment (Ireland) resulted in a maximum axle load of 146kN for Griffith Bridge. The profile of deflections for Griffith Bridge, with the vehicle weight increased by 50%, giving a maximum rear axle load distribution of 155kN per axle, was similar to that of Figure 7 with a peak deflection of 0.72mm. No evidence of distress to the bridge was evident from the finite element model at this level of load.

The ultimate individual axle loads, on a double bogey, for failure of Griffith Bridge, determined from the finite element model, was found to be approximately 264kN. The crack distribution at ultimate load, as predicted by the model, is shown in Figure 11, and it is noted that the first cracks to form were longitudinal cracks near the crown due to transverse effects. The predicted factors of safety against failure for MEXE (UK) and MEXE (Ireland) are thus 3.2 and 1.8 respectively.

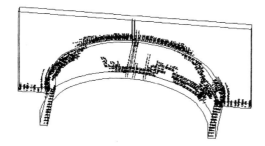

Figure 11 : Griffith Bridge – Crack Distribution at Ultimate Load
(2.56 x Test Truck Axle Loads at Crown)

Conclusions

Three-dimensional finite element models have been demonstrated to be capable of predicting the response of two very different masonry arch bridges under service loading. Both the service load tests and the finite element models demonstrate the importance of transverse effects in understanding the response of masonry arch bridges.

Assessments undertaken using MEXE (UK) were found to be unduly conservative. The maximum allowable axle loads are less than those used during the service load tests and also those that might traverse the bridges regularly. Bridge strengthening or weight restrictions based on these assessments alone would be unwarranted. MEXE (Ireland) gives a more reasonable assessment of bridge capacity with the allowable axle loads being greater than those used in the tests. This is consistent with the repeatability of bridge response, the lack of any evidence of distress to the bridges during testing and also the predictions of the finite element models.

References

ANSYS (1999), ANSYS Manual Set, ANSYS Inc., Southpoint, 275 Technology Drive, Canonsburg, PA 15317, USA

Boothby, T. E. (2001), "Load Rating of Masonry Arch Bridges", *ASCE Journal of Bridge Engineering*, March 2001

Boothby, T.E. and Roberts, B.J. (2001), "Transverse Behaviour of Masonry Arch Bridges", *The Structural Engineer*, May 2001

Department of the Environment (Ireland) (1998), "Report on Inspection, Assessment and Rehabilitation of Masonry Arch Bridges", May 1998

United Kingdom Department of Transport (1997). *Design Manual for Roads and Bridges, Volume 3, Section 4, Part 4. The Assessment of Highway Bridges and Structures.* London, Department of Transport, 1997

Royles, D. and Hendry, A.W. (1991), "Model Tests of Masonry Arch Bridges", *Proceedings of the Institution of Civil Engineers (UK)*, Part 2, 91(6):299-321

Full Scale Testing of High Performance Concrete Bridge Beams with In-Situ Slabs

DAVID J. DOYLE, DAMIEN L. KEOGH
Department of Civil Engineering, University College Dublin, Ireland.

ABSTRACT

This paper reports on a series of full scale load tests carried out on three high performance concrete bridge beams with cast in-situ slabs. Three prestressed concrete beams with strengths of 75 N/mm^2 were constructed. 1m wide slabs with strengths of 25, 50 and 75 N/mm^2 were poured on top of the beams. The development of the concrete mixes for the beams and slabs is discussed. The beams were tested to loads that exceeded their expected capacities. One of the beams was tested to complete collapse while the other two were loaded well past their serviceable limits. Details of the loading arrangements and instrumentation are given in the paper. Moment deflection curves are presented for each of the test beams and for one beam in particular, the moment deflection curve is given for three repeated cycles of loading and unloading. This shows the extent of recovery. The neutral axis location at each load increment is shown for each of the three tests. A moment curvature graph is also shown. Comparisons are made between theoretical serviceability and ultimate capacities and those attained in the tests. In all cases the expected moment capacities were exceeded. A discussion is also presented on the longitudinal and interface shear capacities.

INTRODUCTION

Precast prestressed concrete bridge beams are a popular form of construction for road and rail bridges in Ireland. The use of High Performance Concrete (HPC), however, has been limited to only specialised construction projects, yet the combination of HPC and prestressed structures offers the designer a huge range of opportunities to improve the efficiency of the design.

High Performance Concrete is the term preferred, by many, for concretes which display not only improved compressive strengths but also improved mechanical properties and superior durability as a result of their lower water / binder ratio.[1]

The improved properties of HPC, both mechanical and durability, have been found to be closely related to the same microstructural features of the concrete. The low water / binder ratio, along with the use of very fine secondary cementitious materials results in a much strengthened hydrated cement paste and transition zone between the aggregate and the cement paste. The improved mechanical properties include higher elastic modulus and greater tensile and compressive strengths. Improvements in the durability of the concrete result from a reduction in the porosity of the hydrated cement paste, thus reducing its permeability to chlorides and other harmful substances.

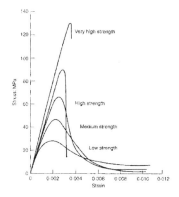

Figure 1 Typical stress-strain relationship of concrete

Together these improvements make HPC an advantageous material for use in the production of prestressed bridge beams. The improved mechanical properties allow for longer spans or greater beam spacing than possible with lower strength concretes. For a given span, the use of HPC allows for a reduction in the number of prestressing strands and member depth, both of which result in cost savings. The improved durability characteristics will lead to extended service life of the bridge with reduced maintenance needs.

This paper deals briefly with the development of HPC mixes using locally available aggregates and cements along with state of the art additives and admixtures. The resulting mix was used in the production of three full scale beams which were load tested to determine their behaviour, with particular reference to the ductility of the HPC and possible mechanisms of failure. HPC is known to be a more brittle material than normal strength concrete, showing a far steeper softening slope beyond the peak of the stress-strain curve.[2] Figure 1 clearly shows the lack of ductility displayed by HPC in uniaxial compression. Realistically, however, a prestressed structure is seldom subjected to pure uniaxial compression as the transverse reinforcement in the compression zone induces sufficient confining stresses to set up a triaxial stress field, under which all concretes, including HPC, display sufficient ductility to prevent a brittle or explosive failure.[3]

DEVELOPMENT OF HPC MIXES

Research was undertaken in association with a local precast concrete producer with the aim of developing a number of trial mixes from which the test beams could be made. Mix variables included size of coarse aggregate, cement type and secondary cementitious material. The producer was keen to use locally sourced aggregate which had proved satisfactory over many years with lower strength concretes and other specialised concretes produced. It is well documented that smaller aggregate, 10 or 12mm, is necessary to achieve very high strength concretes.[1,3] It was decided to compare the 12mm aggregate with the larger 20mm coarse aggregate. Five types of readily available cements were compared, each blended for specific uses such as rapid hardening, low alkali, high strength as well as ordinary Portland.

The role of the secondary cementitious material is vital in the production of high performance concretes, two pozzolanic materials Microsilica and Metakaolin were compared. Microsilica, a slurry form of silica fume, has been widely available for a period of time and its advantages are well documented. Metakaolin is a newer material, available in ultra-fine powder form, originally used to suppress alkali-silica reaction but also found to produce higher compressive strength.

Seven trial mixes were finally compared with combinations of the cements and secondary cementitious materials. The results of compressive strength tests are shown in figure 2. All mixes displayed very similar results, with excellent early strength development and relative stability between 28 and 56 days.

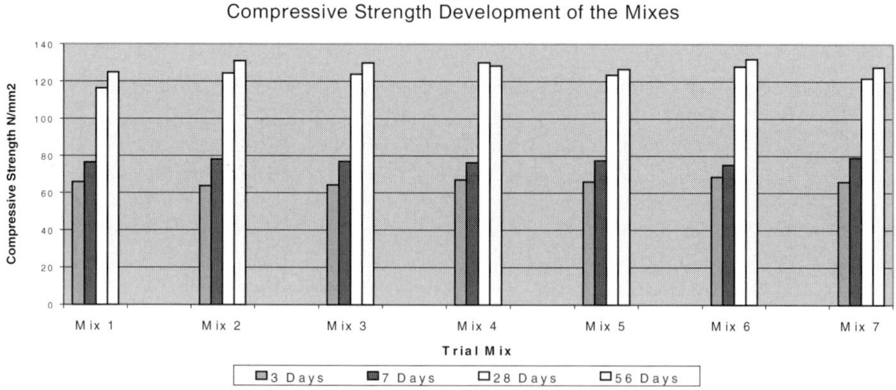

Figure 2 Compressive Strength of Trial Concrete Mixes

The producer had many other deciding criteria apart from strength characteristics. Storage and handling of the secondary cementitious materials, ease of placement and workability of the fresh concrete, ease of revibration and the appearance of the finished product were all important. The microsilica was found to be difficult to batch consistently as the slurry required vigorous agitation to prevent sedimentation and unequal concentrations. The metakaolin mixes were found to be generally superior on most of the decision criteria and produced a more consistent, brighter finished surface. The microsilica mixes produced a darker concrete, with inconsistent colouring due to problems encountered with the concentration of the slurry.

The preferred trial mix was that containing 12mm aggregate, a 42.5 N Low Alkali cement, a new generation superplasticiser and Metakaolin. The 12mm coarse aggregate was found to produce strengths in excess of 90N/mm^2, but was not necessary for strengths between 60 and 90N/mm^2 where the more economical 20mm aggregate was sufficient.

FULL SCALE TESTING OF PRE-STRESSED BEAMS

Detailed analyses of HPC beams being used in bridges with in-situ cast deck slabs were carried out by French et al.[4] It was shown that an increase in the compressive strength of the beam from 48 to 69 N/mm^2 could allow for a 16% increase in maximum span length. The main advantage of using higher strength concrete in precast prestressed bridge beams is that it allows the development of a greater prestress force by increasing the number of prestressing strands. The effectiveness of adding more strands eventually diminishes due to reduced eccentricity. It was found also that the use of higher strength in-situ concrete in bridge deck slabs can increase the maximum span by up to a further 25%.

Full scale testing of HPC I-girders with in-situ deck slabs was carried out by by Ozyildrim and Gomez.[5] This work produced some interesting results with the beams' attaining ultimate moments which were 15% greater than their predicted values. The testing also revealed a pattern of strand slippage well within the service range, which was contributed to poor consolidation of the concrete at the ends due to highly congested shear reinforcement making vibration difficult. The poor consolidation resulted in poor bond strength and led to exceptionally long transfer lengths in all test beams. The slippage did not contribute greatly to any reduction in the flexural capacity of the beams.

TEST BEAM DESIGN

Three standard 'TY5' beams were cast with a target compressive strength of 75 N/mm^2 and spans nearly 50% longer than typically used for this size of beam. The span of the beam was maximised so as to develop as large a moment as possible. The beams were 600mm deep with an overall length of 19.3 metres, giving a span to depth ratio of 31. To simulate actual bridge construction a composite concrete slab was cast on each beam, 1 metre wide and 180 mm deep, giving a span to depth ratio for the composite section of 25. The three slabs were cast on profiled GRC permanent formwork propped from the beams, as would be normal practice in bridge construction. The strengths of the slabs were varied so as to investigate the effect of the slab strength on the flexural capacity of the composite section and to determine how to make the most of the advantages of high strength concrete beams. The geometry of the beams is shown in figure 3.

The test beams were designed with a target compressive strength of 75 N/mm^2 and produced average 28 and 56 day cube results of 79 and 83 N/mm^2 respectively. The beams were topped with reinforced concrete slabs of low, medium and high target strengths of 25, 50 and 75 N/mm^2. Specimens were taken from each mix to test for compressive strength, elastic modulus, flexural tensile strength and splitting tensile strength in accordance with BS 1881.[6] The results of these tests are summarised in Table 1.

Member	Compressive Strength, f_{cu}, N/mm^2		Elastic Modulus, E, kN/mm^2	Splitting Tensile Strength, f_t, N/mm^2	Flexural Tensile Strength, f_{ft}, N/mm^2
	Target	28 Day			
Beam	75	79.2	40.1	4.91	9.04
Slab 1	25	33.5	32.7	4.52	3.40
Slab 2	50	65.7	34.3	4.68	5.48
Slab 3	75	84.0	41.1	5.06	6.68

Table 1 Results from Concrete Tests

The beams were prestressed with 33 no. 7 wire 12.9mm superstrands with a nominal cross sectional area of 100mm^2 each. The strands have a nominal ultimate capacity of 186kN each and were stressed to 75% of their capacity, giving a total prestress force of 4603.5 kN with an eccentricity of 79mm below the centroid of the beam section. The layout of the strands is also shown in figure 3.

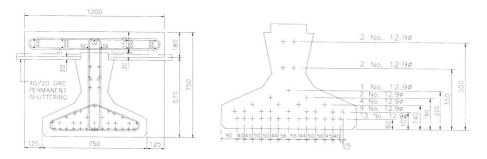

Figure 3 Test Beam Layout

Shear reinforcement was provided by T10 vertical links which also acted as interface shear reinforcement by protruding into the slab. The spacing of these links was varied at each end, with 200mm at one end and 300 mm at the other. This was done to allow investigation of the vertical and interface shear behaviour with different link spacings. Design calculations for vertical shear showed that only nominal links would be required and the design was based on interface shear requirements.

EXPERIMENTAL SET UP

The three beams were tested over three consecutive weeks at the precasters load testing facility. The beams had reached an age of between 186 and 207 days by the time of testing, while the slabs were cast in reverse order of strength with the strongest first, such that the weakest wouldn't gain too much strength and to allow the strongest develop its maximum strength. The order of testing was the medium strength slab first, followed by the weakest and finally the strongest. Each beam was initially tested within its serviceability limit state and released to measure the level of recovery. It was then tested to its design ultimate moment capacity and beyond.

The load testing facility comprised of two steel frames with a pair of hydraulic jacks which are fed by the one oil supply so as to ensure equivalent loading in both. The jacks were spaced 6 metres apart and 6.375 metres from the end supports. This form of four point loading ensured an area of near constant moment across the midspan and equal shear forces at each end. The load set up can be seen in figure 4.

Figure 4 Experimental set up and Instrumentation

INSTRUMENTATION

The beam was extensively instrumented, using load cells, surface mounted vibrating wire strain gauges and linear voltage displacement transducers (LVDTs). A data acquisition system was used to continuously log the readings of the load cells and LVDTs, scanning at a rate of 10 readings per second. The vibrating wire strain gauges were read using a meter are recorded by hand. The load applied by the jacks was monitored by the load cells.

The LVDTs were used to monitor displacements at various locations. Vertical deflection of the beam at midspan relative to the ground was measured by 2 LVDTs, one each side. An LVDT at each end, by the supports, measured any displacement due to compression or settlement of the supports, this was then subtracted from the midspan values to give the true displacement. To monitor any vertical displacement of the slab relative to the beam an

LVDT was attached to each side of the beam at midspan and placed in contact with the outer edge of the underside of the slab. An LVDT attached to the beam, in a horizontal direction, just below the slab near the end recorded any horizontal movement, or slip, of the slab relative to the beam due to interface shear. Two further LVDTs were used to monitor strand slippage at one end of the beam, one was placed on the middle strand of the bottom row, the other attached to a centre strand three rows above. The instrumentation on the beam at midspan can also be seen in figure 4.

Surface strains were measured at ten different points on the beam and slab at midspan using surface mounted vibrating wire strain gauges (VWSGs). The gauges were positioned at a single vertical cross section so as to get an accurate record of the strain distribution due to applied loading in the full composite section. VWSGs were chosen for their reliability, accuracy and relative ease of application in an outdoor environment. The gauges were mounted to the surface using a rapid hardening high strength waterproof resin and read manually with the VWSG reader.

RESULTS

Figure 5 shows the moment deflection relationships for each of the three tests. The moment shown is the total moment including that due to self weight. The deflection shown is due only to applied load. Consequently, zero deflection corresponds to a moment of approximately 500 kNm which is the self weight moment. The small 'dips' in the lines resulted from the load dropping off slightly while it was being held for observing the beams.

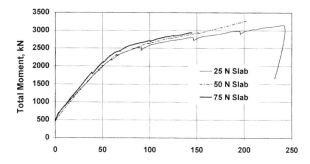

Figure 5 Moment Deflection Relationship for Three Test Beams

It is clear from each of the tests that the beams behaved in an elastic manner up to a moment of about 2000kNm. Flexural cracks began to appear once this moment was exceeded as the beams began to behave in-elastically. The beam with the 25N concrete slab was tested to failure. The point of collapse is clearly evident from the graph, this occurred at a moment of 3150 kNm at which stage the beam had deflected by about 240mm and large flexural cracks were evident. The graph shows that the failure was ductile as large increases in deflection resulted from small increases in load just before failure. This ductility was observed visually as a very large sag in the beam before collapse. Figure 6 shows the sag before collapse and the localised failure at midspan after collapse. The final failure mode resulted from crushing of the concrete in the slab at midspan. The reinforcement in the slab provided confinement to the concrete thus avoiding an explosive compressive failure. The concrete appeared to 'spall' initially which was quickly followed by collapse.

540 RESEARCH

Figure 6 Sag in 25N Test Beam before Failure and Failure at Midspan after Collapse

The beam with the 50N slab was tested up to a moment in excess of the failure moment of the 25N beam but it did not collapse. The deflection was in excess of 200mm and large flexural cracks were evident. It was not possible to increase the load further due to limitations of the loading rig. The beam with the 75N slab was not tested to collapse either but once again large flexural cracks were evident. It is clear from figure 5 that each of the three beams behaved in a similar manner but that the stiffness increased slightly with increase in slab strength.

The beam with the 75N slab was tested three times in succession. The first time within its elastic range and released, then into its in-elastic region, close to its ultimate strength and released again then finally back up to this level again. The moment deflection relationship for these loading cycles is shown in figure 7. This shows almost full recovery after the first and second load cycles as all of the curves follow a similar path. A small permanent set of less than 5mm is evident after the second load cycle.

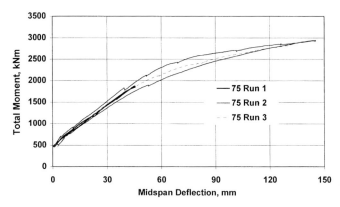

Figure 7 Moment Deflection Relationship for Beam with 75N slab

The neutral axis location of the composite beam and slab was investigated by looking at the strain distribution from the VWSGs at midspan. The neutral axis was taken as being at the bottom of the compression zone. Figure 8 shows the neutral axis location for each of the three tests as the bending moment increases. It can be seen that the neutral axis follows approximately the same path for each test. It remains straight initially, this corresponds to the elastic region. It then moves up towards the top of the beam, this corresponds to the in-elastic region where the concrete in the bottom of the beam has gone into tension and cracked and the lower layers of the prestressing strands begin to yield. Figure 9 shows the neutral axis location for the beam with the 75N slab for the second two loading cycles. This follows a similar pattern to figure 8 initially but some deviation is evident after the elastic region. It is felt that this is a result of the concrete having been cracked in tension from the previous load cycle.

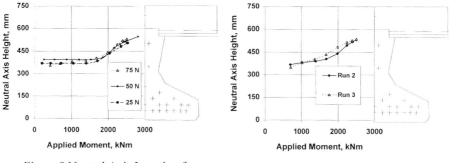

Figure 8 Neutral Axis Location for Three Test Beams

Figure 9 Neutral Axis Location for Beam with 75N Slab

Figure 10 shows a moment curvature graph. This is developed from the strain distribution of each beam. The slope of the moment curvature graph at any point is equal to EI, the product of the elastic modulus and the second moment of area of the section at that point. It is a measure of the stiffness and condition (with regard to cracking etc) of the section at a load. Three distinct regions can be seen in this graph. Initially, the three beams behave in a similar manner with a constant slope up to an applied moment of about 1500kNm, this is close to the load where first cracking was observed, and is consistent with elastic behaviour. Secondly, between 1500 and 2000kNm the concrete begins to crack which causes a reduction in the I value. This results in a substantial change in slope of the curves but occurs at different rates for each of the three beams. Finally, as the beams approach failure, the concrete no longer has any tensile capacity so the I values, and slopes of the curves, do not change any further.

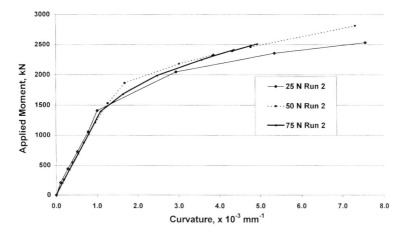

Figure 10 Moment Curvature Graph for Three Test Beams

ANALYSIS OF RESULTS

BS 5400[7] specifies 3 classes of member which may be designed for. These are, class 1: no tensile stresses allowed, class 2: tensile stresses must not exceed design flexural strength of concrete, and class 3: controlled crack widths. Figure 11 shows the moment deflection curves for the three beams along with the limits for class 1 & 2 specification. The British Concrete Society Technical Report 49[3] specifies a design flexural strength of 4.1N/mm^2 for 75N concrete, however flexural testing of a sample made with the beams indicated a flexural strength of 9.04N/mm^2. The Class 2 limits derived with both of these values are shown on the graphs. Despite the large difference between the specified and experimentally determined flexural strength, the actual cracking moment based on first visual evidence of cracking, is much greater than both. The point at which first cracking was observed in two of the tests is also shown in figure 11.

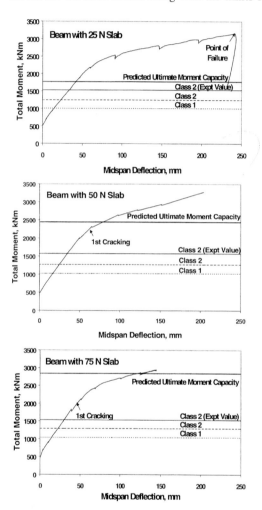

Figure 11 Moment Deflection Curves

Calculation of the design ultimate moment capacity was carried out in accordance with BS 5400[7] and Clark[8] using the strain compatibility approach. The values determined are displayed on each of the graphs. It can be seen that this method is highly sensitive to the compressive strength of the slab as that was the only variation between the three tests. The variation of the predicted values is not matched by the true behaviour of the beams. Although the second two beams shown in the figure were not tested to failure, it can be seen from the shape of their moment deflection curves, and by extrapolation, that their failure moments would not have exceed the predicted values by the same margin as the first beam.

The variation of spacing of the shear reinforcement at the two ends was considered when looking at the horizontal interface shear behaviour. The interface shear resistance is provided by a combination of the bond of the in-situ concrete to the precast surface and by the vertical reinforcement. The capacity of the interface was calculated according to BS 5400[7] and it was anticipated that it would be exceeded and horizontal movement of the slab relative to the beam would occur. This was monitored using 2 LVDTs at the ends. No movement of the slab relative to the beam was observed for any of the three tests. This would indicate that the values given in the design code for the shear capacity of the bonded surfaces may be conservative. It was not anticipated that any of the beams would suffer from cracking as a result of vertical shear. However, in all three beams, diagonal cracks occurred in the web close to the jacks. The cracks occurred at shear forces of 335kN with coexistent moments in excess of 2750kNm. It is felt that it was the combination of these that caused the diagonal cracking in the web, rather than vertical shear alone.

CONCLUSIONS

Three full scale bridge beams made from HPC and slabs of different strengths were instrumented and load tested successfully. Moment deflection curves showed a region of elastic behaviour followed by an in-elastic region of reduced stiffness. Large deflections and flexural cracking was evident in all of the beams before failure. The beam with the lowest strength slab was tested to collapse. The final failure involved spalling of the concrete slab followed by collapse. Neutral axis locations were found from the strain distributions for each beam. This remained straight during the elastic region and then rose towards the top of the beam as the concrete cracked and the strands began to yield. A moment curvature graph was plotted which showed three distinct zones. Design serviceability and ultimate moment capacities were exceeded in all cases. This was most evident in the beam with the lowest strength slab. The design capacities were seen to be very sensitive to slab strength, this was not found to be the case in the tests. Interface slippage between the beam and slab was expected but did not occur in any of the beams. This suggests that the codes of practice may be conservative on this issue. Vertical shear cracking was not expected but this did occur at high levels of load. It is felt that this was as a result of a coexistence of high shear and flexural stresses rather than vertical shear alone.

ACKNOWLEDGEMENTS

The authors would like to acknowledge the support and assistance provided for this work by Banagher Concrete, Co Offaly, Ireland.

REFERENCES

1. Aitcin, P.C., *High-performance concrete,* E & FN Spon, London, 1998
2. Mendis, P., Pendyala, R., and Setunge, S., *Stress-strain model to predict the full-range moment curvature behaviour of high-strength concrete sections*, Magazine of Concrete Research, **52**(4), pp. 227-234, 2000
3. Concrete Society, *Technical report No. 49: Design guidance for high strength concrete*, Concrete Society, Slough, 1998
4. French, C. et al, *High-strength concrete applications to prestressed bridge girders*, Construction and Building Materials, 1998, **12**(2-3), pp. 105- 113, Elsevier Science.
5. Ozyildirim, C. and Gomez, J. P., *High performance concrete in a bridge in Richlands, Virginia*, VTRC 00-R6, Virginia Transportation Research Council, Charlotte, Sept 1999
6. British Standards Institution, *BS 1881: Testing Concrete, Various Parts,* London
7. British Standards Institution, *BS 5400: Part 4: 1990., Code of practice for design of concrete bridges,* London
8. Clark, L. A., *Concrete Bridge Design to BS 5400*, Construction Press, London, 1983

Categorization of Damaged Locations on Concrete Bridge Structures by a Neural Network

LOJZE BEVC, IZTOK PERUŠ
Slovenian National Building and Civil Engineering Institute, SI-1000 Ljubljana, Slovenia

Abstract

The problem addressed in the paper is deterioration of existing concrete bridges. When assessing a deteriorated structure degraded areas on structural components are usually classified into deterioration categories. This categorisation should not be subjective, but based on the results of selected tests carried out as part of an in-depth inspection and/or visual parameters.

Two models will be presented and discussed: visual and test model of deterioration categorisation. In the case of reinforcement corrosion, the following indicative test results were selected and processed to become input parameters, describing the deterioration category in case of experimental model: gas permeability, content of chloride-ions at reinforcement level, alkalinity of concrete, depth of carbonatisation and electrochemical-potentials. Visual model of deterioration category takes into account a number of visual parameters, i.e. surface condition, scalling, delamination and spalling of concrete, reinforcement corrosion, cracking, joints condition and moisture. The output parameter is the deterioration category, a value between 1 and 5.

INTRODUCTION

Bridge structures are very vulnerable part of the road network. Due to the environmental load (marine environment, air pollution, humidity, temperature) or using de-icing salts to provide adequate driving conditions during winter times, a great number of bridge structures suffer deterioration processes caused by reinforcement corrosion. Deterioration can significantly reduce durability of the structure and expected service life as well as the load carrying capacity, if the structures are not properly and on-time maintained. Therefore, it is very important that structures are regularly inspected at appropriate time intervals to detect initiation of damages at an early stage. If no immediate action or in the near future is undertaken, damaged areas will extend as well as the intensity of the damage and the costs of the repair work may increase enormously.

CATEGORISATION OF DAMAGED LOCATIONS

Locations of damages do not depend only of the aggressive media itself, but are mainly governed by the weak points of the structures due to design inhered in construction process, poor detailing or performing work during the construction as well as poor or neglected maintenance. One type of structures with inherent weak points due to construction process are structures made of precast elements.

In early 70's a great number of superstructures of large viaducts in Slovenia were made of precast concrete elements. Longitudinal and transverse post-tensioned girders were made of one to three elements. Bridge deck was assembled of reinforced concrete precast deck elements, which were 2 m wide and half of the viaduct width long. They were interconnected with cast-in-place joints 20cm width. For casting the joints the concrete precast elements or asbestos-cement plates were used as scaffolding. In the cantilever area on both sides of the viaducts wooden scaffolding was used with plastic pipes as a means for fastening scaffolding to the deck elements with steel bolts. After the removing of scaffolding elements the holes were not filled and so became the source of continuous wetting of outer surface of edge girders.

In Slovenia a systematic inspection of bridge structures began in early 90's. During the first major inspections a few viaducts at age about 20 years were found with huge damages and heavily corroded reinforcement and tendons, some of them were even broken. On the edge girders even a few hundreds of damaged locations were found with different intensity as well as extent of the damages. To make approximate estimation of costs for repair of such amount of damaged location with different intensity and extent, the methodology of categorisation of damaged location was provided. It is based on the results of two types of assessment. The first is based on the visual assessment of visible signs on concrete surface, the second is based on the results obtained during on-site measurements as well as in the laboratory. Visual assessment (visual categorisation) is made on all damaged spots, while test assessment (test categorisation) is made on a smaller number of selected damaged locations chosen among several different assessed visual categories. No single type of test is sufficient to determine the state of corrosion and combination of some of them is required. In this methodology it is also assumed that combined results of sufficient number of different tests and measurements is more accurate estimation of the state of the reinforcement than only a visual assessment. By comparing the results on the selected spots of visual categorisation with results of test categorisation and how do they comply the assessment can be made for all damage location on the structure.

When methodology of categorisation was first applied[1], it was made by engineering judgement. To avoid subjectivity in the categorisation as much as possible and to use results of past categorisation analysis in the future work, a CAE (Conditional Average Estimator) Neural Network model was developed to predict visual and test categorisation and to use real results of categorisation for learning process of the model. For this model a more refined categories for visual categorisation as well as for the test categorisation were selected with respect to the model based solely on engineering judgement.

Damage Categories

Determination of damage categories is based on the relation between the damage intensity and basic repair procedure, which are shortly described in table 1. Damage categories are divided into five classes with respect to the damage intensity with gradation from damage of minor intensity (minor or no repair) to very severe damage with huge intensity (very demanding repair work).

Table 1: Description of basic repair procedure for damage category

Category	Repair procedure
I	Do nothing or only minor local superficial repair work (filling voids, levelling of local shallow spalls, ...)
II	Cleaning concrete surface and removing locally cracked shallow concrete cover or delaminated concrete cover above the stirrups; cleaning locally corroded stirrups if visible; reinstate the alkaline environment around the stirrups if necessary; make a new concrete cover to original level of concrete surface or upgrade the concrete cover depth if it is originally made too thin; apply protective coating if necessary (to prevent or retard further carbonation and/or chloride ingress)
III	Cleaning concrete surface and remove delaminated concrete cover as well as contaminated and/or carbonated concrete around the reinforcement; clean the reinforcement and reinstate the alkaline environment around the reinforcement; make a new concrete cover to original level of concrete surface or upgrade the concrete cover depth if it is originally made too thin; apply protective coating if necessary (to prevent or retard further carbonation and/or chloride ingress)
IV	Cleaning concrete surface and remove delaminated concrete cover as well as contaminated and/or carbonated concrete around the reinforcement; clean the reinforcement and add reinforcement in the case of reduced cross section (if needed); reinstate the alkaline environment around the reinforcement; make a new concrete cover to original level of concrete surface or upgrade the concrete cover depth if it is originally made too thin; apply protective coating if necessary (to prevent or retard further carbonation and/or chloride ingress)
	If protective ducts of post-tensioned concrete elements are corroded, they must be removed as well as contaminated grout; reinstate the alkaline environment around the prestressing reinforcement; reinstate appropriate concrete cover;
V	Cleaning concrete surface and remove delaminated concrete cover as well as contaminated and/or carbonated concrete around the reinforcement; clean the reinforcement and add reinforcement and/or external prestressing; reinstate the alkaline environment around the reinforcement; make a new concrete cover to original level of concrete surface or upgrade the concrete cover depth if it is originally made too thin; apply protective coating if necessary (to prevent or retard further carbonation and/or chloride ingress)
	If protective ducts or prestressing reinforcement are corroded, they must be removed as well as contaminated grout and broken or heavily corroded prestressing reinforcement; reinstate the alkaline environment around the prestressing reinforcement; reinstate appropriate concrete cover; add external prestressing if necessary;

Visual Categorisation - Categorisation based on visual assessment

Visual signs of concrete deterioration due to reinforcement corrosion caused by chloride ingress and/or carbonation have many stages, which primarily depends on the quality of the concrete and thickness of the concrete cover. Parameters of concrete surface, which are used in categorisation based on visual assessment, are shortly described in table 2. The assessment of each parameter is divided into four categories. The assessment of the crack width on the concrete surface and in the joints between the precast segments is primarily based on the influence the crack has on easier access contaminants to the reinforcement and further corrosion process. The categorisation of surface wetting is based on the experience gained during in-depth inspections of viaducts. The assessment of the corrosion of the reinforcement during visual categorisation is made only if the reinforcement is exposed due to spalls or removal of delaminated concrete, otherwise it is made after testing on-site is finished and inspection windows are made to inspect the condition of the reinforcement.

Table 2: Visual categorisation

Visual assessment	Category	Description
Surface condition	I	Shallow voids
	II	Deep voids
	III	Honeycomb in the concrete cover
	IV	Honeycomb beyond the level of reinforcement
Scalling, Delamination, Spalling	I	Superficial scalling; spalling of concrete surface a few mm thick
	II	Scalling and spalling of concrete cover to the level of stirrups
	III	Scalling, delamination and spalling to the level of the main reinforcement
	IV	Delamination beyond the level of the reinforcement; delamination to the level of prestressing reinforcement
Corrosion	I	Superficial corrosion, no reduction in the cross section
	II	Reduction of the reinforcement cross section up to 10%
	III	Reduction of the reinforcement cross section more than 10%; corroded protective sheath and minor to moderate corrosion of prestressing reinforcement;
	IV	Broken reinforcement; heavily corroded or broken prestressing reinforcement;
Cracking, joints	I	Crack width $w \leq 0,1$ mm
	II	Crack width $0,1 < w \leq 0,3$ mm
	III	Crack width $0,3 < w \leq 1,0$ mm
	IV	Crack width $w > 1,0$ mm

Table 2: Visual categorisation - continuation

Visual assessment	Category	Description
Moisture (wetting)	I	Very light
	II	Light (only a few signs of rust stains may be visible; thin deposits of efflorescence may be visible)
	III	Heavy (moderate signs of rust stains may be visible; moderate deposits of efflorescence may be visible)
	IV	Very heavy (a lot of signs of rust stains may be visible; thick deposits of efflorescence may be visible)

Test categorisation – categorisation based on the results of on-site and laboratory test

On a few selected location on-site measurements and laboratory tests are carried out. On each selected location concrete cover depth is measured by rebar locator. Depth of carbonation front is measured by fenolftalein test. On-site are also performed gas permeability as well as electrochemical potentials measurements. The interpretation of electrochemical potentials is made upon the standard ASTM C876, although for the convenience of dividing categorisation into four categories the range of electrochemical potential in the range between –200mV and –350 mV is divided into two classes. Concrete cores diameter of 50 mm or more are taken to determine chloride profile and pH value of the concrete at the level of the reinforcement. Parameters, which are used in the test categorisation as well as category classes are presented in table 3.

Table 3: Test categorisation

Test	Category	Description (boundary values)
Gas permeability (on-site test)	I	$g_p \leq 10^{-16}$ m^2
	II	$10^{-16} < g_p \leq 5 \times 10^{-16}$ m^2
	III	$5 \times 10^{-16} < g_p \leq 10 \times 10^{-16}$ m^2
	IV	10×10^{-16} m$^2 < g_p$
Chloride profile (level of critical chloride content = 0,4% b.w.c.) Laboratory test	I	0,4% b.w.c. at the middle of concrete cover
	II	0,4% b.w.c. at the level of stirrups
	III	0,4% b.w.c. at the level of main reinforcement
	IV	0,4% b.w.c. beyond the level of the main reinforcement
pH value at the level of the reinforcement Laboratory test	I	pH $\geq$ 11 at the level of the main reinforcement
	II	$10 \leq$ pH < 11 at the level of the main reinforcement
	III	$9 \leq$ pH < 10 at the level of the main reinforcement
	IV	pH < 9 at the level of the main reinforcement

Table 3: Test categorisation - continuation

Test	Category	Description (boundary values)
Depth of carbonatisation front On-site test	I	Less than half of the concrete cover depth
	II	At the level of the stirrups
	III	At the level of the main reinforcement
	IV	Beyond the level of the main reinforcement
Electrochemical potentials On-site test	I	Ep > -200 mV
	II	-200 mV ≥ Ep > -275 mV
	III	-275 mV ≥ Ep > -350 mV
	IV	-350 mV > Ep

CATEGORISATION BY CAE NEURAL NETWORK

Theoretical background

Reliable treatment of natural phenomena is always based on measurements and on the description of relations between the observed test results. From the theoretical point of view, these relations can be most appropriately described in terms of abstract mathematical models representing mathematical laws. However, from the practical point of view, simulated analogue models based on electronic devices are sometimes more convenient[2,3,4]. The CAE model of deterioration category takes into account a number of parameters, which in mathematical sense corresponds to the components of the model vector

$$\{X_1, X_2, ..., X_N\} \qquad ... /Eq. 1/.$$

When formulating the modeller of a phenomenon of categorisation $C = C(p_1, p_2, p_3, ..., p_L)$, it is assumed that one particular observation of the phenomenon can be described by a number of variables, which are regarded as components of a ***model vector***:

$$X = \{p_1, p_2, p_3, ..., p_L, C\} \qquad ... /Eq. 2/.$$

Vector **X** is composed of two truncated vectors:

$$P = \{p_1, p_2, p_3, ...; \#\} \text{ and } R = \{\#; C\} \qquad ... /Eq. 3/,$$

where # always denotes the missing portion. Vector **P** is complementary to the vector **R** and their concatenation will yield the complete model vector **X**. The problem to be solved is the estimation of the unknown complementary vector **R** by the given truncated vector **P** and by the model vectors $\{X_1, X_2, ..., X_N\}$. Using the conditional probability function the optimal estimator for the given problem can be expressed by:

$$r_k = \sum_{n=1}^{N} A_k \cdot r_{nk} \qquad ... /Eq. 4/$$

where

$$A_k = \frac{a_n}{\sum_{j=1}^{N} a_j} \quad \text{and} \quad a_n = \exp\left[\frac{-\sum_{i=1}^{L}(p_i - p_{ni})^2}{2w^2}\right] \quad \text{.../Eq. 5/.}$$

Coefficient a_n can be written explicitly for both categorisation models, visual and test respectively. For example, expression

$$\exp\left[-\frac{(Pe-Pe_n)^2 + (Cl-Cl_n)^2 + (pH-pH_n)^2 + (Ca-Ca_n)^2 + (Ep-Ep_n)^2}{2w^2}\right] \quad \text{.../Eq. 6/,}$$

stands for coefficient a_n in case of test model. Terms Pe, Cl, pH, Ca, Ep denote gas permeability, chloride ingress, alkalinity at the level of the reinforcement, carbonisation and electropotential parameters, respectively. In similar expression for visual model

$$\exp\left[-\frac{(Su-Su_n)^2 + (Ds-Ds_n)^2 + (Co-Co_n)^2 + (Cr-Cr_n)^2 + (Jo-Jo_n)^2 + (Mo-Mo_n)^2}{2w^2}\right]$$

... /Eq. 7/,

Su, Ds, Co, Cr, Jo, Mo denote surface condition, scalling/delamination/spalling, corrosion, cracking, joints and moisture parameters, respectively. w is so called smoothing parameter.

The output parameter is the deterioration category, denoted as C. This result is a uniform variable which can have any value between 0.5 and 5.5. Individual deterioration categories have been determined by convention in the following ranges:

- category 1:   0.5 - 1.5 (mean 1.0),
- category 2:   1.5 - 2.5 (mean 2.0),
- category 3:   2.5 - 3.5 (mean 3.0),
- category 4:   3.5 - 4.5 (mean 4.0),
- category 5:   4.5 - 5.5 (mean 5.0).

Basic model

Database for basic model was obtained by questionnaire, which was filled out by experts. Questionnaire was filled out separately for visual and test categorisation. It was prepared in the form of locations with randomly generated categories for each parameter of the categorisation model and the experts have to give their assessment of damaged categorisation for each generated location. Prior to fill out the questionnaire all illogical combination of parameter categories for damage location were eliminated (e.g. high potentials with no wetting surface and no chlorides, etc…). By CAE method the assessment results can be presented either on a set of tables or on a set of simple diagrams, showing the iso-lines of equal deterioration or the boundaries between two deterioration categories. For better visual impression the results can be presented also as 3D-surface graphs. Only a few typical results for deterioration category based on visual model and test model are presented on figures 1 and 2.

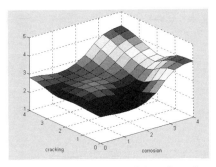 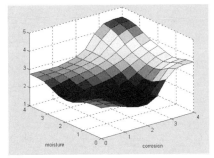

Fig. 1: 3D presentations of deterioration category as a functions of two parameters – "visual model".

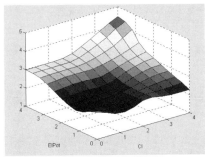 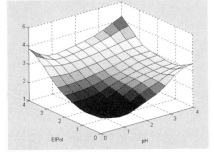

Fig. 2: 3D presentations of deterioration category as a functions of two parameters – "test model".

Example of categorisation using CAE NN

Data of one viaduct were selected to demonstrate the applicability of the model. Eleven locations were selected on which on-site measurements and laboratory measurements were carried out. The average concrete cover depth on the measured locations is 26.1 mm to the level of stirrups and 39.8 mm to the level of main reinforcement. The concrete cover depth to the post-tensioned ducts is about 52 mm.

Table 4: Results of visual categorisation

Location	surface	delmn/spall	corrosion	cracking	joints	moisture	BM	BM+Learn
1	1	2	1	0	0	3	2,50	3,22
2	1	2	2	0	0	3	2,50	3,81
3	1	3	2	0	0	4	3,00	3,98
4	1	0	0	0	0	1	1,00	1,19
5	2	3	2	0	0	4	2,75	3,99
6	1	3	2	1	0	3	3,00	3,89
7	1	0	0	0	0	2	1,33	1,68
8	1	3	1	0	0	2	3,00	3,92
9	1	0	0	0	0	1	1,00	1,19
10	1	0	1	0	0	1	1,00	1,67
11	1	0	0	3	0	2	3,50	1,50

Results for visual categorisation are presented in table 4 and for test categorisation in table 5. In column BM are given the results for basic model, in column BM+Learn are given the results after the model was improved with additional learning. Through additional learning process a real categorisation was added to the database, which was evaluated after on-site measurements were finished. The results of visual categorisation for basic model show that of 11 locations 36% of them were put into category 1, 55% into category 3 and 9% into category 4.

Table 5: Results of test categorisation

Location	gas perm.	Cl	pHx=reinf	carbonization	ElectPot	BM	BM+Learn
1	0	3	1	1	4	4,07	3,23
2	2	4	1	1	4	4,21	4,01
3	3	4	1	3	4	4,98	4,02
4	2	1	1	1	1	1,55	1,18
5	0	4	1	1	4	4,86	3,87
6	2	4	1	1	4	4,21	4,01
7	2	3	1	1	1	2,60	2,05
8	2	4	1	1	3	3,83	3,99
9	3	0	0	0	0	2,44	1,50
10	0	3	1	1	0	3,63	2,29
11	0	3	1	1	2	3,18	1,61

Comparison of these results with the results of basic model for test categorisation revealed that 75% of locations put in put into category 1 by visual categorisation will change into category 2 and 25 % into category 3. The analysis of locations, which were put into category 3 based on visual categorisation show that 67 % of these locations will change into category 4 and 33% into category 5. The analysis also shows that 25% of locations put into category 4 will change back to category 3. This result is primarily the consequence of experts' assessment of the influence the crack width has on the progress of corrosion reinforcement.

After the additional learning of the basic model, results (column BM+Learn) of visual categorisation changed with respect to the results of basic model and they comply quite well with results of test categorisation, which confirm the results of visual categorisation. Visual categorisation put 18% of 11 locations into category 1, 27% into category 2, 9% into category 3 and 46% into category 4. These results are finally applied to all damaged location on the structure.

CONCLUSION

A brief description of categorisation methodology of damaged location of concrete structures by using CAE Neural network has been presented. This methodology provides a tool to make a preliminary (approximate) estimation of costs for repair of damaged location with different intensity and extent. Categorisation consists of two models, visual and test categorisation model. Visual model is based on the categorisation of visual signs of deterioration on the concrete surface and test model is based on the categorisation of result of on-site and laboratory measurements. The final estimation of the categorisation is made by comparison the results of visual and test categorisation, where results of test categorisation prevail over the results of visual categorisation. A future work is needed to improve the model by additional learning of the model as well as to include another parameters into the model (e.g.

inclusion of the results of on site measurements of resistivity of concrete in test model of categorisation) as well as how to improve the estimation of categorisation if some data for test model at a particular locations are missing.

References

[1] Bevc L., Capuder F., Terčelj S., Kuhar V.: *"Damage Assessment of precast post-tensioned viaducts after 22 years exposed to harsh environment"*, FIP Symposium on Post-tensioned concrete structures, pp.305-312, 1996.

[2] Grabec, I., & W. Sachse, *"Synergetics of Measurement, Prediction and Control"*, Springer-Verlag, 1997.

[3] Grabec, I., *"Self-organization of neurons described by the maximum entropy principle"*, Biol. cybern. 63, pp.403-409, 1990.

[4] Peruš, I., Fajfar, P. & Grabec, I., *"Prediction of the seismic capacity of RC structural walls by non-parametric multidimensional regression"*, Earthquake Engineering & Structural Dynamics 23, pp. 1139-1155, 1994.

A Qualitative and Quantitative Comparative Study of Seismic Design Requirements in Bridge Design Codes

MOURAD M. BAKHOUM
Associate Professor, Structural Engineering Department, Cairo University, Egypt

SHERIF S. ATHANASIOUS
Structural Design Dept., Arab Consulting Eng.(Moharram - Bakhoum), Cairo, Egypt

Abstract:

The paper presents quantitative and qualitative comparisons of Seismic Bridge Design codes. It is observed that there are many similarities but also there are quite large differences between the code provisions in their definitions, response spectra, behavior modification factors to consider ductility, load combinations, etc. Discussion of these differences is presented. Moreover, in order to assess how these differences affect the design of bridges, numerical examples are given, related to computation of simulation of input seismic forces (Elastic and Design Response Spectra), straining actions (moments and shear in bridges).

1. INTRODUCTION:

In order to assure safety of structures after an Earthquake, in order to satisfy the required level of functionality, and in order to organize the design process, design codes give provisions to satisfy these requirements. The paper presents Quantitative and Qualitative comparisons of four international Bridge Design Codes. It is very interesting to observe - through the discussions and the numerical examples in the paper - how different codes in different countries of the world tackle the problem of Seismic Design of Bridges. A comprehensive comparison of the codes needs several volumes. Due to space limitations, the paper will focus only on some of these aspects, particularly those which could be of higher priority in practical design, and code developers.

A very brief review is presented first. To assess the effect of an Earthquake on a bridge, and design against it many parameters need to be defined. Some of the most important parameters relevant to structural engineers are mainly related to: (A) Seismic activities (magnitude, recurrence, distance from bridge, frequency content,...) and their definition as seismic loads, in a format to be usable by structural engineers to design the bridges. (B) Consideration of soil parameters at the bridge site. (C) Aspects related to physical dimensions of the bridge materials supports the structural system. (D) Aspects related to the design assumptions, such as how structural engineers model the bridge, combine seismic loads with other loads (own weight, traffic). (E) Aspects related to the users who are benefiting from the bridges, what the bridge crosses over, availability of substitute bridges in the vicinity. (These could be expressed in terms of Bridge importance). Also, as expressed recently in performance based design codes: the allowable level of damage in the bridge after an earthquake. Figure 1 shows schematically the effect of Earthquake on Bridges, and some of the main parameters related to the definition of seismic actions on a bridge. Table (1) summarizes how some of these parameters are considered in the codes.

Table (1): Basic Requirements of Seismic Design

Item	AASHTO	CANADIAN	ENV 1998	JRA
Seismic input: Seismic response (1)	Elastic seismic response (C_{sm}) 1 equation for R.S. 3.10.6	Elastic seismic response (C_{sm}) 1 equation for R.S. 4.4.7	Elastic response spectrum (S_e) 4 equations for R.S. Part 1 [4.2.2]	Several spectra are given for: Two E.Q. levels. For higher level, two E.Q. types (Near & far field). 3 equations for each R.S. Ch. (4,5,6)
Return Period (2)	2500, 475 years	1000, 475 years	950, 475 years	
	Related to bridge Importance			
Acceleration (3)	Acceleration coefficient (A) Contour maps 3.10.2	Zonal acceleration ratio (A) Contour maps 4.4.3	Design ground acceleration (a_g) Maps given in NAD Part 1 [4.1]	Modification factor for zone (c_z): Zone maps 3.5
Soil/ Site effect (4)	Soil profiles (I,II,III,IV) Site coefficient.: table 3.10.5.1-1	Soil profiles (I,II,III,IV) Site coefficient.: table 4.4.6.1	Subsoil classes (A,B,C) Soil parameter: table 4.1, Part 1-1	Ground classes (I,II,III) Characteristic value $T_G(s)$: table 3.6.1
Importance category (5)	Critical bridges, Essential bridges, and Others 3.10.3	Lifeline bridges, Emergency bridge, and Others 4.4.2	Greater than average, Average, and Less than average Part 2 [2.1(3)]	Bridges class B and bridges class A 3.4
Modification factors (6)	Response modification factors (R) Tables 3.10.7.1-1,2	Response modification factors (R) Table 4.4.8.1	Behavior factor (q) Table 4.1 Part 2 [4.1.6]	Allowable ductility factor of the bridge pier (μ_0) 9.2
Analysis methods (7)	Uniform load, Single-mode, Multi-mode and Time history method 4.7.4	Uniform load, Single-mode, Multi-mode, Time history and Static push-over analysis 4.5	Fundamental mode, Power spectrum analysis, Time series analysis and Non-linear time domain analysis Part 2 [4.2]	*Equivalent static: Seismic Coef., Ductility design *Dynamic analysis 4,5,6
Combination of seismic response in different directions (8)	100% of the force in one direction + 30% of the force in perp. Direction 3.10.8	100% of the force in one direction + 30% of the force in perp. Direction 4.4.9.1	Apply (SRSS) $E=(E^2_X+ E^2_Y+ E^2_Z)^{1/2}$ OR $A_{EX}+0.3A_{EY}+0.3A_{EZ}$ $0.3A_{EX}+A_{EY}+0.3A_{EZ}$ $0.3A_{EX}+0.3A_{EY}+A_{EZ}$ Part 2 [4.2.1.4]	It is assumed that the inertial forces in two perpendicular horizontal directions act separately 3.3.1
Combination with other loads (9): Refer to table [4]				
Special precautions: Minimum support length (10): Example calculations are given in table [5]				

Note: ( ) number in parentheses in the first column, relate Table 1 to Fig. 1

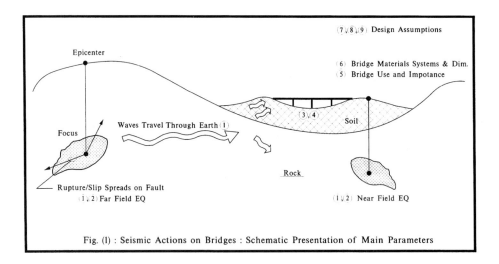

Fig. (I) : Seismic Actions on Bridges : Schematic Presentation of Main Parameters

2. QUALITATIVE COMPARISON

2.1 Response Spectrum: The Canadian code gives the following method to define the above mentioned parameters for seismic actions on Bridges. Note: Equations defining the R. S. for other codes are not given here due to space limitation but numerical comparison are given in the next section (Fig. 2, 3, 4, 5). Table 2 below summarizes JRA requirements.

$$\text{Design Response Spectrum (DRS)} = \frac{1.2 \text{ ASI}}{R\,(T^{2/3})}$$

Where: (A): Acceleration at the rock base at the bridge site. (A) should take into consideration the Earthquake expected at the site, and its recurrence. Small Earthquake could occur several times during the life of the bridge, while large level earthquake occur less frequently (may be never during the use of the bridge). (S): Takes into consideration the possible amplification in seismic waves due to soil nature at the site. (I): Takes into consideration the importance of the bridge. (T):   Depends on the structural system of the bridge. Deduced using elastic analysis methods. (R): Related also to the structural system of the Bridge and its details. It acknowledges the fact that it is not economical to resist large Earthquake by an elastic system, and that some yielding, and controlled level of damage could be permitted.

Comments: (1) Large differences are observed in the elastic response spectra in the different codes. However, these differences are much reduced in the design response spectra as will be seen in Fig. 2, 3, 4. (2)All the response spectra are dependant on seismic zone, soil conditions and structural system of the bridge. (3) The response spectrum of both AASHTO and Canadian code are given for 5% damping, other values of damping can be considered in ENV 1998 by factor (β0), and in JRA damping is represented by the factor (CD). (4) A sample of the response spectra of JRA is given separately in Fig. (5) A direct comparison with the other codes is not possible, because acceleration are included in the R.S. of the three zones, and not given as PGA for a specific location. From table (2) it is noted that JRA gives two levels of seismic action. For the higher magnitude, it gives two types of Earthquake & (far and near fields).

Table 2 : Summary of JRA requirements

Level and Type of Design Ground Motions	Design Objective	Design methods	
		Regular Bridges	Irregular Bridges
		Equivalent static lateral force methods	Dynamic Analysis
Level 1: Ground motions with high probability to occur	Prevent Damage and yielding. Elastic behavior	Seismic coefficient method	Step by step or RSA
Level 2: Ground motions with low probability to occur. Considers two types of EQ. Type I: (Plate boundary earthquake) Type II: (Inland earthquakes)	Class -A: Standard Br: Prevent critical damage. Class-B: Important Br.: Limited damage after EQ	Ductility design method	Step by step analysis or Response Spectrum Analysis

2.2 **Bridge Importance:** The codes agree in using three different categories for Bridge importance (except JRA uses only two), refer to Table 1. However, there are large differences in the contents and description of these classifications. A discussion would be necessary Taking the highest classification or maximum importance category, for example (others are not mentioned here for lack of space). The codes mention:

AASHTO-LRFD : Critical bridges:
- Must remain open to all traffic after the design earthquake. Must be usable by emergency vehicles and for security/defense purposes after a large earthquake (2500-year return period event). Map given in AASHTO is for a return period of 475 years. To consider 2500 years, changes are made to (R). Note: Map for 2400 years is given in AREMA.

Canadian Highway Bridge Design Code: Lifeline bridges: Importance factor (I=3.0).
- Carry or cross over routes that must remain open to all traffic after the design earthquake (15% probability of exceedance in 75 years, equivalent to a return period of 475 years). Must also be usable by emergency vehicles & for security/defense purposes immediately after a large earthquake (7.5% prop. of exceed. in 75 years & return period of 1000 years).

Eurocode 1998-2 : 1994: Greater than average: importance factor ($\gamma_I = 1.3$)
Bridges of critical importance for maintaining communications, especially after a disaster.
- Bridges whose failure is associated with a large number of probable fatalities. Major bridges for which a design life greater than normal is required. Map to be given for return period 475 years in NAD (National Application Documents).

Japan Road Association Specifications: Class B:
- Bridges of national expressways, urban expressways, designated city expressways, Honshu-Shikoku Highway, & general national highways. Double-section bridges & over bridges of prefectural highways & municipal roads & other bridges, highway viaducts, etc. especially important in view of regional disaster prevention plans, traffic strategy, etc.

Comments on the Application of the bridge importance:(1)AASHTO-LRFD: Bridge importance is included in the design response spectrum by selecting the (R) value.(2) Canadian Code: Bridge importance is represented in the response spectrum by introducing the importance factor (I=3, I=1.5, I=1, for the three classes respectively). (3) ENV 1998: Bridge importance is represented by multiplying the design response spectrum by the importance factor ($\gamma_{I=1.3}, \gamma_{I=1.0}, \gamma_{I=0.7}$ for the three classes respectively). (4) JRA: Bridge importance defines the level of allowable damage accepted after the EQ, and also is included in the allowable ductility factor (μ_0) of the bridge pier which is used to evaluate the equivalent horizontal seismic coefficient.

2.3 **Response Modification Factor (R) or Behavior Factor (q):** Table (3a, 3b) summarize some the (R) factors for bending as given in the codes. It can be seen that – in most of the cases- there are a large difference between the values given. There is even difference in the terminology. (The drawings below the table present a possible correlation between the codes). It is noted that the Eurocode gives two levels of ductility. Moreover, it gives a numerical definition for wall piers, which could be useful in practical design. It is noted that these codes agree on (R) factor = 1 for Shear.

Table (3a): Response Modification Factor According to AASHTO and Canadian Code:

Substructure	AASHTO			Canadian Code
	Critical	Essential	Other	
Wall-type piers – larger dimension	1.5	1.5	2.0	2.0
Single columns (ductile R.C. or steel)	1.5	2.0	3.0	3.0
Multiple column bents	1.5	3.5	5.0	5.0
Connections: (for example): • Superstructure to abutment • Columns or piers to foundations	 0.8 1.0			 0.8 1.0

Table (3b): Behaviour Factor for Concrete According to ENV 1998-2:

Ductile Elements	Seismic Behaviour	
	Limited Ductile	Ductile
Reinforced concrete piers: • Vertical piers in bending (H/L$\geq$3.5) • Squat piers (H/L=1.0) • Inclined struts in bending	 1.5 1.0 1.2	 3.5 1.0 2.0
Abutments	1.0	1.0
Arches	1.2	2.0
Columns carrying elastomeric bearings	1.0	1.0

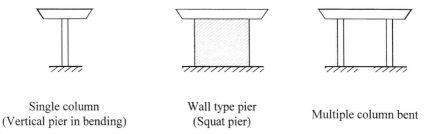

Single column
(Vertical pier in bending)

Wall type pier
(Squat pier)

Multiple column bent

2.4 Load combinations: Table (4) shows the load combinations used in the three codes. The notes below the table show the combination for JRA. Comments: It is interesting to note from the table that the codes agree on load factor equals to one for the Earthquake action. This is logical because the way defined in the codes, earthquake loads are considered as ultimate loads. However, for DL, the load factor = one might need reconsideration. Differences in weight of structures, distribution of permanent loads between the members, modeling of point of application of DL could all include some levels of uncertainty, requiring a load factor larger than one.

Table (4): Combinations of Seismic Load Effects with other Loads (ultimate):

Code	DL	LL	PS	WP	FR	EP	EQ
AASHTO-LRFD	1.0[a]	0.5[b]		1.0	1.0		1.0
Canadian Code	1.25:0.80		1.05:0.95	1.10:0.90		1.25:0.50	1.0
ENV 1998	1.0	0.2:0.3[c]	1.0				1.0
JRA [d]	See notes below						

(a) The load factor for permanent loads γ_p ranges between 0.45 to 1.80 according to type of loading and its effect. (b) The possibility of partial live load, i.e., $\gamma_{EQ} < 1$, with earthquake is considered. Application of Turkstra's rule for combining uncorrelated loads indicates that $\gamma_{EQ} = 0.5$ is reasonable for a wide range of values of average daily truck traffic. (c) live load should be multiplied by 0.2 for road bridges with intense traffic, 0.3 for railway bridges.
(d) Part III of JRA provides instructions for designing concrete members using both service loads and ultimate factored loads. Service loads for both superstructures and substructures are as follows: (Principal load + Earthquake Force) (allowable stresses are increased by 1.50) (Principal load + Earthquake Force + Thermal) (allowable stresses are increased by 1.65) where principal load includes all the loads other than live load and impact.
For ultimate factored loads, consider: 1.3(D+EQ)   and   1.0D + 1.3EQ

3. QUANTITATIVE COMPARISON:

Three different types of quantitative comparison between codes are given in this section.
3.1 Elastic and Design Response spectra: Elastic and Design response spectra (ERS, DRS) are compared assuming a PGA = 0.15g. Different types of soil, different levels of importance are considered as shown in Fig. 2, 3, 4. It is interesting to note from Fig. 2a, that while the ERS is much higher for the Canadian code than the AASHTO, they come very close together in the DRS. On the other hand, while the EC and AASHTO are very close together in the ERS, they are different in the DRS. EC gives higher results than AASHTO for case of limited ductility (EC1), and lower for the case of ductile bridges (ECd). Now we consider effect of soil. Figure 3 shows the comparison also for bridges of maximum importance, but on very weak soils. In this case, Canadian code comes with the highest response. AASHTO and EC1 are very close in the high period region. ECd gives the lowest response as in all the comparisons considered. Now, comparing Fig. 3, 4 gives an assessment of the Bridge Importance categories. First, we note that the maximum response is about the double for bridges with the max. importance, except the Canadian which is almost three times higher. The Canadian and Ec1, have shifted their positions as the code with the highest response in Fig. 3b, 4b. In all the ERS and DRS, it is assumed that damping equals 5%. Damping needs to be further studied.

3.2 Straining actions in a Bridge Example: The example shown in Fig. 6a was previously analyzed according to AASHTO (William 1998). Here it is additionally analyzed using the provisions of the Canadian code and EC8 (limited ductile & Ductile behavior). The results are shown in Fig. 6b. The first column gives the straining actions (response) due to seismic load in the longitudinal direction, combined with 0.3 of transverse. The third column includes the effect of behavior modification factor (R or q). The sixth column presents the design moments and shears due to seismic actions, combined with dead load. Similarly, for columns 2, 4, 7 but for main seismic action in transverse direction. It is interesting to note that while the largest value of elastic bending moment occurs due to Canadian code, however, after dividing by R, the difference between the three codes is much reduced, except -as observed previously- for the Eurocode with limited ductility, which gives more than double the value of bending moment than the other codes. The above observation applies for longitudinal and transverse moments. For shear, the Canadian code gives highest response.

3.3 Minimum support length: An important requirement that is given in all the codes is the provisions of minimum support length for Bearings, to prevent collapse of bridge spans due to column movement. The table below gives a comparison for the example shown in Fig. 6a.

Table (5): Minimum Support Length:

Code	AASHTO	CANADIAN	ENV 1998-2	JRA
Minimum Support Length (cm)	57.3	56.8	69.2	106.9

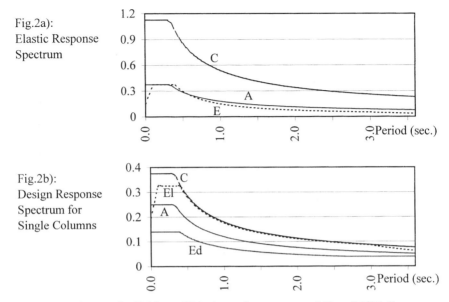

Fig.2: Response Spectra for Bridges of Maximum Importance and Very Stiff Soil (Type I or A) amd Seismic Coefficient=0.15g

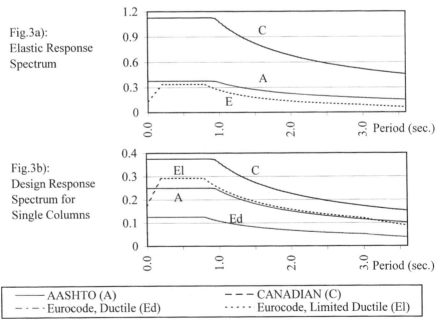

Fig.3: Response Spectra for Bridges of Maximum Importance and Very Weak Soil (Type IV or C) amd Seismic Coefficient=0.15g

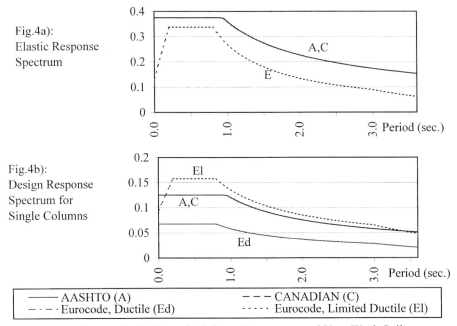

Fig.4: Response Spectra for Bridges of Minimum Importance and Very Weak Soil (Type IV or C) amd Seismic Coefficient=0.15g

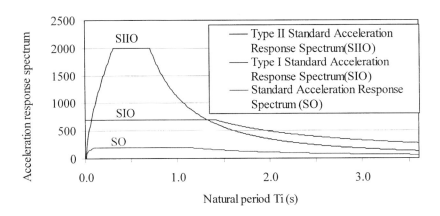

Fig. 5: Acceleration Response Spectrum for JRA, Soil Class 1 for the Dynamic Analysis Method (Ch. 6)

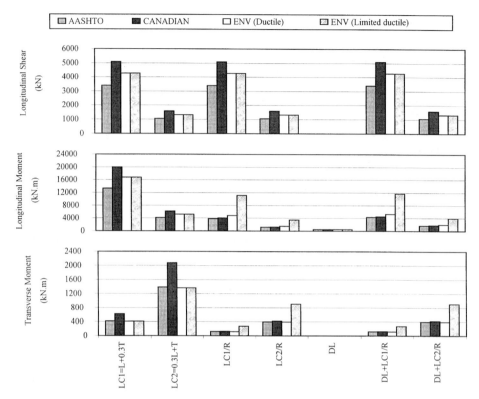

LC1 = 100% of effect of seismic action in long. Dir.(L) + 30% of the effect in trans. dir. (T)
LC2 = 100% of effect of seismic action in trans. Dir.(T) + 30% of the effect in long. dir. (L)
R = Response modification factor (Behaviour factor q)

Fig.6b: Straining Actions in the Top of the Exterior Column

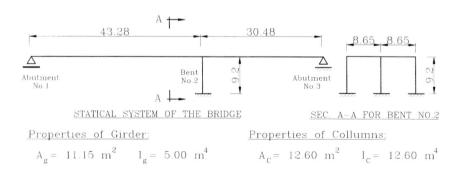

Fig.6a: Description of the Example used in the Quantitative Study

4. SUMMARY AND CONCLUSIONS:

The paper presented Quantitative and Qualitative comparisons of the seismic provisions in four international bridge design codes: AASHTO (USA), Canadian Code, Eurocode (EC8), and Japanese Code (JRA). The following conclusions are given:

(1) It is interesting to note that there are quite large differences in the definition of seismic actions on bridges, classification of bridge importance to users. There is also a wide variation in the consideration of ductility in the different codes (R factor). (2) Despite these large differences, the values of bending moments in an example bridge do not differ so much (except for the case of limited ductility in EC). (3) For the cases of low ductility and high ductility in the Eurocode, there are large differences in the Bending moments induced in the columns. It is important to recognize these differences, but it is important to keep these two levels of ductility (for some structures), till all the reinforcement details needed for achieving high ductility are developed, experimentally tested, and there is sufficient experience of their construction possibility without congestion of reinforcement. (4) Comments concerning load combinations and damping were given in the paper. (5) The Japanese code presents explicitly in its provisions two EQ types (Inland, plate boundary). Also, it presents two EQ levels: level 1 functional and level 2 safety. This is the trend for recommendations of code improvements (ATC-18, 32), and performance based design codes. (6) The use of two earthquakes levels: one with a high probability of occurrence during the life of the bridge, and another with a lower probability of occurrence is a reasonable and logical process (the first is related to check allowable stresses, the second to ultimate strength), but it could prolong design process.

References:

Athanasious, S. S., (2001) Master Thesis under preparation at Cairo University

(AASHTO) American Association of State Highway and Transportation Officials, LRFD *Bridge Design Specifications,* 2nd. Ed., Washington, D.C.,1998.

(JRA) Japan Road Association (1996) *"Design Specifications of Highway Bridges*, Part V: Seismic Design Japan Road Association, Tokyo, Japan. Also, New Seismic Design of Highway Bridges (http://iisee.kenken.go.jp/html/highway.htm)

Applied Technology Council (1996). ATC-18. & ATC-32. Redwood City, Calif, USA.

Canadian Highway Bridge Design Code 2000

(EC8) Eurocode: Design provisions for Earthquake resistance of structures. ENV 1998:1-1, 2

Hausammann, H., Wenk, T. (1998) *"Seismic Bridge Design According to Eurocode 8 and SIA 160"* XIth ECEE, Eleventh European Conference on Earthquake Eng. Sept. 6-11, France.

William, A.,(1998) *"Seismic Design of Buildings and Bridges"*, Engineering Press, Texas.

Yashinsky, M. *"Seismic Design and Retrofit Procedures for Highway Bridges in Japan"* (www.dot.ca.gove/hq/esc/earthquake_engineering/research/seisjap/seisjap.html)

Yen, W. P., Cooper, J. D (2001) *"A Comparative Study on US-Japan Seismic Design of Highway Bridges Part I: Design Methods"* draft paper to be submitted to E.S. (EERI). http://incede.iis.u-tokyo.ac.jp/ctoc/digitalbook/dgb_docs/bridgecriteria.html).

Symbols Frequently used:
(EQ): Earthquake forces, (DL): Dead load, (LL): Traffic load, (PS): Prestress force, (WP): Water pressure, (FR): Friction, (EP): Earth pressure. (ERS, DRS): Elastic and design response spectra. (R): Response modification factors, (q): Behavior Factor.